LEAN ENGINEERING
THE FUTURE HAS ARRIVED

BY

J.T. BLACK

DON T. PHILLIPS

AUBURN UNIVERSITY

TEXAS A&M UNIVERSITY

"Lean Engineering," by J.T. Black and Don T. Phillips. ISBN 978-1-62137-343-8.

Published 2013 by Virtualbookworm.com Publishing Inc., P.O. Box 9949, College Station, TX 77842, US. ©2013, J.T. Black and Don T. Phillips. All rights reserved. No part of this publication may be reproduced, stored in a retrieval system, or transmitted in any form or by any means, electronic, mechanical, recording or otherwise, without the prior written permission of J.T. Black and Don T. Phillips.

Manufactured in the United States of America.

Dedication

This book is dedicated to Carol Strom Black who passed away in 2012. You can't hardly find this kind of woman anymore.

J.T. Black 2/19/13

This book is also dedicated to Candyce Jean Phillips, who endured countless hours of composition, editing and changes by a usually frustrated old college professor.

Don T. Phillips

The Toyota style is not to create results by working hard. It is a system that says there is no limit to people's creativity. People don't go to Toyota to work, they go there to think.

Taiichi Ohno
February 29, 1912 – May 28, 1990

Acknowledgements

There are many individuals who have contributed to this book on Lean Engineering. Shovan Mishra worked tirelessly on the manuscript, figures and problem sets, Kavit Antani wrote the bulk of Mixed Model Final Assembly in Appendix A as part of his PhD research; and Dr. David Cochran wrote Chapter 18 and made significant contributions to Chapter 6. Thanks go out to Dr. Guy L. Curry and Dr. Richard Feldman of Texas A&M University, who have served as a constant source of encouragement and created a deeper understanding of Queuing Processes. Dr. Hiram Moya provided a great deal assistance in creating the Index, Bibliography and Table of Contents. The pioneering work of Dr. Ward Whitt should not go unnoticed. Dr. Mark Spearman and Dr. Wallace Hopp first awakened interest in the field of Factory Physics and without their pioneering work of establishing a scientific body of knowledge to unify Manufacturing Systems Analysis, the content of this book would not be placed in its proper perspective. Their work in awakening the engineering community to the corrupting influence of variance has not gone unnoticed. Of Course, there are numerous pioneers of the Toyota Production System which must be acknowledged including Taiichi Ohno and Shigeo Shingo. Early lean pioneers such as Bob Hall, Kenichi Sekine, Dr. Dick Schonberger and Yasuhiro Monden must also be recognized. The contributions of former students Dr. Steve Hunter and Dr. Joseph Chen are gratefully acknowledged, along with the work and publications of Dr. Bernard Schroer and Michael Baudin. Each were lean engineers in their own way. Concepts and ideas have been modified, presented and summarized which stem directly from Dr. Rajan Suri, Dr. John M. Nicholas, Dr. Dave Sly of ProPlanner and of course Mark Spearman and Wallace Hopp. Admiration and respect are offered to James P. Womack and Daniel T. Jones for their landmark publications in Lean Thinking, and to John Krafcik who coined the term *Lean*. We join these pioneers in believing that Lean thinking and Lean Manufacturing are here to stay. We only hope that in some small way we have captured the DNA of a new engineering discipline called ***Lean Engineering*** in this book.

Henry Ford
Pioneer, Visionary, Genius….. Lean Engineer
July 30, 1863 – April 7, 1947

Table of Contents

Chapter 3 Design Rules for Lean Manufacturing Systems 75

Chapter 8 Lean MSP Implementation Methodology 245

Chapter 12 The Disruptive Effects of Variance **381**

Chapter 15 Integrated Production and Inventory Control 477

Chapter 16 Integrated Preventive Maintenance (IPM), Reliability and Continuous Improvement 511

Preface

*"Perhaps we should stop looking for the sharpest needle in the haystack
And find one that we can sew with...."* March & Siman, Organizations

The motivation for constructing this textbook evolved from over 30 years teaching and industrial experience in manufacturing systems design and analysis. As the years passed, it became abundantly clear that traditional mass production lines were not suited to the flexibility and versatility required to competitively compete in an evolving international marketplace. In order to frame this observation and describe how it came about, we need to briefly describe the evolution of modern manufacturing systems.

Manufacturing is not a new endeavor; it has been around since ancient times. Biblical and historical records record extended battles between ancient empires. These encounters required thousands of bows, arrows, knives, swords and shields. We know that the Romans manufactured these items in Rome to equip their legions. In those days, manufacturing was a craft activity. Skilled individuals made these and other things using crude tools and no automation. This continued for over 3000 years. In the late 18th century and the early 19th century all of this began to change. Exploiting the wheel and the concept of leveraging power by using gears, a new manufacturing paradigm slowly emerged. Using the power of water from streams, manufacturing plants began to automate by generating power from falling water. Pulleys and gears were connected by crude homemade belts and the cottage industries of individual skilled craftsman gave way to groups of powered machines and the job shop emerged. These initial efforts at automation were pioneered and perfected in the Eastern United States where water power was plentiful and creative men were out to make their fortune. The world came to see......and the world adopted these methodologies. Water power reached its zenith and its destiny when James Watt (1726-1819) invented the first modern steam engine. This paved the way for the locomotive, ocean liners and powerful machinery. The old had passed away, and the new had emerged. This was the *First Industrial Revolution,* called the American Armory System by the historians.

The first true Manufacturing engineer was undoubtedly Henry Ford. Henry Ford pursued a vision of creating an automobile that almost any American could afford, and at the same time operate without failure for long periods of time. Initially, Henry Ford was influenced by two key principles: (1) standardization and interchangeability of parts and (2) specialization of labor (Turner, Mize and Case). As far as we can tell, Adam Smith (1776) and Charles Babbage (1832) were among the first to recognize the efficiency and productivity of division of labor. The advantages and superiority in mass production of part standardization and interchangeability were pioneered by Eli Whitney (1785-1793) when he standardized parts which were used to build a cotton gin.

Fueled by these two key principles, Henry Ford became the progenitor of the *Second Industrial Revolution.* Almost single handedly, he designed and built the first Mass Production Assembly line and built millions of Model T (1908-1927) and Model A (1927-1931) Ford vehicles. As a testimony to his genius and engineering skills, many of these vehicles are still running today.

"I will build a car for the great multitude. It will be large enough for the family, but small enough for the individual to run and care for. It will be constructed of the best materials, by the best men to be hired, after the simplest designs that modern engineering can devise. But it will be so low in price that no man making a good salary will be unable to own one – and enjoy with his family the blessing of hours of pleasure in God's great open spaces" Henry Ford, 1908

Although Henry Ford was very successful, he had a fatal flaw in his thinking. Obsessed with standardization, he purposely eliminated any variety or customization of mass produced vehicles. One of his famous quotations was: *"You can buy any color of Model A that you choose, as long as it is black"*. Reluctant or unable to respond to the individual customer, his pioneering efforts were soon to come to an end. Almost simultaneously, entrepreneur name William C. Durant arrived on the scene.

In 1908, Durant recognized that the automobile was here to stay, but that a customer was going to demand variety and customization. He formed General Motors Corporation, and the *Second Industrial Revolution* was underway. Durant set up several automotive divisions (Buick, Pontiac, Cadillac and Chevrolet) and within each division he produced one line of automobiles, but with variety in several different models. General Motors pioneered the Mass production of mixed-model automobiles and the concepts of high speed assembly lines. A new engineer emerged which we now call a manufacturing engineer, but they were mostly Mecanical engineers. This new breed of manufacturing engineers and manufacturing practices were fueled by the work of Frederick F. Taylor (Scientific management) and Frank & Lillian Gilbreath (Motion and time Study). Both were concerned about the best way to perform a task (time and cost). Henry Gantt developed charts to schedule these factors. These people can rightly be considered to be the first Industrial Engineers. The world's first Industrial engineering Degree was established at Penn State University in 1909 in the Department of Mechanical engineering.

While mass production lines were being pioneered by General Motors, there was a dark cloud which hung over world peace. In 1914 WWI started and lasted until 1918. Mass production of war machines such as guns, tanks and planes were built upon existing knowledge and much success was achieved. However, modern mass manufacturing and paced assembly lines were not begun until 1941 when the United States entered WWII (1939-1945). Between 1941 and 1945 there were enormous strides taken in the mass production of war machines. After WWII, these methodologies and production innovations began to find their way into general American Manufacturing. Following WWII, America emerged as the world leader in manufactured goods and services. Led by the aerospace and automotive industries, mass production continued to evolve and produce at higher rates. During this era, Industrial Engineering played a large role in finding better ways to do things, designing more efficient systems and optimizing key parameters. The work of Charles Babbage, Henry Metcalf, Henry Gantt, Frederick Taylor and the Gilbreaths (Frank & Lillian) established Industrial Engineering as the *productivity people*. Between 1923-1932 Elton Mayo conducted his now famous *Hawthorne Experiments* at the Western Electric Company. Mayo directed a series of studies which scientifically established that there were close and undeniable connections between the physical conditions, mental attitudes and welfare of the workforce to productivity, quality, and a host of other factors. These factors were directly related to corporate profits. These experiments expanded the scope of Industrial engineering to include Human factors and Ergonomics, and had a dramatic influence on Taylor and the Gilbreaths. George Danzig developed Linear Programming at Stanford University in 1939, and proved its usefulness in solving a wide range of practical problems. During WWII, the field of Operations Research emerged in England under a group called *Blackett's Circus*. The Industrial Engineering community adopted Operations Research into their program of study, and modern Industrial Engineering emerged.

The Second Industrial Revolution was not instantaneous, but spanned a period of over 50 years, including two world wars. The evolution of assembly line manufacturing started with Henry Ford's production system and is currently best typified by the high speed assembly lines of General Motors Corporation, Ford Motor Company and the Chrysler Corporation. In this era, the Mass Production System reached unprecedented capabilities to produce a wide variety of automobiles. However, as previously discussed, the Achilles heel of American Automobile Manufacturing is the lack of flexibility and versatility to rapidly change from one set of finished products to another. This may seem paradoxical in contrast of the previous statement that a wide variety of automobiles are produced, but manufactured variety is actually built into the mass production systems. A typical American automobile manufacturer will spend up to 3 years designing future automobiles, and to change mass production from one year to the next requires a complicated and expensive changeover sometimes spanning 6 months. Modern Mass production Lines often involve massive *supermachines* and highly automated processes, blended with some manual tasks. This is a highly automated form of Henry Ford's flow line for building and producing modern automobiles in extremely large volumes (400,000 to 500,000 per year). The enabling technologies are repeat-cycle, dedicated machines with automated material handling devices and sophisticated control systems. Although a wide variety of automobile configurations are produced, within any one family of cars there is a large degree of standardized basic configurations and interchangeability of parts.

As the modern Mass Production System evolved, it became necessary to develop and deploy highly sophisticated Production Planning and Control Systems. In 1964 Joseph Orlicky introduced Materials Requirement Planning (MRP) as a tool to combat emerging international competition. In 1983 Oliver Wight transformed MRP into Manufacturing Resource Planning or MRP II. As the modern computer matured into an incredible computational platform and the PC emerged, the race to implement computerized production planning and control was underway. By 1975, MRP was implemented in 150 companies. By 1981, this had risen to over 8000 companies. MRP II brought master scheduling, rough-cut capacity planning, and material requirements planning under one computational umbrella. By 1989, about one third of the software industry software sold to American industry was MRP II ($1.2 billion worth of software). However, major problems began to surface. First, modern MRP II systems can require a large support staff to manage its capabilities. Second, there has been a failure to fully integrate factory floor status into the planning framework because of cost and time. Third, such a wide degree of computer control tends to dehumanize the factory worker. Skilled labor with years of experience began to be supplanted by computerized instructions. When things went bad (frequently) and work flow priorities disrupted the planned schedule, expeditors were sent to resolve the problem. Worse, because expensive high speed mass production lines were driven to 100 % utilization, any disruption to product flow caused a corresponding failure in planned production. The stage was set for a new Industrial Revolution to emerge.

The *Third Industrial Revolution* came from an unpredictable and surprising source. After WWII ended, the Japanese Toyota automobile manufacturing company was ready to take on the world supremacy of American Automobile Manufacturing; but there was a lack of sophisticated computer software and hardware to build upon. Further, labor was plentiful and comparatively cheap. Finally, the Japanese economy was not ready to support the sales required to support another General Motors Corporation. In the early 1950s Taiichi Ohno conceived and designed an entirely new manufacturing system concept that was destined to change the entire manufacturing world. His system became known as the *Toyota Production System* (TPS), and was rooted in five(5) basic concepts: (1) Just-in-time production… only manufacture what is needed, in the quantity needed, at the correct point in time (2) Implement pull vs push production strategies (3) Control production with simple, manual control strategies which became known as Kanban (4) convert all Mass Production lines into a Linked-Cell Manufacturing System (L-CMS) with Mixed Model Final Assembly (MMFA). Finally, and most important (5) Aggressively and continuously vigorously attack any form of waste (Muda) in the entire manufacturing system from the supply chain to final product delivery. The methodologies, strategies and tools to accomplish this transformation are contained in this textbook.

Ohno found perhaps the one man in Japan that could make the TPS concept a reality. His name was Shigeo Shingo. When history records the Third Industrial Revolution, the name of Shigeo Shingo will rank with those of Henry Ford, Frederick Taylor and Lillian Gilbreath. Shingo was a Japanese Industrial Engineer. If Ohno was the conceptual force behind TPS, Shingo was the driving force behind its implementation. Every Industrial Engineer should read his books on the Toyota Production System. This new design took years to develop and implement, but by the mid-1960s the world began to take notice. In 1982 Shoenberger wrote a landmark textbook called *Japanese Manufacturing Techniques* after visiting Japan. In 1990, Black published his first book describing manufacturing cells called the Design of the Factory with a Future. In 1991 Womack, Jones and Roos published a book called The *Machine that Changed the World.* The term lean production introduced here was coined by John Krafcik. After extensive trips to Japan, Womack and Jones rocked the entire industrial world by publishing a textbook called *Lean Thinking* in 1996. The term *Lean* stuck, and TPS became known as the *Lean Production System.* Academicians such as Hall, Monden and Black soon joined the chase; and understanding the transformation to Lean Production began to emerge. The concepts of U-shaped, Lean Production Cells were published by Black and his colleagues (Black and Hunter, 2003). Lean concepts are now being applied to service systems, hospitals, insurance companies, small job shops and large manufacturing companies. One of the first companies to jump on the Lean bandwagon was Harley-Davidson, who reported spectacular results (1980s).

Around 2008, J.T. Black and Don T. Phillips met and discussed the need for an exciting new breed of manufacturing engineer… one who would be thoroughly trained and equipped to lead Lean implementation across a wide variety of applications. It was agreed that perhaps the best engineering discipline from which this new engineer might emerge was Industrial Engineering. This decision was by no means self-serving: The modern Industrial Engineer is trained to continuously improve the intersection of people, scarce resources and equipment. The TPS and current Lean transformations require extensive knowledge of this intersection, with specific training in systems design and analysis, statistics, human factors and ergonomics, production planning and control, facilities layout and design, supply chain management and warehousing/distribution/material handling. However, Black and Phillips soon realized that while modern Industrial engineering provided a strong foundation, there were significant educational changes which needed to take place. Traditional IE skills need to be augmented with the methodologies of aggressive variance reduction, the principles of Factory Physics, Lean cell design, setup time reduction, decoupler design, Poka-yoke insertion, Kanban control, full preventative maintenance, total quality management and zero defects, 5 S methodologies and WIP reduction strategies: Worker paced production and not machine paced production, push versus pull strategies and many other new skills including the intersection of 6-sigma methods. This new breed of Industrial Engineer deserved a new name: We call this new engineer a *Lean Engineer*. We will be quick to state that the new Lean Engineer does not supplant or replace the traditional Industrial Engineer, but rather creates a new career path and game-changing engineering discipline with specialized training in Lean Concepts. This concept prompted us to construct this textbook, which represents only an initial effort to define and equip this new engineer with a core set of tools. In a real sense, we seek to convince the discipline of Industrial Engineering to embrace, support and educate this new Lean Engineer. Succinctly stated, this new baby needs a home. We propose…..now we plead….that Lean Engineering and Lean Thinking be unified under Industrial Engineering and taught as a new, emerging discipline.

Many academic institutions, including Texas A&M University and Auburn University, have introduced stand-alone Lean course(s). While this has created a general awareness of Lean concepts, a unified body of knowledge and the unique design aspects of Lean Systems design have not yet emerged. The truth is also that Lean Thinking and Lean Concepts are no longer confined to automobile manufacturing. These concepts have been found to be universally effective in reducing wastes, improving system performance, increasing quality and simply resulting in higher profits. Lean is now surfacing as a transformative theory in all forms of production, service and mass production systems. In a recent informal survey by the authors, 61% of all manufacturers reported that they were either planning to implement Lean or were in the process of doing so. This is not a fad nor is it a temporary fix…. It is the Third Industrial Revolution. It demands to be unified under the title of Lean Engineering. The authors are convinced that when some Engineering discipline decides to do this… the world will beat a path to their door. But much needs to be done; this text is only a meager beginning.

As efforts to integrate Lean concepts, Industrial engineering and 6-sigma emerge into a unified engineering discipline; it will become clear that the main benefit will be to create a deeper understanding of the unified power in these now mostly disjoint but complementary disciplines. The first step is not simply to integrate these concepts into a cohesive program of study, but to transform and change the very culture of manufacturing engineering. If 60%-70% of all manufacturing companies have begun this cultural change, how can we not also do so?

We are simply hopeful that this book will help articulate and define the very unique nature of Lean Transformation and Lean Systems Design, creating a pathway to *Factories with a Future* and *Companies with a new Concept*.

J.T. Black and Don T. Phillips
2013

Chapter 1
Introduction to Lean Engineering

Changing any system will always take longer than you might think.   J.T. Black

KISS...Keep It Simple Stupid, or at least as simple as possible.   Don T. Phillips

1.1   What is Lean Engineering?

In the second decade of the 21st Century we are witnessing in America the restructuring of manufacturing systems. Manufacturing is on the decline, and assembly is on the rise. Extensive outsourcing to operations outside the United States border is becoming more commonplace. The entire manufacturing and assembly Enterprise is now the focus of intensive cost saving activities and waste reduction. The restructuring of the manufacturing enterprise and the supply chain which supports that enterprise is underway to produce goods and services at less cost, with higher quality in less time with higher customer satisfaction. Underlying these goals are principles of lean engineering, six-sigma engineering and Industrial Engineering. Leading these innovative practices is a Japanese company called Toyota, who has risen from meager beginnings to the world leader in automobile manufacturing. Toyota pioneered a new manufacturing system design and operation called the *Toyota Production System* or TPS.

Since its origin in the early 1950s this new system has acquired many names; including Just-in Time (JIT), World-Class-Manufacturing, Stockless Production, Zero Inventory and Integrated Manufacturing Production System (IMPS). A full suite of common names for Lean Production is given in the following. The most commonly used name for this new system of operation is the *Lean Production System* which originated in a book called *The Machine Which Changed the World* by Womack and Jones in 1990.

The term Lean Production stands in opposition and contrast to traditional *Mass Production* systems. In the early 1950s, Toyota Motor Company restructured its business with the goal of being the number 1 automobile manufacturer in the world. To achieve this goal it invented and implemented a new manufacturing system design which it called The *Toyota Production System* (TPS). In any manufacturing system, products, goods and services are produced by transforming raw materials to finished goods and products by a set of manufacturing process involving people, materials and machines. The manufacturing system systematically transforms *raw materials* into *final product(s)* through a system of *Value Added* activities. It is important that we distinguish the *Manufacturing System* from the *Manufacturing Enterprise*.

Table 1.1
Name and Origin of Lean Manufacturing System

Names and Origin for the New LEAN Manufacturing System	
Lean Production	James P. Womack and Daniel T. Jones (In book The Machine That Changed the World)
The Toyota Production System	Mondon, Ohno and Shingo (Toyota Motor Co.)
Integrated Pull Manufacturing System	AT&T
Minimum Inventory Production System	Westinghouse Corporation
MAAN-Material As Needed	Harley-Davidson
Just in Time System	Dick Schonberger (In book Japanese Manufacturing Techniques)
World Class Manufacturing	Dick Schonberger (In book World Class Manufacturing)
ZIPS-Zero Inventory Production System	Omark and Hall (In book Zero Inventory)
Quick Response Manufacturing	The Apparel Industry (The Toyota Sewing System)
Stockless Production	Hewlett - Packard Corporation
Kanban System	Many Japanese and American Companies
The New production System	Suzuki
Continuous Flow Manufacturing (CFM)	Many American Companies
The Linked Cell Manufacturing System (L-CMS)	JT. Black - Auburn University
One - Piece Flow	Sekine
The Integrated Manufacturing Production System - IMPS	JT. Black (In book The Design of the Factory with a future)
The Lean Engineered System (LES)	JT. Black and Don T. Phillips (Texas A&M University)

It is interesting to note that the term *Lean Thinking* has come to be synonymous with Lean Manufacturing. Lean thinking is simply the mental process of aggressively and continuously reducing *Muda* or waste. While this is certainly a necessary and continuous goal of Lean Manufacturing, it fails to capture the whole range of design, implementation and sustainment activities required to create a true *Lean Enterprise System*.

The Lean Engineer does not represent a completely new body of knowledge, but a unique combination of several traditional and some new areas of knowledge. This book is an attempt to capture the tools, techniques and operational activities necessary to transform a traditional system or enterprise into a Lean System or enterprise. The force required to do this is embodied in what we now call a *Lean Engineer*.

An *Enterprise System* supports and manages the *Manufacturing System* where the actual value-added activities take place. Figure 1.1 provides a high level view of the management and control functions which constitute a typical *Enterprise System*. The manufacturing system is a self-contained or distributed system which includes all of the manufacturing and operational components to manufacture a final product. In today's global economy, this system may be distributed among various plant locations or international subassembly operations. These operational components are coordinated, managed and controlled by a *Supply Chain Management System*. The elements in a typical manufacturing system are automated or semi-automated machine tools, workstations, tooling and fixtures to support the workstation, human resources to support the system, material handling, warehousing and distribution systems to support product flows. The arrangement, functions, physical locations flexibility and versatility of these elements are largely determined by the *Manufacturing System Design*.

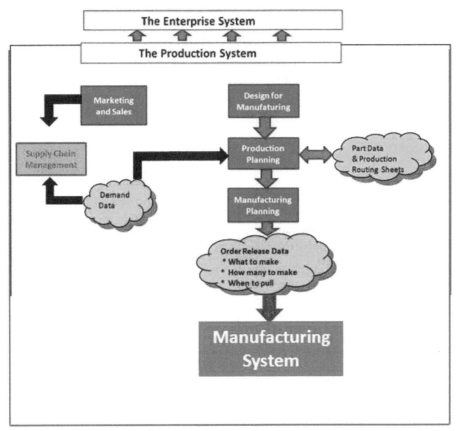

Figure 1.1
The enterprise system takes product demand data and product definition data to plan the work in the manufacturing system.

1.2 Manufacturing Systems

The collection of processes and people that actually produce a final product is called a *Manufacturing System*. A typical manufacturing system is shown in Figure 1.2, and is characterized by a complex arrangement of physical <u>elements</u> characterized by measurable operating parameters (Black, 1991). Typical operating parameters are machine availability, through put time, cycle time and output rates (or production). The relationship of necessary manufacturing components and their complexity determine how efficiently a system can be operated and controlled. System control not only involves individual process steps, but also involves the entire manufacturing system. How well each process and element of the system harmoniously exists and supports one another largely affects the profit margin, efficient use of scarce resources and satisfaction of stated system goals and objectives. The entire manufacturing system must be continuously coordinated and controlled to efficiently move raw materials (work-in-process) throughout the system, schedule people, processes and customer orders, manage inventory levels, insure high product quality, maintain target production rates and minimize system costs.

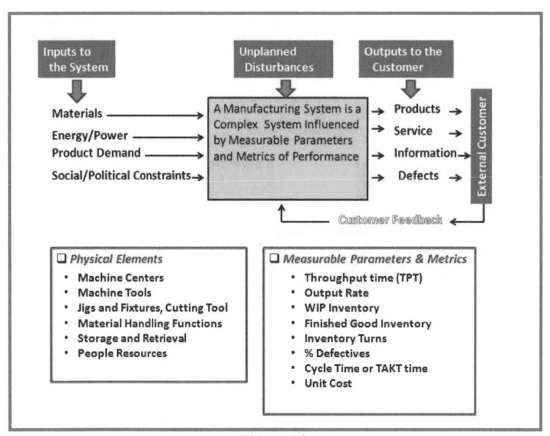

Figure 1.2
Definition of a Manufacturing System with Inputs and Outputs

The manufacturing system receives inputs that include materials, information and energy. The integrated system is a complex set of elements that include a wide variety of machines, tooling, material handling equipment and people. The entire system exists to serve *customers*; both internal and external. Individual machine centers within the system represent *internal customers*. The *external customer* is the person or entity who consumes goods and services produced by the

$$\text{Revenue} = \text{Sales Price} - \text{Cost of Production}$$

manufacturing system. Internal customers create value as raw materials are transformed into final product. External customers (usually) pay money for the delivery of these final goods and services. There is (usually) only one reason for a competitive, production system to exist... and that is to generate profit (revenue). In a traditional manufacturing system, revenue or profit is usually determined as a reasonable corporate goal, and is accomplished by calculating the total cost of production and establishing a sales price to guarantee revenue. This traditional way of thinking is a recipe for corporate disaster in a competitive sales environment. If the sales price is too high, the consumer will look elsewhere for a comparable product. In the Lean Engineering world, a subtle but powerful alternative is used.

$$\text{Sales Price} = \text{Revenue} + \text{Cost of Production}$$

4

In this world view, if desired revenue is set at say 30% profit, then the path to lower unit cost to the consumer….and hence higher sales if a quality product is produced…. is to lower the cost of production. This is the real driving force behind Lean Engineering. Lean Engineering and its associated methodologies are directed to *waste reduction*. In the Lean world, all forms of waste are aggressively attacked and systematically reduced. The result is reduced through put times, increased production rates, tighter system control and a higher quality product. This book will provide a framework and definition of how to achieve these goals, which we call *Lean Engineering.*

All manufacturing systems are complex, dynamic and subject to variation. This means that they must be designed and built to constantly accommodate change. Many inputs previously described cannot be fully controlled by management, and the typical response to variance and disturbances are counteracted by constantly manipulating, changing and modifying order release, sequencing, scheduling and material availability. Constantly changing these functions of a manufacturing system can only create more variability in a system. Lean Engineering seeks to level production and create a stable manufacturing environment.

In order understand the difficulty in designing or changing existing systems using modeling and analysis tools the following list will provide insight.

- A Manufacturing system is usually very complex, difficult to rigorously define and exhibit different goals in different areas
- Accurate data describing system operational behavior is usually either nonexistent or difficult to easily obtain. Even when system data exists, it may be inaccurate, out of date or too obscure to analyze and use without considerable effort.
- Interactive behavior and operational relationships may be awkward to express in analytical terms and exist in nonlinear form. Hence, many standard modeling tools cannot be applied without oversimplification.
- The physical size of large manufacturing systems may inhibit detailed modeling and analysis
- Real-world systems rarely reach "steady state" behavior, and dynamically change through time. Many external forces such as environmental parameters can change system behavior.
- All forms of accurate systems analysis are subject to modeling errors (inaccuracy or lack of sufficient detail); errors of omission (missing data and operational logic) and errors of commission (failure to properly use all of the data).

Because of these and other difficulties, systems simulation analysis has emerged as the most important tool for detailed systems analysis and for manufacturing system design. Of course, higher levels of modeling and analysis such as queuing networks often provide extremely valuable insight. As new Lean Engineering modeling and analysis capabilities emerge, systems analysis will be more effective.

1.3 Critical Control Functions in a Manufacturing System

Ideally, manufacturing systems control should be vertically integrated, with each level of system control being composed of horizontally integrated subsystems dedicated to appropriate tasks.

Each level of control is vertically integrated to support the goals and objectives of the entire system. The most critical control functions are *production control, inventory control, quality control and machine tool control.* While the system as an integrated entity will have a number of goals and objectives (cost control, profit, etc.) these system goals are often sought to be optimized by optimizing selected bits and pieces of existing subsystems. This will never work in a complex, interactive system. Real system control functions require complex information analysis from the lowest levels of operation to the highest. All decisions and operational policies must reflect both local (machine level) and global (corporate level) impact. To control any system:

- The boundaries, constraints and interactive dependencies must be clearly identified.
- System response and behavior to any system change must be identified.
- System behavioral objectives (Cost, profit, span times, throughput rates, etc,) must be linked to subsystem operational rules

The rules and laws which describe and capture system behavior are difficult to discover and describe. Recent landmark work in this area is beginning to produce meaningful methodologies and procedures (*Factory Physics* by Hopp and Spearman for example). Accurate rules and laws of behavior must accurately link the behavior of a system to perturbations or changes in both input and behavioral characteristics. The entire field of Lean analysis is directed to this goal.

1.4   Trends in Manufacturing Systems Design

In Chapter 2, we will devote considerable time to describing historical, current and future Lean systems design of manufacturing systems. The behavior of both *greenfield systems* and existing systems can be greatly influenced by the principles of Lean Engineering which will be presented in this book. However, the operational field of systems engineering has been partially clouded if not destroyed by academic researchers; principally those in Operations Research. Academicians have historically looked at manufacturing systems as an application *playground* for optimization theory. In order to sustain this approach, many important nonlinearities, complex systems interactions and stochastic behavior have been *assumed* away. The exception to this rule seems to be a subset of researchers applying simulation modeling techniques. In almost all cases, the operational and production system has often been assumed to be given or fixed to a large degree, and not something which requires change. As the reader proceeds through this book, it will become clear that Lean thinking and Lean operation is a path that *requires* and *demands* system behavioral, operational and managerial change in the fundamental way that a system is designed and operated. This new way of systems design and operation has been fueled by the following trends.

- Many manufacturing systems now operate on an international, global scale. There are two aspects of this trend that should be noted.
  - As manufacturers and producers of goods and services build plants around the world, the new manufacturing systems design must function in places with different cultures, currency ad languages.
  - Outsourcing United States manufacturing functions to other high-tech, low-cost countries like Indonesia, Brazil and China will continue.

  These two facts do not negate the reason to embrace Lean Thinking and to apply the principles of Lean engineering .... Rather they demand that waste elimination,

WIP control and more effective system control policies be adapted to survive and be cost competitive.

- The proliferation of variety in products has resulted in a significant decrease in manufacturing build quantities. Smaller production lot sizes are necessary to facilitate increases in product options and variety; level production; quickly respond to customer demand; and support manufacturing versatility and agility.
- The use of modern, exotic varieties of raw materials such as composites and plastics have created a need for old manufacturing processes and equipment to be seamlessly replaced by new, modern processes. New manufacturing systems must be designed to support rapid system change, and existing systems must be reengineered to support versatility and agility. In fact, many manufacturing systems such as those in the semiconductor and automobile industry feel that they must respond to new technologies to even exist.
- The United States over the past decade has experienced a continual decrease in both the skills and desire for workers to embrace a career in advanced manufacturing (See Figure 1.3). This trend has resulted in increased automation and the emergence of "workerless warehouses". As a consequence, direct labor hours are no longer considered the basis for calculating cost per unit and direct labor cost has dropped to 5%-10% of total cost in almost every modern factory today.

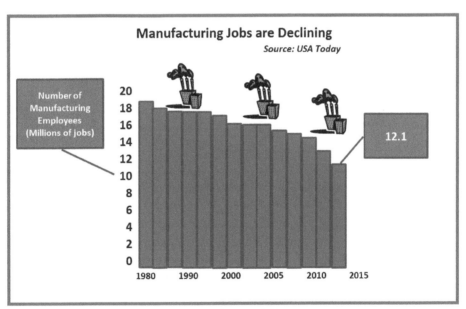

Figure 1.3
The Number of Manufacturing Jobs in the USA

- Product reliability has increased in response to competition. Product liability (lawsuits) from use, misuse and abuse of products has also increased.
- To remain competitive, the total time from product design to finished product has had to decrease. This requires a more flexible manufacturing system design.

7

- The continued growth of computer usage has led to more tightly integrated product design and manufacturing system design. The use of 3D simulations, CAD/CAM application and rapid virtual prototyping has significantly impacted the design/manufacturing cycle time.
- Ergonomics and worker safety continue to grow in importance as worker absenteeism and compensation costs have escalated.
- Green Manufacturing philosophies and social pressures to become environmentally friendly are rapidly becoming driving forces. Zero wastes and disposable products will continue to present themselves as significant product design issues.
- Typically, 50% of total manufacturing costs come from the materials used (Direct cost) and almost 35%-40% from overhead and indirect cost. The Total Cost for a typical unit of product is shown in Figure 1.4. For example, if a small model car cost $20,000 in a modern automated factory, then $10,000 would be for direct cost and $7,000-$8,000 in overhead and indirect costs This would result in $2,000

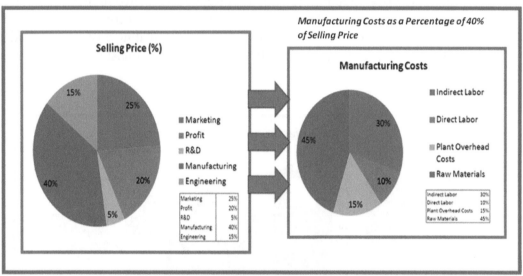

Figure 1.4
Manufacturing Cost as a Percentage of Sales Prices

- (10%) to$3,000 (15%) profit per car. Viewed another way, on an assembly line producing 400 cars per 8 hour shift, we might have 1600 workers. Each worker works on each car for 1 minute. If the worker is paid $20 per hour, the cost per laborer per car is $.30. Direct labor cost per car would be $480 or about 5% of the manufacturing cost.

The response to customer demands for cheaper-better-faster is translated into lower profits and higher costs unless Lean Manufacturing principles are applied throughout the product development life cycle. Reducing unit cost while increasing value; delivering superior quality products; meeting on-time delivery and literally guaranteeing customer satisfaction has forced American companies to adopt Lean strategies to become more nimble and responsive. Either using original TPS strategies; adopting TPS strategies to system design and operation; or developing new Lean methodologies are enabling companies to bring competitive products to market cheaper, faster and on-time by matching production rates to final assembly, employing just-in-time strategies, implementing pull systems and smoothing production. These same strategies have resulted in increased flexibility and the capability to quickly respond to varying product demands and product mix.

1.5 Four Basic Manufacturing System Designs

All factory configurations may look different, but there are only four basic manufacturing system designs (MSDS); four classic configurations and a new linked cell Lean manufacturing system design. The four classic system configurations are (1) the job shop (2) the flow shop (3) the project shop (4) continuous processing. The new design is a (5) Lean Linked cell system. The linked cell design is a new Lean manufacturing concept composed of manufacturing, assemblies, and subassembly lines configured in U-shaped cells; all feeding a mixed model final assembly. It is conceivable that some might view the high speed assembly line as another type of manufacturing system which is a specialized and typically highly automated flow shop. Table 1.2 lists examples of each system and Table 1.3 compares each type of manufacturing line in terms of operating characteristics. These two tables are not meant to be an exhaustive enumeration but representative.

Table 1.2
Types and Examples of Manufacturing Systems

Type of Manufacturing System	Service System	Product Sales
Job shop	Auto Repair	Machine Shop
	Hospital	Metal Fabrication
	Restaurant	Custom Jewelry
	Insurance	Machine Tool Producer
Flow Shop or Flow Line	Hospital	TV Manufacturer
	Cafeteria	Auto Assembly Line
	Custom Shirts	Pre-Fab Housing
		Hot Water Heaters
Project Shop	Movie Set	Aircraft Carrier
	Broadway Play	Lamination Chamber
Continuous Processing	Telephone Company	Chemical Plant
	Cable Television	Oil Refinery
Linked-Cell Lean Shop	Subway Sandwiches	Automobile Manufacturing
	Car Wash	Clothing Manufacturer
	Upholstry Shop	Pump Manufacturer

9

Table 1.3
Characteristics of Basic Manufacturing Systems

Characteristics	Job Shop	Flow Shop	Project Shop	Continuous Processing	Lean Manufacturing
Types of Machines	Flexible, General Purpose	Single Purpose, Single Function	General purpose, Mobile, Manual & Automated	Specialized, High Tecnology	Simple, Customized & Single Cycle Machines
Layout	No Particular Order or Pattern	Product Flow	Fixed Position	Immovable Locations	Linked-U_Shaped Reconfigurable cells
Setup & Changeovers	Long, Variable & Frequent	Long, Expensive & Complex	Product Specific, Virtually None After Design	Rare & Expensive	Short, frequent
Operators or Skilled Personnel	Highly skilled, Multifunction, Versatile	Specialized, Highly Trained, Dedicated to process step	Specialized, Highly Skilled	Skill level varies; Processing, Packaging and Engineers	Multifunctional, Cross Trained, Mobile and Skilled
Inventory (WIP)	Very low (specialty Shop to Very High (Piece Part MFg.)	Large with Buffer Storage	Variable, but Usually Large	Very large but in Tanks, Trucks, Rail Cars or Ships.	As Small as Possible
Process Routings	By order and by machine. No Fixed Routing	Fixed and Prioritized	No such Concept, but Stages of Work are Critical	Many Pipelines and Different Storage Locations/Modes to Route through.	Fixed within Cells by part type and at Final assembly
Production Control	Once Work is Released, flow according to Due Date and Machine Availability	Follows a Travellor or a Process Routing Sheet	Work accomplished in planned stages	Release and Process as soon as Possible. Usually capacity limited	Controlled by a Kanban Signal Strategy
Order Release	Release Order as Soon as Possible Upon Demand. No Forecasts.	MRP or an Order Scheduling System Based upon Forecasted Demand	No such Concept Except to control Subassemblies & Module	No such Concept as Compared to Piece-Part Manufacturing.	Based upon a Combination of Actual Demand and Forecasts.
Unit Flow	One at a Time	In Large Lots	No Analogy to Traditional Manufacturing	Continuous in Liters or Gallons	One-At-A -Time or Small Lots
Basic: Push or Pull	Push	Push	Push & Pull	Push	Pull

1.5.1 The Job Shop

The distinguishing feature of a job shop is its flexibility. In a typical job shop, small and even unit lot sizes are produced to specific customer order specifications. Because the job shop must perform a wide variety of manufacturing operations, several different manufacturing processes and basic machine types are usually found, sometimes in a random fashion as new capabilities are added. General purpose and highly flexible machines are required, and each manufacturing operation is usually done by highly skilled labor. Process steps are frequently characterized by a wide range of work content, and standardization is usually difficult if not impossible.

As the demand for certain parts increase, machines are grouped according to the general manufacturing capabilities and characteristics represented by each work group. For example, one area might contain lathes; other milling machines; other grinders…etc. A finished product (part) will have a unique sequence of operations to be performed in a serial fashion, which usually results in and scheduling rarely exists. As the demand for certain parts increases, jobs may be manufactured and moved in medium sized lots or batches.

A typical Job shop is shown in Figure 1.5. A particular part may enter or leave the same functional area many times in a re-visitation sequence. The process flow chart for products in a job shop often looks like what is sometimes called *spaghetti flow*. *Routing sheets or travelers* are used to specify particular routing sequence with operating instructions. Material handlers, fork lift trucks or *pusher dogs* are often used to manually transfer a part from one machine to another. Sophisticated sequencing and scheduling rarely exists. A typical Job shop is shown in Figure 1.5.

A job shop becomes extremely difficult to manage as it grows, resulting in long throughput times, large in-process inventory and unpredictable delivery dates. Charles Carter, an engineer who worked for Cincinnati Milacron, performed a study where he tracked several parts and part types as they moved through the factory. He found that parts moving from start to finish spent about 95% of their time waiting and only about 5% of their time in actual value added processing (raw material transformation). Further studies revealed that even when a part was at a machining operation; only 30%-50% of *value added* time was actually material transformation.

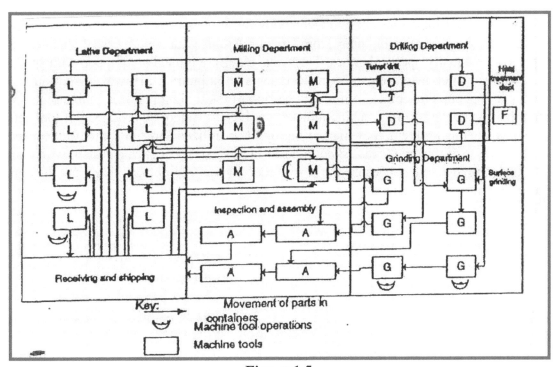

Figure 1.5
The job shop collects machines into functional groups of processes, uses highest skilled labor and produces products in small lot sizes

The remaining percentage of time was spent waiting on set-ups, tool changing, loading and unloading parts and inspection. Thus, the total time actually adding value was only 2%-3% of the through put time. These figures have since been verified many times, and even seem to hold in other type of manufacturing or assembly operations.

In recent years, programmable controllers and CNC/DNC capabilities have created standardization within certain classes of manufactured parts. However, the modern job shop still produces a lot of low volume parts, but by grouping orders and building make-to-stock, larger lot sizes or batch production can be achieved. Manufacturing lot sizes are often determined by well-known Economic Lot Size (EOQ) calculations. This is due to the simplicity of EOQ model calculations and not due to the efficacy of these models. The basic assumptions of EOQ simply cannot be met in a job shop. In later Chapters, we will show that a basic pull- strategy with load leveling yield far superior operating systems.

Basic batch and large lot production is usually used to satisfy continuous customer demand or relatively stable yearly demand patterns for an item. Because the production capability of any one machine often exceeds customer demand over a given period of time, most job shops build-to-stock for each item. The machine is continually being reconfigured to build value into a product at a particular step in its routing sheet. The corruptive effects of this policy are insidious and often financially large. It is necessary to build large quantities of work-in-process from which products are pulled as needed. Perhaps worse, each part will experience delays at each processing step as the machine needed is being set up and being changed over.

In the case of DNC or CNC machines, these tools are often designed for higher production rates, which necessarily use higher speeds and feeds with multiple cutting tools. For example, automatic lathes capable of using many stored machine tools with computerized changeover will far outperform manually operated lathes. These machines are typically equipped with specialized jigs and fixtures, which increase precision, accuracy and output rates. However, such superiority over manually operated machines comes with a price. If the tooling required to manufacture a particular part is not part of the machine tool set on, the part cannot be produced. If tool sets are added to mitigate this problem, they may be highly underutilized and subject to obsolescence. One of the authors of this book was asked to perform an analysis of such a system with many automated machining centers because the cost of tooling was beginning to bankrupt the company and other alternatives were needed.

Many domestic products are made in small to medium sized job shops, such manufacturing systems may be called machine shops, specialty shops, foundries or press working shops to name a few. It is unfortunate that a wide variety of names exist, because they all exhibit similar serious problems. The old adage is true… a rose by any name is still a rose.

It is estimated that as much as 75% of all domestic manufacturing in the United States is done in job shops with between 5 and 100 machines with production lot sizes of 5-100 pieces per order. Hence, traditional and modern job shop operations constitute a large part of our gross national product. All forms of job shop operations can greatly be benefitted by the allocation of Lean thinking and Lean engineering.

1.5.2 The Flow Shop

When product demand is fairly stable and is of sufficient volume to support large lot or batch sizes, a flow shop configuration becomes a superior manufacturing option to a job shop. A flow shop emerges when a single product or a group of products can be made on identical or similar dedicated machines at each step in a routing sequence. Traditionally, the flow shop is arranged such that the manufacturing line is serial and sequential. Flow shops arranged in such a fashion can be very long with many different types of machines arranged such that the output of one machine proceeds directly to the next machine in the manufacturing sequence as shown in Figure 1.6.

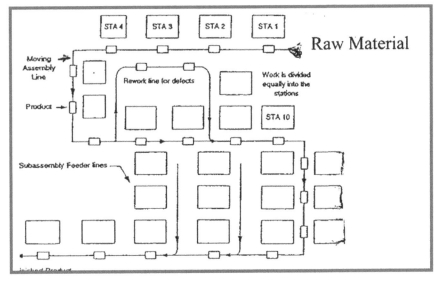

Figure 1.6
A Flow Shop is where Men, Machines and Materials are Assigned to a
Particular Product or a Product Family and Arranged for Sequential Flow

Parts are transported between machines by material handlers or conveyors depending upon material volume being transferred and the manufacturing lot sizes. This form of manufacturing or assembly system can exhibit very high production rates and higher levels of automation than a job shop. A single flow line may have highly specialized equipment dedicated to the manufacture of a particular product. A flow shop might be designed to produce parts from cradle to grave, subassemblies or intermediate part types. The typical capital investment in machines or machining operations is usually very high. The extra cost of a flow line is justified by higher production rates and a large return on investment. In most cases, the sophisticated manual skills in a typical job shop are transferred to automated or semi-automated assembly/manufacturing operations. Ideally, items or parts flow through the system one at a time. The time that each item spends at each process station is the same, and corresponds to Takt time; which is the inverse of the required production (output) rate. Such a production line is said to be *balanced or line balanced*.

As product demand continues to increase and hopefully stable across a particular time horizon the flow shop will give way to the modern *mass* or *high speed assembly line*. A good example of a high speed assembly line is the modern automobile assembly line. A typical rate of production might be 60 cars per hour or one car per minute. We will later show that a car cannot spend more (or less) that one minute per car or one car per minute output rate from any one machine or a parallel group of machines. To be economically feasible, high speed production lines must operate continuously or over long periods of time. The capital investment for an automobile assembly line is in excess of $1.5 billion. These manufacturing systems are designed to produce very large volumes per year, and design changes usually follow a 12 month cycle. Part of this phenomenon is the insatiable appetite of the consumer for more sophisticated and attractive vehicles on a yearly basis. It is unlikely that due to capital investment costs that the process could be radically changed less frequently.

Paradoxically, automobile consumers expect and demand many variations in color, add-ons and style. So to meet the demands of a free economy and be competitive in an international market, automotive assembly lines need to be designed to exhibit flexibility and versatility, while retaining the capability to mass produce. In the flow line manufacturing system, the processing and assembly workstations are arranged in accordance to the product's sequencing of operations. Work stations or machines are arranged in a serial production line with only one type of operation being performed at each processing step. Duplicate workstation (parallel machines) might be added to balance the line, but they all perform the same function. The entire line is designed to produce a product or subassembly of the same type or from the same family of products. In most cases, there are little or no changeover and setup operations, and if there are they may be long and complicated. Hence, basic line integrity and functionality is difficult and expensive to change.

1.5.3 Mass Production

A mass production line is an extension of a flow line to produce more products in shorter periods of time due to extensive automation and control. Mass production facilities require a steady demand of product over relatively short periods of time. As demand dictates a shift from batch mode to high volume production, mass production lines emerge. Mass production lines are usually fed by other high volume flow lines which insert key subassemblies into the main line.

A variety of approaches have been used to develop machines and machine centers which are highly effective in mass assembly lines. Line efficiency and effectiveness usually depends in a large degree to how closely coupled engineering and product design are to the mass production line. Standardization of products, methods and methodologies are key to balanced line operation. If a particular part is fairly standardized and can be manufactured in large quantities, specialized machines with a minimum of *touch labor* can be designed and implemented. It should be noted that the emergence of mass production lines with little or no skilled labor intervention has created havoc in many old corporate planning systems. Key performance metrics and product costing can no longer be based upon direct labor hours per part. This has given rise to *Activity Based Costing* and other new accounting schemes.

1.5.4 The Product Shop (Fixed Position Layout)

In some types of manufacturing systems, the product must remain in the same geographical area, and product transformation must be done in a fixed location. All raw materials, people, processing including machining, assembly and inspection operations must be brought to the product site as shown in Figure 1.8. In the late 1920's Henry Ford built his automobiles using a product shop. Skilled laborers with specialized tools and raw materials were brought to a single point of assembly. Some years later, Henry Ford used a skid drawn on by horses to move the car-in-progress along a straight line to the same (now stationary) set of workers and raw materials.

This is when the modern assembly line was born. The product shop might be considered an archaic, historical production strategy today but in fact it still exists in many forms. Large ships, aircraft, dams, machines and skyscrapers are just a few product shop examples. A typical product shop may in fact contain elements of job shops, flow shops and mass production within its boundaries. Subassemblies and components may even use mass production lines. All product shops are fertile ground for Lean engineering principles.

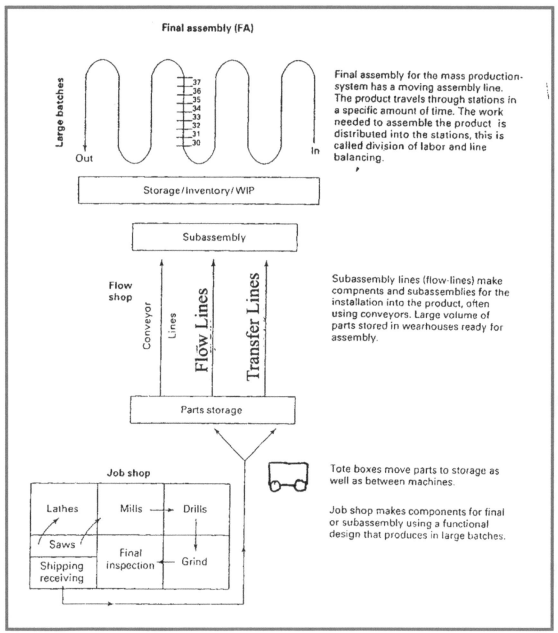

Figure 1.7
The Manufacturing System Design called Mass Production Produces Large Volumes at Low Unit Cost but Lacks Flexibility

Clearly, the product shop is used to produce very large and often one-of-a-kind products. Work is usually scheduled by project management techniques like the *Critical Path Method (*CPM) or the *Project Evaluation and Review Technique* (PERT). These methods use precedence diagrams and specialized analysis algorithms to sequence work and maintain project schedules (time and cost). Project shops are usually very complex, labor intensive and very expensive, often producing only a single product for millions of dollars, like the space shuttle.

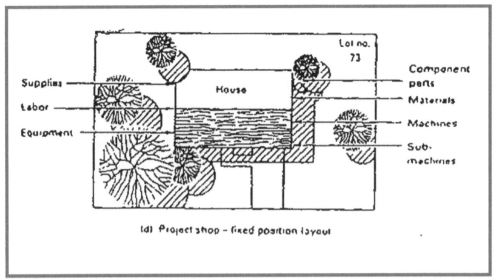

(d) Project shop - fixed position layout

Figure 1.8
Building a House is an Example Product Manufacturing

1.5.5 Continuous Processing

Lean Engineering and Lean practices have not been widely used in the continuous processing industries, but significant cost savings and waste reduction have recently been reported (King and Floyd). In continuous processing, products usually flow in large quantities measured in barrels of product or thousands of gallons. Hence, these systems are often referred to as *Flow Production Systems.* Oil refineries, chemical processing plants and Biofarma production facilities are good examples of Continuous processing, although each of the previous examples of continuous processing actually have mass production components associated with many final products (cans of oil, plastic products, bottles of pills, etc.). In reality, most continuous processing facilities are not strictly processing products continuously, but rather in a mixture of high-volume flow lines. Movement of liquids, gasses and powders do flow continuously and require extremely complicated transformations of raw materials (distillation columns and separators). See Figure 1.9. It is also true that facilities such as oil refineries are the least flexible of all mass production systems, but paradoxically very efficient with large output rates. The control systems are very complex, and work-in-process is hard to measure and discretize. Lean Engineering in these types of systems requires specialized knowledge in Chemical engineering, Materials engineering and Metallurgical engineering in many cases. There is an interesting intersection between the Lean engineer who is attempting to execute waste reduction and the Chemical engineer who is usually assigned the task of managing and controlling the system

Figure 1.9
An Oil Refinery is an Example of a Continuous Flow Process

1.5.6 The Lean Linked-Cell Manufacturing System (L-CMS)

The Toyota Production System (TPS) has developed a radical new approach to the design and operation of almost any piece-part manufacturing system and many other types of manufacturing systems as well. In a linked-cell manufacturing system, three major parts of mass production facility and flow shops are reconfigured to greatly improve product time in the system, WIP levels and output rates. (Figure 1.10). The p*roduct dedicated final assembly lines* are changed over to *mixed model final assembly line* so that the daily production rates for erratic customer requirements can be *leveled* and *smoothed.* Production leveling, also called *production smoothing* is a set of Lean Engineering principles for reducing waste and guaranteeing that each process step produces the same output rate. When demand is relatively constant production leveling can easily be accomplished, but when customer demand fluctuates, flexible and versatile manufacturing strategies with optimal WIP must be developed. The subassembly lines that feed final assembly or intermediate assemblies are also reconfigured from linear, straight-line, flow lines, using conveyers into volume-flexible U-shaped cells with tightly controlled WIP as shown in Figure 1.11. The operations are all manual which require smaller capital investment, and are performed by cross-trained walking workers. Multiple workers are assigned to multiple tasks to achieve worker equality and line balance; often self-balanced, ergo patterns based upon preference and skill sets. Any job shop layout like that shown in Figure 1.5 can be redesigned into a U-shaped manufacturing cell such as that one shown in Figure 1.12.

U-shaped Lean manufacturing cells allow each operator access to multiple machine tools with a minimum of transit or walking time. The operators usually walk in pre-assigned clockwise or counterclockwise loops with no crossing patterns. This U-shaped design lends itself well to standard operations sheets and worker flexibility. The implementation strategy to form these cells and the methodology used to rapidly adjust to changing production rate requirements will be thoroughly discussed in Chapters 3, 4 and 5.

After U-shaped cells are implemented, the cells can then be linked to other sub-assembly points or final assembly with a production control system called *Kanban* which is based upon a *pull* strategy versus a *push* strategy. Similar production and implementation design rules are used to configure final assembly. Key proprietary processes are easily protected and hidden within cell structures.

In a typical linked-cell system, a predetermined level of inventory is maintained within each cell and between cells to minimize work-in progress. These inventory levels are tightly controlled by *Kanban* cards to match the desired output rate and quantity of goods produced. The system is also configured to compensate for any disturbances or variation which cannot be controlled or eliminated. Kanban card systems are discussed in Chapters 13 and 15.

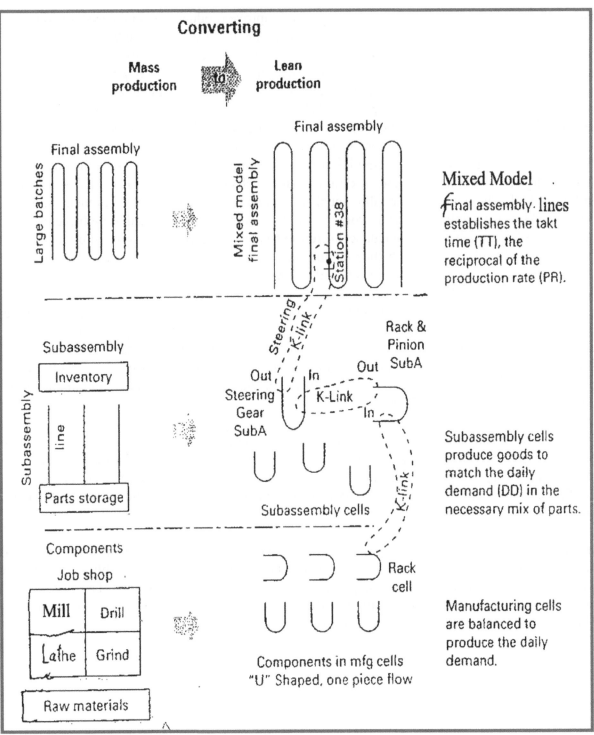

Figure 1.10
Converting the Mass System to a Lean Systems Design

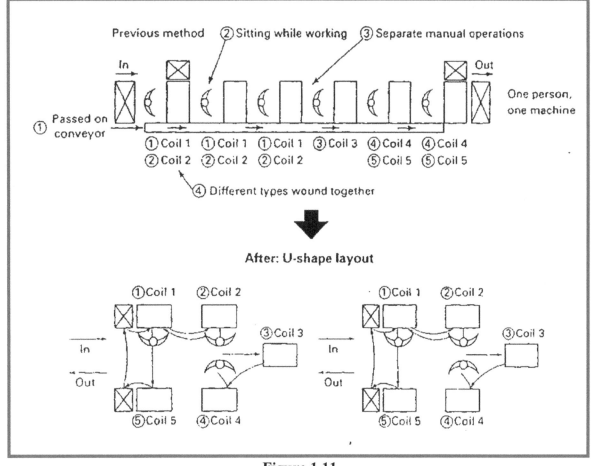

Figure 1.11
Subassembly Lines are Re-designed into Unit Flow Subassembly Cells.
(Source: Sekine, 1990 and Black and Hunter, 2003)

Let us try to quickly summarize the key characteristics of a Lean Engineering designed production system with L-CMS production cells and compare these characteristics to other alternative configurations. It may be useful to once again examine Table 1.3. Lean cell design produces parts in a highly flexible way, meaning that cell output and product mix can be altered quickly to respond to customer demand and market fluxuations. One of the myths of Lean manufacturing is that it can only be applied successfully to production lines which manufacture a single product with fairly steady demand. Nothing could be further from the truth. U-shaped flexible cells are *not* dependent upon either machine location in a U-shaped line nor on machine cycle times. Indeed, families of similar parts can easily be produced in cells using one piece flow, cross trained work, simple single-cycle machines, flexible work-holding devices, and rapid changeover tooling systems. These capabilities lead to small or unit lot size production and product-mix load leveling. Inter-cell inventory is tightly controlled by Kanban strategies and small *inventory supermarkets* (See Chapter 15).Quality is controlled within the cell, and less worker error is experienced due to training and worker experience in executing assigned tasks. Worker pride and morale is increased. It should be noted that no intra-cell sequencing or scheduling required, and no attempt is made to maximize the utilization of the machines… only the workers and a balanced workload. Cell throughput time and output rate is determined by the number of workers (See Chapters 4-6) once adequate production capacity is created.

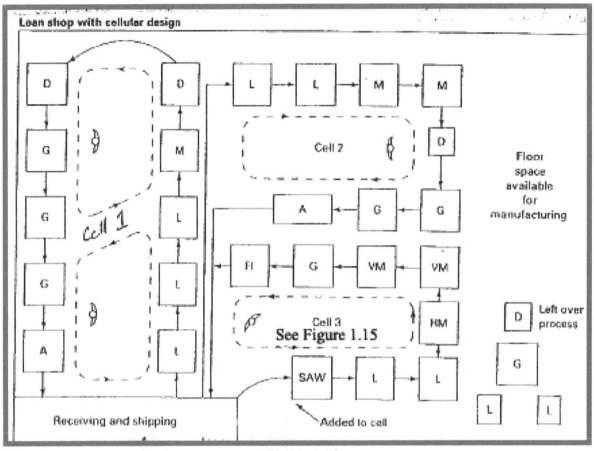

Figure 1.12
Lean Shop Composed of Three Lean Manufacturing Cells
Designed for One-piece Flow

Putting process steps together in a U-shaped cell allows both design engineers and skilled workers to see how parts are made and to immediately uncover production limitations and operational problems. Cell arrangement and multifunctional workers also create an opportunity to uncover any quality problems. In fact, *zero defects* are a mandate rather than a choice. The design engineer can more easily configure and implement future design changes. This is truly designing for manufacturing.

A major advantage of Lean linked cells is that the basic design facilitates *one-piece flow* (OPF) and production flexibility. OPF is the capability to move parts through the system one-at-a-time. Each machine in a cell executes a single processing step in the production sequence. Each part is checked carefully before moving to the next processing step (machine) in sequence. Volume flexibility is achieved by worker-machine separation and the instant reallocation of workers to machines. Contrary to popular belief, mixed model production in L-CMS is simple to implement as long as setup and changeover times are minimized (SMED). Simple or complex parts can be made in a properly configured cell.

20

1.6 Evolution of Factory Designs

There was a time when there were no factories as we define them today. From earliest times until about 1850, most manufacturing was done by single, highly skilled workers in home workshops. This is often referred to as *cottage industries*. These early artisans were watch makers, blacksmiths, tool makers and wagon makers, etc. Any machines used for manufacturing were manually powered. In the mid-1800's the first integrated factories began to emerge. Automated machinery was being discovered, and large quantities of raw materials could be moved by locomotives or through rivers and lakes. These were typical of a wide range of *enabling technologies* which emerged to drive larger scale production. The first factory designs also evolved because power was being made available by new methods. According to Amber and Amber (1962) the major enabling technological innovation that led to modern factories was the ability to power machines on a reliable and consistent basis. The evolution of modern power systems was fueled by the emergence of steel mills. Producing and machining hard metals such as steel required significantly more power than wood or softer metals. Initially, factories were placed in locations dictated by where water could be used to turn large water wheels that drove large gears.

1.6.1 The First Factory Evolution

Closely following the production of unlimited power was the concept of interchangeable parts by Eli Whitney and others. The first factories were built in New England next to rivers, where the water flowing downstream would turn large waterwheels. Machines were run by connecting waterwheels to large gears that drove long shafts which powered machines by connecting leather belts to gearing mechanisms (See Figure 1.13). Later, the waterwheels were replaced by a steam engine, which enabled factories to be built away from waterways. Eventually, steam engines were replaced by large generators and electrical lines connected to the factory. Finally, electric motors were put on each machine in different areas or locations. The initial manufacturing configuration was job shop built around manufacturing functionality. The earliest factories were steel mills and gun manufacturers. Companies like Colt and Winchester invented high speed production and the race was on. Initial American gun manufacturing

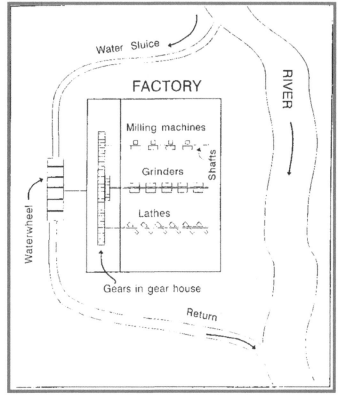

Figure 1.13
Early Factories Used Water Power to Drive Machines Using Overhead Shafts

factories became famous worldwide ad people from all over the world came to see what was called the *American Armory System*. These were the beginnings of the *First Factory Evolution*.

1.6.2 The Evolution of the Job Shop

The job shop concept flourished because of its simplicity. Machines were simply added as demand called for more production capacity. Modern day versions of the job shop exist today in almost the same type of functional layout. It is estimated that 50%-60% of all manufactured products in the world are by job shops, which produce finished goods in lot sizes between 50 and 200 pieces. Processes are still functionally grouped together. Similar types of machines are usually grouped in common areas separated by walls or in separate buildings. It is usually difficult to track and locate work in process without modern data collection systems, and almost all job shops are characterized by having large amounts of work-in-process using an order release system that releases orders to the shop floor as they are received.

Figure 1.14
Singer Sewing Machine Plant in 1854 Showing Men Along the outside Walls Finishing the Parts by Hand. (Source: Invention and Technology Spring 1986)

1.6.2.1 Controlling a Job shop

Throughout the late 19th century and well into the 20th century, there was little or no sophisticated tracking and control systems for typical job shop operations. Engineering drawings, shop floor travelers and manufacturing specifications were introduced but most of the manufacturing control was by individual, highly skilled workers. Machines and machine tools were introduced which was single cycle, repeat cycle automatics. Some were integrated into a single platform such as the Bessemer steel making process. Products moved in unit lot sizes (custom products) or in small batches through the job shop. Then as is today many products were unique and one-of-a-kind. Little theory or scientific management principles existed and little integrated thought was given on how to change capacity or processes within a factory.

Figure 1.15 Frederick Talor was President of ASME When he published "On the Art of Cutting Metals"

This all changed in the late 1890s and early 1900s when a Mechanical engineer named Frederick W. Taylor developed the concepts of *Motion and Time Study*. His pioneering work in improving worker productivity also gave rise to the concepts of *Scientific Management*. Taylor developed "preferred practices" and "standard times" for all sorts of manufacturing activities. Taylor also discovered and invented high speed steel which

**Figure 1.16
Frank Gilbreth
Developed the
Principles of Time 7
Motion study**

doubled cutting speeds.

A colleague of Taylor named Henry Gantt developed charts and procedures for scheduling and controlling work in a factory; he is famous for the Gantt chart which bears his name. Almost at the same time, Carl Barth developed the *slide rule*, which was the first modern computational device for producing shop calculations. Post, and Keffel and Esser, both produced slide rules which soon hung on the belt of every engineer. In this same era, Frank and Lillian Gilbreth began scientific studies based upon *micromotion analysis*. Lillian Gilbreth was formally trained in psychology and sociology, and along with her husband Frank soon developed theories directed to the motivation and attitudes of the workforce. Frank and Lillian more than any other investigator(s) pioneered the *best way* to do a particular task.

**Figure 1.16
Henry Gantt
Developed
Technologies to
Schedule and
Control the Job
Shop**

The pioneering work of Frederick F. Taylor, Henry Gantt, and Frank and Lillian Gilbreth was the innovative and scientific spark which spawned an entirely new engineering discipline called **Industrial Engineering** *which grew up in the era of mass production.* Modern Industrial and Systems engineering provides the foundation for the new and modern Lean Engineer introduced in this textbook.

1.6.3 The Second Factory Revolution

The second industrial revolution was development of the *flow shop*. This factory revolution was fueled by the genius of Henry Ford. Henry Ford defined the concepts of *economy of scale* and *work specialization* in his Model T (1908-1927) and Model A (1927-1931) Ford automobile production systems. His innovative approach to large scale production was built upon division of labor, the moving assembly line and the principles of time and motion study. Ford was the first to attempt to minimize work-in- process on a large scale, and he was the first to fully integrate the *supply chain* with the production system. He exploited standard setup and tooling procedures used with interchangeable parts and product standardization. His most famous saying might have been: *You can have one of my Model A motor cars in any color as long as it is black.* Ford and his manufacturing engineers, led by Charles Sorensen, were responsible for many product design innovations and large scale manufacturing strategies. For example, Ford developed a streamlined process for manufacturing engine blocks. The basic Model A design was a single cast iron engine block instead of pieces bolted together. The use of this engine block led to more power and decreased the weight of each car.

**Figure 1.18
Henry Ford with his Model
A Assembled on a Movig
Assembly Line**

However, almost without question the single most enabling technology Ford pioneered was that of interchangeable parts based upon rigid standards of measurement. Ford demanded that every car produced met standard specifications. He pioneered *gage calibration* to reduce variance. In concept, the second factory revolution was founded upon the principles of the first, but the second factory revolution was rooted in standardization and specialization of labor.

What we now call *flow manufacturing* began in the early 1900s and matured with the first Ford motor Company moving assembly line around 1913. By 1915 Henry Ford and Charles E. Sorenson had achieved the first integrated manufacturing system design. All elements in the Ford Manufacturing System....raw materials, machines, machine tools, people and inventory had been integrated into a single, remarkable system. This new factory design and production system attracted people from around the world; all came to see how this new system of manufacturing worked. So what was the problem?

Henry Ford was obsessed with people productivity and minimizing cycle time at each workstation. He implemented specialists who worked only one job. But how would you like to *install shock absorbers* 40 hours a week every week, every year? Ford soon had critical labor problems. He paid his workers an unheard of $5.00 an hour and still could not maintain a critical, trained labor force. But this was not his most serious problem. Obsessed with efficiency, production rate and minimizing costs, his philosophy of standardization soon caught up with him. How would you like to have nothing but a black car with black seats for years and years? Enter Alfred P. Sloan.

Ford refused to change his system and create more variety. His customers wanted other colors, other interiors and other options. This did not resonate with either Henry Ford or his Ford production system. Alfred P Sloan recognized the superiority of Henry Ford's manufacturing system, but he adopted the basic concepts of supplying variety to the customer. In retrospect, Henry Ford made a cardinal error…In his obsession to standardize and reduce costs, he failed to recognize customer satisfaction. Sloan reconfigured General Motors Corporation. Not only did he produce a wide variety of specific automobiles, he did so by adopting the basic mass production concepts of Henry Ford. Exploiting Ford's pioneering work; he formed production lines for Chevrolets, Buicks, GMC vehicles and others, adopting flow line and mass production concepts within product lines (car and truck) or car families. The strategy worked, and by the mid 1930's General Motors had passed Ford Motor Company in both volume of cars sold and corporate profits.

General Motors Corporation designed a hybrid manufacturing system, based upon the best manufacturing strategies of job shops, flow shops and mass production. Subassembly lines were designed to feed into main assembly lines. Subcontractors were developed using the same production strategies. This integrated system design enabled General Motors Corporation to produce a wide variety of cars and trucks, while maintaining standardization within product families at lower unit costs. As technology innovations grew, individual machines and processes became highly automated and autonomous. Subassemblies and assemblies were produced in a fixed cycle time on paced, moving assembly lines. The principles of division of labor were still used but unskilled laborers with the proper training replaced the traditional craftsman.

The mass production system pioneered by Henry Ford reached its zenith in the war years between 1928-1944. America was at war and war machines (airplanes, tanks, guns, clothing, etc.) had to be manufactured in great quantities and delivery in a minimum amount of time. During the war, all kinds of new methods, machinery, jigs, fixtures and tooling were developed to support mass production. During this era, the modern *transfer line* evolved. The transfer line is a modern version of the flow line for products like car engines produced in large volumes (200,000-400,000 engines per year for example).The enabling technology is automated, using repeat cycle, mechanized machines. The system demanded interchangeable and standardized parts based upon precise standards of measurement. The transfer line is designed for large volumes of identical parts. These specialized flow lines are very expensive and inflexible.

Work-in-process, which feed the transfer line, usually come from a flow shop in large lots; held in inventory for long periods of time; and then moved to the line in large quantities where they are automatically loaded. A new product could typically only be made by going through a long changeover and retooling cycle. This type of mass production line was pioneered and implemented during WWII to produce large quantities of fighter planes, tanks and jeeps.

After WWII, this form of production line continued in the form of manufactured automobiles such as Ford, Dodge and Chevrolets. Just as it appeared that this new system would dominate automobile production for years to come, a new manufacturing system emerged, which would eclipse all known production systems at that time. This new manufacturing paradigm was called the *Toyota Production System* (TPS), and it would revolutionize the way all manufacturing systems were perceived, designed and implemented. The emergence of TPS ushered in the Third Factory Revolution.

1.6.4 The Third Factory Revolution

The third industrial revolution was pioneered by the Japanese Toyota Motor Car Company, and was led by a genius called *Taiichi Ohno*. The evolution of TPS was generated by the desire of Toyota Motor Company to produce small quantities of Toyota automobiles, at an affordable price, for the emerging Japanese market; and later to overtake Ford Motor Company, Chrysler and General Motors as the leading car manufacturer in the world. Both of these goals were accomplished. The first step that Ohno took was to redesign the job shop into *manufacturing cells*. These cells were configured to hold minimum work-in-progress, and produce at an output rate determined by consumer demand The capacity and output rate of these cells were be determined by the *number of laborers* in the cell and not by the *number of machines*. This was a radical design concept. Machines were upgraded to single-cycle, automatics, so operators could run many machines. Final assembly was accomplished by merging assembly lines for different models of cars into one final assembly line synchronized by leveling production across all assembly. Toyota called this smoothing of production and it is now called Mixed Model assembly, discussed in Chapter 7. This Mixing of products (any model, any color) results in leveling the demand for subassemblies on the supply chain. Workers on final assembly were empowered to stop production at any time using pull cards if a quality or operational problem was detected. Next, restructure subassembly lines into U-shaped cells with balanced workloads achieved by using roving workers with fixed boundaries and operation assignments to meet output requirements.

**Figure 1.19
Taiichi Ohno was
the Architect of TPS**

Methodologies were developed to achieve flexibility, including controlling output as a function of the number of workers in the cell (See Chapter 6).

These innovations were soon followed by a *Kanban system* to control production and reduce inventory, WIP. *Production* and *move* Kanban cards were configured to control the movement of work-in-process and to trigger production runs. On the supply side, external suppliers were integrated into the production system by implementing just-*in-time* rules. Internally, each *downstream* production cell was viewed as an internal customer of the *upstream* production cell. Material handling and delivery was controlled by sending empty carts to trigger upstream production and full carts to satisfy downstream demand. The invention of *autonomation* by Shigeo Shingo replaced pure automation. Autonomation separated workers from machines through mechanisms that automatically detected production abnormalities and automatically shut down the machine. Many traditional, complex control and operational functions were redesigned to reflect *human reasoning*. This became known as autonomation; automation with a human touch. These efforts to fully exploit human reasoning and actions into a production system did not come easy. Shingo said later that there are twenty-three stages between purely manual and fully autonomated work. To be fully autonomated, machines must be able to detect *and* correct their own operating problems, which were, at that time, too complex and not cost-effective. However, Toyota attests that ninety percent of the benefits of full automation can be gained by autonomation. The concepts of Just-in-time were fully addressed and exploited by Toyota. For Just-in-Time (JIT) systems to work, it is absolutely vital to produce zero defects. Defects and rework always disrupt the production processes, create variance in the system, and disrupt the orderly flow of work. Taiichi Ohno considered JIT & Autonomation the pillars upon which TPS is built. The Japanese word for autonomation is *Jidoka, the decision to stop and fix problems as they occur rather than pushing them down the line to be resolved later*; is a large part of the difference between the effectiveness of Toyota and other companies who have tried to adopt Lean Manufacturing. Therefore, Jidoka is an important key element in successfully implementing Lean Manufacturing. If Autonomation and JIT are the pillars upon which a lean manufacturing system is built, the *foundation* upon which these two pillars rest is the aggressive and relentless pursuit of *waste elimination*. In fact, waste elimination has been the single most concepts identified with *Lean Thinking* (Womack and Jones). Lean thinking demands that direct labor utilization be maximized; waste of time and waste of materials be minimized; and machine utilization adjusted to accommodate fluxuations in demand. These goals are achieved by implementing U-shaped manufacturing cells, one-piece flow and the systematic removal of inventory and WIP by carefully designing a *Kanban control system*. We call this new design a Linked-Cell Manufacturing System (L-CMS): Figure 1.20**.**

Part Description				Part Number	
Smoke-shifter, left handed.				14613	
Qty	20	Lead Time	1 week	Order Date	9/3
Supplier	Acme Smoke-Shifter, LLC			Due Date	9/10
Planner	John R.		Card 1 of 2		
			Location	Rack 1B3	

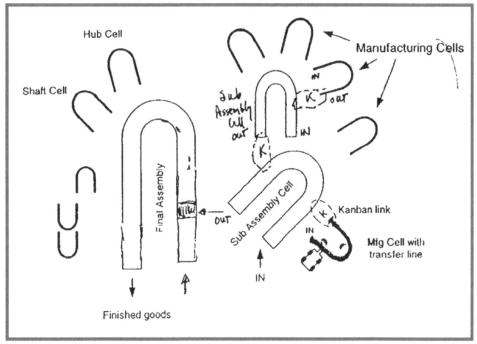

Figure 1.20
The Linked-Cell Manufacturing System has Many Manufacturing and Subassembly Cells Connected by Kanban Links to Final Assembly. These Links Hold the In-process Inventory.

Finally, a key element in the TPS was the ability to produce final product *only when needed, in the right quantities at the right time.* This goal was accomplished by realizing that scheduling production based upon a *push philosophy*, where systems abased entirely upon forecasted customer demand such as MRP II or ERP could only result in excess WIP, long lead times, long throughput times and inflated finished goods inventory. This gave rise to a new production control system called a *pull system.* Pull systems *ideally* only produce goods and services when the customer demands goods and services, in only the quantities ordered. The procedure is often misunderstood. Suppose that a production line consists of three workstations, WS1, WS2 and WS3. The entire system is idle. A customer asks for delivery of 100 units of finished product in lots of 10. When the order is received, it is immediately passed to WS3, who is authorized to produce 100 units. Of course, WS2 must send 100 units of work-in-progress to WS3 before the final production operation can be started. A Kanban (card) signal is then sent to WS2 to supply 100 units of in-process work to WS3. WS2 repeats this procedure and sends a Kanban signal to WS1 to produce 100 units as soon as possible. Suppose that the line is balanced, and each workstation takes 1 hour per part. Units are moved between stations on pallets in lots of 10. It is clear that this pure pull system has a major problem. WS 1 can begin work immediately, but WS2 is idle for 10 hours waiting on a pallet of 10 items. WS3 cannot begin for 20 hours. The first part will not be produced until 30 hours have passed from when the order was received and the first order will be filled in 31 hours. Of course one could immediately say, put the stations side by side and movement of parts in unit (1) lot size. Now the first part will come off the line in only 3 hours. The optimal procedure is to immediately start production at all workstations as soon as the order is received. After 1 hour the first unit from WS2 will reach WS3. Meanwhile, WS2 will have immediately begun work at time zero and so would WS1. The entire order will be filled in only 100 hours. However, WS3 must have 1 unit of in-process work finished by WS2 to immediately begin; and WS2 must have 1 unit of work completed by WS1. If raw materials are available when the order is received, WS1 can immediately begin work (assuming zero delivery time and no setup time for all three Workstations): It is obvious that the line would never be idle. To make sure things go smoothly, 1 unit of raw material might be stored for WS1; Note that the system will be restored to its original state after 100 units are produced by Workstations 1, 2 and 3. Of course, the price to pay is at least 2 units of WIP at each workstation at all times during the manufacturing cycle, and 1 unit at all other times. So much for the *myth* of zero WIP! In fact, this is an example of a *production line law* which we will prove in Chapter 9: In an N station serial, balanced production line, the optimal system WIP is N units when the line is operating, which is exactly what we determined in this example. This is called producing *Make-to-Stock inventory* to minimize non-value added (idle) production time. The work-in-process needed to immediately start the line is stored in what is referred to as a *Supermarket*. These concepts for both single product and Mixed Model Production will be further discussed in Chapter 7.

1.7 Lean Transformation, Kaizen Events and Continuous Improvement

The relentless reduction of waste and WIP is the primary goal of any Lean Production System. Driving both waste and WIP to lower levels will *always* reduce unit cost and improve customer satisfaction. It also usually results in better quality. Toyota properly recognized that if the entire production system is to be optimized (or run as efficiently as possible), every subsystem must be optimized, but only within the context of every other system component. The entire manufacturing enterprise should (ideally) run like a fine Swiss watch. Smoothing output and supporting balanced production is facilitated by one-piece flow with minimum set up and changeover times (See Chapter 3-6). Of course, initially switching to a Lean Production system cannot be accomplished by instantly transforming the entire factory; it must be done in pieces. See Chapter 8. Piecewise improvements are accomplished by what is called *Kaizen events*. It must be stated in the strongest terms that one *CANNOT* just pick areas to transform without knowing how the entire system will be affected. This requires system modeling techniques to choose the correct sequence of Kaizen events, such as a Systems Simulation analysis or Queuing Network Analysis (Chapters 9, 10 & 11). Finally, and possibly most important, is that the system is staffed and operated by workers empowered to make continuous improvements and understands how the entire system works. It should be clear that by reducing all setup and changeover times to as small as possible (SMED) eliminates wasted time, creates great flexibility and facilitates smaller lot sizes, balanced production lines and faster throughput times.

It has been almost 40 years since the TPS was introduced, and we are just now beginning to develop a scientific approach to Lean transformation. We have learned that the transformation to Lean is not based strictly upon computer hardware or sophisticated software, but the implementation of a set of Lean Transformational Activities. These set of principles and activities are collected into a body of knowledge which we call ***Lean Engineering***. The presentation and explanation of these principles is the subject of this book.

1.8   The Lean Engineer

The new engineering discipline that we call a *Lean Engineer* is closely aligned with Industrial and System Engineering (ISEN). ISEN has always been tied to manufacturing and factory design. One might conjecture that Henry Ford was the first Industrial Engineer by practice, building upon the work of Frederick Taylor, Henry Gantt, Frank and Lillian Gilbreth and other pioneers. Over the past 85 years, Industrial Engineering has grown into an engineering discipline which embraces Operations Research, Production Planning & Control,

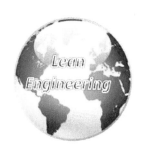

Sequencing and Scheduling, Inventory Control, Human Factors & Ergonomics, Methods and Time Studies and Total Quality Assurance. The modern industrial engineer is well schooled in Probability and Statistics to address complex system modeling using Queuing Theory and Systems Simulation. In addition, over the past 20 years many Industrial Engineering departments have broadened their scope of application to *Systems Engineering* and a discipline called *System of Systems Engineering*. The core tools of Industrial and Systems engineering are ideally suited to adopt the principles and Lean Engineering methodologies in this book, and become a new breed of Industrial Engineer called a *Lean Engineer*. Although not thoroughly covered in this book, we must stress that Industrial Engineering is also ideally suited and trained to use the modern tools of *6-Sigma Engineering*. Six-sigma engineering forms a core discipline to execute *variance reduction*…the prime enemy of Lean transformation. An entire Chapter in this book is directed to the impact and sources of variance (Chapter 4). We would be negligent and short sighted to not discuss another emerging cornerstone of Lean Engineering: *Environmental and Green Engineering*.

1.9   Environmental and Green Engineering

While all forms of service and manufacturing industries are becoming more efficient, and saving time and cost by embracing Lean concepts, there is an emerging if not mandated need to integrate green and environmental issues into Lean transformation. The growth of the human population, which is now proceeding on an exponential rate, has resulted in more and more scarce resources being consumed. In direct correlation to these phenomena has been

the need to consider both product disposal and environmental impact. Both issues are beginning to change the face of factory design, mass manufacturing and Lean engineering. New factory designs, product design and manufacturing methodologies are needed for recycling, reuse, recovery of scarce resources and zero waste. It will not be long before a new *Energy Engineer* will also enter the scene…if not already. The need for higher gas mileage, lighter and more disposable cars and zero pollution is already changing the way automobiles are designed, manufactured and disposed. Who will lead this new wave of engineers? Again, the *Lean Engineer* is better suited than any other engineer to embrace these needs because of their systems engineering training and perspectives.

Green Engineering focuses on the design, manufacture and commercialization of processes and products that can be produced quickly and economically with a minimum impact on *renewable resources*. The term renewable resource is somewhat misleading, since almost any resource consumed at a faster rate than it can be reclaimed or renewed can become a scarce resource, and eventually non-renewable. Feasible and economical solutions to Green engineering problems are logically the only set of solutions that will be adopted by any industry or service sector. We will offer only 3 alternative groups of potential benefits, although there are many more.

> ➢ Eliminate waste through more efficient and wasteful methodologies and designs. This is clearly within the scope of Lean Engineering and intersects one of its main platforms (waste elimination). This usually means going beyond regulatory mandates and embraces green principles at the source.
> ➢ Engineer out costs associated with environmental, energy and green compliance.
> ➢ Exploit marketing and sales advantages associated with this new breed of products

Since about 1995, the automobile and processing industries have initiated several efforts in this area…some because of legislative mandates and some to reduce Life Cycle Costs. Concurrently, Lean transformation activities and new management structures have been implemented to support these new initiatives.

Toyota has attempted to embrace these principles into TPS and has made significant impact. Toyota's 5R program was initiated to reduce pollution, reduce energy consumption and address *environmental impact*. Tools such as *Green Stream* and *Energy Mapping*, following the success and principles of Value Stream Mapping, IDEF0, IDEF1 and IDEF2 (See Chapter 17) need to be developed. Currently, green methodologies, principles and applications are surfacing all over the world. In 2003, a summit meeting on green engineering produced 9 fundamental green principles from a group of 65 participants. Applications and extensions of these 9 principles are now beginning to surface in the technical literature, but actual factory transformations are slow to develop. This is an exciting new field of research for the emerging Lean Engineer.

Both a wide variety of Environmental Management Systems (EMS) and ISO standards have been developed to guide this transformation. Sadly, no one engineering discipline has elected to champion Lean, Green, Environmental and Energy efficient transformations. We contend that this new breed of engineer should be called a Lean Engineer, and should be spawned out of Industrial and Systems Engineering; just as Industrial Engineering emerged out of Mechanical Engineering during the First Industrial Revolution. This book will introduce the emerging concepts of a *zero-waste, Green and environmentally sensitive factory*. The fundamental goal of zero waste is to recycle all finished products back into some renewable resource or back into the market place as a derivative product; just as a caterpillar morphs into a butterfly. At this time, a unified engineering discipline, its fundamental tools and education required and its professional identification is waiting to be defined and exploited. The best we can hope for in this book is to energize and stimulate such an engineering discipline. For now, we call this ***Lean Engineering***.

Review Questions

1. Define a Manufacturing System.
2. What do you understand by internal customers?
3. List five basis manufacturing system designs.
4. Explain Job Shop Design.
5. How is Flow show different from Job Shop?
6. List the differences between continuous processing and lean manufacturing.
7. What is kaizen event?
8. What is a Green engineer?
9. What is an Environmental engineer?
10. What is a Lean engineer?
11. Who was Henry Gantt?
12. Who were Frank and Lillian Gilbreth?
13. Who was Frederick Taylor?
14. What role did henry Ford play in the evolution of Manufacturing Systems?
15. Who was Taiichi Ohno?
16. Who was Shigeo Shingo?
17. What is a L-CMS Lean manufacturing system?

Thoughts
   and Things……..

Chapter 2
Manufacturing System Design

"Perhaps we should stop looking for the sharpest needle in the haystack, and find one that we can sew with"   *March Siman, Organizations*

2.1 Introduction

In any factory, *manufacturing processes* are assembled together to form a *manufacturing system* (MS) to produce a desired set of goods. The manufacturing system takes a set of orders from customers, translates these orders into resources and raw materials to support production, adds value through a series of manufacturing processes and transforms these raw materials into final goods. It is important to distinguish between the manufacturing system and the *enterprise system* which is comprised of marketing, sales, production planning, supply chain management, finished goods inventory, product delivery and a myriad of other support functions. Figure 2.1 provides a composite view of the enterprise system.

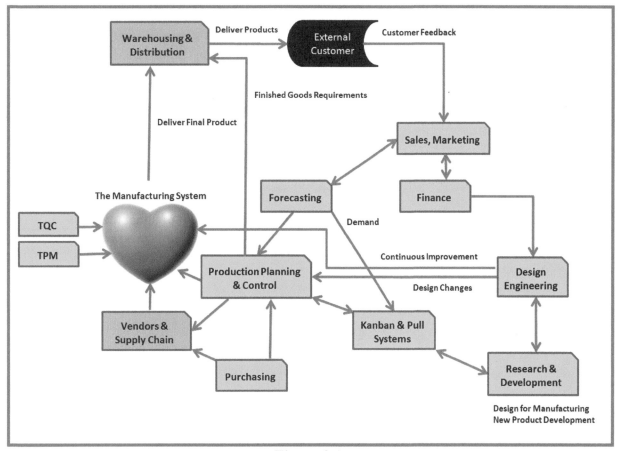

Figure 2.1
Functions of the Production Systems

The enterprise system exists to support the manufacturing system, which is where raw materials are transformed into finished products. The enterprise system is composed of people and information used in product design, analysis and production control functions. These functional areas exist for only one reason…to make money. The entire enterprise system contains all aspects of the business, including design engineering, manufacturing engineering, sales, advertising, production control, inventory control and most importantly the actual manufacturing system.

2.2 The Manufacturing System

The total number of people and processes that actually produce a desired finished product or provide goods and services to produce a finished product is called the *Manufacturing System.* The total manufacturing system is a complex arrangement of physical elements which are guided by measureable parameters (Black, 1991). Physical raw materials are transformed by a series of *process steps or workstations* which add value to the work in process. The degree to which these measurable parameters dictate both the individual process steps and the entire sequence of operations largely determine how well the system can be managed and controlled. The entire manufacturing system must be continuously monitored to enable the movement of materials, people and the management of value added processes.

The goal of modern scientific management is to design a manufacturing system that operates harmoniously as one living organism, and to include all the subsystems as functional components. In nature, there is a law called the *constructional law (*Bejan and Zane) which identifies two basic properties of anything that flows as the substance that flows and the system through which it flows. For example, a river will evolve into a system that resembles the main trunk and the branches of a tree through which water flows. This is a natural structural evolution that supports the movement of water. In a properly constructed manufacturing system, the same type of principles should exist. For a system to operate effectively and efficiently over time, its configuration (design) must evolve in such a manner to fully support the harmonious movement of parts and raw materials through the system.

This phenomenon exactly describes the design, control and operation of a manufacturing system constructed around *lean principles*. Just as tributaries to a river constantly erode and change to support water flow, a lean system will continuously change to support cost effective and efficient production of goods and services. The constant search to remove problems enhances product flow and decreases the product throughput time while increasing output.

The control functions for a manufacturing system are production control, product quality control, inventory control, RAM (Reliability-Availability-Maintainability), supply chain management…etc.

In a manufacturing system, there are two main components that drive many performance indicators: (1) Work-in-Process, including made to stock and finished goods inventory and (2) Variance. Work-in –Process (WIP) is manifested in three main categories.
  (1) Finished goods inventory (product waiting to be shipped to the customer)
  (2) Raw materials inventory (raw materials waiting to be used)
  (3) Parts or goods flowing through the system waiting (queued) either to be moved or to be processed); Partially finished goods stored to buffer imbalances in production and demand (Make-to-stock inventory)

Major portions of this book will be devoted to how WIP can be reduced or eliminated. A wide variety of studies across a wide range of manufacturing systems has confirmed that parts which flow through a manufacturing system spend about 95%-96% of the total manufacturing throughput time waiting for a value added transformation. The two performance measures that are most affected by WIP levels are the *production Rate* (PR) and the Time in the system or *throughput time* (TPT) per unit. One of the invariant laws of manufacturing systems was derived by John D.C. Little and is called *Little's Law*. Little proved the following relationship exists for *any* manufacturing system.

$$WIP = PR * TPT$$

We will fully exploit and use this law in later Chapters of this book in a variety of applications. For now simply note that the production rate is directly related to and dictated by customer demand. If it is fixed, then as WIP increases for any reason whatsoever, the throughput time directly increases. We will see in subsequent Chapters, that there is an optimal level of WIP which exists for a lean manufacturing cell design.

The second, and perhaps the most important factor in poor system performance, is *variance*. Using the mechanics of queuing theory and the principles of *Factory Physics* we will analyze various components of variance which manifest themselves in manufacturing systems and analyze their causes and effects. It will be shown that increasing variance in *any way* will *always* increase WIP and hence increase throughput time. A major goal in Lean Thinking is to systematically and aggressively reduce variance in any way possible at every step in the manufacturing processes.

2.3 Typical Manufacturing System Production Facilities

As previously discussed, there are three primary production configurations which dominate current manufacturing philosophies: (1) The job shop (2) The flow shop and the (3) High volume mass production line. The job shop production system has existed since the 1800's, and still dominates both stand alone and integrated manufacturing system designs. Approximately 60%-70% of and goods and services manufactured are in small lots with customized specifications by skilled labor. This manufacturing system design is based upon flexibility and functionality. Flexibility is traditionally achieved at the expense of high in-process inventories and long throughput times. A part is scheduled through multiple processing steps which usually involve multi-purpose machines manually operated. Production control has the responsibility for order release to the shop floor and a material handling function (often manual) is responsible for moving work-in-process. The traditional flow shop evolves as product demand increases, and involves *machining centers* where different types of machines are clustered together (Lathe, Mill, Grinder, Drill etc.) The process layout is more efficient than typical job shop layouts, but might still result in spaghetti-type flow with multiple set up operations required. The high speed, mass production line evolves as demand for a single part or part family increases to the point that dedicated manufacturing cells or production lines can be configured to support mass production. Such manufacturing lines are used in the automobile industry and in commodity production such as chainsaws .The modern automobile assembly/production line is divided into hundreds of stations sequenced to systematically build an automobile. There are usually main assembly lines and subassembly lines coordinated and balanced to sustain smooth, uninterrupted flow. This strategy is referred to as *one piece flow*. Figure 2.2 shows a mass production line which integrates all three manufacturing system designs into one production system.

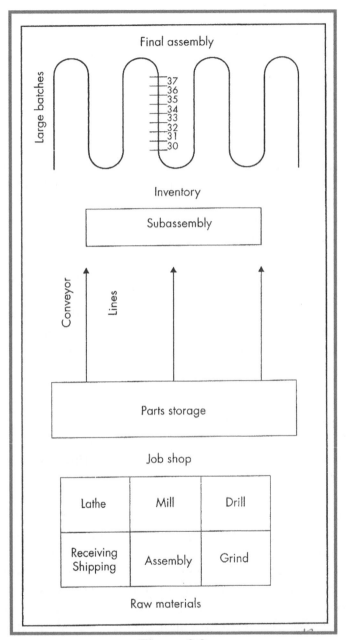

Figure 2.2
The Mass-Production System has a Moving Assembly line (stations 30- 37) Fed by Conveyorized Subassembly Lines and Job Shops

2.4   The Toyota Production System (TPS)

After World War II, the Toyota Motor Company, led by the genius of a manufacturing engineer called Taiichi Ohno, developed a new manufacturing system design and a new manufacturing philosophy simply known as the Toyota Production System (TPS). In the 1980s this system was referred to as Just-in-Time, World Class Manufacturing, and the Pull System to name a few. In

1990, this system was given a name that would become universal called the *Lean Production System* (Womack and Jones). Lean production focuses on waste production and is often characterized by the *two pillars of TPS* shown in Figure 2.3. These two pillars are Just-in-time (JIT) and Jidoka. JIT implies that the system is designed and operated to produce only what is needed, in the right quantities at the right time. Jidoka provides the means to achieve these goals and is multifaceted in nature. The primary mission of Jidoka is to produce product with zero defectives and with no in-process quality drop outs. These objectives are partially achieved by: (1) Enforcing quality-at-the source (2) Empowering workers to stop production when poor quality develops and (3) Training each worker to do the jobs assigned to him/her. Equipment is

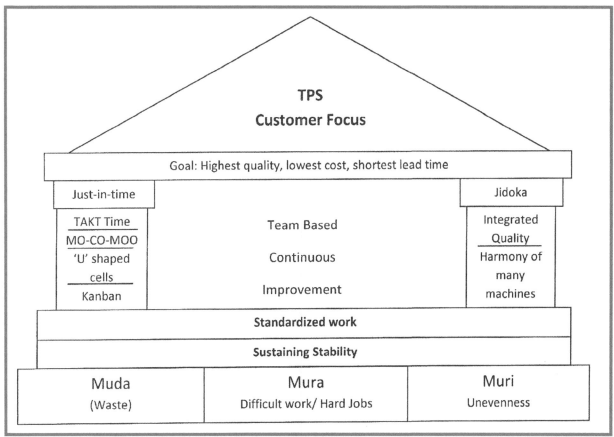

Figure 2.3
The Toyota Production System is based upon the Twin Pillars of Just-in-Time and Jidoka

redesigned to execute self-diagnosis, detect anomalies when they occur and to automatically stop when a problem arises. Operators are also given the power and authority to shut down a line or process if necessary on their own. Stopping a production machine instantly raises *red flags* and highlights a problem immediately. *Jidoka* is called *human engineering* because it empowers a worker to gain complete control over a process machine and not vice-versa. In a real sense, every operator is also a quality inspector. These things are an integral part of Lean manufacturing, but what was truly different about Lean production when contrasted to traditional mass production or flow shop designs?

The conversion of the Mass System to the lean system occurs at three levels, as shown in Figure 2.4. First, in a lean production system the final assembly area is converted to mixed model. In automobile production (for example), this means that different models of cars within a car family come down the production line to final assembly in any order. Two door sedans, four door sedans and convertibles all get final assembled on the same line in any order. Really food lean companies can run different models (Civics and Accords) on the same time. This is achieved by creating level (not constant) demand for assemblies, subassemblies and main components. Final assembly is run at a mixed production rate based upon customer demand (parts/unit time) and this levels raw material and in-process requirements. The reciprocal of required production rate is Takt time and is dictated by customer demand.

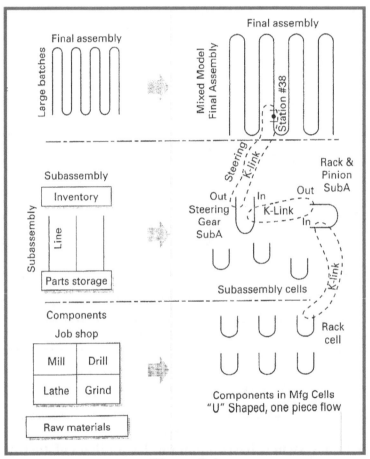

Figure 2.4
The Mass Production System is restructured into a Lean Manufacturing System Design to Achieve Single-piece Flow

Second, as shown in Figure 2.5, linear subassembly conveyor-paced lines are converted into *Lean Self-balancing subassembly Cells*, often eliminating the need for complex material handling and conveyors. The production rate of each Lean cell is determined by the number of workers in each cell and not strictly the number of machines (See Chapter 6). Each Lean production cells are balanced and produce to down-stream requirements required to achieve final assembly Takt time goals. There is virtually no sequencing and scheduling in these cells, and balance is achieved by allocating different workers to different sets of tasks; very similar but not identical to line balancing schemes usually employed in non-lean production facilities.

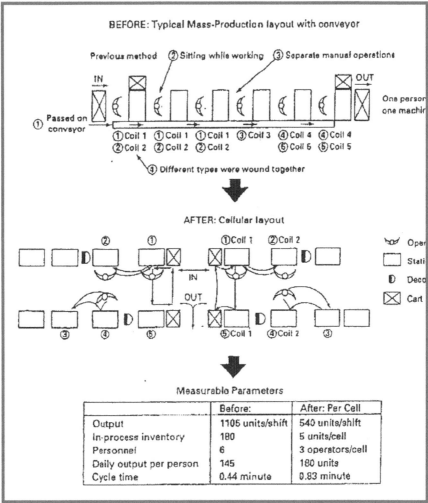

Figure 2.5
A Subassembly Flow Line, with Conveyor, can be redesigned into two cells Using Walking Operators, with Stations Arranged in Parallel Rows. The Main Advantage of the Cell Design is its Flexibility and the Ability to Easily Change the Output on a Daily Basis.

Third, any job shops are converted into U-shaped manufacturing cells, as shown in Figure 2.6. The redesign of job shops into manufacturing cells requires that an engineer trained in Lean thinking (*The Lean Engineer*) convert all machine tools into *single-cycle automatics* with near Zero Setup and Changeover Time.

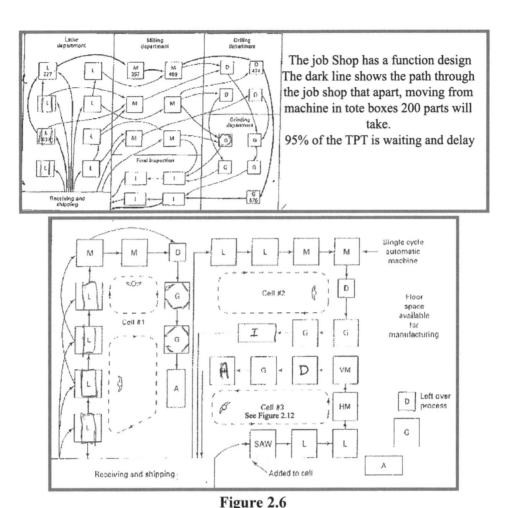

The job Shop has a function design The dark line shows the path through the job shop that apart, moving from machine in tote boxes 200 parts will take.
95% of the TPT is waiting and delay

Figure 2.6
The Functionally Designed Job Shop can be restructured into U-shaped Manufacturing Cells to Process Families of Parts at Production Rates that Match Demand Rates

It is critical that Lean cells minimize set up and change over times, equip machine tools with work and tool holders that can be rapidly reconfigured, and design and install mechanisms to connect the flow of parts between machines. Devices called *Poka- yoke* are designed and installed to monitor performance, realize zero defects and automatically shut a machine down if it malfunctions.

All manufacturing cells are designed to provide a *WIP-Cap* within the boundary of each cell. A Wip-cap ensures that work-in-process will not exceed a predetermined level. This Wip cap is usually no more that than twice the number of machines in the cell, and is strictly enforced (See Chapter 5). All component, subassembly, make-to stock WIP locations and final assemblies are linked together by a system called *Kanban* which pulls material through the system. The Kanban system creates an integrated material movement, production control and demand patterns for the entire system. Ideally, these philosophies result in a smooth flowing system with minimum WIP and desired output rates characterized by superior quality products and on-time delivery to the customer (s). This new system design can and should operate in ways that the old mass production or job shop system cannot. Each manufacturing cell in the system operates using one-piece flow, just as the final assembly lines. All manufacturing components are designed for mixed model production. The system is most efficient and totally superior to traditional manufacturing operations when demand is fairly stable and repeatable. However, we wish to debunk and dismantle the concept that Lean Manufacturing strategies cannot be applied to other systems, such as low volume, high product mix systems. Manufacturers and service organizations around the world are implementing Lean Engineering principles in every type of manufacturing system with spectacular success.

It should be repeated that perhaps the most important characteristic of a properly designed and operated Lean system is *flexibility*. Flexibility and the ability to produce a wider variety of basic products is the key to higher profits and deeper market penetration. Systems designed using the Lean engineering principles in this book allows a company to rapidly adapt to changes in product demand and product mix, and provides a framework by which changes in product design requiring new processes can be implemented. Notice also that a Lean manufacturing system integrates all of the quality assurance, production planning, materials handling and machining /processing requirements into one cohesive, coherent operating framework. The Lean production system is in every way designed to produce superior quality products. Toyota believed in Total Quality Management and Control, and enforced this philosophy at every level of the company from president to truck drivers. Toyota was transformed from a company which had a reputation for poor quality into one with superior quality products. It is no surprise that the Toyota line of automobiles is generally recognized today as the most reliable, defect free automobile in the world; and it leads the world in sales. This was made possible by enforcing quality upon both the raw material supply chain and upon every step in the manufacturing system.

We cannot overemphasize enough that the key to both superior quality products and efficient system performance is to aggressively and systematically reduce all sources of *variance*. This is also the key to *continuous improvement*. The kinds of variance that must be considered are:

- Variation in raw material characteristics
- Variation in methods and process activities
- Variation in processing times
- Variation in cost
- Variation in customer demand
- Variation in material delivery times

These and a myriad of other sources of variation can be systematically reduced by applying good, solid Industrial Engineering tools. Toyota made the seven tools of quality control standard knowledge for its workers and even Doug Montgomery has identified and promotes the use of 7 basic tools. He calls the *Magnificent 7*. These are shown in Figure 2.7. Of course, there is nothing magical or magnificent about any one of them; they simply represent a set of tools that when properly applied, can be used to eliminate variance and improve quality. The interested reader is referred to his book on Statistical Quality Control.

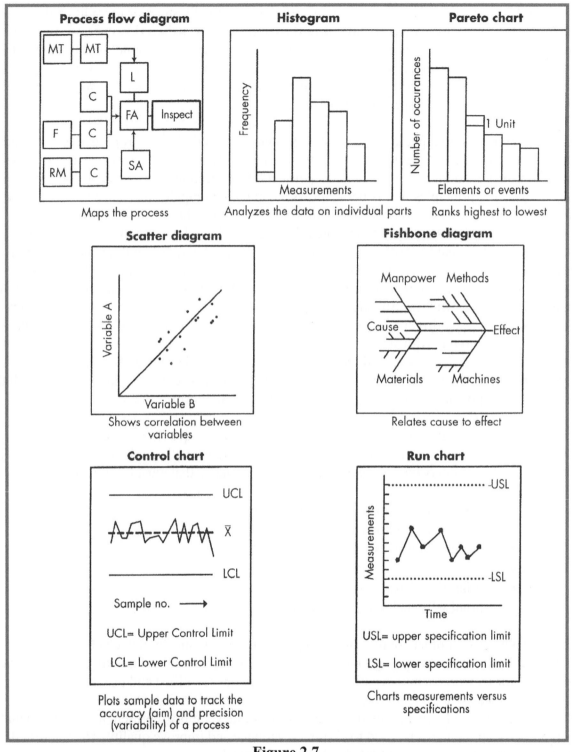

Figure 2.7
Seven Tools of Quality Control Used by a Lean Engineer to Control the Process and Achieve Zero Defects (Montgomery)

Cost reduction is achieved by reducing any form of waste, including the efficient use of space, materials, machines, people and storage. This is why the system is called *Lean*. Unique application of Lean thinking is achieved through the creativity and work of the Lean Engineer, who is constantly looking for ways to reduce waste, minimize WIP, improve processes, reduce set up and changeover times, and redesign the workplace. In restructuring traditional manufacturing systems into Lean manufacturing systems, the final assembly line and all subassembly cells are redesigned to support one-piece flow. Workers are assigned to unique, independent tasks and workload balance is enforced (See Chapter 10, or Black & Hunter, 2003). In redesigning manufacturing check lines into manufacturing cells, the axiomatic principles of work cell operation (See Chapter 4) are used to *decouple* or separate one task from another. Cell cycle times are rendered independent of individual machining times. Thus variation is removed from the supply chain and the order release process. Hence, scheduling work to the cell is rendered very simple by the use of *Kanban* and a strictly enforced *pull* strategy. The WIP level in each cell is held constant throughout the entire production period. The output rate is controlled by worker utilizations and not machines. Cell workers are *multifunctional* and cross trained on each machine. Every cell operator can perform every operation and also execute routine machine maintenance duties. Lean cell manufacturing strategies completely eliminate the concept of one worker per one machine, except in rare capacity-limited situations. Since each cell operates autonomously of every other cell, the Lean Cell is ideally suited to hosting mixed model production from a common family of parts. Part families can be identified by applying *group technology* or other well-known clustering algorithms. Proper identification of part families help to reduce setup and changeover times.

2.5 Four Key Terms in Lean Engineering

A Lean production system must be continuously refined and improved to achieve success. This refinement must involve four key aspects of Lean thinking: (1) Leveling and Smoothing (2) Balancing (3) Sequencing and (4) Synchronization. Knowledge of these four terms is critical to the success of Lean transformation.

2.5.1 Leveling Production

Leveling production is simply stated: *the distribution of both the type of product produced and the volume of product produced over time*. Leveling is the process of planning and executing a level production schedule. Leveling production for a single product assigned to a single line amounts to leveling the product demand to be manufactured over a period of time: usually by weeks or by month. For example, suppose that Product A is to be produced according to the following demand schedule over the next 5 weeks.

Week 1	Week 2	Week 3	Week 4	Week 5
180 units	110 units	205 units	150 units	135 units

One way of scheduling production is to *produce to demand*. In this case, the weekly production schedule would vary from 180 units to 135 units. If this is a final assembly schedule, all of the upstream processes are going up and down in the same manner. This is called *chasing demand* and the violent, frequent changes in production requirements is called the *Bullwhip Effect*. Clearly, the amount of time devoted to this product and the operation of a production cell would vary from week to week. Inventory and raw material requirements would change from week to week and cost overruns might occur.

An alternate strategy would be to set production at 180 units per week, but at the end of the first week would see that demand has significantly decreased, so production would be reduced accordingly. At the end of week 3, suddenly demand has almost doubled, so production is doubled (overtime, additional shift, etc.). The bullwhip effect is still very evident. Suppose the following production schedule is used.

Week 1	Week 2	Week 3	Week 4	Week 5
156 units	156 units	156 units	156 units	156 units

Clearly, this schedule would be easier to manage. Note that to accommodate this schedule, a certain amount of WIP will be required; which seems to violate one of the main objectives of Lean Thinking. However, note the following.

- Customer demand is fully met on time (possibly both cases)
- Predictable work schedules
- Fixed production requirements
- Make-to-stock inventory need be accomplished only one time

The principle behind load leveling is simply to regulate production in all subassembly and final assembly such that constant production rates are achieved. This avoids *spikes* in production schedules which occur when production follows demand. Final assembly should not pull production from upstream cells or subassemblies as demand changes.

In almost all Manufacturing companies, production and assembly lines must be designed and implemented to accommodate the manufacture of multiple products in the same facility. This is called *mixed model* production. Mixed model Lean production requires application of the same Lean principles as single product production, implemented in multi-dimensions. The key to equivalent mixed model production is to reduce set-up and changeover times to an absolute minimum (SMED) and enforce a zero defects production program. A myth in changing over to Lean production with leveling is that integrated forecasting systems such as those contained in MRP II shop floor schedulers can be abandoned. Forecasted demand is still needed, but must be smoothed to achieve one piece flow. Traditionally, fluxuations in demand rates were accommodated by scheduling setting production rates on upstream processes to maximum expected rates, causing overproduction and excess inventories.

Suppose the demand pattern is as shown in Figure 2.8-a. There are 1000 units which need to be made over a 4 week planning horizon. The schedule calls for 100 units to be produced in weeks 1 & 2; 200 units in week 3 and 600 units in week 4. This calls for a typical rush to produce at the end of the month. To level this schedule, two-2 week schedules are created to divide a typical month into two parts. Next, the two week periods are divided into one week periods, and finally one week periods into daily production requirements. This is called a leveled production schedule and will greatly influence and improve upstream system behavior. Daily production requirements are set at 50 units/day; working 5 days per week. Of course, if the shipping

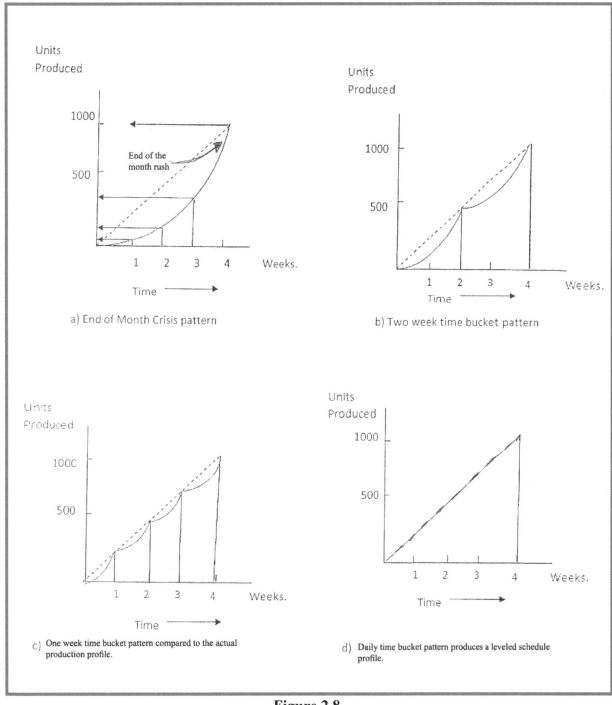

Figure 2.8
Leveled Production Means a Plot of Units Produced Versus Time Would be Linear (dashed lines in the Figure)

schedule corresponds to the original manufacturing schedule; this will require an average WIP of 360/units per day in make-to order inventory. If finished goods are shipped every day, this WIP becomes only 25 units/day. The correct policy must be determined by balancing holding costs against shipping costs.

Advantages and Disadvantages

Leveling production has the following *advantages*.

- Raw material and part suppliers can build and ship on a smoother schedule
- Highs and lows in production system demands are smoothed out
- WIP is reduced if shipping is also leveled
- Leveling production can facilitate automated , with standardized containers
- Overall productivity usually increases and waste is decreased
- Direct labor hours may be decreased
- Quality is greatly improved

Leveling production may also have the following *disadvantages*.

- Material handling requirements will increase
- Setup and changeover times must be greatly decreased or eliminated in Mixed model production
- Workers should be trained to be Quality inspectors and maintenance people (these are also advantages!)

Workers must be thoroughly indoctrinated in the principles, techniques and reasons for Lean production and leveling. Their approach to manufacturing must include continuously seeking ways to improve production, reduce costs and eliminate waste (*Muda*). Successful leveling strategies must be fully supported by every factory worker, or the results will be resistance and mass confusion.

2.5.2 Mixed Model Leveling

Almost all manufacturing companies manufacture and sell products of several different types. A company can produce products of different type (General Motors: Chevrolet, Buick, Etc.) or variations of individual products (F250 Ford Truck: Diesel, 4 wheel drive, 2 or 4 door, Etc.). Consider the following demand profile for products A, B, C and D.

Product A	6000 units/mo.
Product B	4000 units/mo.
Product C	2000 units/mo.
Product D	8000 units/mo.

It is not uncommon to observe that many companies would traditionally run this monthly production profile in units of A=6000, B=4000, C=2000 and D=8000. The sequence of production might be chosen to minimize setup and changeover times. The reason for this line of thinking is that to minimize the time lost to setups and changeovers this is the "thing to do". If in fact, these times are very large; this might indeed be the right solution. Early in the formation of the Toyota Production System, a major emphasis was placed upon reducing setup and changeover times to as near zero as possible. This initiative became known as SMED, which stood for Single Minute Exchange of Dies. Indeed, after much time and effort these times were reduced from hours to minutes. SMED is absolutely necessary to implement Lean Manufacturing and Just-in-time strategies.

Continuing with the example, assume 4 weeks per month and 5 working days per week. This would result in production requirements of A=300, B=200, C=100 and D=400 units per day. This is still not a level production schedule. Maximum level production is ideally achieved by using *one piece flow*: Produce one unit of A, then one unit of B, etc. Any sequence of producing multiple products is called a *Mixed Model Production Schedule*. As a general rule, first calculate the largest integer divisor that would divide evenly into all products. Next, determine how many of each type of product will be included in a partial production sequence by dividing each product demand in the lowest time bucket (called a *Planning Horizon*) selected by that integer divisor. Third, determine a *basic sequence* which would produce that set of products. Finally, determine how many times that sequence would need to be repeated over the Planning Horizon to satisfy the time bucket over which total production is planned; called the *Planning Period.* For this example:

- The planning period is *monthly:* Production Requirements are: A=600, B=400, C=200 D=800
- The planning horizon is *daily:* Production Requirements are: A=30, B=20, C=10 and D=40
- The largest integer divisor is 10
- The Basic Sequence is: A=3, B=2, C=1 and D=4.
- A possible production sequence is ABDDABCDDA

The production sequence would need to be repeated 10 times per day to achieve monthly requirements. There are many other possible sequences, including one that would result in maximum leveling: DABDABDCBD. Here we must now determine the maximum amount of available time per day that would likely be required to produce the desired sequence of products. Assume that the manufacturing facility is working two shifts per day 7 hours per shift. This will give 840 minutes of available production time per day. Using the sequence of A=3, B=2, C=1 and D=4 if we will assume that the following production times per unit are required: A=240 sec; B=120 sec; C=400 sec; and D=180 sec. Each sequence will require 140 seconds, repeated 10 times or 1400 seconds per day or 23.33 min/day.

Migrating to Lean production also requires maximum flexibility in final assembly lines so that production can be scheduled around firm customer orders rather than chasing demand. So basically, how is flexibility achieved?

2.5.3   Smoothing Final Assembly

Lean manufacturing companies have yearly production plans that forecast how many items need to be produced over the next planning period. For example, a yearly production plan might employ a running two month plan. Products are scheduled into production and a detailed routing sheet is constructed built around a one month planning horizon and the amount to be produced on a daily or weekly basis is placed on the order release schedule. The production plan should ideally support minimum time in finished goods inventory and just-in-time delivery. Leveling production is an important piece of the actual production schedule.

For example, suppose that a final assembly line is required to produce a total of 10,000 vehicles monthly according to the plan shown in Table 2.1 for a mixture of 4-door sedans, 2-door sedans and mini-vans.

Table 2.1
Mixed-Model Final Assembly: Sequencing to smooth the system

Assembly Sequence 4DS, 2DS,4DS, MV, 4DS, 2DS, 4DS, MV, 4DS, 2DS			
Model	Monthly Demand	Daily Demand*	Takt Time
4-door sedan (4DS)	5,000	250	1.92 minutes
2-door sedan (2DS	2,500	125	3.84 minutes
Minivan (MV	2,500	125	3.84 minutes
		500 = Daily Demand	

$$Takt = \frac{480\ min.\ x\ 2\ shifts}{500} = 1.92\ min.\ per\ vehicle$$

*Assuming 20 working days in the month

Table 2.2
Characteristics of Basic Manufacturing Systems

Characteristics	Job Shop	Flow Shop	Project Shop	Continuous Process	Lean Shop
Types of machine tools	Flexible, general purpose	Single purpose, single function	General purpose, mobile, manual automation	Specialized, high technology	Simple, customized single-cycle automatic
Layout of processes	Functional or process	Product flow layout	Project or fixed position	Product	Linked U-shaped cells
Setup or changeover times	Long, variable, frequent	Long and complex	Variable, every job different	Rare and expensive	Short, frequent, one touch
Operators	Single function, highly skilled: one worker – one machine	One function, lower-skilled: one worker – one machine	Specialized, highly skilled	Skill level varies	Multifunctional, respected standing
Inventories (WIP)	Large to provide for large variety	Large to provide buffer storage	Variable, usually large	Very small	Small
Lot sizes	Small to medium	Large	Small lot	Very large	Small
Manufacturing lead time	Long, variable	Short, constant time	Long, variable	Very fast, constant	Short, constant

Using for example a 20 day work month, this would require a daily production of 500 vehicles; 250 4-door sedans, 125 2-door sedans and 125 minivans to satisfy demand. The factory is scheduled to work two 8-hour shifts per day. Each shift would contain 480 working minutes. In addition to the three basic types of vehicles, suppose that due to options there are 300 different variations to the basic vehicles. The production of these automobiles is a good example of *mixed model* production. For this example, there will be no set up or changeover times between models or variations. Inclusion of these times is not hard to accommodate. Once the production sequences are established, all of the setup and changeover times are simply deducted from the available working time.

The assembly line receives a production schedule at the end of every month. The average daily production time required to meet the schedule is then calculated. The process step(s) are then examined to see if sufficient capacity exists for the schedule: If not, capacity is adjusted. If a Lean, U-shaped manufacturing cell is being used, then capacity is easily adjusted by assigning the appropriate number of workers to the appropriate number of machines to match required Takt time (See Chapter 4). All operator tasks must be standardized and all workers cross trained to operate any machine in the cell. Overtime can usually be used temporarily when demand exceeds capacity. If this condition persists, then parallel machines could be added to increase capacity.

If demand decreases, the reverse strategy is used. Workers take on more machines in their assigned work cycle and machine utilizations decrease. Excess workers are moved to other cells or other duties until needed. A goal of Lean manufacturing is to produce goods and services with a minimum of skilled laborers (See Table 2.2 for Characteristics of basic manufacturing systems). Conversely, it is not necessary or desired to keep machines running all of the time or configure a cell with the minimum number of processes. For this example, the Takt time for this cell is to produce a 4-door sedan every 1.92 minutes, a 2-door sedan every 3.84 minutes and a Minivan every 3.84 minutes.

2.6 Mixed-model Final Assembly

To perform leveled production in a factory that manufactures several types of final products, Mixed-model final assembly lines must be designed and implemented. Mixed model subassembly lines may or may not be needed. Every U-shaped Lean manufacturing cell might be dedicated to a particular type of subassembly, but some may need to be Mixed Model cells. The decision depends largely upon the production quantities that need to be produced over a fixed planning horizon. Here we will constrain our thoughts to Mixed Model final assembly, although the same principles apply to both Subassembly cells and component manufacturing cells.

- Eliminate or bring to a minimum all setup and changeover times.
- Standardize all material handling as much as possible.
- Determine the required Takt time(s)
- Determine the sequencing and dependency of all final assembly operations
- Level the Mixed model production schedule by properly sequencing and grouping parts
- Design workstations to balance line workstation times

2.7 Determining Takt Time(s)

In final assembly, the required cycle times between part releases are called Takt times. The production rate is mathematically the inverse of Takt time. Takt time is determined by dividing the operating minutes per day (scheduled time-breaks-allowances, etc.) by the daily demand. This sets the pace of manufacturing at final assembly, which in turn sets the pace for every operation that precedes final assembly.

Let: TT=Takt time (Minutes/Part)
H=Number of operating minutes per day
DD = daily demand (Parts)
OM= Required output/month

M= Number of operating days/month
TPT=Throughput time per part
PR= 1/TT=Production Rate

Then: TT = H/DD in minutes/part
And: DD = OM/M in parts/day

As usual, Little's Law can be written as:
$$WIP = TPT *(1/TT) = TPT* PR$$

In Lean production systems, final assembly Takt time is held constant over the planning/production period. By Little's Law, any reduction of WIP will result in a corresponding reduction of throughput times. This is a major goal of Lean manufacturing.

Contrary to popular belief, a Lean system does not strictly react only to customer demands. We have already pointed out the fallacy of this policy. The required output from final assembly comes from a combination of actual orders-in-hand and a sales forecast. Here we kill another widespread misconception of Lean. Sales forecasting is a necessary part of Lean manufacturing. So how is this done? A common belief is that Lean thinking and Lean production will replace MRP, ERP and other shop floor control systems. While this is partially true, these legacy systems can still play a major role in production planning. A major contribution can be sales forecasting.

Once the required cycle time is determined, the Lean Engineer develops standard procedures and best practices for every worker on the final assembly line. The line is designed for flexibility to be reconfigured as demand/product mix changes. Likewise, workers must be cross trained to execute many process steps. Standard procedures and best practice is the best arrangement of men, materials and machines coupled with the best methods used to produce a defect-free final product. It is Takt time which determines equipment capacity, work breakdown structure, work balance, leveling and product mix.

2.8   Determining the Proper Sequence of Operations
The sequence of operations in final assembly is easily determined by the process routing sheets. This simply identifies which operation must precede another. Closely associated with sequencing, is the granularity of processing steps. In general, each processing step should be as short as possible. This will make it easier to balance production on the line.

2.8.1   Determining the Mixed model Sequence Schedule
The sequencing problem at final assembly is the order and run quantities that need to be introduced to smooth and balance production. Sequencing involves two main goals.

- *Balance* total assembly time at each workstation by product (Workers must *equally assemble 2 doors or 4 doors coming down the line*)
- Keep the rate at which each part is produced constant. This is called *leveling demand.*

50

In Mixed model assembly, not all products have to exactly meet Takt time requirements. Some may be lower or higher than Takt times (Just as in traditional Line Balancing). Often one is forced to average Takt times. For this reason, products introduced into the line with assembly times which exceed Takt time(s) should be followed by products which are less than Takt time(s). Workers are often trained to assist across station boundaries, thereby achieving workload balance. When Mixed Model production is in place, inventory (WIP) levels are also affected. Generally, WIP levels in Mixed Model assembly lines are larger than individual product lines. WIP levels are required to prevent starvation and to accommodate temporary overproduction.

2.8.2 Scheduling

Scheduling a Mixed model, final assembly line to achieve good line balance is a key component in designing the entire production system. If a station on the final assembly line starves or is blocked, the entire system can be starved or blocked. All upstream cells and subassemblies must be tuned to final assembly. Great flexibility is required to seamlessly move between different types of products with different cycle times. Lean Mixed model final assemblies usually result in higher on time delivery, lower finished goods inventory and customer satisfaction.

After setting weekly and daily production schedules, determination of the daily sequence is critical. The sequence of products dictates the order of final assembly, the just-in-time resources required and the balance. It is important to realize that if three products are being made (A, B & C), and if the final assembly Mixed model sequence is 1000 replications of AAABBCC, this sequence is important not only to final assembly, but also to the upstream processes or cells. In order to synchronize upstream cellular manufacturing with final assembly, upstream production must mirror the final assembly schedule. There is only one way to avoid this: it is to store upstream mass production in an interim Supermarket, which in turn will ship to final assembly. While this will work, it will create excess WIP, longer throughput time and violate lean manufacturing principles. This is the most important and fundamental aspect of linked-cell, Lean production system, and it distinguishes Lean Mixed model final assembly from traditional Mass manufacturing systems. In general, upstream component and subassembly lines receive only monthly or weekly production schedules. Workers are assigned to cells and processes based upon this schedule. In the Lean linked-cell system, Mixed Model final assembly builds a final product (s) by pulling components and subassemblies directly from upstream production cells with only a small buffer to keep from starving final assembly. This is accomplished by issuing a Kanban withdrawal signal to the upstream subassembly cells. This in turn causes a production Kanban signal to upstream cells to deliver subassemblies/components to final assembly. Hence, upstream production does not need short term or detailed production schedules. Kanban signals provide all of the scheduling orders to trigger upstream production as parts are pulled toward final assembly.

Another distinction between final assembly and upstream suppliers is that workers on the final assembly line must know what final part (vehicle) they are building next. This information is usually generated to them from a central computer via a local computer terminal. This same terminal also provides bar code stickers/plates and other needed final product information. In most cases, detailed assembly instructions are available also, but after a short period of time experienced workers need little or no help. We will again warn that the best final assembly sequence for Mixed Model production can be difficult to optimize or even determine for smooth production. The goal is, of course, to minimize variation in both the processes and in the required output rates. Consumption rates, subassembly deliveries and processing times must be as consistent and constant as possible. This requires great ingenuity, much work and technological innovation as product demand and variety increases.

2.8.3   Long Range Forecasting

The Toyota Motor Company advertises an outstanding 70 or more inventory turns per year. This means that there is only about 3 days of inventory in the system at any one time. Toyota calls this their *instant delivery system*. Toyota spends a great deal of time and effort to support forecasting methods to predict short and long term demands for their automobiles. How does Toyota do this? Do they know something about forecasting methods that the rest of the world does not know? To answer these questions requires a deeper understanding of how the TPS actually works.

Toyota claims that they can produce customized vehicles in two days or less, but the production throughput time is much longer. How can two day delivery be accomplished? The answer lies in how they schedule production, manufacture subassemblies and make-to-stock certain key components. The body, frame and other key parts are manufactured to a relatively fixed production plan, and each has been standardized for a variety of vehicles. The real customization of populating the interior, painting and other custom add-ons can be accomplished in two days. There is no real magic here, only good design for manufacturing and good manufacturing planning.

Toyota determines to a large degree its basic production plan by constantly performing sales forecasts and tracking monthly trend; and surveying over 60,000 potential customers twice a year. As a result, Toyota establishes fixed monthly production plans for key parts and then breaks those plans into daily manufacturing orders. Precise daily production plans are produced to level production, producing the same amount every day until circumstances dictate otherwise. This daily schedule is only communicated to Mixed Model final assembly. Specific demands for all components and parts are then passed upstream to precedence activities. For example, the subassembly line that puts together custom dashes with gages and trim only has 2 hours to respond to an assembly line order. In contrast, the typical American mass production system is a combination of the functional job shop and flow lines that build-to-stock for final assembly, creating large banks of in-process inventories and longer throughput times.

Many American companies, recognizing the superiority and advantages of the TPS, have launched efforts to become Lean, but they have failed to convert job shops and flow shops into cellular manufacturing. This is a sure recipe for chaos. Small lot production with minimum WIP requires that the traditional job shop be eliminated (possibly not totally) and replaced by a linked-cell system in which zero defects are produced, setup and changeover times drastically reduced or eliminated, and final assembly is converted to Mixed model production. All of this is achieved by driving the system with accurate forecasts of potential sales, and the versatility to react to forecast errors. The ultimate goal is to produce all parts one-at-a-time, or at least using minimum lot sizes. Coupled with new design for manufacturing innovations, this new system can quickly and easily respond to fluxuations in demand and support shorter forecasting windows.

2.9   Line Balance

Line balancing is making sure that each worker in final assembly works the same amount of time on the processes assigned to him/her. Task times assigned to each workstation must, of course, be less than the Takt time required by final assembly. The traditional line balancing problem is an old Industrial engineering problem that has now been solved using computer methods. The goal is to meet Takt time while minimizing idle time and imbalance between workstations. When the final assembly line is Mixed Model, both Takt time and line balance are subject to constant change. This characteristic is a problem at final assembly that must be solved by line flexibility.

2.9.1   Balancing Subassembly Cells to Final Assembly

As the level of inventory in the factory shrinks and processes become more tightly linked, production rates for each component and subassembly cell become closely aligned to those of final assembly. Ideally, the cycle time of each component, part, and the subassembly would also be close to the final assembly line's Takt times. Balancing the output of the subassembly cells with final assembly carries over to balancing labor and processes. In a traditional factory line, balancing entails shifting people and tasks along the assembly line. This creates a balanced flow line. In a lean production factory, balancing extends upstream to subassembly cells and manufacturing cells for the component parts.

As a production job shop converts to a lean shop, different types of machines are rearranged into manufacturing cells. The machines in these cells are arranged to move parts between them one piece at a time. Inventory between machines disappears when only one part flows between machines, which led some people to call this a zero inventory system. Close attention to machine setup time, quality (no defects) and total preventive maintenance (TPS) is essential.

The very critical reorganization task in building a lean system is to redesign subassembly lines into U-shaped Lean cells. Operations and processes are placed near one another, and arranged in U-shaped designs. Subassembly cells consist of small, simple pieces of equipment. Workers walk in loops completing precise time schedules. If possible, the cells are designed so that their cycle times are slightly less than the Takt times of final assembly lines. Balancing for the cells means that it can meet the daily demand even if the cycle time is not yet about equal to the Takt time. Output is balanced with assembly line needs. That is, cell output over short periods of time is matched to the rate of part use by subsequent operations and ultimately to final assembly. The Kanban link between cycle and Takt times absorbs any mismatching of cell production and usage rates by subassemblies or final assemblies. The fewer mismatches, the smaller the volumes are in Kanban inventory links.

Suppose a cell is making two products: A and B, which is subsequently used downstream in equal quantities. Part of the time a cell is making Product A. While it is making Product A, there must be a sufficient amount of Product B in the Kanban link to meet downstream demand for Product B, assuming there are no other problems in the manufacturing cell. Of course, if setup and changeover times have been reduced to a minimum, the feeder cell might make one unit of A, then one unit of B in repeating patterns. If SMED cannot minimize or virtually eliminate setup/changeover times; it will be necessary to maintain a small amount of WIP using a Kanban controlled link between the upstream and downstream manufacturing operations.

In Machining cells, different part families produced in the same cell need not be produced on the same machines in identical machining times. Parts usually require different processing times, but parts made in one cell should require labor times in the same general range for each part produced in the cell. This is a general requirement so that a different number of workers are not required to be in the cell for each different part in the family. This does not mean that total machining time between two parts must be balanced or equal because machining times are decoupled from the cycle time. The cycle time depends on the number of machines a worker visits on a trip around the cell.

Adding or subtracting workers changes the output rate from cells. In addition, workers often cross cell boundaries when they perform tasks on different parts. During any fixed schedule period, the number of workers within a cell should remain constant. However, when the daily demand changes, personnel and personnel loop assignments might both change. Occasionally a cell is designed to add or subtract machines. This may be acceptable if a cell is reconfigured and staffed to support long production runs, but if machine capacity or modification is required frequently, it will be better to let some machines sit idle for a particular part. In any event, a change in either demand patterns or product mix will usually cause the work pattern in the cell to change. These changes are necessary to match cell cycle times to those required by a new downstream final assembly schedule (FAS).

Workers who have performed only a single function or operated a single machine may find that Lean manufacturing requirements are difficult to accept. Cells are designed with simple, single-cycle machines and equipment modified for flexibility. A properly trained operator should be able to operate and set up every machine in a cell. This is a goal but not a prerequisite for starting up a cell. Higher management, finance and accounting functions may also resist Lean philosophies which often result in lower utilization of machines.

Supervisors are responsible for maintaining data needed to determine how many workers are required for different cycle times in a cell. This job does not require large amounts of data, but rather a set of rules based on past performance, trial and error, and perhaps a calculation or two. The lean engineer can assist in solving complex balancing problems.

Balancing a cell is easier when the required cycle time is greater than 30 seconds. If the needed cycle time is less than 30 seconds, then replication of an existing cell should be considered. If there are two different parts being made in a Mixed Model cell, each cell should be dedicated to a single part. This requires twice as much equipment. However, since machines are simple, single-cycle automatics this may not be a large capital investment.

2.9.2   Balancing the Entire Factory
The steps for balancing an entire factory are:
- Balancing the rate of parts production to match the final assembly-line rate (overall cycle times are the inverse of production rates).
- Adjusting work content and cycle times at each cell or station until times match system cycle times are nearly as possible; and
- Trying to off-load work content of selected stations until selected stations are no longer needed.

The final assembly line Takt time requirements should line side subassembly and point of use storage when possible. The commonly held and proclaimed notion that there are no storage areas on a factory floor is incorrect. However, the idea is to minimize material in these storage areas. The best places to minimize storage areas are near points of use and close enough to producing areas so operators have visual signals of part usage.

Linked subassembly cells must be balanced to final-assembly Takt time. For example, if vehicles are assembled with a Takt time of 60 seconds and each vehicle needs a steering gear, then steering gear subassembly should have cycle times of around 60 seconds. If each steering gear needs a rack and pinion assembly, then these assembly cells should also have cycle times of 60 seconds as well feed by manufacturing cells machining the rack and another machining the pinion and a third machining the housing. Most lean plants run the subassembly and manufacturing cells a few seconds less than the Takt time to buffer any unexpected variation or unusual events.

In a factory, events rarely work out exactly as planned. Therefore, a continuous effort should be made to reduce deviations between the rate set by final assembly and the production rates of upstream elements. Real production improvement results from matching upstream processes closely to the rhythm set by the final assembly Takt time.

In the mass system, traditional line balancing refers to balancing the amount of work or lab or at each station, regardless of cycle time. There is no attempt to achieve overall factory balance. In lean manufacturing, balancing the line or plant also refers to balancing material flow. The pull system of material control balances material flow. Material balance and labor balance are dependent on each other. The primary indicator that labor is out of balance is an excess or shortage of material. The amount of work in process between cells results from unbalanced cycle times. The objective is to set cycle times as required by the schedule and then to shift tasks and operations accordingly. This will result in less and less work for one worker, and eventually this worker can move to another area. It must be remembered that systems are designed to minimize labor content as well as improve labor efficiency.

Plant balancing is a dynamic, ongoing process. Normally, a plant must be rebalanced whenever the production rate changes. The fear of losing line balance is the major reason for reluctance to stop a balanced assembly line once it is up and running. This is also true of an entire factory. There are countermeasures to this difficulty including:

- Visible signaling systems, like *andon*, allow a system to respond to temporary variation in part-usage rates and to changes in product mix.
- Flexibility built into cells permits a system to quickly adapt to requests for increased or decreased production rates or to changes in product mix.
- Less-then-full-capacity scheduling means keeping a little slack time, perhaps 20-30 minutes per shift at each cell, so a system can respond to variations from the planned schedule. This adds flexibility to a system.

Since a plant's schedule changes periodically, the operations also may have to change. Manufacturing cells and subassembly cells can usually continue working without rebalancing if the cycle times required do not vary more than about 10%. Cycle-time variations beyond 10% usually require rebalancing. Detailed planning for production must take this into account.

2.10 Standard Operations (Also called Standard Work)

Standard operations are designed to allow manufacturing cells and subassembly cells to use a minimum number of workers. The *Standard operations routine* is a standardized form for standard work in manufacturing cell. It shows the sequence of operations, the manual times for the operator, the walking times and the machining cycle times. Workers are expected to write down operations, and this listing is compared with the standard operations routine. This procedure helps ensure that new workers are performing the correct steps in the correct sequence. A standardized operations routine sheet like that shown in Figure 2.9 is used for this purpose. The cycle time is 2 minutes, greater than the longest machining time of 95 seconds. The cycle equals the sum of the manual times (98 sec) plus 22 seconds of walk time for 120 sec.

Standard operations try to achieve a balance among operations so the cycle time is in compliance with the final assembly cycle time. Producing the daily demand in a given period supports the final assembly tact time. Therefore, the cycle time concept should be incorporated into standard operations.

With the use of cams, micro-switches, and similar mechanical devices, different types of machine tools can be set up to have a processing time less than the necessary cycle for the cell. At completion of each machining cycle, the machine is stopped automatically and its components and attachments are returned to the start position. If there is space to hold a finished work piece on the output side of a process (the downstream decoupler), the work piece can be automatically ejected from machine to the decoupler. If an empty space is not available, the machine must wait until the operator arrives to unload the machine. An empty decoupler provides a signal to the upstream machine to make another part. If an unprocessed work piece is available in an upstream decoupler, it may be automatically fed to the machine using guides to position it without human assistance. Once the unprocessed work piece is located and clamped into the machine, the next machining cycle is started.

Leveling quantities and synchronizing processes can significantly reduce delays, thus greatly reducing manufacturing throughput time. For example, using one-piece flow to eliminate lot delays for two serial processes reduces throughput time. The multiplier effect of eliminating process delays and lot delays with one-piece part movement in cells may improve throughput time by 98% if 10 or more processes are involved.

The final part of standard work is achieving a minimum quantity of material in the cell. This standard quantity is called *stock-on-hand*. In other words, only the minimal quantities of parts necessary to complete standard operations are kept on hand in the cell. Therefore, standard work consists of cycle time, standard-operating routines, and minimum quantities of materials. Concurrently, the elimination of accidents, breakdowns, and defects is also a major component of standard operations.

2.11 Internal Customers

An important principle of lean manufacturing is that the internal customer is the most important resource of a manufacturing organization, as well as one of the most limiting factors.

In the lean production factory, processes produce only the amount of product required. Lean design concentrates on worker utilization. Using processes to make more than is required violates the basic principles of lean production. Overproduction means that eventually inventory builds in a manufacturing system. This excess must be stored, tracked, and retrieved – all costly and wasteful operations. Frequently, these wasteful operations lead to the purchase of automatic storage-and-retrieval systems that require a maintenance person to keep running, a computer system to track materials, another worker to program the automatic storage, retrieval-system computer, etc. This scenario is a total waste of valuable resources and it lowers an organization's ability to compete.

Lean production uses a minimum number of workers to achieve daily demands by balancing factory operations. Modifying cycle times by laying out cells close to other cells makes it easy to balance tasks among workers. Elements of work are shifted from one worker to another until operations are balanced. This may result in a situation in which a worker is no longer needed in the cell and can be assigned to another cell.

What can be done at full capacity if the production rate must be increased (and the cycle time lowered)? This situation presents a lack of flexibility at full capacity. One problem-solving approach would be to develop two cells with half the number of workers. This process doubles the cycle time while producing the necessary number of parts. For instance, five workers might be able to produce 57-58 parts per hour, just two or three less than the required 670. A little overtime would permit workers in cells to meet system-level requirements. Notice the objective of improvement is not to reduce cycle time. Changing cycle time only is necessary for meeting schedule changes dictated by final assembly. The objective is to minimize the amount of required labor while producing at a rate that yields parts needed for final production. Changes resulting only in excessive inventory stored in the manufacturing system are deceptive and really do not improve productivity.

Replication or duplication of a manufacturing cell can add variation in both production rates and throughput times because no two cells will operate exactly the same. This may seem to violate the basic objective of minimizing or eliminating sources of variance, but if Cell A is dedicated to Product A and Cell B to Product B, then variability issues should be resolved.

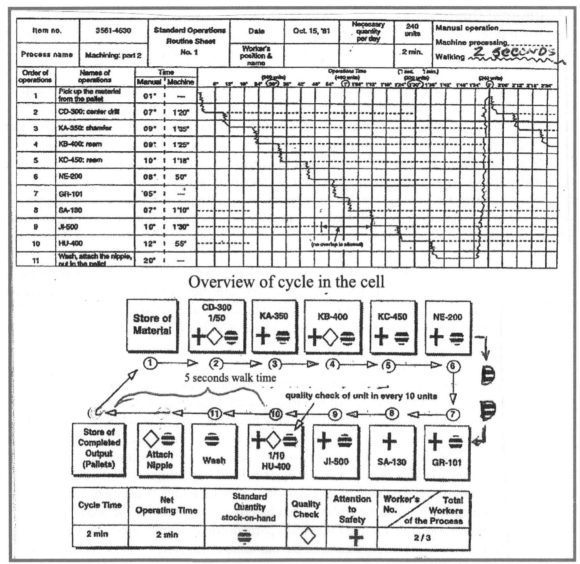

Figure 2.9
Example of a Standard Operations Sheet and the Associated Cell Layout for a Lean Manufacturing Cell

2.12 Inventory: Not zero but Minimized

Two historical names for lean manufacturing have been *stockless* production and zero inventory. Nothing could be farther from the truth. Every manufacturing system has inventory. It is the life blood of the factory. No factory can run with zero inventories or have a stockless production system. In the TPS, the inventory is held in the Kanban links. The WIP is in each link while the material in the cells is called *stock-on-hand*, and it is not considered inventory, hence the term zero inventory has some validity. In the TPS, the inventory is minimized and controlled (reduced) in order to expose problems in the system. Taiichi Ohno's most famous schematic called *Rocks in the River*, where the river represents the WIP inventory in the system (flowing or stagnant) and the rocks represent problems. Lowering the level of the river (the level of inventory) exposes the rocks (problems) which are worked on to improve the system and permit further lowering of the WIP inventory.

2.13 Synchronization

Synchronization refers to the process of timing movement of material between portions of the assembly line and the major subassemblies. Even when material quantities have been leveled and balanced, unnecessary storage of in-process material can occur between unsynchronized operations. However, once operations are leveled, synchronization is just a matter of efficient, integrated scheduling. Leveling must precede synchronization because it helps eliminate process delays that make synchronization difficult.

Subassemblies that are synchronized for final assembly in the automobile industry are large items such as seats, panels, headliners, cockpits, doors, and engines. These subassemblies are all specific to certain vehicle models. They are made in sequence with the final assembly and delivered to designated workstations on the final assembly line at the same time as the vehicle needing that subassembly. Only the best lean manufacturers can accomplish synchronization, since any subassembly-line failure can also stop a final assembly line. Conversely, if final assembly is stopped, then synchronized subassembly cells and lines also must stop so elements stay in sync. Production processes that have stopped must restart together.

2.13.1 Yo-i-don Synchronization

Yo-i-don in Japanese means: *ready, set, go*. It is the name given to a method of synchronizing startup of manual manufacturing processes or operations. This method is not used with a mechanical transportation mechanism such as a conveyor line. In these situations, the final assembly line would pace the line door assembly. For example, suppose a left front door is removed from a vehicle early on the final trim line at Station 26, goes by overhead conveyor to the door subassembly line, and gets fully stuffed with components (windows, door handles, speakers, etc.) It then returns to the final assembly line at Station 110 where the door is reinstalled on the same car body from which it was earlier removed. Thus, the door line must be kept in sync with the final assembly trim line, stopping and starting with the final assembly line.

The Yo-i-don method is also used in body welding. Operations for the body may be divided into four primary processes such as underside, side bodies and top body or roof (Figure 2-10). The underbody and side body processes can be divided into six processes and three sub-processes. U_1, U_6 and S_1, S_3, respectively. The top, sides, and bottom pieces come together at B_1 in a large jig and are spot welded.

Suppose a final assembly line is producing one unit per minute for the system Takt time or factory output rate. Operations at each of the subassembly, processing, and main assembly areas must be completed in one minute or less. Since each car needs a body, the car-body welding area produces one body every minute. Workers and spot welding robots in each area must complete tasks and pass welded parts to subsequent workstations at the end of prescribed one-minute-cycle times. After completing assigned tasks and passing the welded part downstream, each operator presses a job completion button.

The job completion button turns on a green light on the *andon*, indicating that certain tasks have been completed (Figure 2.11). The andon is a signal board with the colored light and designation relating to stations on the line. The andon hangs over the line so it can be easily seen. At the end of each cycle, a yellow light illuminates if there are any incomplete tasks. When this happens, adjacent workers and supervisors provide assistance. A red andon light illuminates and the entire line comes to a halt and if a task has not been completed, once it is completed the red light is turned off and the process cycle starts again with all processes beginning in sync. This method synchronizes plant operations by getting them all to start together in each cycle. The andon board is called a *bingo board* in some factories.

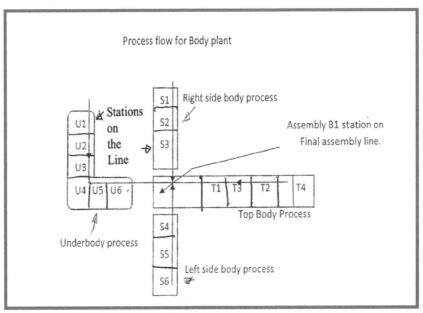

Figure 2.10
Synchronization of Body Parts in a Spot Welding Line (Monden, 1983)

Many plants serving as a tier 1 supplier to final assembly are also synchronized with final assembly. They receive an electronic signal from the final assembly area when a car comes out of paint and enters the final assembly line. The synchronized supplier begins to build the subassembly at that time to match the final assembly item (car). The subassembly is built, shipped, unloaded, and sent to the correct station *just-in-time* to be installed on the final assembly item.

2.14   How Lean Manufacturing Cells Work

An example of a lean U-shaped manufacturing cell that has seven machine tools and a final inspection station is shown in Figure 2.12. In the cell, one worker moves from machine to machine, unloading, checking, and loading parts. Each time the worker completes a loop (the dashed line) one part is completed and all parts are advanced to the next machine in the sequence. Each machine is a single cycle automatic capable of completing some machining operation on its own and then shutting off. The next machine may have a work holding device

Figure 2.11
Andon on a Toyota Assembly Line: A Signal Board Indicating Progress at Various Stations

that checks to make sure that the previous machine produced its *step* correctly before the next machine performs its processing steps. Sometimes the checking occurs in the part/holding transporting device called a decoupler between the machines. Decouplers are shown in the layout with a large D symbol. The decoupler device may simply check a dimension, and turn on a red light when it finds a defect. Decouplers that prevent defects are called *Poka-yokes*.

The operator is critical in this design. He is handling every part and checking the output from every process. In the manufacturing cells, the operators are considered to be the company's most important (fixed) asset, a point of view not traditionally held in U.S. factories. People are much more flexible than computers. Computers are viewed only as a useful tool in the process. Cells that use equipment that was initially designed for stand-alone applications in the job shop are sometimes called *interim cells*. When such equipment is grouped into U-shaped cells and properly modified, the cell can be easily operated by one standing, walking worker.

If demand on the cell increases, it will be necessary to add one or more workers. Figure 2.13 shows two-workers in the cell. Part flow between each machine and the two workers is connected with a decoupler. If a third worker is added, the cell output will increase again so long as it is not constructed by a machining cycle time.

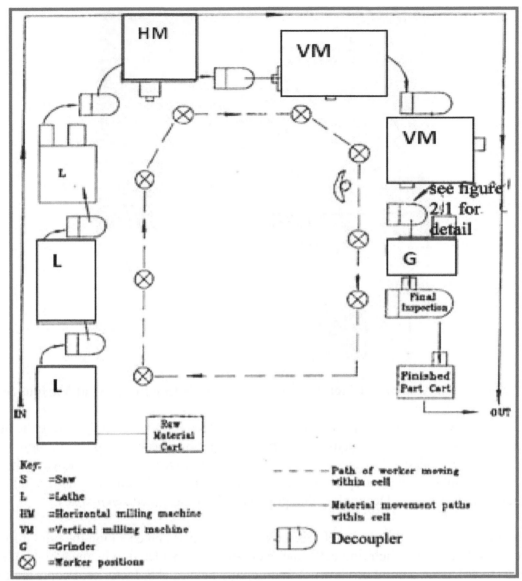

Figure 2-12
Interim Manufacturing Cell with Seven Machines Operated by One Worker
Width of aisle = 4 feet. (Black, 1991)

Decouplers are also placed between the processes, operations, machines and workers to provide cell flexibility, part transportation, quality control, production control, and process delay. Figure 2.14 shows a decoupler that is a slide or chute. This transfer device also presents the part to sensors and laser measuring devices that conduct automated quality control inspections. Upon completing the QC check, the decoupler then orients the part correctly and moves it to the grinding machine.

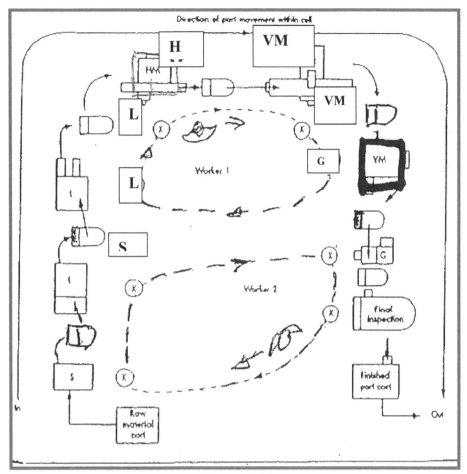

Figure 2.13
Interim Manufacturing Cell Operated by Two Workers Connected by Decouplers

2.15 Final Assembly

The final assembly line is a product-oriented layout and is often a highly automated version of a flow shop. When volume gets large, especially in an assembly line, it is called *mass production*. These systems may have a high-production rate, typically 200,000-400,000 per year in the automotive industry. Specialized equipment is dedicated to the manufacture of a particular part or model. The integrated systems of men and machines interact to create a manufacturing flow line. One machine of each type is typical, except where duplicate machines are needed to balance the flow. The entire plant often is designed exclusively for production of

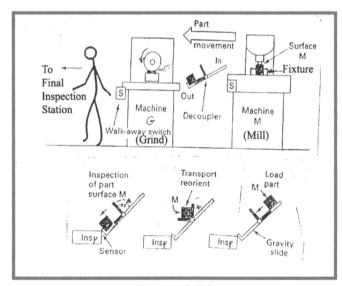

Figure 2.14
A Decoupler That is a Slide or Chute

63

a particular product, with special purpose process equipment rather than general purpose equipment.

Capital investment costs of specialized machines and tools are very high, as are the risks. If design changes are necessary, highly automated and dedicated machines are often not flexible enough to make the transition. Many productions skills are transferred from operators to machines resulting in lower levels of manual labor skills than in a production job shop. Items flow through a sequence of operations by material-handling devices (conveyors, moving belts, transfer devices, etc.) Items usually move through operations one at a time.

Flow shop layouts are either continuous or uninterrupted. A continuous flow line basically mass produces one complex item in great quantity and nothing else. For example, a transfer line for the assembly of a Camry engine is shown in Figure 2.15. An interrupted flow line manufactures large lots, but periodically changes over to run similar but different components. Changing over a complex flow line may take hours or even days. For automobile engines, transfer lines are usually dedicated. A flow line for manufacturing facilities is arranged according to a product's sequence of operation. A line is organized by the processing sequence needed to make a single or regular mix of products. The transfer line advances the work (transfers it) forward one station at a time with each station performing some operations, all done within the cycle time. Such lines cost millions of dollars and may operate for years to make enough units to recover.

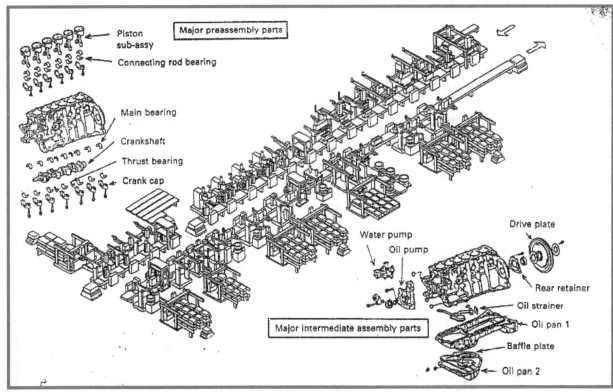

Figure 2.15
A Highly automated and Integrated Flow Line

The automated-transfer machine that assembles engines (at a rate of 100 per hour) is an example of a super machine for mass production. Transfer machines are specialized, expensive to design and build, and usually not capable of making another product. These machines must be operated for long periods of time to spread the cost of the initial investment – typically $20-40 million over many units. Although highly efficient, they usually make products in large volume.

Desired design changes in a product must be avoided or delayed, because it would be too costly to scrap the machines. Such systems are usually not flexible enough for product or process-design changes. Smaller versions of transfer machines developed for smaller-sized products made in large volumes range in cost from $2-3 million.

A hybrid form of the flow line produces a batch of products moving through clusters of workstations or processes organized by product flow. This is called batch flow. Garment or apparel manufacturing is traditionally done this way. For instance, a batch of shirts moves through a sequence of different sewing operations. Usually, setup times to change from one product to another are long, and often the process is complicated.

Mass production factories are mixtures of job shop and flow line systems. Demand for products can precipitate a shift from batch to high-volume production, and much of production is guided by that steady demand. Subassembly and final assembly line are further extensions of the flow line, with the former usually being more labor-intensive.

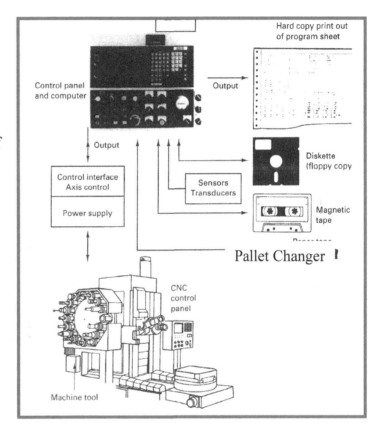

Figure 2.16
A CNC Machining Center

Since the advent of the mass production system, various approaches and techniques have led to the development of machine tools that are highly effective in large-scale manufacturing. Their effectiveness is closely related to produce-design standardization and the length of time permitted between design changes. A machine producing a part with a minimal amount of skilled labor can be developed if the part can be highly standardized. The part can then be manufactured in large quantities. An automatic screw machine (a complex lathe) is a good example of a machine for the manufacture of small parts.

The development of the numerical-control machine tool in the late 1950s and early 1960s permitted programmable control of the position of cutting tools in relationship to the workpiece. By the late 1960s, automatic tool changers had been added to the numerical control machine, making the birth of the machining center. Computers and material handling devices (pallets) were added next. Today, the computer-numerical-control machine tool is readily available to all manufacturers. The machine shown in Figure 2.16 is an example of a CNC machining center.

Products manufactured to meet demands of the free economy and today's mass-consumption markets must include changes in design for improved product performance, as well as style. Therefore, automation systems must be as flexible as possible while retaining the ability to mass produce materials. In the late 1960s, this lead to a coupling of the transfer line with the numerical-control machine and the flexible manufacturing system (FMS) was born.

The primary components of the flexible manufacturing system are computer-numerical-control machine tools, material-handling systems to transport parts, cutting tools, machine-work holding devices (pallets), and computer-control networks. The machine tools are usually horizontal or vertical spindle-milling machines or multiple spindle CNC lathes. The flexible manufacturing system (FMS) design shown in Figure 2-17 has four computer-numerical-control machining centers, each equipped with large cutter-tool magazines and a parts-pallet-changing system. The system designers claim this FMS is capable of making more than 100 different parts. Such a system must be programmed and scheduled for each part it produces. It can be much more complex as the job shop it replaces.

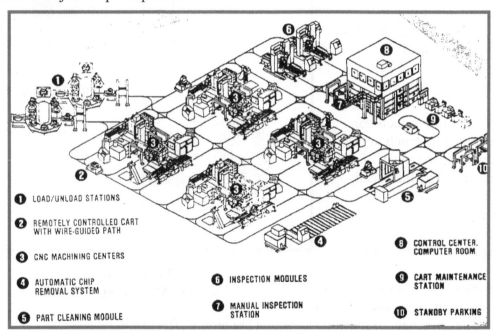

Figure 2.17
A Flexible Machining System

The modern computer-numerical-control machine can be programmed to automatically change tools, work pieces, and cutting parameters. It seems logical that the versatile machine be joined with the transfer line to expand manufacturing-part variety. Most flexible manufacturing system installations have a system manager who is responsible for supervising workers, including material handlers who perform loading/unloading tasks, a roving operator who presets tools and reacts to unscheduled machine stops, and a mechanical/hydraulic technician who repairs transfer devices, machine, work holders, and pallet changers.

Much has been written about flexible manufacturing systems. The systems are expensive to design and often require months or years to implement. They are complex to program, analyze, debug, and control. Currently there are about 1,000 FMSs existing world-wide. In the United States, most of these systems are found in large companies than can afford large capital outlays or that receive governmental backing, such as military defense contracts. The flexible manufacturing system represents the super-machine philosophy at its ultimate. Fundamentally, it is an attempt to blend the job shop's flexibility with productivity. Parts may require two or three passes through a flexible manufacturing system. Fixturing in the flexible manufacturing system is costly and complex. The system's control computer must control the conveyor, maintain the computer numerical control library of programs, download these to the machines, handle scheduling of the system, track tool maintenance, track performance of the system, and print management reports. Not surprisingly, the system's software development often proves to be a major limiting factor.

2.16 Group Technology (GT)

Manufacturing cells are key building blocks in linked-cell manufacturing systems, composed of directly connected manufacturing cells or cells linked by Kanban. There are two basic types of cells that feed final assembly: manufacturing cells where most processes and machines are single-cycle automatics (i.e., complete the processing cycle untended); and self-balancing, subassembly cells, where most, if not all, operations require an operator to be present to do tasks. One way to form a cell is by using group technology but most companies prefer instead to simply form cells based on product families.

Most group technology methods ignore the worker and simply find machines that will process groups of parts. Group technology is a philosophy of grouping similar parts into part families. See Chapter 17. Parts of similar size and geometry can often be processed by a single set of processes. A part family based on manufacturing –process type would have the same sequence of manufacturing processes. The set of processes is arranged to form a cell. Thus, with group technology, job shops can be systematically restructured into production cells, with each cell specializing in a particular family of parts. Usually no new machines are needed. Floor space is often freed up by GT conversions. The machines have at least the same utilization as in the job shop, but throughput times are greatly reduced, and workers are more effectively utilized.

Cells are typically manned, but unmanned cells with robots replacing workers have been developed. A robotic-cell design (Honda) is shown in Figure 2.18 with one robot, one operator, and six metal forming machines. If the operator is removed, the machines must have higher levels of automation. Decouplers are designed to connect the part flow to and from the operator.

For the robotic cell to operate on a pull basis, the removal of a part from the last decoupler initiates a string of commands back to upstream decouplers to replace the part removed. The first robotic pull cells were built at Auburn University in 1988 to manufacture gear couplings.

2.17 Kaizen and Continuous Improvement

Kaizen provides the means for continuous improvement of the MSD and the motivation to encourage operators to take part in redesigning, improving and managing their own jobs. Kaizen improvements in standard work help maximize productivity in the cells and in the workstations.

Because standard work within the cells involves following work methods consistently, problems in the working sequence occur repeatedly and the team members can easily identify problems and note them in the daily meetings. Team leaders then develop kaizen events to implement improvements. Similarly, monthly changes in production volumes may require changes in the standard work, so team leaders and team members devise new standard work procedures to accommodate monthly changes in production volumes.

Kaizen activities include measures for improving equipment as well as measures for improving methods of work. But work-kaizen tends to be easier, faster, and less expensive than equipment-kaizen. So teams usually start with work-kaizen when trying to resolve a problem. If changing the working sequence or methods is insufficient to resolve a problem, solutions through equipment changes are considered.

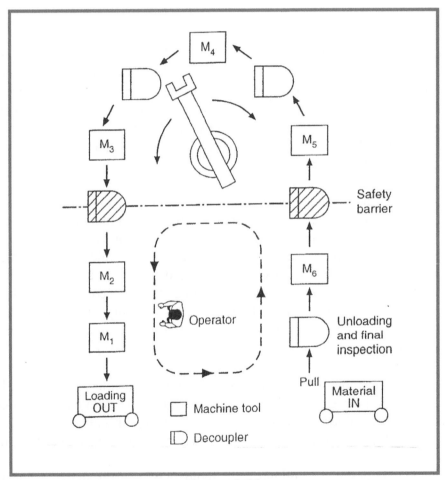

Figure 2.18
An Example of a U-shaped Cell Manned by an Operator and a Robot with Part Flow Connected by 2 Decouplers.

Figure 2.19 illustrates an example of a kaizen improvement of a small subassembly line into a U-shaped cell. In this subassembly example, the transition to a new design involves the following steps:

- Conversion from a straight-line using a conveyor to a U-shape subassembly cell
- Add simple automation steps (driving screws and applying glue)
- Reduction from 3 operators to 2 operators

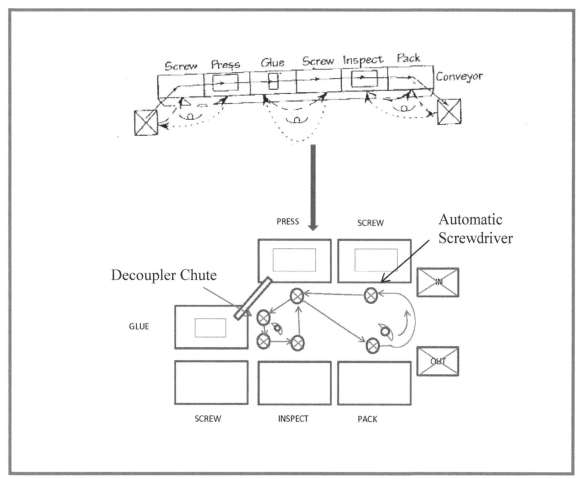

Figure 2.19
Kaizen Example of a Linear Sub-Assembly Conversion to a U-Shaped Cell Design (Baudin)

When flow shop elements within a factory are redesigned into U-shaped cells, long setup times typical of flow lines must be vigorously attacked. It is imperative that flow lines be able to change over quickly from the manufacture of one product line to another. The U-shape design eliminates the need to line-balance a flow line every time it changes to another part.

2.18 Design for Flexibility

Flexibility (the capability of the MSD to respond quickly to changes) is a premier design feature. Once implemented, the L-CMS can react quickly to changes in customer demand (volume and mix), product design (ECO's and new product designs) and lead time changes in custom orders. Manufacturing cells are linked to subassembly cells or assembly lines at the point of use or they are connected by Kanban links. This makes a factory product oriented and flexible in:

69

- Its operation of equipment,
- Changeover,
- Process, and
- Capacity or volume

The key to flexibility is there should not be any adjustments in equipment settings, automatic-error detection must be in place, and rapid tooling changeover is a must. Changeovers must be easy, with the speed of the setup and exchange of tooling and dies smooth and rapid. In a flexible, lean environment, the ability to increase or decrease production output, rate, and volume is significant, so the ability to handle a different mix, a different order in the mix, and a different volume in the mix is critical. The key points in forming a manufacturing cell are:

- Machines are arranged according to the sequence of operations.
- The cell is designed in a U-shape (with a 4-foot aisle).
- Parts are made within a cell on a one-piece flow basis.
- Workers attend to more than one process.
- Takt time dictates the cell cycle time.
- Workers perform their jobs while standing and walking,
- Machines are modified for quick changeover, zero defects, high reliability and walk-away switches.
- Often smaller, slower, dedicated, less-expensive machines are used.
- Unique manufacturing process technology is often developed in-house.

2.19   Control of Manufacturing Functions - Revisited
Every manufacturing system has certain control functions that must be carried out. Table 2.3 summarizes eight functions and lists techniques that both the lean production system and mass production system use to aid these functions.

Regardless of the type of manufacturing system, the same control functions are performed for all. However, system tools used in lean manufacturing differ greatly from mass production tools. Table 2.4 summarizes control in Lean Manufacturing vs. control in traditional mass manufacturing. In the linked-cell system, many tools are manual or physical such as Kanban cards, Andon lights, Poka-yoke checks, and oral orders. In the job/flow shop factory under material requirements planning, the most important tool is the computer. If the system being computerized is well-designed and robust, the computerized version will have a good chance of succeeding. However, if your mass system is having quality, delivery or cost problems, or you don't understand how your MRP system works or your computer fails during a power outage, you will no longer have control over the life blood of your factory and fighting fires will be a way of life.

How eight manufacturing functions are controlled			
Functions	Categories	L-CMS – Lean	Job shop/flow shop - Mass
How many to make per day	Families of products	Leveling the manufacturing system	Production plan-orders plus forecast
What mix of products to make each day	Finished goods for make-to-stock, customer orders for make-to-stock	Master production schedule	Master production schedule
Getting materials required to the right place at the right time	Components – both manufactured and purchased	Pull system WLK cards	Push system – material requirement planning (MRP)
Capacity of the system	Output for key work centers and vendors	Controlled by number of workers	Capacity requirement planning (CRP)
Executing capacity plans	Producing enough output to satisfy plans	Meet downstream needs	Input/output controls, route sheets
Executing material plans – manufactured items	Working on right priorities in factory	POK cards-pull system	Dispatching reports, route sheets
Executing material plans – purchased items	Bringing in right items from vendors at the right time	Kanban cards and unofficial orders	Purchasing reports, invoices
Feedback information	What cannot be executed due to problems	Immediate and automatic	Anticipated delay reports

Table 2.3
The Control of Manufacturing Functions

2.20 Comparing Lean Production to Other Systems

Table 2.2, 2.3 and 2.4 provide brief comparison of lean production versus mass production in a flow shop. A popular idea a few years ago was to find the functional job shop system's bottlenecks or constraints and try to eliminate them. This approach views material queues as necessities that permit downstream operations to continue, when there was actually a problem with the upstream or feeding operations. The linked-cell manufacturing system approach recognizes the job shop design as a retardant to flow. Better management by constraints only results in small improvements in productivity and quality in the job shop. It is the manufacturing system that must be redesigned. Figures 2.20 and 2.21 summarize manufacturing systems by comparing different methodologies based on production rates and flexibility (the number of different parts the system can handle). Continuous process systems are not shown. Cells provide a wide middle ground between job shops and dedicated mass-production flow lines. These figures were developed using factory-floor data. Note that the widely publicized flexible manufacturing system lies between the job shop and the transfer line. This is expected, since the flexible manufacturing system was developed from a merger of these two systems.

As in a job shop layout, it is necessary to schedule any job shops within a flexible manufacturing system. This function, however, makes FMS designs difficult to link to other manufacturing systems. The flexible system often becomes an island of automation within a job shop, which is unfortunately a characteristic of super machines. Because both the accounting functions and upper management (incorrectly) dictate that these expensive machines must run 100% of the time, and work is *pushed* to these machines by atypical order release system such as MRP or MRP II, Work in process tends to accumulate around these types of flexible manufacturing systems.

Table 2.4
Control in Lean Manufacturing vs. Control in Traditional Mass Manufacturing

HOW LEAN MANUFACTURING PHILOSOPHY DIFFERS FROM THAT OF A TYPICAL US COMPANY		
Factors	**Lean Manufacturing**	**Mass Manufacturing**
Inventory	Inventory is wasteful. It hides problems. It is a liability. Every effort must be extended to minimize inventory	Inventory is a necessity. It protects against forecast errors, machine problems, late vendor deliveries. More inventory is "safer" and necessary.
Lot sizes	Keep reducing lot sizes. The smallest quantity is desired for both manufactured and purchased parts	Formulas. Keep revising the optimum lot size with some formula based on the trade-off between the cost of carrying inventories and the setup costs.
Setups	Eliminate/reduce setups by extremely rapid changeover to minimize the impact. Fast changeover permits small lot sizes and allows a wide variety of parts to be made frequently.	Low priority. Maximum output is the usual goal. Rarely does similar thought and effort go into achieving quick changeover. Use EOQ to determine lot size.
Vendors	Procure from a single source. Vendors are remote cells, part of the team. Daily, multiple deliveries of active items are expected. The vendor takes care of the needs of the customer, and the customer treats the vendor as an extension of the factory.	Adversaries. Multiple sources are the rule, and it is typical to play suppliers against each other to get lower costs but multiple vendors increase the variability in the components.
Quality	Zero defects. If quality is not perfect, then improvement can be made. Continuous improvement in people and process is the goal.	It costs money to make high quality products. Tolerate some scrap. Track what the actual scrap has been and develop formula for predicting it. Plan extra quantity to cover scrap losses.
Equipment maintenance	Constant and effective. Machine breakdown and tool failure are eliminated or reduced by routine maintenance.	As required. Not critical because inventory is available.
Lead times	Keep them short. This simplifies the job of marketing, purchasing, and manufacturing as it reduces the need for expediting.	The longer the better. Most foremen and purchasing agents want more lead time, not less.
Workers	The internal customers are the experts. Changes are not made until consensus is reached. Employee involvement is critical, especially in the design of the cells. Managers are coaches who serve workers in teams.	Engineers provide ideas and are the experts. Management is by edict. New systems are installed in spite of the workers, not thanks to the workers. Measurements are used to determine whether or not workers are doing as directed.
Cost reduction	Cost reduction comes by non-stop like water through the pipe type manufacturing, thus reducing the TPT.	Cost reduction comes by driving labor out of the product and by having high machine utilization.
Production Control	Material should be "pulled" through the factory, using kanban.	Material should be coordinated by MRP and "pushed" out into the factory.
Overhead	Any function that does not directly add value to the product is waste.	Overhead functions are essential.
Accounting's view of labor	Labor is a fixed cost. The internal customers are one of the system's primary resources.	Labor is a variable cost.
Equipment maintenance	Machines are distance runners, slow but steady and always ready to run.	Machines are sprinters, and pulled hamstrings are to be expected.
Automation	Autonomation is valued because it facilitates consistent quality and prevents overproduction.	Automation is valued because it drives labor out of the product.
Expediting	Expending is a manufacturing sin.	Expediting is a way of life.
Cleanliness	Housekeeping is everyone's responsibility.	Work means getting your hands dirty.
Evaluation (measurable parameters)	Multiple performance criteria based on cost, quality, ontime delivery, and flexibility.	Evaluation is based on quantified direct cost.

2.21 Closing Remarks

In summary, Lean manufacturing cells produce parts one at a time in a flexible manner. Cell capacity and cycle time can be quickly altered to respond to customer demands for change. For manufacturing cells, the cycle times does not depend on machining times or upon the number of process steps involved, but upon the number of workers assigned to each manufacturing cell. Identifying families of parts with similar designs, flexible-work holding devices, and tool changes in programmable machines using Group Technology concepts allow rapid changeover from one component to another.

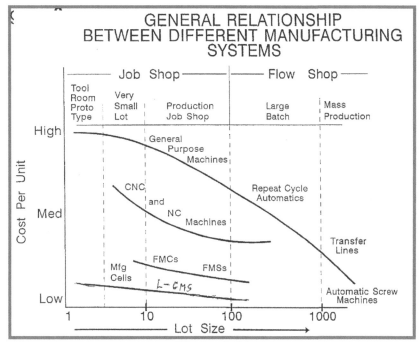

Figure 2.20
General Relationship Between Different Manufacturing Systems

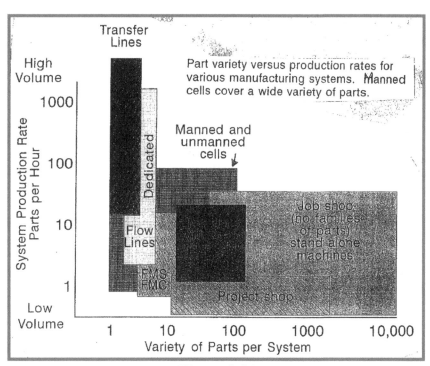

Figure 2-21
Production Rates (parts/hour) Versus the Number of Parts in Various Manufacturing System Configurations

Rapid changeover means quick or one-touch setup, often like flipping a light switch. Significant inventory reduction between cells is possible with rapid changeover when the inventory levels can be directly controlled. The cell operators are directly responsible for quality in a cell. Specialized quality control statistics and skills rest primarily with the trained Lean engineer working directly with the cell operators. Equipment is routinely maintained by the workers as part of their assigned duties. Only major maintenance tasks are relegated to the Maintenance (TPM) organization.

For robotic (automated) cells, robots typically load parts into computer numerical-control machine tools, and sometimes are used to unload parts into a decoupler following the machine cycle. To the casual observer, a traditional machining center appears to be similar to a Lean cell consisting of one machine. However, a machining center is not a cell. It is not as flexible or productive as a manufacturing cell, which uses many simple machines. Production rates are dictated by the number of machines and their machining cycles, compared to a Lean cell in which output rates are controlled by the number of workers. These characteristics of Lean cells will be discussed in some detail in Chapters 3-5. Multiple spindle machining centers can have overlapping of machining times and are the logical choice for large complicated parts in smaller quantities. Cellular layouts facilitate the integration of critical production functions while maintaining flexibility and producing superior-quality products. For production workers, cells provide opportunities to perform more tasks and to experience a sense of overall job enrichment.

In the L-CMS, product designers easily see how parts are made since each processing step is usually manned by a single worker. The workers sense of how he or she fits into the grand scheme of things is often clouded or unknown. Design changes are often very difficult in traditional, highly automated manufacturing systems, but flexible Lean cells are specifically designed and operated to accommodate change.

The linked-cell system is designed for one-piece flow. In essence, each piece of a final product is made or assembled by a series of single processes, steps, or operations. When something goes wrong with the product it is easy to identify the problem and process. Problems can be quickly fixed. Make-one, check-one, and move-one-on is the operational watchword for lean manufacturing system design. Chapters 4 and 5 will address the design and operation of U-shaped Lean cells. Before we do that, Chapter 3 will present the basic design rules for Lean systems.

Review Questions

1. Explain Little's Law
2. What are the main advantages of cell design compared to typical mass production system?
3. List seven tools of quality control used by workers to control process and achieve zero defects.
4. What are the advantages and disadvantages of leveling production?
5. What do you understand by standard work?
6. How does lean manufacturing cell work?
7. Briefly explain Kaizen and continuous improvement.
8. Why is flexibility important in a manufacturing system?

Chapter 3
Design Rules for Lean Manufacturing Systems

Build a better mousetrap and the world will beat a path to your door   Ralph W. Emerson

3.1 Introduction

The Toyota Motor Company has risen to lead the world in automobile manufacturing. In a relatively short period of about 30 years, Toyota has moved from last place to first place in both sales and profits. How was this amazing feat accomplished? Under the leadership of Taiichi Ohno and Shigeo Shingo, they redesigned traditional Mass manufacturing systems and implemented a new paradigm of manufacturing called the Toyota Production System (TPS). TPS is now known worldwide as Lean manufacturing. In implementing TPS, they changed all final assemblies and key subassemblies into Mixed Model production lines; they balanced and leveled production; converted linear flow and mass production lines into U-shaped manufacturing cells; almost totally eliminated Job shop manufacturing; reduced Work-in-Process (WIP) to a minimum and drove production with Kanban cards, pull strategies instead of push, and produced according to a measure called Takt time. Starting at final assembly, every component of production is designed to respond only to authorized or forecasted demand. Final assembly is configured to have an output rate only slightly higher than demand (Takt time) and to operate using one piece flow. Single cycle machines are designed to operate with flexible jigs and fixtures and to automatically check for quality and defective items (Poka-yoke). Between each machine a single item of WIP is maintained designed to render each machine independent of any other machine in the cell. These are called *decouplers*.

When the Toyota Motor Company invented and implemented TPS they introduced an entirely new way of thinking about manufacturing systems design and operation. This new way of thinking was called Lean Thinking (Womack and Jones) and focused on only one paramount and overriding goal: *Eliminate all forms of waste*. The details of TPS are well documented by early proponents of TPS (Ohno, 1988; Sugimori, 1977; Hall, 1982 & 1983; Shingo.1985; Mondon, 1983 and Schonberger, 1982). These authors described how Toyota changed their final assembly lines into Mixed-model production lines and leveled demand from the manufacturing shop floor through their supply chains. Other authors focused on the redesign and operation of traditional Mass production lines into U-shaped manufacturing cells (Black, 1986; Black, 1991; Black and Cochran, 2000), this new design Black (1986) called a *Linked Cell Manufacturing System* (L-CMS). The L-CMS is a radical departure from traditional Mass production lines. The output rates of L-CMS are controlled by allocating workers to machines in a self- balancing paradigm. L-CMS cells are flexible by design, they execute one-piece flow, they can easily accommodate changes in product mix or customer demand, and they are easier to accept changes in technological innovation and product designs. Figure 3.1 shows a typical U-shaped cell manned by two workers.

Each worker completes the tasks (machines) assigned to them, and repeats the assigned sequence of operations. The cell output is the maximum cycle time of either worker. Ideally, this cycle time will match Takt time production requirements and every worker will work exactly the same amount of time per cycle (Balanced production). How is this *magic* accomplished?

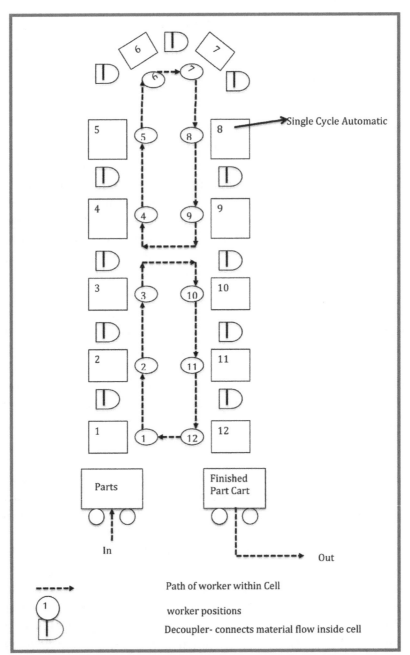

Figure 3.1
Example of Lean Manufacturing Cell Operated by 2 Workers

3.2 Design Rules for the L-CMS

The design and operation of a L-CMS is based upon a set of simple axiomatic design rules. Axiomatic design rules are based upon *tautology*; they are operationally necessary and true but they are based more upon operational knowledge rather than mathematical proofs. The force of gravity is a tautology: Everyone knows that gravity exists and holds things in place, but the theory of gravity is extremely difficult. Tautologies evolve out of practice and observation. A minimum set of design rules for a L-CMS are given below.

- *Axiom 1*: Independence
  - ➤ All functional requirements should be independent of one another

➤ An optimal design will always maintain independence of functional requirements

This Axiom states that each machine in a U-shaped cell operates independently of any other machine. This does not imply that there is not a precedence relationship between machine centers. It does imply that the work assigned to each individual processing step is a complete and uninterrupted operation.

- **Axiom 2**: Information
  - ➤ Minimize any needed information
  - ➤ Provide any information needed in the simplest form possible

The best cell design is one that has the minimum set of specifications and instructions. A design that satisfies Axiom 1 by maintaining functional independence is called an *uncoupled design*. Any design that does not satisfy Axiom 1 is called a *coupled design*. Any coupled design can always be decoupled by separating the coupled functional requirement(s).

- **Axiom 3**: Cell cycle time is dictated by the immediate downstream Takt time requirements. The output rates (Takt time) from all cells upstream from Mixed Model final assembly are synchronized by the Takt time of final assembly, and whether or not there is a one-to-one coupling relationship or a many-to-one coupling relationship (four table legs are shipped to final assembly to be bolted on to a single table top).

The third axiom is critical to regulating and bounding cell output. The output(s) of a Lean manufacturing system are dictated and set by the final assembly line(s), which produce at a rate determined by Takt time. Takt time is based upon daily demand (DD), which is a smoothed weekly and monthly demand. Contrary to popular belief, Takt time for final assembly does not *chase* demand but is based upon projected and forecasted demand. Leveling the actual demand enables all final assemblies, subassemblies and component production to produce the same amount every day.

- **Axiom 4**: Balance the amount of time required by each worker in the cell.

We will shortly demonstrate that the output of Lean manufacturing cells is controlled by creating enough capacity to meet demand, and then controlling the output rate by appropriately assigning machining tasks to one or more workers in the cell. Each cell requires at least one worker, and no more than one worker at each machine. Assignment of processes to workers should be such that all workers work the same amount of time. This is called *intercell* balance.

- **Axiom 5**: Worker balance should equal Takt time as close as possible.

It should be obvious that if each worker assigned to a U-shaped cell works the same amount of time, this time should be as close as possible to Takt time. It is usually impossible to satisfy this axiom exactly. In most real-world cell designs, the individual worker's cycle time should be as close to one another as possible, with cell Takt time based upon the maximum cycle time of any worker in the cell.

- **Axiom 6**: All equipment in the cell should be arranged in a U-shape in such a way as to guarantee that precedence constraints are satisfied and to minimize walking time.

Just as in a serial production line, machines or sequences of machines in a U-shaped cell, should be ordered from first to last reflecting precedence relationships.

- **Axiom 7**: Any item that is processed on SCA machine should follow a Make-one, Unload-one, Check-one, Load a new one, and Move-one on sequence.

Each machine in a U-shaped cell is operated independently of every other machine in the cell (Axiom 1). The sequencing of work at each machine should be the same at all machines. When a worker reaches any one machine (processing step) and begins an assigned work cycle, it should always be: (1) Unload the machine (2) Inspect the part. If the part is good; (3) Hold in left hand or drop in decoupler (4) Load the next waiting part to the machine (5) Start the machine with a simple walk-away-switch, which will execute the processing step (6) Walk to the next machine in sequence and begin this sequence again. We will soon see that this same basic sequence of steps is followed by both *assembly* and *machining* operations/cells; except in an assembly cell the operator manually executes an assembly operation (provided it is not totally automated. In a machining cell, the operator simply begins an automated machining operation and walks away. Any cell can contain both machining and assembly operations in any order.

- **Axiom 8**: Minimize as much as possible all set-up, changeover, load and unload times. It should be obvious that since all setup, changeover, mount and dismount times at each machine (if manually executed) can be a large part of the cell cycle time. These times should be minimized as much as possible (SMED). This is particularly important in Mixed-model production cells.

- **Axiom 9**: Enforce a Zero defects policy and execute Total Preventative Maintenance at each process step (machine)

It will shortly be made obvious that the utilization of every machine in a Lean cell is usually considerably smaller than that of a corresponding, traditional Mass production line. While each machine can absorb some unscheduled downtime, Total Preventative Maintenance (TPM) should be introduced to try and guarantee 100% availability. Each machine must finish its assigned manufacturing operation, and shut itself down before the worker assigned to that processing step returns. Steady, uninterrupted flow of parts through a Lean cell cannot tolerate any rejected or scrapped part in the production cycle. A policy of *zero defects* must be strictly enforced. The worker serves as a preventative maintenance engineer.

- **Axiom 10**: Redesign as many machines as possible to be single-cycle automatics with start/stop mechanisms at each process step

Each machine in a U-shaped Lean cell should be retrofitted with a simple single cycle, start switch. This is called a *walk-away switch*.

- **Axiom 11**: Introduce no parts into the Manufacturing cell unless they are authorized and waiting in WIP before the first workstation in the sequence of workstations. All parts are removed from the cell to an output station and are not moved downstream except when authorized.

One of the most important characteristics of a U-shaped Lean cell is that the amount of stock on hand allowed in the cell is strictly regulated and enforced. A properly designed cell for one worker will have one part residing on each machine. It may be necessary to make the work cycle of multiple workers assigned to a cell independent of every other work cycle. For example, if there are three workers assigned to a single cell, their work patterns resemble three separate *loops*. Each loop is made independent of every other loop by holding one part in a decoupler at the loop interfaces. A part is never introduced into the cell unless a part has finished. Cells are buffered (made independent) from every other cell by placing a limited amount of WIP before the cell loading station. Parts in this loading area are replaced by a Kanban, pull mechanism. In a similar manner, the output station will load a limited amount of parts (WIP) onto a conveyance device (pallet, box). This device (temporary storage) is removed according to a similar Kanban, Pull system triggered by downstream requirements. If the output device is full and cannot be moved, all cell production ceases.

- **Axiom 12**: Decouple all operations/machines contained in the cell with a single item of stock between machines.

A single item of stock-on-hand between before each workstation *decouples the* machining operations. It can also help to buffer any irregularities or variance in the processing cycle. Decouplers can also perform quality control checks on the most recently completed part.

- *Axiom 13*: All workers should be cross trained to operate any machine in the cell.

The key to responding to different demand patterns or product mix is the ability to add workers or to change worker assignments. This is only possible by cross training each worker to run any machine in the cell. One worker is often assigned to cover the input and output processes.

- *Axiom 14*: Any individual machining time in the cell must be strictly less than the Takt time which is driving cell output.

This is one of the most important design principles in U-shaped Lean cell design. Since the cell cycle time is based on the required Takt time (MIN/part), every machine in the cell must be able to meet the following rule:

$$MT_{ij} < \text{Takt time} \quad i=1, 2\ldots\ldots M \text{ machines}$$
$$j= 1, 2\ldots\ldots \text{part families}$$

Where: MT_{ij} is the processing time on machine I for part j.

This Axiom insures that cell production will meet the Takt time requirement. Once it is started, a machine will automatically complete the machining time assigned to that process step. Upon completion, it will automatically cut off. Meanwhile, the worker that started that machine cycle is working his/her way through all other machines in his/her assigned group of machines (his work *loop*). The machine cycle must be completely finished before the worker returns, or the machine will cause a *bottleneck* and cause the worker to wait.

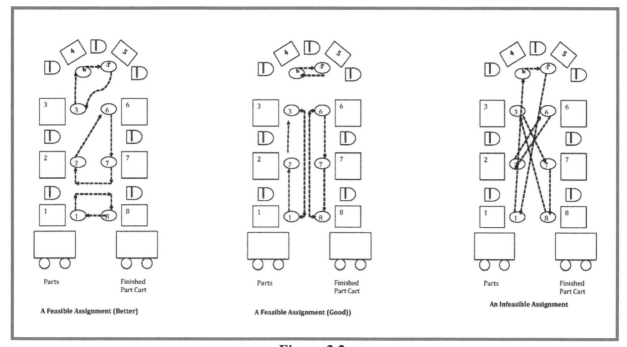

Figure 3.2
Feasible and Infeasible Worker Assignments

79

- *Axiom 15*: In cell with multiple workers and/or workers operating multiple machines/operations, his or her walking pattern in the assigned loop cannot cross that of any other worker.

This axiom is given to prevent any interference between workers. It is a *rule of convenience* and will constrain the maximum number of total worker assignments which can be made between N workers and M machines. The operating principle is not an absolutely hard and fast rule, but the idea is that as each worker walks through his/her assigned route, that rout cannot cross another worker's route.

Using these 15 Axiomatic design rules, a U-shaped Lean manufacturing cell can be designed and implemented. Beneath each rule are a lot of innovative and difficult design activities. Naturally, the cells will be operated by people who will use and improve *autonomation* principles. Everyone who works in the system must understand and buy into Lean thinking. The result of following these design rules will result in a U-shaped Lean cell which is robust, flexible and simple to operate with a minimum of complex calculations and operating data.

3.3 The Evolution of Lean Factory Design

The Lean linked-cell factory design (L-CMS) is an outgrowth of the job shop and the flow shop, but it is neither. The Job shop as a manufacturing system emerged in the early 1800's. Job shops replaced craft or cottage manufacturing when it became necessary to replicate individual part types and produce them in larger quantities. The functional design initially evolved because of increasing power requirements to automatically run many machines simultaneously. The source of power was *water* from streams and rivers, which was routed to fall upon a water wheel; which in turn was used to drive machines using metal drive shafts, gears and leather belt drives. Groups of functionally similar machines were placed under a common drive shaft, with each machine connected to gear drives by leather belts. Historians named this type of manufacturing system the American Armory System. The birthplace of these new manufacturing systems was in New England. People from all over the world came to see and build these systems.

In the early 1900s, the first vestiges of the flow shop began to emerge with flow lines designed for small part manufacturing. The flow line reached maturity at the Ford Motor Company around 1913. Once again, the world came to see how this new system was designed and operated. Over the next 20 years, hybrid shops with Job shop and Flow shop components emerged using the pioneering work of Henry Ford and Frederick Taylor. Taylor was the first to define and implement the principles of scientific management. New system designs enabled factories to produce large volumes of identical products at low unit cost. Subassembly lines were designed for make-to-stock inventories creating large in-process-inventories but sustained continuous production of the main flow lines. These systems helped America win World War II and become the world leader in manufacturing. After WWII, the Toyota Motor Company, led by an engineering genius called Taiichi Ohno, pioneered a new manufacturing system design called the Toyota Production System (TPS). In 1991 it was given a name that promoted its universal application to any form of service or manufacturing system: The *Lean Production System* (Womack, Jones and Roos).

3.4 Defining a Manufacturing System

The Manufacturing System is the beating heart of any Enterprise system. The manufacturing system is defined as a complex arrangement of physical entities which are characterized by measureable parameters. The physical entities are machines, tooling, material handling resources, raw materials and most important; people. Manufacturing systems systematically transform raw materials into finished goods which are sold for profit. Companies and individuals who buy finished goods are *external customers*. The workers performing the processing steps and machining operations which add value to the finished product are *internal customers*. In designing a manufacturing system, this is a key concept. Any manufacturing system must satisfy the needs of both external and internal customers. Both types of customers respond to measurable parameters that define how well a company is operating. There are many measurable parameters which must be carefully controlled. Profits, costs, cycle time, throughput time, the amount of work-in-process, scrap and rework rates and customer satisfaction to name a few.

Different system designs can greatly influence the response of measureable parameters. Time is one of the most critical drivers, measuring throughput, cycle and customer response. Lean systems have as one of its main goals the reduction of throughput time. Lean systems are continually seeking ways to reduce WIP levels, improve quality by eliminating rework and scrap rates and eliminating non-value added activities. Using Little's Law, if the desired output rate is fixed by Takt time, then systematic reduction of WIP will directly reduce throughput time. This is one of the main reasons to design and operate U-shaped cells. We have already pointed out that cells have a rigid SOH or WIP Cap. Work in process held between cells is minimized by invoking Kanban control and is also held at a minimum. Inter-cell inventories are determined by the needs of *internal customers*. Kanban control is based upon minimum inventory levels which provide important production control functions to upstream suppliers and customers. Final goods inventory are set to smooth product delivery requirements caused by varying demand. The entire Lean, linked manufacturing system constantly tells all manufacturing components *what* to make, *how much* to make and *when* to make or store product. Parts being worked on at machines within cells or which reside in the decouplers are referred to as *stock-on-hand* (SOH). SOH is not considered to be WIP inventory, but part of *standard work*. The design of the factory and the manufacturing cells are tightly coupled to product design (work sequencing) and required production rates (upper bounds). System design includes close attention to ergonomic and human factors issues.

Final assembly is converted to mixed model and dictates how all subassembly and component cells operate. Leveling and smoothing of all production components follow Takt time specifications. All subassembly and component production is accomplished in relatively independent U-shaped Lean cells, with Takt time requirements linking all downstream cells to upstream cells. Cells operate on a one piece flow basis just like final assembly. All subassembly and component cells are linked to one another and to final assembly by Kanban and pull philosophies. This usually eliminates the need for paced conveyor lines.

One thing often misunderstood is the fact that a linked-cell Lean manufacturing system has inherent, integrated control functions built into the system design and operation. This type of built-in production control is quite different from traditional manufacturing systems that operate by pushing orders into the manufacturing system. Traditional manufacturing systems require a constant dialog between order release systems like MRP II and production control systems with Computer Aided Manufacturing (CAM) modules.

Every Lean system is designed and operated built upon a *zero defects* manufacturing philosophy. Toyota believed in company-wide Total Quality Control (TQC), and taught zero defects operational procedures to everyone in the company from the president to the shop floor worker. The enabling technology to support TQC was to redesign the manufacturing system to support one-piece flow, where work-in-process is inspected after each processing step. Every worker in the factory became a quality control inspector.

A second equally important concept was to aggressively and continuously work on *variance reduction*. In Chapter 12 we will focus on quantifying the *corrupting influence of variance*. Variance reduction is the key to consistent, steady production and is a key component of continuous improvement, waste (*Muda*) elimination, and the efficient use of people, machines on-time delivery, and supply chain integration. Components of variance to be eliminated include variation in quality, variation in hourly/daily output rates, variation in processing time and variation in standard practices. We will continually stress the importance of variance in any form.

Lean systems can and should be changed to promote waste elimination and variance elimination. A Lean system is designed to be versatile and flexible to respond to uniqueness and creativity. It is the responsibility of every shop floor worker to develop and implement manufacturing processes, reduce setup and changeover times, improve quality, and continuously redesign the workplace (methods improvement). Of equal importance is the placement of machines and processes in Lean cells to bring processing times in line with required Takt times. Many innovative solutions become company secrets, closely guarded to gain competitive cost and time improvements. The result is to evolve into a Lean Manufacturing Solution that produces lower cost, superior quality products that promote customer satisfaction. A Lean system operates in ways that the old mass production system cannot. It is responsive and easily reconfigured to design changes, product mix and changing demand patterns. So what steps need to be taken to begin a Lean transformation?

3.5 Steps Necessary to Implementing Lean Manufacturing Strategies

Changing traditional manufacturing system design and integrating the control functions into a new Lean system design requires commitment from top level management to the lowest factory worker. Every employee must be fully committed to system change. Total employee involvement is absolutely necessary. If the manufacturing enterprise is unionized, the union(s) must be fully briefed and involved in the transformation. In general, it is not the labor union or the shop floor workers who raise barriers to Lean production, but it is usually middle management (MM). Middle management will have the most to lose in implementing Lean manufacturing strategies. Many MM functions get integrated into the operational strategy and design of a Lean manufacturing system. We have identified some basic steps which must be taken to implement Lean production.

1. All levels of the plant from the president to hourly workers must be fully educated in Lean production philosophy and concepts. There must be a clear understanding of how a Lean manufacturing system differs from traditional Mass manufacturing systems.
2. Top down support and solid commitment is critical. The entire company must be involved and committed to change, particularly the existing design and manufacturing engineers.
3. Top management must realize two fundamental truths. The first is that once the company begins to embrace and switch over to a lean system design, things will get worse before they get better. Initially, it may appear that Lean principles simply do not work. There will be a gradual recovery back to where the company was operating prior to change, then a steady improvement in system performance measures, cost and customer satisfaction. This cycle will be repeated many times as technological and operational changes take place as a result of continuous improvement. Second, there will be resistance to change: equipment utilization will be lower in general (by plan), workers roles and responsibilities will be broadened, and corporate costing strategies will need to be changed to abandon the traditional cost per labor hour.
4. The selection and monitoring of system performance measures to reflect change is critical. Everyone in the company must be aware that Lean thinking is directed to waste elimination, WIP reduction, quality improvement and throughput time reduction. Cost and not profit margins determine profits.
5. All engineers and shop floor workers must be trained to find waste, and to eliminate all forms of waste as a part of their job description. There must be a relentless pursuit of eliminating waste and continuous improvement. New and more comprehensive educational programs are essential. Lean cell operators must understand why change is necessary and how change takes place. Operators need to be involved in cell design; cross trained to operate all machines in a cell; be proactive in methods improvement; and function as Quality control inspectors empowered to shut a system down if problems arise. A critical function of all cell operators is to continuously find ways to reduce setup and changeover times.
6. Shop floor personnel and cell workers should be rewarded for reducing waste and improving system performance. Innovation awards and bonus payments are commonly used to reward innovative workers.
7. Middle management and corporate executives should take an active role in Lean transformation and avoid passive participation.

Table 3.1 gives an overview of a successful implementation strategy. The suggestions in Table 3.1 were extracted from several American companies who have successfully implemented Lean Engineering principles. This implementation strategy is covered in more detail in Chapter 8.

The remainder of this Chapter will provide a high level presentation of some important Lean design rules. The existing manufacturing system must be redesigned, simplified and integrated before any computerized systems are put into place to support production. *Autonomation*, which is automation with a human touch, becomes a large part of final transformation after system integration has been achieved. *Concurrent engineering* becomes the last step in the effort to restructure the entire manufacturing enterprise. A word of caution: piecemeal and isolated implementation of Lean manufacturing will never realize the full potential of Lean transformation. Isolated pilot projects may, in fact, prove to be counterproductive. The order in which Lean transformation occurs is also important. Many companies have attempted WIP reduction first by implementing Kanban control and pull versus push methodologies. These companies have, in fact, experienced immediate WIP reduction even before U-shaped cells were implemented. However, because this is only a part of Lean production, many such approaches have resulted in failure. Even worse, many companies have resisted continuing to run the company with large production control software systems such as MRP and ERP. Such an approach fails to realize the full potential of total Lean transformation.

Finally, it should be stressed that the main manufacturing facility cannot be isolated and transformed by itself. The Supply Chain and the final distribution system must also be changed. The principles and methodologies used to change logistics and the total supply chain are just now beginning to emerge. This book introduces the Kanban system and sole – sourcing supply chains while supply chain management is the subject of the book by Ravindran.

3.6 Takt Time Design Rules

A linked cell manufacturing system is shown in Figure 3.3. The subassembly and component manufacturing cells are linked to Mixed Model final assembly by a Kanban control system. Each manufacturing cell is designed to produce product as near as possible to customer demand; which coupled to downstream requirements defines Takt time for each cell. Within cells, a serial set of processes is put into place to both balance the cell and to meet the Takt time requirement. Takt is a German term that historically refers to a symphony conductor, who is responsible for keeping every member of a large orchestra in time and sequence. Takt time (TT) is available production time divided by demand. For example:

$$TT = \frac{\text{Minutes available for Production/Day}}{\text{Daily Demand (DD)}}$$

Table 3.1
Successful Implementation Strategies

1. **Level and Balance the manufacturing system: Smooth the material flow (Monden, 1983)** Leveling involves the design of Mixed-model U-shaped lean cells and the implementation of Mixed-Model final assembly. Balancing is making sure that every set of processing steps are equalized in terms of time required to execute each step.	• Calculate Takt times and daily demands • Implement Mixed-Model final assembly • Balance output from each supplier • Develop one-piece flow • Sequence products and subassemblies with other final assembly
2. **Design and reconfigure the Manufacturing system** Design and implement component, subassembly and final assembly cells. System design must consider the manufacturing specifications of each part and product, and the requirements of both internal and external customers	• Set standard work for operators in cells • Implement U-shared lean cells • Implement self-balancing subassembly cells

3. **Initiate SMED to reduce or eliminate all setup and changeover times** All setup and changeover times are non-value added and increase throughput time. By Little's Law, this also creates WIP, Redesign to accommodate small lot sizes.	• Teach everyone SMED • Develop one touch setup procedures • Train operators to decrease these times and to perform all setups and changeovers.
4. **Initiate Total Quality Control throughout the manufacturing system(Shingo, 1986)** Eliminate all defective parts, rework and scrap parts. Develop automated Poka-Yoke quality controls	• Eliminate formal quality inspection by attacking poor quality at the source • Turn every process operator into a QC inspector • Use the "7 tools for quality control" • Make zero defects the goal • Implement Poke-Yoke devices
5. **Fully implement Total Preventative Maintenance – TPM (Nakajima, 1988)** Redesign machines for Poka-yoke and to be self-diagnostic. Make maintenance practices proactive and not reactive.	• Design/redesign all machines for high reliability and availability • Train operators to solve reliability problems • Train operators to perform daily, simple preventative maintenance
6. **Integrate all production cells; link cells; pull to final assembly** Implement just-in time strategies; ask where, when and how. Install Kanban controls to link all manufacturing functions (Black, 1991)	• Link all cells • Pull parts to final assembly • Use Kanban to replace scheduling and order release
7. **Integrate Inventory Control** Reduce all WIP to minimum levels. Set WIP by leveraging Kanban, pull and unit flow principles (Black, 1991)	• Standardize WIP material handling devices • Reduce lot sizes • Minimize WIP between cells • Use Kanban/Pull to schedule and control production
8. **Integrate all Sub-tier Suppliers** All Suppliers should be treated as if they were part of the factory. Suppliers become simply "remote cells". Suppliers are trusted partners.	• Make all suppliers sole source • Install lean practices • Implement steps 1-7 throughout the external supply chain
9. **Focus on Autonomation; not strictly Automation** Automate when practical, but do it with the operator in mind	• Simplify….simplify….simplify • Upgrade machines to prevent defects • Upgrade machines to prevent overproductions
10. **Design the entire enterprise system around lean**	• Execute lean concurrent engineering • Lean purchasing, sales and distribution

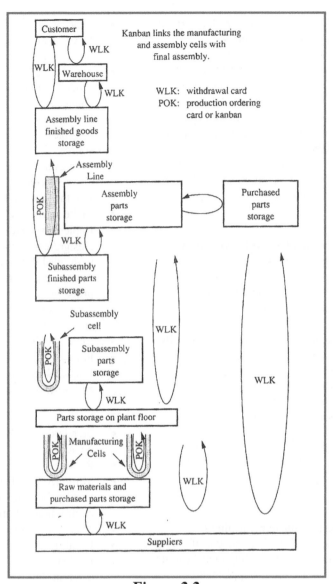

Figure 3.3
A Linked-Cell Manufacturing System has Many Manufacturing Cells Linked by Kanban to Other Cells and Final Assembly.

Takt Time is the *heartbeat* of a Lean production system. Takt time calculation is easy to understand, but requires a lot of Lean Engineering to properly calculate. Final assembly (FA) lines must be converted to a Mixed-Model (MM) final assembly that can make multiple products in any order. TT is simply the required cycle time for final assembly. The inverse of Takt time is *production rate* for final assembly. While the required production rate is mathematically the inverse of Takt time, there is a subtle difference in these two measures. A typical production rate might be 5 parts per hour. The reciprocal of this is one part produced every 12 minutes. In Lean manufacturing, the production rate is a long term average of output per unit time, but the time per part is a goal which must be met. Takt time sets the pace or goal for the entire factory. Mixed-model final assembly sets the requirements for balanced and level production for the rest of the factory and for any Kanban-linked supply chain. Producing to final assembly Takt time based upon customer demand sets the functional requirements of producing the right quantity, at

86

the right time in the right place. All operations in final assembly are designed to be changed over rapidly to Mixed Model requirements, so that finished products can be made with great variety in small batches.

3.7   Subassembly Cells vs. Manufacturing Cells

In the Linked Cell Manufacturing System (L-CMS) there are multiple subassembly and U-shaped cells, all aligned to the Takt time of the Final Assembly Schedule (FAS). Some assembly lines are synchronized with final assembly and stop and start if final assembly stops and restarts. In manufacturing cells with automatic repeat cycle machine tools, each operator only remains at each machine through the loading/unloading cycle. Once the machine is restarted, the operator moves on to the next machine in sequence. Cells are manned by multifunctional, cross trained workers who move in circular patterns between machines arranged in a U-shape to minimize walking time (unless only one machine is assigned to one worker). So how do Lean subassembly cells operate differently from flow lines and manufacturing cells? Figure 3.4 shows a typical linear flow line.

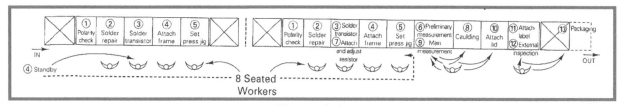

Figure 3.4
A Typical Subassembly Line for the Mass System has Conveyors and Seated Workers.

This type of linear flow is representative of most mass production lines. In this type of flow line, multiple operations are grouped according to precedence constraints and duration to meet required line production rate. These stations are then assigned to individual workers. Ideally, every machine/process grouping consumes the same amount of time. This is called *line balancing*, and is well-known Industrial and Manufacturing engineering procedure.

In a Lean, U-shaped subassembly cell, walking workers are assigned to one or more machines in such a way as to achieve the Takt time requirement as shown in Figure 3.5. Multiple workers are responsible for one or more operations, and ideally the worker cycle time of each worker is identical. Task times at each process and walking times between machines can be very different. Different output requirements or changes in product mix are easily accommodated by reassigning tasks (mostly manual) and processes to workers, and by increasing or decreasing the number of workers in a cell. There are three variations of this basic scheme: (1) Rabbit chase (2) subcell and (3) The Toyota sewing system also known as Bucket brigade. These worker configurations will be explained in Chapter 6.

In the manufacturing cells, almost all the machines are single cycle automatics (machine completes cycle untended and stops). The operator unloads, checks, loads and starts the machine, carrying the part to the next machine.

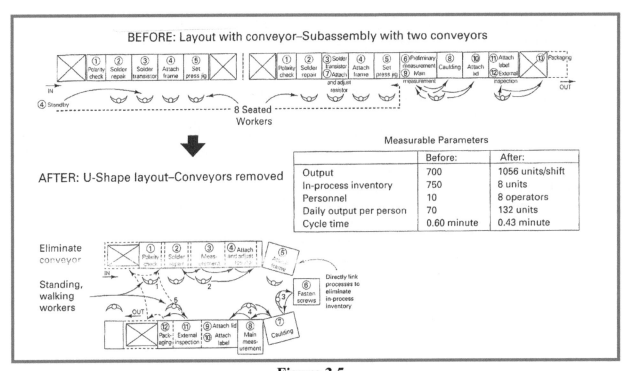

Figure 3.5
The Traditional Assembly Line Using Conveyors can be Redesigned into a U-shaped Cell with Walking Operators

3.8 Decouplers

Decouplers are placed between adjacent processes or at the point where the work cycle of any one worker interfaces to the work cycle of another worker. This point of interface is called the *Baton passing zone*. This zone usually contains a small amount of stock-on-hand. A critical function of the SOH in a baton passing zone is to absorb unavoidable differences in the worker cycle times. In this way, variations in processing times are absorbed and each worker operates independently of every other worker. Decouplers, both between adjacent machines and at the interface of worker sub-cells, are so named because they physically separate one process from the other. The term decoupler originated by Black based on pioneering work by Nam Suh in axiomatic design. Decouplers can play many critical roles in the operation of U-shaped Lean cells (Black and Schroer).

- Decouplers greatly enhance the flexibility of U-shaped cells by eliminating any codependency between machines or worker sub cells.
- Decouplers facilitate flexible worker assignments or reassignments as demand and/or product mix changes
- Decouplers promote continuous, single piece flow by buffering disruptions
- Decouplers provide the opportunity to perform automated quality control inspections without adding to the cell cycle time which is determined by the workers and not the machines. Decouplers can be automated to perform part orientation for machine loading and unloading functions.
- If two machines are located some distance apart, part of the decoupler functionality might be to execute material handling responsibilities.

88

- If a particular machine is functionally coupled to another feeder machine or cell in an assembly operation, the decoupler can execute orientation and staging capabilities.
- Decouplers can be useful in smoothing production if a cell is a Mixed Model production cell.
- Simple process transformation requirements such as heating, cooling or drying can be part of decoupler functionality. Figure 3.6 shows a paint drying decoupler. The drying decoupler in this example has six different fixtures with different surface orientation. Each fixture has a 2 minute drying cycle, so the total drying time per part is 12 minutes. Of course, the required Takt time for the entire cell cannot exceed 12 minutes.
- Decouplers control the level of SOH within a cell, and are part of the cell WIP cap. The capacity of a decoupler should never exceed two units except in rare cases, or in Mixed Model cells.

The cell shown in Figure 3.7 operates using one-piece flow. Any of the machines can quickly switch from one part type to another. A decoupler is shown between each machine. This particular cell produces a family of four parts. Each decoupler holds one part from each type of part processed to that point in the cell. This rapidly enables the cell to switch from one part type to another. Parts are manufactured one at a time in the sequence shown. The number of parts to be made by type is determined from product demand, and might change frequently. In this case, the workers walk opposite to the part flow. The triangular part does not use machine four, but as the worker walks around the cell opposite to part flow, there is no confusion or ambiguity as to which part is to be loaded to the next machine in sequence.

Once the Mixed Model sequence has been determined, there is no sequencing or scheduling or routing sheet needed whatsoever in the cell. As parts are removed from each decoupler, the next part in line is loaded onto the machine. It makes no difference at all to the worker if machines are not in the routing sequence of a particular part. The operator will rapidly know the machine sequence for each part type, but it is common practice to color code each part's routing sequence on the floor and at each machine. Clearly, the cell must be initially loaded with parts, processed one part to one machine at a time, and properly staged in each decoupler. Ideally, all parts are properly staged and sequenced in a flow-through rack at the input station. Properly staged and sequenced, any mix of parts can be processed without error. There is one vitally important design principle which must be followed.

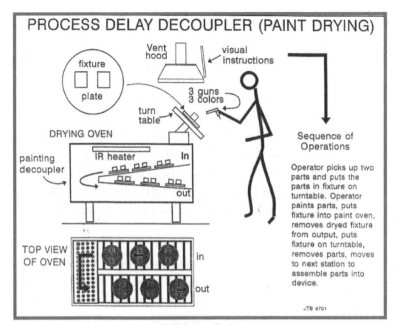

Figure 3.6
A Decoupler for Paint Drying from a Toyota Supplier

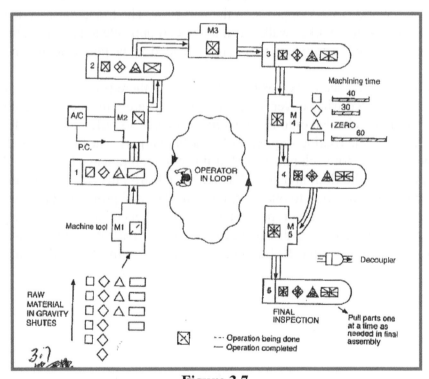

Figure 3.7
**A Typical Mixed Model Manufacturing Cell Run by One Worker Moving
Counterclockwise while Product Moves Clockwise**

The processing time for each part in each machine must be less than the cycle time. Remember that CT is determined by the maximum amount of time the worker takes to complete his loop around the cell. The operation of this type of cell requires that setup and changeover times are reduced to near zero; every machine produces no defective parts; and TPM practices make the availability of each machine 100%. It makes no difference if any one machine is fully automated and each machine can be an assembly operation or a machining operation. Cell/subcell cycle times are determined as usual. For example, Machine 2 is a CNC machining center that has been equipped with adoptive controls (AC) with a feedback loop to monitor and control the process. Note that AC capabilities are not the same as a Poka-yoke device. AC controls attempt to stop defects from occurring, while a Poka-yoke device will immediately shut the machine down if a defect is detected. Other ingenious and innovative uses of decouplers can be found for specific Lean cells.

3.8.1 Classifying and Defining Decouplers

In examining the various uses and functions of decouplers, a more comprehensive definition of a decoupler might be appropriate.

> *Any functional component in a U-shaped, Lean production cell that holds, delays, orients, inspects or transports parts is operationally subject to Pull process flow initiated by an operator, and its design is critical to cell control.*

In Mixed Model Lean cells, one of its most useful functions is to introduce flexibility into cell operation and facilitate smooth flow of parts through the cell. In other words, decouplers can be classified as an *operational control device* and not simply a material storage device. If we use this broader definition of a decoupler, it is possible to classify a decoupler by function as follows. A *Primary function* is a device that is essential to achieve controlled flexibility. A *Secondary function* is any other task or use that the decoupler executes.

Primary Functions	*Secondary Functions*
Decouple process steps	Poka-Yoke (Defect prevention)
Control cell WIP levels	Inspection/Quality checks
Buffer for Mixed Model production	Part transportation
	Part orientation
	Part registration
	Process delays for Inspection, cooling, Etc.
	Part transformation :Heating, cooling,etc.

Subassembly cells and assembly cells can use simple and efficient *cell decouplers*. The cell shown in Figure 3.7B was designed and built by operators who worked for Hewlett-Packard. The application was the assembly of disk drives for one of their computers. They called this decoupler a *Kanban Square*, even though were not part of a Kanban-pull system. When a part was removed from the decoupler, this was a signal to the upstream worker to make another disk drive.

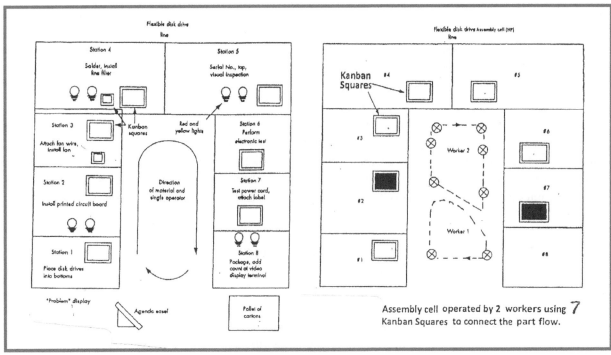

Figure 3.7B
This Subassembly Cell Developed at Hewlett Packard had Decouplers called Kanban Squares and Could be Operated by One or More Workers

Finally, it should be noted that decouplers can often relieve humans of boring, simple repetitive tasks and improve cell efficiency. Historically, the important role and functionality that decouplers provide has been overlooked. In Lean Engineering, they now play extremely important roles in designing and implementing Lean production cells. They are vital to Lean systems with Kanban control and pull philosophies.

3.9 Integrating Production Control

The traditional mass manufacturing system is very complex. Large, multiple function software systems such as MRP II and ERP interface to literally hundreds of functions and sometimes thousands of workers. People who execute production control functions sequence and schedule the entire manufacturing facility, which means that they literally determine *where* purchased parts and subassemblies go; *when* they should be scheduled to go there; and *how many* should be moved. Many companies have attempted to completely computerize these functions and eliminate production control personnel, but the general result has been frustration, wasted time and money, disappointment and largely failure. Why has this been so difficult? There are many reasons, but we believe that the main problem rests with the software systems being used. Software packages are basically designed for planning, and not control. Many software systems being used today were developed and installed by consultants and outside people who do not use these systems. One of the missing ingredients is that manufacturing control involves people and not strictly things. The other major problem is the inherent capability of hard rule-based systems to intelligently respond to uncertainty, change, variations and disruptions.

The Lean manufacturing approach greatly simplifies and redesigns traditional system control functions. Many system and subsystem control functions are integrated into the system design using versions of Kanban. *Physically*, production control and synchronization is achieved by creating Lean, U-shaped assembly, subassembly and component cells linked by Kanban control and pull philosophies. *Logically*, nothing is moved or manufactured without some form of signal built into the Kanban system. Lean layout and product flow rigidly define the paths that parts can take as they move through the plant. Each manufacturing component is precedent linked to every other component; eliminating the basic need for a routing sheet.

3.10 Production Control in a Linked Cell System

Production in upstream U-shaped Lean cells is pulled by the downstream demand. For discussion purposes we will refer to the diagram in Figure 3.8. Assume that cell C1 is providing parts to only cell C2. The output station of cell C1 unloads parts into carts at a part storage area. These parts will be moved to an input station in front of the receiving cell (C2). We have previously indicated that the WIP in carts at both stations is tightly controlled by Kanban signals. Further, nothing can be moved from C1 to C2 without authorization. In Chapters 13 and 15 we will describe several types of Kanban control systems that can be used. For now, we will describe a simple pull system that does not require any Kanban controls at all. Let us briefly address the following questions. (1) How many units should be stored in the output station of cell C1? How many units should be stored in the input station of cell C2? What triggers part movement from C1 to C2? How many parts should be moved from C1 to C2? How many units should be placed in a full cart?

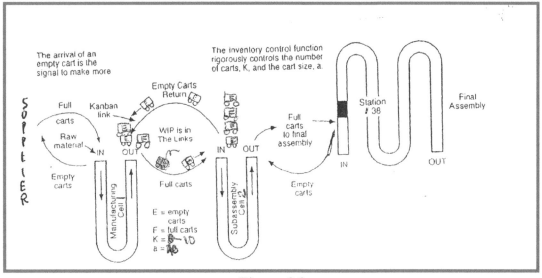

Figure 3.8
Cells are Linked by a Kanban system. The Design Rule for Determining the Maximum Inventory in a k-link is an Operational Form of Little's Law.

The system shown in Figure 3.8 is typical of how two Lean cells are linked to one another. A design objective is to minimize the total amount of WIP at the output of Cell C1 and the input of Cell C2. The most important decision to be made is *when* to ask Cell C1 to deliver more parts and *how many* parts to deliver? Recall that the Takt time at Cell C2 and the Takt time at Cell C1 are both the same by design. Assume that Cells is producing one part every 60 seconds, so the output rate is 1 part per minute. The order is placed at Cell C2 for more parts from Cell C1 by the return of an empty cart. Assume that the total amount of time from when the order is placed until a cart is received is 10 minutes. During these 60 minutes, 40 subassemblies will be built in Cell C2. Hence, an order for 40 parts will be placed. Each material handling cart has been designed to hold 40 parts. An equivalent policy is then to request full cart from Cell C1. If Cell C1 is producing at the same rate as Cell C2, then Cell C1 will need to have full cart ready to go when an empty card arrives. Notices that if everything is synchronized, Cell 2 will deplete full cart (40 parts) while full carts are being shipped to Cell C1. Using this information, the cells will operate and communicate as follows

(1) Assume that 4 full carts are sitting at the Cell C2 input station, and full cart 15 sitting at the output station of Cell C1. Other carts are in transportation or being loaded or unloaded.
(2) Both Cell C1 and Cell C2 begin production at the same time.
(3) At Cell C2 an empty cart arrives, this is a signal to order full cart from Cell C1. The order is placed, meanwhile Cell C1 responds by shipping full carts to Cell C2. After the order is placed, full cart arrives at Cell C2. Meanwhile, Cell C2 has worked off full cart and is just beginning to use the next cart of parts from the one full carts left in queue. During the last 40 minutes, as the full cart is unloaded at Cell C2, empty carts are moved to Cell C1 for refilling.
(4) This cycle will repeat itself until the daily production run is finished. The arrival of an empty cart is the signal for cell I to make more.

Notice the elegance and simplicity of this scheme. There are no real production control functions in this system… no order release, no sequencing and no scheduling. The total number of parts in the system is fixed and can never increase. Parts are delivered to Cell C2 *just-in-time*, at just the right quantity, just when needed. The astute reader may have wondered why the link has 10 carts in it. After all, the system as described can operate with only three. This is true, as long as everything goes exactly as planned. Here we invoke *Murphy's Law*, which states that*: Anytime something can go wrong, it will.* The extra carts protect against any uncertainties, and it will prevent any shut down at Cell C2. The reverse is true at Cell C1; carts will guarantee that there is a full cart available when a shipping request is received from Cell C2. It should be noted that this simple scheme can also be used in Mixed Model production. In that case, the sequence of different parts arriving at Cell C2 (Load Leveling) and the amount in each cart will be determined by type and sequence. It would be a good idea to color code each cart for immediate visual verification and identification.

Finally, It should be noted that Cell C1 is likely linked to its immediate upstream cell, and Cell C2 is linked to its immediate successor cell, a final assembly station. All the linked cells are synchronized and controlled in exactly the same way just described.

3.11 Kanban Signals
There are many different types of Kanban signals which can trigger a multitude of events (Chapter 15), but the basic design rule for Kanban control is:

$$K = \frac{L*DD+SS}{a}$$

Where:
    L=Lead Time (time to complete the link)
    DD= Daily demand
    SS=Safety Stock
    a=the number of items put into a standard container

Ignoring safety stock, (K*a) is the total WIP in all containers or 400 in our example. L is the Lead time which includes throughput time; and DD is a production rate or a form of demand. Hence, the basic Kanban equation is just an application of Little's Law. This will be discussed in Chapter 15.

Lean manufacturing cells and subassemblies are linked to final assembly by some form of Kanban control. In a Lean system that has been built to manufacture only one product, the lowest number of each item in-process should be transported between cells. Unit lot size is the ideal situation, but subsequent assembly operations sometimes dictate the number of items in a container. For example suppose that a furniture manufacturer is making kitchen tables. To support final assembly, a Lean cell is configured to make table legs for final assembly. Each table requires four legs. If final assembly produces a table every 16 minutes, the cell that manufactures a table leg must produce one leg every 4 minutes. It seems reasonable to ship a set of 4 legs every 16 minutes. Ideally, the U-shaped leg production cell would be placed as near as possible to the point in final assembly where the legs are attached. A decoupler holding two sets of 4 legs should be sufficient to buffer any delays or variance in production rates.

3.12   Mixed Model Final Assembly

Now consider a Mixed Model final assembly line. The company has contracted to make 3 different types of tables, which use 3 different sets of legs. A possible Mixed Model assembly sequence which meets a particular demand pattern might be AABBBC. This sequence might be repeated 100 times a day. Note that the FAS determines the sequence of production in the leg assembly cell. For simplicity, assume that every set of legs can be attached in 5 minutes. In this case, the lean manufacturing cell would make and deliver 8 "A" legs every 30 minutes, or one leg must be made about every 3.5 minutes; 12 "B" legs every 30 minutes; and 4 "C" legs every 30 minutes. The reader should verify that the above Kanban formula will yield the same result with no safety stock. In general, safety stock will not be moved between cells as part of a standard container.

Taiichi Ohno said that *Kanban is a visual control that is only good for Lean production systems.* Linking cells together with visual controls provide information as to where the parts must be delivered and in what quantities, hence making routing sheets and travelers unnecessary.

3.13   More Kanbans?

A second basic type of Kanban is called the *dual card* system. In a dual card system there are two basic types of Kanban: *Withdrawal* or conveyance Kanban (WLK) and a *Production Ordering* Kanban (POK). In this type of system, the output (order of production) of each cell is directly controlled by the input of another cell. The output of the upstream cell contains a fixed (one or more) number of material handling carts, and each cart contains the same number of parts. Cart capacity should be as small as is practical or possible. We will call the upstream cell a production cell, and the downstream cell a consumer cell. Assume that the upstream has carts are sitting on the output side ready to be delivered downstream. Each of these carts have POK in them. Downstream, the consumer cell is pulling parts from a container sitting on its input side. When a container is emptied, the downstream cell puts a WLK card in the empty cart. The empty cart is now moved upstream to the output side of the production cell. The WLK card is attached to a full cart and the POK is now a signal for the upstream cell to begin production to fill the cart. The quantity to be made is on the POK card. Note two important aspects of this simple visual control system: (1) Nothing can be made in the upstream cell unless authorized to do so (2) Nothing can be moved downstream unless authorized to do so. The number of carts in the system multiplied by the number of parts per cart put a *WIP Cap* on the participating cells. Note further that the ability to respond to changes in demand in both cells are still dependent only upon available (bottleneck) capacity, smoothed production (number of stations and work done at each station) and the number of workers active in each cell. The same kind of linkage is implemented between all cells.

The logic in specifying Kanban links to execute synchronized production is in reality just an operational implementation of Little's Law (Chapter 15). Recall that Little's Law states that in any type of manufacturing system, there is an invariant mathematical relationship between WIP (Work-In-Process inventory), Throughput time (time spent in any system) and a specified production rate (the inverse of Takt time). Little's Law is as follows.

WIP = TPT*PR

> *Where*: WIP = Work-in-process
> TPT = Throughput time or Time in the system
> PR= Production rate or demand rate (1/Takt time)

A *system* can be any manufacturing component which has boundaries. A single machine with waiting space, a group of machines with individual or pooled waiting space, a Lean cell or even a total factory. Another useful version of Little's Law.

$$WIP_{Queue} = Time_{In\ queue} * \lambda$$
$$\text{where } \lambda = \text{demand rate}$$

This form of Little's Law mathematically relates the amount of inventory waiting to be worked on in front of a manufacturing component (WIP_{Queue}) to the amount of time per unit spent in queue ($Time_{In\ queue}$), multiplied by the production or demand rate, These two forms of Little's Law are coupled by the following relationship's.

$$WIP_{System} = WIP_{In\ queue} + WIP_{Currently\ in\ Manufacturing\ component}$$

$$Time_{In\ System} = Time_{In\ queue} + Time_{In\ Manufacturing\ component}$$

These two equations and the two coupling relationships will be extensively used in Value Stream Mapping (Chapter 17) and in Process Flow Modeling.

3.14 Characterizing and Defining WIP

It should be recognized that WIP is just another way to characterize inventory in a system. Inventory may be waiting, blocked or delayed parts; parts being worked upon or raw materials being transformed; Make-to-Stock, Make-to-Order, intermediate storage or final finished goods inventory. Parts residing in decouplers are actually WIP also. All material in a manufacturing system is considered to be inventory. We will define three basic categories: (1) Raw materials (2) Work-in-Process and (3) finished goods inventory. WIP has often been analogized as *water in a river*. When water (inventory) is high, any rocks (problems) in the river cannot be seen. If the water level in the river (Manufacturing facility) is reduced, then the rocks (problems) are exposed. If inventory (work-in-Process) is high then every machine and every worker feels the need to produce whether needed or not. This type of traditional manufacturing behavior is typical of manufacturing control systems that *push* work orders into the system with little or no consideration of current WIP levels or machine status (utilization). This causes workers and machines to just work harder to try and catch up. Overproduction, subcontracting and overtime are then employed to bring the system back under control. This results in high make-to-stock and make-to-order inventory levels. Inventory turns radically decrease and customer delivery times cannot be met. The result is disaster. On the other hand, if WIP (inventory) levels are drastically reduced instead of increased, the problems (rocks) are more easily exposed and can be addressed. Taiichi Ohno and Shigeo Shingo both steadfastly maintained that inventory is the *root of all evil*: lower the inventory levels (the river) and expose problems (rocks). Their analogy is absolutely correct. In Lean manufacturing, any problems which are exposed receive immediate attention when exposed. Figure 3.7 illustrates this analogy.

We must point out that many Lean strategists loudly proclaim that one of the main objectives of Lean implementation is *Zero Inventory*, The goal is to reduce all WIP to zero… or as near as possible. We must in the strongest possible way expose this type of thinking as dangerous and incorrect. The river analogy is that if the river goes dry, there is no flow at all. Of course, the same Manufacturing engineer will then claim to be misinterpreted. It will be explained that the goal is minimum WIP outside of any work directly being manufactured. In particular, the material which resides in a Lean cell is called *stock-on-hand* in most of the Lean literature; so technically it is not WIP. No matter what it might be called; a rose by any other name is still a rose and any material between entry into the system and exit from the system should be called work-in-process. To placate the zero WIP engineer; we suppose that there is no WIP in any material handling functions; no WIP to buffer any unexpected delays; no WIP in manufacturing cells and no WIP to smooth and balance production in Mixed-model manufacturing. To be fair, if a manufacturing system could be constructed in which every processing step took exactly the same amount of time; material handling and supply chain support was instantaneous; and there was no setup and changeover times this philosophy might be worthy of exposition. However, goals are fine but the Lean engineer lives in *Realville, USA*. The real problem as we will repeatedly point out is the corrupting and disrupting effects of *variance*. When variance components of any kind are present in a manufacturing system, the system MUST pay for it in degraded system performance. With all due respect and gratitude, we will quote from the work of Hopp and Spearman in their landmark book *Factory Physics*.

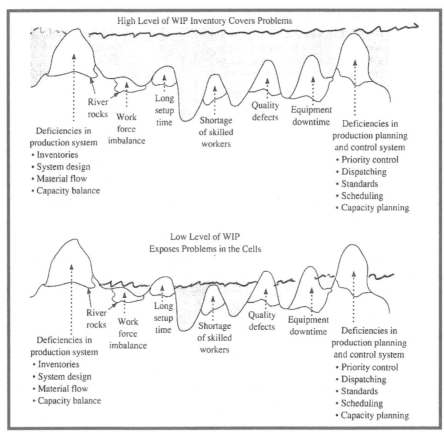

Figure 3.9
The Ohno Rocks-in-the-River Analogy is Very Revealing

Any form of variability causes congestion. This is stated as one of the *Laws of Factory Physics.* Congestion means more inventory, and more inventory in a system with fixed demand causes a corresponding increase in Throughput time by *Little's Law.* This implies that variability reduction is a key for improving system performance. Unfortunately, when proposing policies or solutions to reduce variance it is common to encounter a response such as: *Well, that might work somewhere else, but our problems are unique; variability is just an unfortunate part of our business and we simply cannot control it.* Sometimes…rarely…this could be true, but is often used as simply an excuse for not attacking a difficult problem; one that they may not be (statistically) equipped to solve. Hopp and Spearman categorically state that solving variability problems might cost time and money, but if these problems are ignored they will *always increase WIP* and will cause more problems. Granted, reducing variability anywhere it is found might be expansive and time consuming, but Hopp and Spearman formulate another *Factory Physics Law* that states what will happen if variability is tolerated or ignored.

Factory Physics Law: *Pay me now, or Pay me later*
  If you cannot pay for variability reduction, you will pay for it in one of three ways.
  1.0 Long throughput times and higher WIP levels
  2.0 Wasted machine capacity and lower utilization of scarce resources
  3.0 Lower production or output rates

Hopp and Spearman continue:

"This law does not imply that variability reduction is always the preferred course of action. In a system where capacity is inexpensive, low utilization may be perfectly acceptable. The point to be made is that there is no free lunch. Since we will pay for variability in one way or another, the decision of what to do (or tolerate) should be made with full knowledge of the consequences. Let the engineer who has never examined variability or its effects be aware…*You will pay in one way or another"*
Factory Physics, 2nd Edition.

In light of these penetrating observations, it should be obvious that regardless of how it is done, minimizing WIP whenever possible is a *Design* Rule and a goal of *continuous improvement.* It should also be clear that one of the primary goals in designing and implementing Lean, U-shaped cells is to standardize the stock-on-hand within the cell. Simultaneously, the WIP levels between cells and in final assembly are controlled and set by line foreman and the Lean Engineer. Here is an example of what Lean Engineers do.

Suppose that in front of final assembly there are 10 carts and that each cart holds 40 parts. Empty carts signal upstream production to refill and deliver a full cart. The maximum inventory is therefore 400 parts in front of final assembly. Suppose that one cart is purposely removed: the WIP cap is now 360 parts. The foreman now waits to see if and where a problem might arise: (Lower the river to expose rocks). To prevent a negative impact on final assembly, when a problem surfaces it is noted and identified; the full cart just removed is placed back into the WIP link. A Kaizen event is then launched to fix the problem, Since WIP is at the previous level, the positive effect of this *fix* may or may not be noticed. When the new procedures are implemented, a cart is again removed. If the problem has been properly identified and properly solved, the system should now function with no problem; but *some* improvement in system performance should occur by the Laws of Factory Physics and Little's Law. After allowing the system to stabilize, the foreman will now remove another cart, dropping WIP to 320 parts in front of the cell. This procedure is repeated several times. If solutions are formulated and are effective, suppose that final assembly is operating with only 5 carts (WIP is now 200). Now something interesting occurs! Over a weekend, the number of carts is restored to 10, but now each cart contains only 20 parts. Note that both upstream and in final assembly as unit lot sizes are approached, this new policy helps to smooth production.. a primary goal of Lean Engineering. By continuing in this fashion, WIP and the number of carts can be systematically reduced. Note carefully that this necessitates changes in both material handling and in setup time reductions in all areas. Any problems which surface, any material handling problems and any setup time reductions required are solved one by one. Following these procedures, WIP (inventory) between cells is continuously reduced. The effect on system performance is to both reduce throughput time and to increase flexibility.

Minimum levels of required inventory can be achieved in many ways. The *zero defects* and *zero rework* goals must be rigorously pursued. Unscheduled machine downtimes must be eliminated. Transportation and material handling functions must be redesigned to guarantee *Just-in Time* delivery. Workers must engage in educational programs to eliminate mistakes, become cross trained on many machines and understand their role in the system. Finally, we cannot stress enough that six-sigma engineering must be coupled to Lean Engineering principles to aggressively and continually reduce variance whenever and wherever it occurs. This is how continuous improvement is executed in a Lean factory. Considering the importance of WIP and variance reduction, it must be realized that inventory should not be considered a dependent, uncontrollable factor, but a dependent controllable factor.

3.15 Lean Manufacturing Cells Replace the Flow Shop

Most companies that design their first Lean manufacturing cell use their existing job shop or flow shop equipment. Flow shops or job shops produce unique or one-at-a-time finished goods. Lean cells are built around part or product families. Rearranging, converting and using existing equipment is a natural precursor to the implementation of mature, Lean cells. This stepwise conversion to Lean manufacturing and U-shaped Lean cells is good, because the company can learn how Lean cells work, how to drastically reduce setup and changeover times, enforce in-line Quality assurance and Preventative Maintenance, implement MO-CO-MOO machine cycle time procedures and operate U-shaped Lean cells. As conversion to Lean systems mature, lean machine tools can be designed and specialized to Lean practices. The initial use of existing machines is a relatively low-cost, risk-free approach. We will call these initial Lean manufacturing cells *interim cells* to distinguish them from complete, mature Lean cells.

Consider how a traditional job shop might be used to manufacture a drive shaft, made in four different lengths for different pickup models. The sequence of operations shown in Figure 3.10 might be followed in a traditional job shop. Note what is called *spaghetti flow,* which is typical of a job shop.

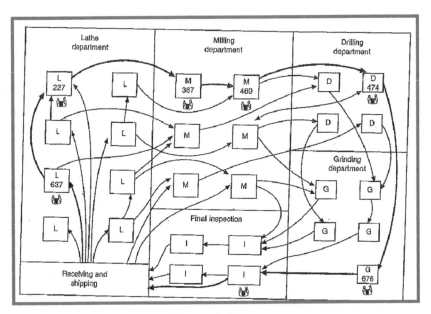

Figure 3.10
Typical Part Flow Patterns for a Job Shop Producing Four Parts

In the job Shop, single skilled operator would man each machine, and parts would move from machine to machine in tote boxes. A process plan which gives the sequence of operations (machines) to be used in manufacturing the drive shaft is crafted by either a Manufacturing engineer or by production planning. This sheet is sometimes called a *routing sheet* or a *traveler*.

An operations sheet is prepared for each machine that contains information about how to set key processing parameters at each step (speeds & feeds), the tooling required, type of raw material, etc. Sometimes standard processing throughput time or machining time is also shown for each operation: These are usually Industrial Engineering stop-watched, sampled times or artificial engineered time standards. These times are also used to execute process and product cost. The routing sheet is accompanied, preceded or augmented by machining instructions, manufacturing/move lot sizes and set-up instructions. All of this information and much more is used to produce one or more parts in a job shop (or a process shop) traditional manufacturing system. Now let us see what happens if we convert this manufacturing system over to a Lean manufacturing system.

All machines in the job shop are arranged into U-shaped configurations as shown in Figure 3.11. In this case, the drive shaft is completed manufacturing cell #2, but the cell is feeding a subassembly cell that is feeding a Mixed Model final assembly line. It should be emphatically stated that the conversion to a U-shaped manufacturing *interim cell* is much more than just moving machines around as suggested by many group technology gurus and value stream mapping technical papers might suggest. A lot of work is required by the lean engineer to make this transition involving everyone in the factory. For example, a new sequence of operations might be implemented adding a horizontal band saw added to the existing six machines. The saw is used to cut bar stock into the correct length, depending upon which of 4 different pinion gears is required. This modification eliminates the need to perform facing and cutoff operations at the first lathe and reduces the cycle time for that machine. Every machine is then modified to execute autonomous, single cycle operations unattended (single cycle automatics). Jigs, fixtures and cutting tooling are designed or modified to drastically reduce setup and changeover from one type of drive shaft to another (SMED). Machines are all equipped with walkaway switches, safety devices, QC and loading/unloading devices. All machines are arranged according to required precedence relationships in a U-shaped layout with a narrow internal isle. Between each machine, a decoupler is added which holds one unit to support one-piece flow through the cell. A small WIP area is designed to hold incoming parts before the first workstation, and following the last workstation in the cell.

The cell is now ready to produce the family of four drive shafts; say A, B, C and D. Assume that the monthly demand is for 1000 drive shafts of type A, 2000 of type B and 500 of type C and D. Depending upon the required demand and delivery schedule, the cell Takt time and output rate can be determined. The cell is then *capacity planned* to meet this schedule. Using 20 working days a month, suppose that the daily demand is then 100 of type A, 200 of type B and 25 of types C and D. To smooth the finished goods delivery and minimize make-to-order final storage; assume that the final *leveled* production sequence is set to be executed 25 times: AAAACBBBBBBBB. Starting at the first operation and moving clockwise, we end up at the second grinder. This is the interim cell design. The machining time at any one process step must satisfy the following design rule. The machining cycle time for every machine or parallel machine configuration in the cell must be strictly less that the necessary required *cell cycle time* (Time between part completions). The next step is to determine (1) how many workers are needed in the cell and (2) the number of machines to be assigned to each worker (workers loop) to meet the required Takt time for the cell. Each worker will spend a certain amount of time at each machine, determined by (a) whether or not it is a machining or assembly operation (b) the changeover/setup times required (c) any visual inspections and (d) the walking times between machines. The time it takes each worker to complete his/her loop of assigned processes is the sum of all these times in the assigned loop. The cell is usually manned by one worker. The maximum output rate is determined by the longest machining processing time. See Table 3.2. For a number of assigned workers less than the number of machines in the cell, the cell cycle time in terms of minutes/part is the *maximum of any of the assigned worker's loop times*.

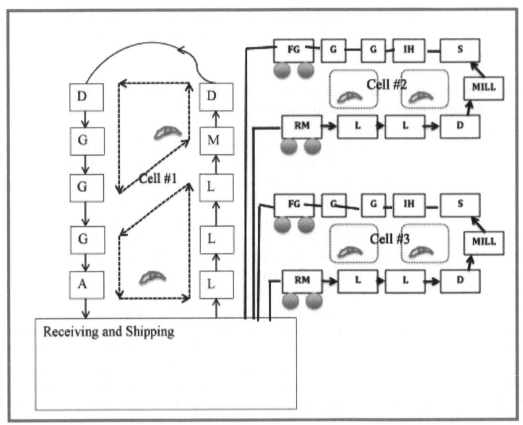

Figure 3.11
The Job Shop can be Reconfigured into Manufacturing Cells

Assume a worker takes 20 seconds at each machine (unloading, checking and loading). Final inspection and cell unloading takes 20 seconds and that about 5 seconds is spent walking from one machine to the next in his/her assigned loop. At each machine, the acronym MO-CO-MOO applies, and QC checks are done manually or in the decoupler. Every processing step is a single cycle automatic machine except for inspection, operation#8, which is a manual operation. This operation takes 20 seconds. The cell required Takt time is 220 seconds. Using one worker the Cell cycle time would be CT_{cell} = 200 seconds, almost 10% less than Takt time. This cycle time meets the Takt time (220 sec). In this example, capacity would be increased by adding a second worker. The cell could be manned by 2 workers. The first worker will be assigned to the first lathe and help with inspection taking 100 seconds.

Table 3.2
List of Operations and Respective Times for Cell 2

Operation Number	Machine Tool	Operation of Process	Machining Times (MT) (Sec)	Operator Manual Time (Sec)	Operator Walk Time (Sec)
1	Lathe 1	Rough turn	90	20	5
2	Lathe 2	Finish turn	40	20	5
3	Horizontal mill	Mill Step	45	20	5
4	Vertical mill	Mill slot	45	20	5
5	Drill press	Drill and tap holes	30	20	5
6	Surface grinder	Grind slot	45	20	5
7	Cylindrical grinder	Grind surface	30	20	5
8		Inspect		20	5

The largest machining time is 90 seconds. Can a third worker be added to the cell? No; the lathe's processing time of 90 seconds exceeds the cycle time of 67 seconds (200 /3).

The cell cycle time and the cell throughput time are based solely upon the number of workers assigned to the cell, the operations in their individual loop(s), the time spent at each machine in their assigned loop and the walking times between machines. Here is the unique aspect of Lean manufacturing cell design. The throughput time and production rate of the cell does not depend upon individual machine times but on the number of workers, as long as there is sufficient capacity to meet Takt time requirements. The machining times have been *decoupled* from the cell cycle time by the way the cell operates. In other words, *the production schedule has been made independent of the processing times at each machine.* This is the magic and genius of the TPS, linked-cell production system. *The job shop and flow lines cannot be operated in this way.* The output of any cell can be easily changed by assigning more or less workers. Generally workers assign themselves to the machines without changing any machining operations (provided capacity exists for expansion).

The number of workers and the machines operated by each worker are determined to meet two critical goals: (1) Balance the workload across all workers and (2) produce to Takt time. Balancing the workload is accomplished by assigning machines to each worker in such a way that the time required for each worker to make one cycle around his/her loop is the same for all workers. In most situations, this goal cannot be accomplished. There is usually a relatively small difference between worker cycle times. The differential can be reduced in a number of ways, including workers sharing one or more machines at the interface of their loops but since both the number of workers and product demand will frequently change, one of the best ways to balance is to reduce long machine cycle times into multiple short, sequential operations. That is why a saw might be added to the cell, to reduce the processing time of the first lathe to something under 67 seconds.

Since a cell's cycle time is dictated by operator and machine assignments, and each machine will automatically cut off and wait for the operator's next arrival, the throughput time per part is not determined by the machine processing time, but the maximum loop time of any workers multiplied by the number of workstations. Hence, machine times and machine cycle times can be changed (within limits) without affecting the production rate of the cell. In traditional manufacturing lines, changing machine times would affect both the output rate and the delivery lead time. Note that the number of output cycles that any one part spends in the cell depends not only upon the maximum loop time of any worker, but also upon the amount of SOH in the cell and whether each machining operation is an assembly operation or a machining operation. Recall that if an operation is assembly, the operator will need to be at the machine during the assembly cycle time. If the operation is machining, the operator will MO-CO-MOO, hit a walk-away switch and go to the next machine. The machine will automatically shut itself down after the machining operation is completed and wait for the operator to return.

3.16 Summary

Lean Engineering is a discipline which is primarily responsible for designing and implementing a radically different design for a manufacturing system. The fundamental premise of Lean thinking first identified by Womack and Jones is still true: Lean production is fundamentally *Waste elimination*, in which all non-value activities are systematically removed. The most important aspects of Lean thinking are to Minimize WIP by enforcing WIP caps at all levels of the Manufacturing system; design and implement Lean, U-shaped manufacturing cells; aggressively reduce all sources of variation in the system; seek one-piece flow in minimum quantities; smooth and balance production; and implement Kanban /Pull control between all functional components. All subassembly, component and final assembly cells should be converted to Mixed-model production, and changeover/setup times reduced to as small a time as possible. Downstream functions, starting with final assembly, dictate the Takt time requirements for all upstream functions. All product is *pulled* through the system using Kanban mechanisms. Transformation to a Lean manufacturing system first requires implementation of lean, U-shaped cells and Mixed-model final assembly, before attempting to implement high technology and innovative solutions such as automation, *Autonomation*, Poka-Yoke, computerization and robotization. We present the following observations based upon experience and observation.

- When implementing Lean transformation, *start at the top*. System level changes demand top level support and strong leadership
- Before starting the journey, *educate everyone in the company* on what you are trying to do. *Involve everyone in the company from top to bottom.*
- Stress that the *purpose* of Lean Manufacturing is *not* to eliminate jobs or to fully automate, but to make more profit and improve customer satisfaction.
- Total Quality improvement and defect free manufacturing is not optional, but *mandatory*. Perfection is the goal. Defects can be prevented. Fix problems at the source. If something goes wrong, shut down the process immediately and fix the problem.
- If ask to choose between cost and quality.... TAKE BOTH.
- Total Preventative Maintenance using proactive and not reactive maintenance is mandatory. The squeaky wheel gets the grease. An hour lost at a bottleneck or critical production station can never be fully recovered without increasing cost and time.

- The *First Key* to Lean production is to *reduce work-in-process* to the lowest possible level. AXIOM: The lowest possible level is NOT ZERO in almost all cases. Determine the optimal WIP level(s) and go there.
- The *Second Key* is to aggressively and systematically reduce all forms of *variance*.
- The *Third Key* is to design and implement Lean, linked, U-shaped manufacturing & subassembly cells. *Link all production cells* with Kanban/Pull control.
- Implement a continuous improvement program to continuously change and improve the Lean manufacturing system design.
- Lowering WIP will reduce throughput times. Do not confuse Takt time, a cycle time with throughput time.
- Build a system design that can be understood by everyone involved. *Keep it simple, or at least as simple as possible.*
- Do not forget that the Supply Chain can drastically affect the performance of the Manufacturing system. Drive Lean principles all the way to the bottom of the supply chain. Implement Lean in second and third tier subcontractors.
- Lean engineering system design must precede concurrent engineering efforts. Eventually include all aspects of the Manufacturing Enterprise, including logistics, marketing, purchasing and sales.

Finally, Lean transformation calls for rethinking and redesigning traditional Mass manufacturing systems. This is difficult to accomplish because most corporate managers are risk averters and are resistant to radical change. Remember, things will get worse before they get better. Stay with the plan, but expect initial resistance at all levels. All companies want to raise productivity, lower cost and make more money, but few are willing to pay the price to facilitate long-term change. The conversion of final assembly to Mixed-model assembly fed by synchronized Lean cells is just the first step in the Lean journey. The entire journey will involve the relentless elimination of all forms of waste, on-time delivery of goods and services; and perfect quality.

The TPS (Lean manufacturing system) is designed so that every person in the system has responsibility to meet goal; understands how the system works; and realizes that people are responsible for making superior products. The L-CMS discussed in this Chapter simplifies the Manufacturing system and seeks continuous improvement. The remainder of this book will mostly address the Lean Engineering design tools and methodologies needed to execute change.

Review Questions

1. Define an enterprise or production system versus a manufacturing system. What are the classic manufacturing system designs (MSD)? How is the mass production system different from the lean production system in terms of their systems design?
2. Review the axioms (Design Rules) for the L-CMS and explain these 4 terms
    Leveling, Balancing, Smoothing, Sequencing, Synchronization
3. In comparing a linear subassembly design with a U-shaped cell design, what measurable parameters were used?
4. What is the main difference between the U-shaped subassembly cells and the U-shaped manufacturing cells?

5. Explain this statement: Cycle times (for manufacturing cells) do not depend on machining times of the manufacturing process.
6. What is the ramification of this independence [MT<CT] on production control (the scheduling of the system)?
7. What is standard work?
8. W. Edwards Deming said that to know when a problem exists we must distinguish normal from abnormal. How does standard work make this idea possible?
9. What does it mean for a system design to be robust or stable?
10. Explain the differences and relationships between cycle time, Takt time, throughput time and production rate.
11. What are some of the modifications you need to make to a machine tool when you incorporate it into a manufacturing cell?
12. If a machine's processing time exceeds the cycle time in a manufacturing cell, what alternatives might the lean engineer consider to ensure that a cell meets the necessary cycle time?
13. One-piece flow in lean cell design has many benefits. Discuss.
14. What is the key role of the worker(s) in the manufacturing cell?
15. Describe a common error-proofing device (poka-yoka) you are frequently in contact with.
16. Why is mistake proofing important in a cell's operation?
17. Name one product that has had a life of 50 years and is produced in volume over 10,000 per year. How is it made? Is it a candidate for lean manufacturing cells?
18. What functions can decouplers do in manufacturing cells?
19. Why do lean companies work toward having sole suppliers for their subassemblies and components?
20. Why are workers 3 and 4 in cell shown in Figure 3.5 going to the same station?
21. Where is the rabbit chase in Figure 3.5?
22. How is the Kanban rule like Little's Law?
23. There are many acronyms in lean manufacturing. Explain these.
    JIT, CT, TPM, SCA, TQC, TPT.

PROBLEMS

Problem 1:
Examine the redesign of the conveyorized layout into a U-shaped cell manned by eight operators seen in Figure 3.5. Examine the "before" and "after" data in the table for the two designs. Discuss cycle time and throughput time.

Problem 2:
Assume that an assembly line produced three products A, B and C. The available operating time per shift is 480 minutes. The cell operates two shifts per day. The average daily demand is 160 of A, 120 of B, 200 of C. What is the Takt time?

Problem 3:
The lean manufacturing cell Figure 3A has the following design: SCA designates a single-cycle automatic machine. There are 8 SCA's in the cell. Assume the walk time between all stations in 3 seconds (allow for 8 walks)

a) What is the necessary cell cycle time for one worker to operate the cell?
b) What is the throughput time for a part going through the cell?
c) Why is TPT not dependent on processing times?
d) What is CT and TPT if two operators work the cell?

Problem 4:
Now assume the cell shown in Figure 3-A has 8 processes (machines) that are not SCA's so the operator has to run the machine to perform the process.
a) What is the CT?
b) What is the TPT?

Problem 5:
One of the classical problems in the flow show is called "line balancing". Explain what this is all about and how it is done today. (This is a student research question)

Problem 6:
Go to your local subway sandwich shop. Define their sandwich manufacturing system ("Sub" assembly process)

Problem 7:
Suppose you added a saw to the cell shown in Figure 3.11 to reduce the machining time on the first lathe to 36 Seconds. The saw's machining time is 30 Seconds but it requires 20 Seconds of operator time for loading and unloading. Layout a new cell, make a new table of operation times and calculate the CT and TPT for one operator. Can a second worker be added to cell?

Problem 8:
If you had a Kanban link with containers that held 50 parts and it took 2 days to get the parts and you used 1000 parts per day, how many containers you recommend we have for this link?

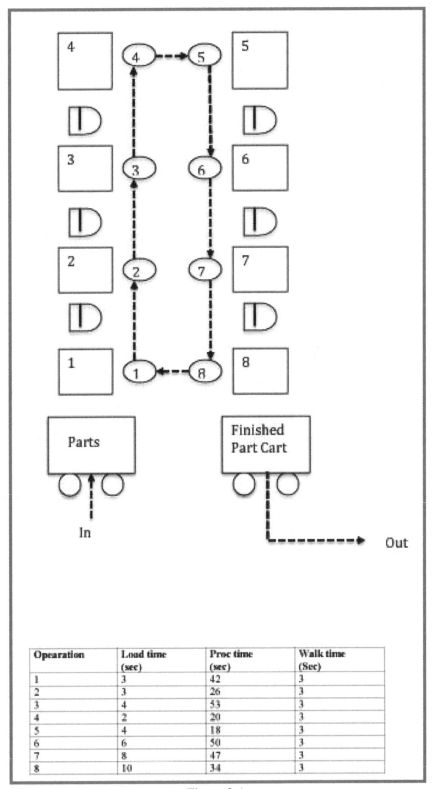

Opearation	Load time (sec)	Proc time (sec)	Walk time (Sec)
1	3	42	3
2	3	26	3
3	4	53	3
4	2	20	3
5	4	18	3
6	6	50	3
7	8	47	3
8	10	34	3

Figure 3-A
A Small Lean Manufacturing Cell

108

Chapter 4
Manufacturing Cell Design

"The superior man, when he sees what is good, moves toward it; and when he sees his errors, he turns from them."   The Book of Changes, China circa 1200 BC

4.1 Introduction

The Toyota Production System (TPS) is a radical redesign of the mass production system which evolved out of Henry Ford's assembly line. Under the leadership of Taiichi Ohno, Toyota invented and implemented a radical redesign of the modern Mas manufacturing system which we will call the *Linked-Cell Manufacturing System* (L-CMS). Schematic of L-CMS design is shown in Figure 4.1.

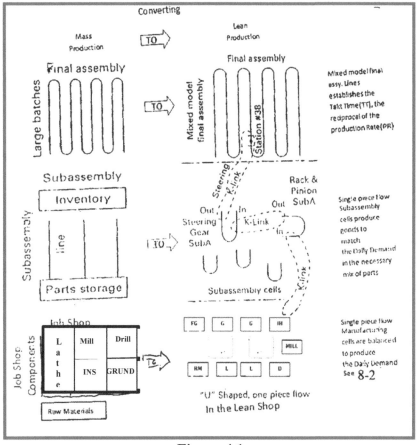

Figure 4.1
Converting Mass Production to Lean Occurs at 3 levels; Final Assembly, Subassembly and Component Manufacturing

For final assembly, the mass line is converted from large batches to a mixed model final assembly (MMFA) which creates a leveled demand for the subassemblies. The subassembly lines are converted from linear lines into U-shaped cells that do not require line balancing as they are self-balanced. The job shop is converted into U-shaped manufacturing cells which can produce components for the subassembly cells as needed in whatever mix the subassembly cell needs. See Figures 3.10 and 3.11 in Chapter 3 for example. Figure 4.2 shows a cell in detail for the manufacture of a gear. The cells are connected to subassembly and to final assembly by Kanban links which hold all the inventory. The material in the cells is called stock-on-hand and the cells are called interim cells. This means they are designed and built using machines that were originally designed and built for the job shop, where machine tools are placed in standalone positions – all in a line in departments. However the machine tools for the cells are extensively modified. Toyota has developed a stepwise strategy for modifying the machines designed and built for the job shop so they can be used in a lean shop. Typically the machines operate on a single cycle automatic (SCA) basis (once loaded, you hit the start button and they complete the cycle and then turn off – like your kitchen toaster). They are equipped with many devices to prevent defects. These devices are called Poka-yokes. The machines are modified so that they can be changed over very quickly from one part in the family to another, and they are maintained so that they do not fail during the shift (using preventive maintenance methodology). The goal is to make perfect parts every time on a one-piece flow basis. The methodology is called one-piece flow or MO-CO-MOO for make one, check one and move one on.

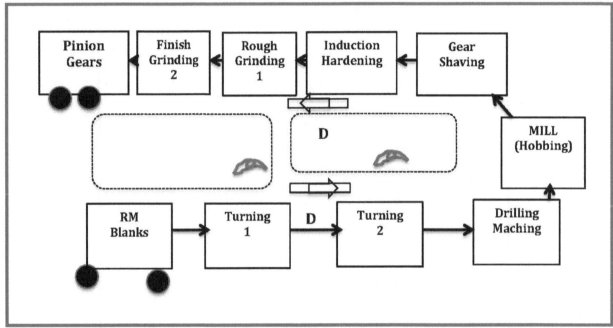

Figure 4.2
Detail of a Cell with 2 Operators and 2 Decouplers Operating on a One Piece Flow Basis

The subassembly cells also work on a one-piece flow basis but here the operations are usually all manual. The cycle time for the cells should be a bit less than the Takt time for the final assembly line to allow for any problems that will occur. The CT for the cell is based on the sum of the manual times for the operator(s) for loading, unloading, checking parts and walking from machine to machine.

True lean manufacturing cells such as that one shown in Figure 4.3 have processes designed and built specifically for the cell. These machines have many features and aspects not available in the interim cell. Notice for example, the narrow frontal footprint which reduces the walk time. The machines are designed ergonomically for easy loading and many feature automatic unloading devices. These cells may have small transfer lines for machining elements of the components that are present on every part in the family. The cell shown in Figure 4.3 can easily be run by two workers, one worker or even three workers, depending on the necessary demand for these components. See Chapter 6 for more details on true lean cells.

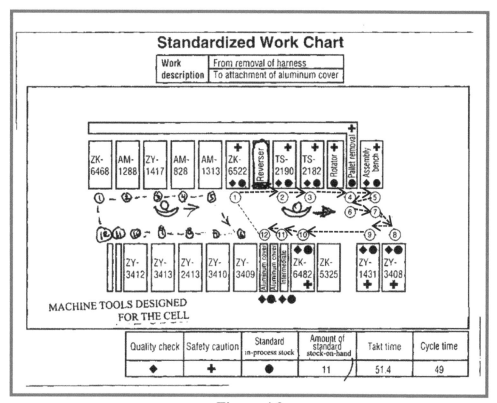

Figure 4.3
Lean Manufacturing Cells Have Machine Tools Designed and Built for the MO-CO-MOO Operations in the Lean Shop

One of the aspects of cells is the manner in which parts are moved from process/station to process when multiple workers are running the cell. The first thing to note is that the station times do not need to be balanced. The operators do the balancing by the way they divide up the work. They can pass work to the downstream worker two ways. They first way uses decouplers between the processes (See large D between processes). The part is placed there and the worker moves across the cell. The downstream worker picks up the part and moves it downstream. In some cases the worker may actually take over the work from another worker at the station or machine. This is baton passing. If decouplers are used, they hold one part and may perform other functions other than transporting the part downstream to the next machine. Poka-yoke devices for the prevention of defects and errors are generally in the machines or processes. Inspection devices in the decouplers can perform inspection on the outcome of the prior process while moving the part to the next station. They may also perform other functions like part cooling or heating or curing. The lean cells contain many processes thought to be job shop monuments, processes like painting or heat treating that can only be done in large batches which generate large queues. In the lean cell, in-line heat treating or painting processes are commonplace. Finally another aspect of the cells is the manner in which robots are used with people to make parts and subassemblies. See Figure 4.4 for an example of a manufacturing cell for sheet metal parts where the decouplers separate the operators from the robots, as a safety measure.

The objective of Taiichi Ohno was to build a simple system that could make complex goods (like cars). Why? Toyota did not have computers and their language was difficult to convey with written instructions, so most of the instructions and controls were visual. Pictures were used at the stations to show the workers what to do at each station.

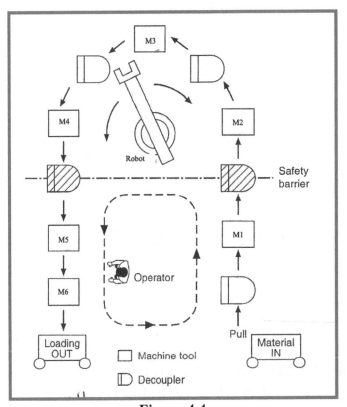

Figure 4.4
A U-shaped Cell Used in Sheet Metal Fabrication is Manned by an Operator and a Robot with Part Flow Connected by 2 Decouplers (Honda).

Because the manufacturing system design is simple, the Kanban system for production and inventory control operates on a few simple rules (when you see an empty container, fill it with what it had in there the last time). The most critical and but poorly undertook aspect of the system design is this. The design of these manufacturing cell makes the throughput times (TPT) for the delivery of the materials independent of the processing or machining times. TPT's depend only on cell cycle times which depend only on manual and walking times for the operators. Thus, TPT (and therefore the schedule) could be easily ramped up and down just by adding and subtracting workers, so changes in demand were easily and quickly handled, making the system very flexible. For a Lean manufacturing system to be successful, the design and operation of the system must always have customer requirements in mind. It must be clearly understood that there are two types of customers: (1) *Internal* and (2) *External*. A good Lean system design should ideally have the following factors in its design.

112

- *Safety*-All machining operations must be designed to the highest safety standards to prevent any possible accident.
- *Human Factors and Ergonomics*- The Lean system might contain many automated components, but it is fundamentally designed to be operated by humans. In a Lean system, the internal customer (the worker) is considered to be a more important asset than a machine. Workstations and daily tasks need to have professionally engineered time standards, and follow methods which have been completely evaluated and developed using ergonomically sound methodologies. No dangerous jobs; no physically challenging manual tasks; and no wasted time or motion.
- *Flexibility* – This term is frequently used to describe how a system will react to changes in product mix, demand patterns and technology innovation. *Flexibility* will be defined as the ability of a Lean manufacturing cell or production line to respond to changes within a particular product family or the ability to quickly change from one product family to another. For example, a particular Lean manufacturing cell might manufacture both Ford 150 pickup trucks and Ford 250 pickup trucks. Within the Ford 150 production facility, there might be well over 100 different F150 pickup configurations (Flexibility). In general, flexibility should be designed into any static cell, but might require cell modification by adding or subtracting machines.
- *RAM: Reliability, Availability and Maintainability* -We have frequently stated that each machine in a Lean manufacturing cell or system must be designed to run continuously with no unscheduled downtimes. RAM represents a broader concept with extended goals. A *Reliable* machine is one that functions without failure during any one production run or scheduled time of operation. *Availability* is the amount of time that any one machine or production line is actually ready to begin its assigned task. Note that if any one machine fails infrequently, but takes a long time to bring back on line, it can have high reliability and low availability. Conversely, a machine can have low reliability but high availability if it fails frequently but can be quickly brought back on line. *Maintainability* is the ease by which a machine can be maintained to prevent failure, or how easily worn or broken machine components can be replaced.
- *Quality Assurance*- Quality assurance is manifested in several different ways. Traditionally, Quality Control involved quality inspections, quality control charts and acceptance sampling. The Lean Engineer will redefine how these traditional quality tools are used but remember in a Lean manufacturing system, quality is designed into the product. Each worker is responsible for in-line quality control; and *Autonomation* devices are designed and implemented on each machine to detect and monitor potential problems. Autonomation is automation with the human in mind.
- *Empowering Employees*- Each employee must be trained in Lean thinking, Lean philosophies and Lean objectives. More systematic and continuous improvement ideas should come from the shop floor than from the ivory towers. However, the Lean Engineer or the design engineer will usually be tasked with system modification and actual implementation. The key is teamwork and transparency from top to bottom.
- *Training and Education*- The training and education function plays a major role in Lean Manufacturing and in cell operation. Each employee must be trained in Lean thinking and Lean philosophies. Each cell operator must be trained to operate any machine in the cell; perform routine maintenance functions; and be trained in the seven basic quality tools.

- *Simplify, Simplify, Simplify-* All operations should be easy to understand, operate, and control with proper training. Operating instructions should be visually displaced, readily available and easy to access.
- *Promote Customer Satisfaction-* Any manufacturing system only exists for one reason: To make money. Profit is generated by reducing production costs, and production costs are reduced by eliminating waste. Waste is called *Muda*. Profits and sales are kept high by keeping the customer happy and satisfying his needs and expectations. Failure to recognize this simple fact is a sure recipe for ultimate failure, no matter how Lean a system might be. Every employee needs to understand their role in providing value added to the external customer and promoting customer satisfaction.

4.2 Case Study: Design and Implementation of a Drive Pinion

A part drawing for a drive pinion is shown in Figure 4.5. The shaft is being machined from 430F; stainless steel bar stock that has been cold rolled and cut to the required length. The bar stock has an outside diameter of 1.780 ± 0.003 inches (45.2 ± 0.08 MM).

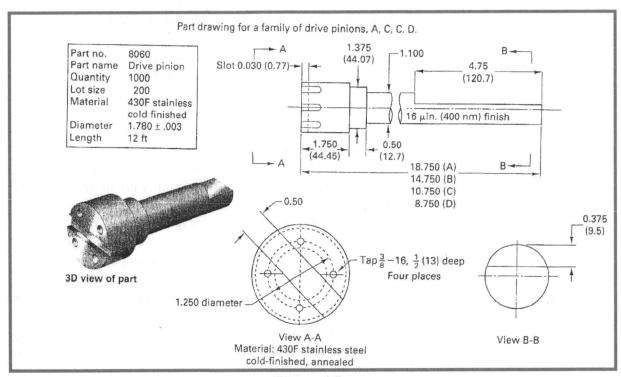

Figure 4.5
Part Drawing for a Drive Pinion

4.3 Traditional Manufacturing:

The drive pinion is currently being produced in a large job shop. Figure 3.10 shows its current route through two turning operations, two milling operations, and a cylindrical grinding operation. The bar stock is first turned on two sequential NC lathes. For the job shop a route sheet has been prepared. See Figure 4.6. The first operation makes a roughing cut (Operation #10) and the second a finishing cut (Operation #11). The turned bar stock is then transferred to a vertical milling machine where a ½ inch slot on the end is machined out (Operation #20). The operations are then followed by a horizontal milling operation (Operation #30) to produce the

114

step on the right end of the shaft. Next, a NC drilling tapping operation (Operation #40) produces the 4 holes on the left end. Finally the shaft is cylindrically ground, operation 50. The drive shaft is made in four different lengths as shown. Each drive shaft uses the same routing sheet, but an entirely different set of machines might be used from shaft to shaft. Hence, a different set of cutting tools, setups and operators might be used. All of this creates variance between drive shafts. Each drive shaft might have to wait outside the job shop for an undetermined amount of time before being introduced into the job shop for an undetermined amount of time before being introduced into the job shop.

Part no. _8060_

Part name _Drive Pinion_

Ordering quantity _1000_

Lot requirement _200_

Material _430F Stainless steel, 1.780 ± 0.003 in._

cold - finished 12-ft

bars = 1000 pieces

Unit material cost _$ 22.47_

Workstation	Operation no.	Description of operations (list tools and gages)	Setup hour	Cycle hour/ 100 units	Unit estimate	Labor rate	Labor + overhead rate	Cost for labor + overhead rate
Engine lathe #137	10	Face A-A end 0.05 center drill A-A end rough turn 1.46 cut off to length 18.750	3.2	10.067	0.117	18.35	1.70	3.65
Engine lathe #237	11	Center drill B-B end finish turn 1.100 turn 1.735 diam	3.2	8.067	0.095	18.35	1.70	2.96
Vertical mill #357	20	End mill 0.50 slot with 1/2 HSS end mill (collet fixture)	1.8	7.850	0.088	19.65	1.85	3.20
Horizontal mill #464	30	Slab mill 4.75 x3/8 (nesting vise HSS tool)	1.3	1.500	0.022	19.65	1.80	0.28
NC turrett drill press #474	40	Drill 3/8 holes-4x tap 3/8-16 (collet fixture)	0.66	5.245	0.056	17.40	2.15	2.10
Cylindrical Grinder #57	50	Grind shaft to 16µm -1.10	1.0	10.067	0.110	19.65	1.80	3.89

$16.58 + 22.47 =

39.051

↗

Estimated mfg. cost per unit

Figure 4.6
A Typical Process Planning Sheet for the Job Shop

4.4 The U-shaped, Lean Production Cell

The drive pinion can be manufactured in a U-shaped Lean manufacturing cell shown in more detail in figure 4.7. The cell is designed to produce drive shafts one at a time. This interim cell has seven machines (processing stations) plus a final inspection station, and is designed to produce the family of four different drive shafts. All workers are trained to be multifunctional and versatile. To be *multifunctional* means any one worker can perform setup, changeover, quality inspections, simple maintenance and continuous improvement tasks. To be *multiprocess* means each worker can operate any machine in the cell, and be equally efficient across all four shafts.

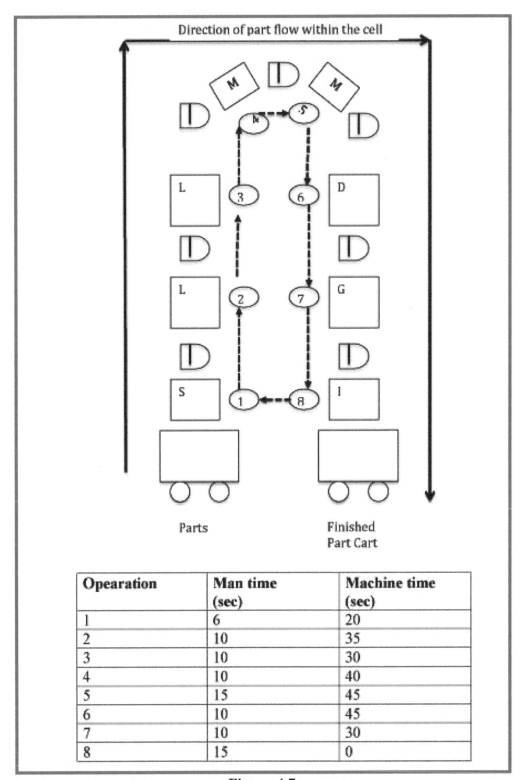

Opearation	Man time (sec)	Machine time (sec)
1	6	20
2	10	35
3	10	30
4	10	40
5	15	45
6	10	45
7	10	30
8	15	0

Figure 4.7
Lean Cell Design for the Drive Pinion
Saw Added to Reduce the Machining Time at the First Lathe

Figures 4.8 and 4.9 show typical, feasible worker assignment loops for two or three workers. For illustrative purposes, we will examine how the cell can be operated by 1-3 workers on only one type of shaft. Mixed model production will be studied in great detail in Chapter later.

Note that the final inspection in this cell design takes about 15 seconds. This task is performed by the worker who has machine 7 in his/her loop. Recall that the four parts are of different lengths, and so the processing times are different for each part. Machining time for part on each machine ranges between 6-45 seconds. Workers spend about 3 seconds walking between machines and from the output storage area to the input station. The isle width is narrow, about 4 feet wide.

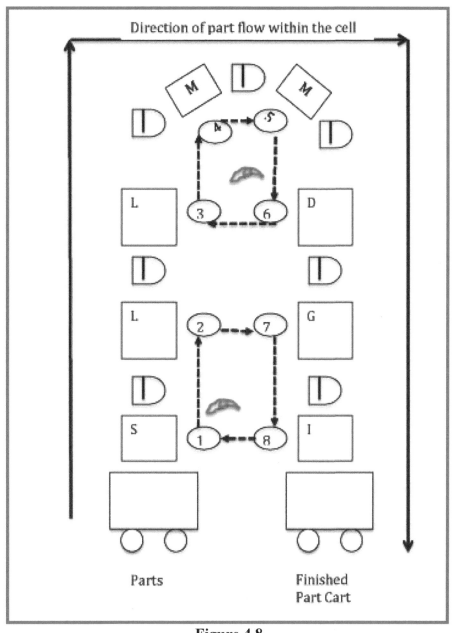

Figure 4.8
An Interim Manufacturing Cell Operated by Two Workers Connected by Decouplers.

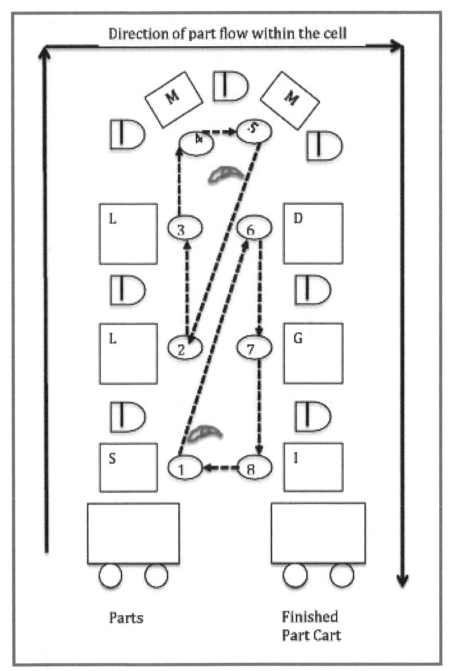

Figure 4.9
An Interim Manufacturing Cell Operated by Two Workers Connected by Decouplers

4.5 Cell Cycle Time
Cycle time for the cell is calculated as follows.

$$CT_{Cell} = \sum Walking\ times\ between\ machines + \sum Manual\ tasks\ at\ each\ machine$$

118

4.5.1 One Worker

The time it takes one worker to complete the cell loop is calculated by adding up the sum of all task times spent at each machine and the walking times. Assume that the time required at all Machines include the loading and unloading times, respectively.

$$CT_{Cell} = 86 \text{ sec (task times)} + 24 \text{ sec (walk time)} = 110 \text{ sec} = 1.83 \text{ minutes/part.}$$

Hence, one part will be produced about every 2 minutes. Is this cycle time greater than any one machining time? Yes it is, so there will be no blocking as the worker reaches each machine in the loop after it has completed its machining cycle.

- The output rate of the cell using one worker is approximately 30 parts per hour
- The minimum throughput time for each part is (No decouplers):

CT_{Cell} * *Number of cycles part was in the cell (8 loops by 1 worker);*
$$TPT = 1.83 \text{ min} * 8 \text{ loops} = 14.64 \text{ minutes}$$

4.5.2 Two workers

What would the cycle time and output rate be using two workers? Clearly, this depends upon the number of machines each worker operates and the manual tasks each worker performs. Figure 4.8 shows the machines each worker might operate at a given time. The loops change as workers can share machines to balance their loop times. See Figure 4.9. The decouplers are used to connect the part flow through the cell. For example, B*aton passing* occurs between the two vertical mills (Worker2 to Worker1) and between the saw and lathe 1 (Worker1 to Worker2) in figure 4.9. In order to synchronize the two workers, it is necessary to hold one extra unit of stock in each decoupler.

The cell loop cycle times for the two workers are as follows.

$$W1_{loop\ time} = 41 \text{ min (task times)} + 12 \text{ min (walk times)} = 53 \text{ sec}$$
$$W2_{loop\ time} = 45 \text{ min (task times)} + 12 \text{ min (walk times)} = 57 \text{ sec}$$

The time to produce one part and the output rate is now:

$$CT_{Cell} = \text{Max } (53, 57) = 57 \text{ Sec}$$
$$Output\ Rate_{Cell} = 0.95 \text{ parts/min} = 63 \text{ parts/hr}$$

As the operators share tasks in the cell, the cycle time will move close to the average 54.5 sec/part or 0.9 min/part or 66 parts/hour.

4.5.3 Three Workers

The cell loop cycle times for three workers are as follows.

$$W1_{loop\ time} = 31 \text{ sec (task times)} + 9 \text{ sec (walk times)} = 40 \text{ sec} = 0.66 \text{ min}$$
$$W2_{loop\ time} = 30 \text{ sec (task times)} + 9 \text{ sec (walk times)} = 39 \text{ sec} = 0.65 \text{ min}$$
$$W3_{loop\ time} = 25 \text{ sec (task times)} + 9 \text{ sec (walk times)} = 34 \text{ sec} = 0.56 \text{ min}$$

The time to produce one part and the output rate is now:

$$CT_{Cell} = \text{Maximum loop time was 40 sec}$$
$$Output\ Rate_{Cell} = 90\ \text{parts/hour}$$

Note that as workers are added to the cell, the output rate increases almost linearly as shown in Figure 4.11. But wait! Machines 5, 6 have processing times of 45 seconds, 5 sec greater than that of the longest loop time. So these machines block the output at 45 seconds and the actual rate is 80 parts/hr. If one worker is assigned to each machine, the maximum output per hour will still be 80 parts/hour. If Takt time cannot be met by workers, additional capacity must be created. We will discuss various ways to create additional capacity in the next section.

Each machine in the cell is a single-cycle automatic, which once initiated by a walk-away-switch will autonomously complete its machining cycle. All switches are located on the machine so as to insure worker safety. Each switch should be convenient, simple and ergonomically designed. It should be clear that the cell can be operated by as few as one worker or by no more than two workers. The output rate of the cell (1/Takt) is determined by the number of workers and their machine assignments (loop of work), and not by the machines. We will call this cell an *interim cell* because it will initially use machine tools originally designed for job shop or flow shop operations. After implementation, the cell will be continuously improved, machine by machine. See Chapter 6. Figure 4.12 shows the interim cell from Figure 4.7 upgraded with a transfer line to do the drilling, tapping operations on the drive shaft. All the drive shafts have the same hole pattern, hole size, hole depth and thread requirements, so the same set of tooling can be used on all the parts. Stainless steel can be difficult to drill and tap and the transfer line eliminated many processing problems and allowed the machining cycle time to be reduced.

The interim cell is a good example of a less-than-ideal cell which can be modified to improve production and eliminate waste (Muda). Each operator will move from machine to machine in his/her loop, performing Mo-Co-Moo operations and starting each machine with a walk-away switch. Each decoupler, both between machines and worker loops, contain one unit. We will assume that there is always a part available at the input station, and that there is always a demand for finished parts.

When machines are first placed into the cell, they are appropriately (time and cost constraints) modified to support Lean cell production. For example, the drilling machine shown in Figure 4.13 might be modified as follows.

- Convert any manual feeds and hand-held tools or work pieces to automated power feeds controlled by specifically designed jigs and/or fixtures.
- Add an automated stop mechanism for machining process,
- Add automation to make the machine return to a ready-to-be-loaded state after the finished [part is removed and a restart button is pushed by the operator.
- Install a walk away switch to activate the next machining cycle once the next part is loaded
- Install an automated unloading device (*Chuku-chuka*).

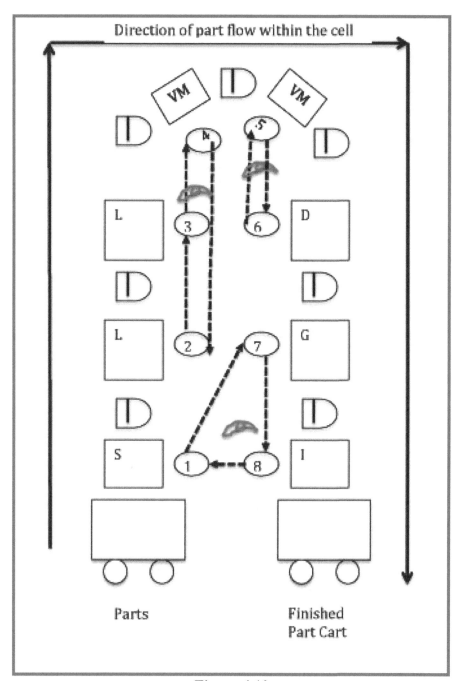

Figure 4.10
Manned Interim Manufacturing Cell for a Family of Parts with Seven Machines Operated by 3 Workers.

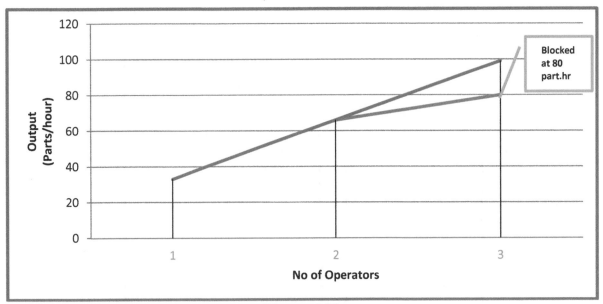

Figure 4.11
The cell design produces a linear output with respect to the number of workers in the cell

- Install an automated loading device for the next part: This is a good robotic application for small robotic arms and transfer lines.
- Possibly develop and install automated Quality control checks and error diagnostics (Poka-yoke). These can be fully implemented later.
- Design a decoupler between each machine to hold and position a part for the next machine. At least design a Quality control check; this can be fully implemented later.

While traditional Mass manufacturing companies historically have designed automated machine loading and unloading mechanisms, few Mass production lines use the concepts of decouplers and single cycle machines with operators moving stock-on hand to subsequent processes. Material handling is usually done by fork lift trucks, pusher dogs, or expensive conveyor material handling systems. It is not unusual for semi-finished parts to be stored in ASRS storage systems and retrieved later when space becomes available downstream. In Lean U-shaped cells the work between stations is tightly controlled. Decouplers usually contain no more than one unit. In interim cells, material handling is often performed by workers placing the part(s) into decouplers. In some interim cells, machines have been modified to both load and unload from decouplers, and almost all second generation Lean cells have some automated machine loading/unloading devices. If automation is used, it is supervised and often activated by workers. Toyota called this *Autonomation*… automation with a human touch. In the lean cell, automated loading, unloading material transfer lines and *decouplers* was employed. Monitoring the machining cycle with *Poka yoke* and sensing the uniformity of machining operations was also necessary to hold tight tolerances. This ensures repeatability, and eliminates variance. Single unit flow replaces lot sizes larger than one.

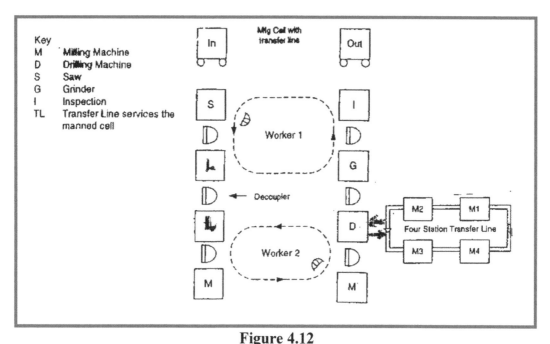

Figure 4.12
The Interim Cell has been Improved to a Mature Lean Cell with a Transfer Line to Reduce the Machining Time at Operation D

Automation of the machine start and stop cycle supports single cycle automation and unit processing, and is essential to the elimination of the *one operator-one machine* syndrome which characterize many Mass production job shops. Partial or full machine automation is accomplished with standard limit switches, sensors and programmable controllers (PLC's). When first initiated by Ohno in Toyota's machine shops in the 1950s, it allowed hands-free operation by replacing manual controls with pneumatic/hydraulic chucks, limit switches, and end-point detection.

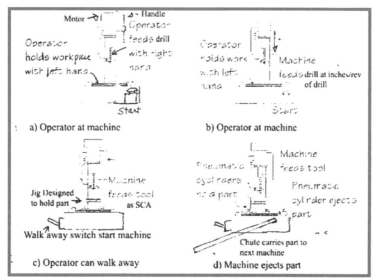

Figure 4.13
The Step by Step Modification of a Drill Press Involves Having the Machine Feed the Drill, Stop and Eject Automatically and Adding Jig and Walk-away Switch

Ohno also developed specialized jigs and fixtures (SMED) to support rapid setup and changeovers for different part types. Finally, Total Preventative Maintenance (TPM) practices were implemented to render each machine highly reliable, and Total Quality Control (TQC) practices were implemented to enable *zero defect* production.

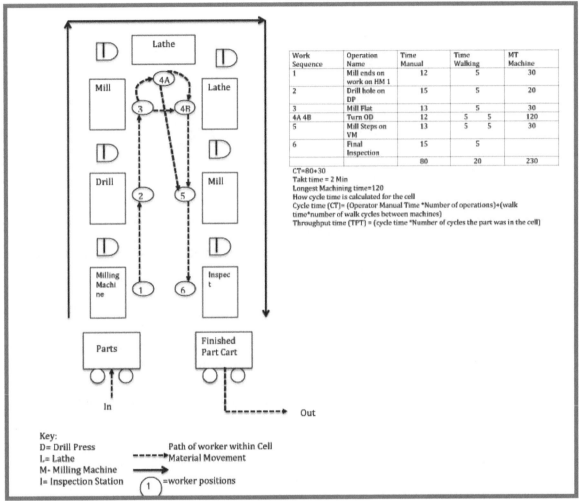

The table within the figure:

Work Sequence	Operation Name	Time Manual	Time Walking		MT Machine
1	Mill ends on work on HM 1	12	5		30
2	Drill hole on DP	15	5		20
3	Mill Flat	13	5		30
4A 4B	Turn OD	12	5	5	120
5	Mill Steps on VM	13	5	5	30
6	Final Inspection	15	5		
		80	20		230

CT=80+30
Takt time = 2 Min
Longest Machining time=120
How cycle time is calculated for the cell
Cycle time (CT)= (Operator Manual Time *Number of operations)+(walk time*number of walk cycles between machines)
Throughput time (TPT) = (cycle time *Number of cycles the part was in the cell)

Key:
D= Drill Press
L= Lathe
M- Milling Machine
I= Inspection Station
Path of worker within Cell
Material Movement
=worker positions

Figure 4.14
Example of a 6-Machine Interim Manufacturing cell with a Final Inspection station (manual) and One Operator

4.6 U-Shaped Lean Cells with Machining Time(s) Greater than Cell Cycle Time

Consider the manufacturing cell shown in Figure 4.14. The Takt time is 2 minutes or 120 seconds. The cell consists of five machining steps and one final inspection station. The table shows processing times in seconds, the walking times and the manual task times at each station. Using a single worker, the cell cycle time will be 110 seconds. The cell output rate is therefore 0.54 unit/minute or 32.7 units/hour. However, machining time for each part on the lathe is 120 seconds. Since machining time at the lathe is 120 seconds, and the worker cycle time is 110 seconds, this operation will delay the worker for 10 seconds each cycle. This violates one of the Axioms for Lean cell design which is $MT_i < CT$

124

MT_i = Machining cycle for machine i = 1, 2…M machines in cell; j = 1, 2…..parts in family; CT = Cell cycle Time

The Lathe operation is not only the longest machining time, and it is also much larger than any other machining time.

This is why the Lean Engineer decided to install a duplicate lathe as shown in Figure 4.14. This will cause the average time to lathe a part to become 60 seconds. Hence, we now have a feasible solution. The machining time of 120 seconds is still at least three times larger than every other machining time, so there is considerable idle time at each machine but the lathe. However, this causes no real difficulty. It simply drives the traditional *bean counters* crazy. The worker will simply alternate between the two lathes. Starting at the input station, the worker will alternate between the lathe *A* and the lathe *B*.

4.7 Creating Capacity

In reality, if a family of several parts is scheduled to be made in this cell, the actual time spent by each part on every machine would likely change as the part type changes. In this case, the Lean Engineer must check the cycle time of each part to make sure that no machining time exceeds the worker's cycle time. In addition, the sequence of parts which flow through the cell will require multiple setups. The total number of changeover setup times must also be added to the cycle time. There are two methods which might be used to determine how long each part should spend at each machine in *designing* U-shaped Lean cells to guarantee feasibility. The first is to use *engineered standards* to determine the time that each part will spend on each machine. Good Industrial Engineering methods analysis should be executed in parallel to achieve consistency and maximum worker efficiency. The second method is to use *mathematical analysis* to calculate machining times. For example, the following equation will predict the drilling time at the drill press.

$$MT_i = (d + \text{Allowance}) / (F*R)$$

Where: d = depth of hole (Inches)
A = Allowance ($\Delta * d$)
F = feed rate (Inches/revolution)
R = Revolutions per minute (RPM's) of the drill

The lean engineer must select the two critical machine inputs of cutting speed and feed. The RPM of the drill is related to the cutting speed by the following equation.

$$R = \frac{12V}{\pi D}$$ Where: D=Diameter of the Drill bit

For the drilling operation in this example, suppose that a 2 inch diameter hole is to be drilled 1 inch deep with a 0.27 allowance. The material to be drilled is cast iron and the drill is made of carbide. The Lean Engineer selects V= 200 feet per minute and F= 0.01 inches per revolution. The drill RPM is therefore R = (V*12) / (π*D) = (200*12)/(2*π) = 382 Rev/min. The feed rate is (382*0.01) = 3.82 inches per minute. Hence:

$$MT_{Station\ 2} = (1+0.27)\ /\ (3.82) = 20\ \text{Seconds}$$

This drilling operation will take 20 seconds to drill a 2 inch diameter hole 1 inch deep with a carbide bit.

As long as the worker loop time is larger than the longest machining cycle in the cell, any part in a part family can be made in the cell. The number of workers assigned to the cell, and the number of machines each worker tends, is determined by the required downstream Takt time requirement(s). The sequence used for multiple parts must be carefully determined. The *sequence* in which parts are presented to each *machine* will influence cell throughput time, cell output rate and the degree to which downstream production can be smoothed. We will revisit this issue in Chapter 7. The cell can continue to function without any physical alterations as long as there is capacity to produce. There is no scheduling required, and the cell is balanced by design. Such a cell is both flexible and *uncoupled* from all other cells except for the required Takt times and the effect of sequencing. In some cases, cell worker balance might require sharing one or more workers between adjacent cells. We will show an example of such a design later in this Chapter.

4.8 Takt Time and Cell Cycle Time

As previously stated, the Takt time requirement(s) for a cell dictate how the cell will be operated. Like all other cells (subassembly and component cells), each cell is designed to produce parts only when required (authorized), and in only the quantity needed (Kanban control) for immediate downstream production. The entire system is paced by Mixed Model final assembly. In the Lean, Linked-cell factory, the cell is operated to produce parts at exactly the required rate and no faster. Cycle times are determined by Takt times with a 10%-20% allowance factor to account for any delays or uncertainties.

An example from a Honda plant in Marysvile, Ohio might be helpful. Suppose the required production rate is 300 automobiles per day. The body for each vehicle requires 24 different sheet metal components. All 24 pieces of sheet metal are stamped out of sheet metal rolls of a set of vertical presses. The presses stamp out 300 hoods, then the dies are changed and 300 roofs are stamped out. Dies are changed again, and 300 right side body panels are stamped out. It takes about 10 minutes to change dies and a press can stamp out one part every 6 seconds, which is a production rate of 600 per hour. One stand of presses operating 16 hours per day can produce every part required. Actual value added time is 12 hours per day, and 4 hours a day are used for setup and changeovers.

4.9 Variance vs. Variation

Each Lean cell operates on the principles previously defined. In implementing interim cells, quality is likely to be monitored by the operators assigned to a cell. In mature, Lean cells automated inspection devices can be installed in a decoupler. Sensors and monitoring devices are installed on machines to monitor the machining cycle. In Chapter 5, we will discuss mature Lean cell implementation where machines are built with customized quality control monitoring devices. Most quality problems are not created by faulty machines, but by our old enemy… *Variance*. Ever since Motorola Corporation first introduced the concepts of 6-sigma engineering, Manufacturing engineers have become acutely aware of the *corrupting influence of variance*. There are two concepts that are almost universally misunderstood: One is *variation* and the other is *variance*. Variation is necessary to sell finished products. Variation corresponds to the number of different ways that a product can be configured (Color, add-ons, appearance, etc.). Variance is created both within production of a particular product, and between production of different product types. For example, Henry Ford was adamant about reducing variance by eliminating variation. He said: *You can have any colored Model-T you want, as long as it is black*. However, he confused variation with variance, because there were always differences in appearance and performance among his black Model-T's. The same phenomenon occurs today. Even within part types, two automobiles that look alike and were built the same way can be vastly different. Some even fail to perform as designed. In the extreme, we call these automobiles *lemons*. Variance is the culprit, not variation... although variation in itself can contribute variance components.

4.10 Eliminating Variance: *A Case Study*

Consider once again the U-shaped Lean manufacturing cell shown in Figure 4.14. The part that finishes machining at HM1 is checked before it goes to the drill press. The hole produced at the drill press is checked for size and location before the part is released to either Lathe1 or Lathe2. However, there is a potential source of variance here. Two machines are being used to produce the same, identical part. However, no two machines can produce exactly the same geometries. This is a problem that must be constantly monitored by cell workers. A possible solution might be sophisticated inspection and measurement devices in the decoupler or on the machines, but a Lean Engineer would try to find and eliminate the root cause of the problem. Perhaps the best solution is to divide the lathe operation into a serial turn and bore sequence. This is shown in Figure 4.15. However, by the two lathes in series, the new cycle time is now 140 seconds, which is 30 seconds over the required time. However, here is where U-shaped, Lean cell design really pays off. Each multifunctional machine must be able to carry out the sequence of operations in less than 110 seconds. This might require new machine parameters, new cutting tools, work holders and/or faster operation cycles. This type of cell redesign represents a migration from an *interim Lean cell* to a final, *mature Lean cell*. The separation of lathe tasks should improve quality by reducing variation in the turn-bore steps. Note that this will add one more process step to the current assigned loop which will increase the manual time to 95 seconds and the walk time to 35 seconds for the total cycle time of 130. Further Lean Engineering work will be necessary to reduce the operator times at machines 2, 4, 5 and 6. Walking times might be reduced by putting the machines closer together. Now the lean engineer walking with the operator was able to reduce the manual time to 75 seconds and the cycle time to 110 seconds.

The lean engineer made other design changes to the cell. He redesigned the cell so the worker travels CCW. He added walk-away switches to the machines. And he changed the sequence of operations, moving the drill press to precede the lathes and the second Mill to follow the lathe operations.

4.10.1 Reducing Variance by Optimizing Machining Parameters

One way to reduce variance is often overlooked by manufacturing engineers who fail to understand how critical machining parameters can be used to both improve quality and to reduce variance. Notice in this example that some of the machining cycle times are significantly less than the required cycle time. As long as the machining cycle time does not exceed the cell cycle time set by Takt, machining speeds and feeds can be changed. The effects of such an adjustment can be dramatic. Lower speeds and feeds can increase tool life, reduce wear and tear on the machine, and improve surface finish.

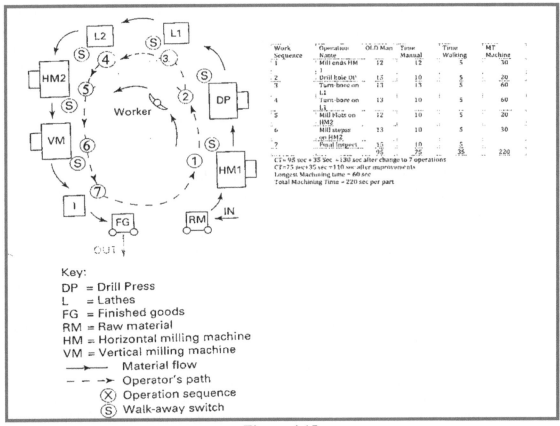

Work Sequence	Operation Name	OLD Man	Time Manual	Time Walking	MT Machine
1	Mill ends HM 1	12	12	5	30
2	Drill hole DP	15	10	5	20
3	Turn-bore on L1	13	13	5	60
4	Turn-bore on L1	13	10	5	60
5	Mill Flats on HM2	12	10	5	20
6	Mill stepss on HM2	13	10	5	30
7	Final Inspect	15	10	5	
		95	75	35	220

CT= 95 sec + 35 Sec = 130 sec after change to 7 operations
CT=75 sec+35 sec =110 sec after improvements
Longest Machining time = 60 sec
Total Machining Time = 220 sec per part

Key:
DP = Drill Press
L = Lathes
FG = Finished goods
RM = Raw material
HM = Horizontal milling machine
VM = Vertical milling machine
——→ Material flow
– – –→ Operator's path
Ⓧ Operation sequence
Ⓢ Walk-away switch

Figure 4.15
Cell Redesigned to Eliminate Duplicate Machines to Reduce Variability

More importantly, there will be less variation between parts. For example, the following relationship exists between cutting speeds and tool life.

$$C = V*T^n$$

Where: V = Cutting Speed (Ft/Min)
T = Tool Life (Min)
n and C are constants that change with tool and material

Typical values for n are 0.14 to 0.18 for high speed steel tools, so a modest reduction in speed results in a large increase in tool life. Reductions in cutting speed will also greatly reduce tool wear, which will improve surface finish. Dull and worn tools cause cutting forces to increase, which can cause tool chatter and tool vibration. Tool wear can lead to failure and a ruined part. It is obvious that tool wear can lead to problems in meeting close specifications, and hence lead to poor quality and slightly different finished parts. Finally, in Lean cells, if a machine can run continuously without changing tools over either a one or two shift operation, quality will be higher due to the consistency between machine cycles.

4.11 Flexibility and Cell Design

Product versatility demands cell flexibility. It should be obvious that in Mixed Model production systems, any Lean cell must be flexible enough to react to changes in both product mix and volume or demand. Lean manufacturing systems should not be confused with Flexible Manufacturing Systems (FMS). U-shaped Lean cells are operationally quite different. In reality, Lean cells are designed for flexibility, but FMS manufacturing systems actually have limited flexibility. Manufacturing flexibility is characterized by four main capabilities.

(1)   The ability to rapidly respond to product design changes
(2)   The ability to rapidly respond to changes in demand and/or product mix
(3)   The ability to rapidly reconfigure a production line (New products).
(4)   The ability to accept technological innovation

First, any manufacturing company must constantly design and redesign the production system to respond to customer demands or gain a competitive advantage. Envision the style, appearance and functional changes in a basic pickup truck such as the F150 Ford. Engineering design changes are simply a fact of life. There are three main design activities: (1) New machine and part design (2) *Concurrent design* (CD) to bring new parts to market and (3) Design for manufacturing (DFM), where existing end products remain basically the same but intermediate production activities are simplified or even eliminated by such methodologies as *net shape* manufacturing.

Second, to be flexible the process/machine must be able to rapidly change to respond to variation in either product demand or product mix or both. There are two different kinds of flexibility, and in traditional mass manufacturing both can create significant problems, often requiring completely different manufacturing configurations and adding or subtracting machines to create capacity to respond. In traditional Mass production lines, these lines sometimes need to be completely shut down and reconfigured. In Lean manufacturing, response to either a change in product mix or a change in demand are both designed and built into U-shaped cells.

Third, although changes in both demand and product mix can usually be accommodated without changing anything but worker machine assignments, the number of workers in a cell, or parallel additions to bottleneck machines. Significant changes might require that the cell to be redesigned and machines be replaced, rearranged or significantly modified. In Lean manufacturing cells, these changes are certainly not minor but because each cell and every machine in a cell is designed to operate independently modification is significantly easier, cheaper and faster than traditional mass manufacturing lines.

Fourth, flexibility is also defined as the ability to accept machine upgrades or new technological innovations. A properly designed U-shaped Lean cell or Mixed Model final assembly line has been specifically designed to easily accept new methods and improved processing workstations.

4.12 Clusters of Lean Manufacturing Cells

Most of the examples so far have focused on individual cell design and operation. However, it is common practice to design a Lean manufacturing system that consists of several linked cells and two or more cells sharing workers. Figure 4.16 shows five different cells (A, B, C, D, E & F) all which are building up components for another subassembly or manufacturing cell or the final assembly line. Cell E has no output WIP area but is directly linked to Cell F. Cell F is feeding the final assembly line. Each cell is being paced by the Final Assembly Schedule (FAS), and the entire area is under Kanban control.

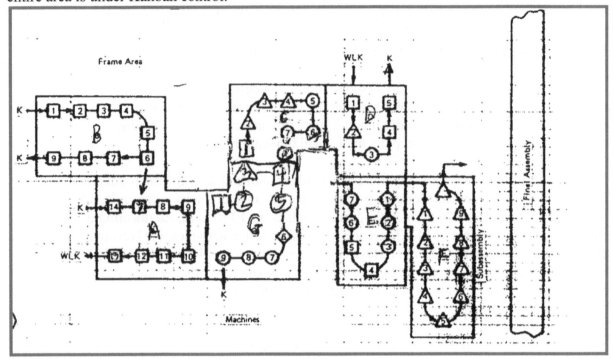

Figure 4.16 a)
Seven Subassembly Cells with Cell F Feeding Final Assembly

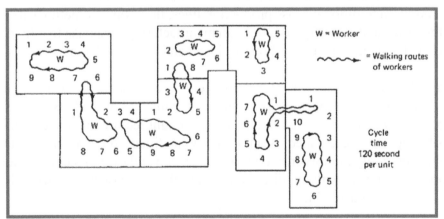

Figure 4.16 b)
Allocation of 8 Workers to the Cells for a 120 Second Cycle Time

Each of the seven cells is linked to one another and to final assembly using Kanban control. In this linked-cell system, work-in-process between cells is controlled by a two-card Kanban system. One card authorizes cell production, and another card authorizes part delivery. The inventory within each cell has a WIP cap on it, with parts residing only on each machine or in a decoupler. In Lean linked-cell systems, WIP within a cell is often called *stock-on-hand,* but any part within a cell boundary is really work-in-process. Parts or inventory held between cells are also actually WIP, but these are usually called *Make-to-Stock* WIP. Final parts storage is called *Make-to-Order*. Although different terminology might be used, from a systems viewpoint, anything which has value and is being worked upon or waiting should be called WIP. Any other terminology camouflages a main objective of Lean thinking, which is to eliminate all non-value added elements from cradle to grave. In this light, anything waiting in a manufacturing system is consuming time and cost without adding value. Of course, we once again point out that some amount of WIP must be designed into the system to promote balancing, leveling and to avoid bottleneck starvation. Decouplers hold stock-on-hand, but a sophisticated decoupler can perform many necessary functions. In some cases, a Kanban link might be completely eliminated if production in both cells is over the same period of time in the same product sequence and the cells are tightly coupled requiring only a conveyor or common linkage.

It is worth noting once again that changes in either product mix or demand will likely alter the cell cycle time, number of workers required, capacity increases, machines required or all of these things. The ability for any one cell to respond to these changes is what we have called *cell flexibility*. In figure 4.16 we see how the allocation of operators can be changed to accommodate a change in cycle time for this cluster of assembly cells. In manufacturing cells, changing a machining time will usually have no effect on the cells production rate, because *the cell cycle time is determined by the time it takes a worker to perform all manual tasks at each machine and walk around his/her loop.* This is perhaps the most important key to understanding cell design and operation. The basic cell design decouples the production rate of the cell from the processing time required by each machine tool (provided capacity exists).

If the same family of parts was made in a traditional Computer-numerical-controlled (CNC) machining center, the cycle times to produce different parts would directly affect product cycle times. It is also likely that the cycle time in a CNC machining center would be longer because machining operations are performed serially by one spindle with possibly multiple tools, other than serial cell production on single cycle spindles.

4.13 Capacity Analysis and Lean Engineering

It should be obvious that when a Takt time requirement is given to any individual cell, the cell may not have enough capacity to attain the required output rate. This will occur in a Lean manufacturing cell when one or more machines *bottleneck* the cell. Machining time is greater than cycle time required. In this case, we offer several suggestions to increase capacity. Some will be more cost effective than others: Some may simply be a temporary fix. The situation and cost will dictate the remedy.

(1)   The process/machine is duplicated to effectively double the output rate. Note that as we discussed earlier, this will introduce variance into the system.

(2)   The process is engineered to reduce machining time(s). We have previously pointed out that depth of cut, speeds, feeds and type of tool material are all important factors to consider.

(3)   Schedule overtime production for the cell. This is only a temporary solution, but one that is commonly used. Working overtime will necessarily increase inter-cell WIP, throughput time and cost per unit.

(4)   Sometimes it is possible to separate the bottleneck processing time into two or more stages on sequential machines. Sometimes it may be possible to shift

operations to another machine with enough idle time per cycle, but this is not always technologically or operationally feasible.

(5)   The product is redesigned by using *design for manufacturing* methodologies.
(6)   Redesign the entire cell (Call for a Kaizen event).

Finally, if demand or product mix cannot be accommodated in any existing cell, build a new cell to produce the parts that are causing a bottleneck. This is a long term solution that usually requires designing and building new machines, so this approach can be costly; but it may be the only feasible solution. If a cell is cloned, then the capacity immediately doubles. In fact, in most cases capacity is more than doubled. To illustrate this observation, suppose that a U-shaped Lean cell is making two parts, A and B. Further assume that business is good, and that demand is twice what the cell is currently producing. Further assume that the cell has reached its output capacity. The first inclination the engineer might have is to clone the existing cell. Enter the Lean Engineer, who proposes that one cell be dedicated to producing Part A and one to produce Part B. The traditional Manufacturing manager will not see the difference. However, by dedicating each cell to a single product all setups and changeovers are eliminated; labor is specialized to one part; machines can possibly use dedicated jigs and fixtures; and result in a number of different time/cost saving advantages. Perhaps more important, dedicated cells will reduce variance in the system. Finally, if demand continues to increase or the mix changes, then both of the cells could return to Mixed Model production.

4.14   Plant Design and Lean Cell Design

Operationally, we have discussed how Lean cells can usually respond to change by using several methods. However, as interim cells are replaced by true Lean cells an opportunity to design Lean into each cell increases. Traditionally, the Industrial or Manufacturing engineer designed a new *Greenfield* facility using Facilities Layout or Plant Layout tools such as Corelap, Aldep, Craft and other techniques. These historical tools do not produce a new Lean production design. Today the modern term is *Facilities Design*, but the general landscape of computerized plant design has not changed very much. Once a plant was laid out, the job was considered finished. Management and production control who doggedly held to this philosophy created *ivory towers* and *legacy systems* which were very inflexible.

The Lean Engineer is fully aware that any manufacturing system must be continuously modified and changed. This is what continuous improvement is all about. The Lean Engineer will design flexibility and the ability to change directly into the facility design. This is the main advantage in using Lean manufacturing cells. Processes and manufacturing production designed from the top down to promote flexibility will improve life cycle costs, quality accommodate product variability and reduce system WIP. The Lean engineer and appropriately educated product designers will know and understand the processing/machining requirements of all parts and the way the manufacturing system operates (Cellular design). This is truly *Design for Manufacturing*.

4.15   How Cells are Linked and Regulated

All U-shaped Lean cells are linked and regulated by Kanban linkage. Recall that the word Kanban means a *signal* or a *control*. Once parts enter the cell, they flow through the cell one at a time. This one piece movement with (ideally) unit decouplers eliminates queues, maintains the order sequence, saves floor space and promotes zero defects. We again cannot overstate the importance of Rapid Exchange of Tooling and Dies (RETAD) which reduces the time for setup and changeovers (SMED). RETAD strategies are presented and discussed in Chapter 17. Each cell has an *input station* and an *output station*. These locations maintain temporary storage between cells, or buffer inventory between component and subassembly cells, or buffer storage between subassembly cells and final assembly. Parts move between these storage points and point of usage on the command of Kanban linkages. There are four factors to consider: (1) The number of carriers in the system, (2) the number of parts per carrier, (3) the number of carriers/parts that can be in any one input/output station at any one time and (4) the signal used to produce and/or move carriers/parts. Carriers between system production components should be standardized to carry the same number of parts. The size of each input/output station is determined by the operational requirements of the downstream cell and how frequently parts arrive to the output station; which incidentally is the same rate that parts enter the system since at steady state production cell input must equal cell output.

The number of parts in each Kanban link is the number of carriers times the number of parts per carrier. The inventory in each buffer has a maximum allowable value. Note that the cell itself does not schedule production or control the level of output WIP. Cell input occurs when a downstream cell or production line needs to receive parts. Cell output occurs when full carts are requested by downstream demand. A simple way to do this is for downstream production to send an empty cart upstream using a withdrawal Kanban card. This request authorizes the upstream cell to deliver a full cart downstream. When the full cart is removed from the upstream cell, a production ordering card is removed from the cart and placed in the production ordering box. This card authorizes production. An upstream operation will not *produce* unless authorized, and product will not be *shipped* downstream unless authorized with a move card. The normally complex, production schedules produced by MRP systems in traditional manufacturing systems are either dictated by a routing sheet or by priority scheduling. Orders are *pushed* into the system and flow from one area to another as they complete process steps; usually in large lots. The critical functions of lot sizing, sequencing and scheduling are integrated and infused into the Lean system and executed based upon current requirements and not a routing sheet pushed into the system. Lean parts are *pulled* through the system in small lot sizes.

4.16 Standard Operations Routine Sheet (SORS)

Standard operations are critical to maintain consistency of production and to make sure that every operation follows a standard procedure. Standard procedures help to reduce variance in any required manual operation. There are three key elements in a standard operations routine sheet.

1.   Cycle time and production rates are all determined by Takt time set at final assembly ($CT \cong TT - Allowance$)
2.   There is a standard sequence of operations needed to complete work in a cell or on an assembly line.
3.   There is a standard quantity of stock maintained in the cell

   In addition,
   - Each manual task is standardized by using approved methods and standards, and standard operating procedures and

- All material handling carriers are standardized

A typical example of a blank standard operation routine sheet is shown in Figure 4.17. The SORS provides a starting point for the manufacturing instructions for each part produced. It can be expanded to include the machining instructions, settings of machining parameters and the expected processing times. All machining times, walking times and manual tasks performed by the worker are usually set using Methods and Standards by the Industrial Engineering group. These standards form the basis for machining and assembly times in each Lean cell. These standards are used not only to standardize work in each cell, but they are also used for product costing. Figure 4.18 shows a completed SORS for a group of 8 single-cycle automatic machines and 3 manual assembly operations in a Lean manufacturing cell.

Date: Part No.: Part Name:		Manual ——— Machine - - - - - Walking
Cycle Time:	Daily Neccesary Quantity	

Work Seq	Name of Operation	Time: Man	WALK	MACH	Operation Time ("": Sec ; ': Min) 4	8	12	16	20	24	28	32	36	40	44	48	52	56	60

Figure 4.17
Example of a Typical Standard Operations Routine Sheet

The cell cycle time of 2 minutes, has been determined by downstream requirements. This SORS shows the tasks or operations of the worker and the sequence of operations. Note that all machining times are less than 2 minutes. As shown, the processing times are not identical at each machine. The planned cycle time consists of 96 seconds worker machine tasks and 24 seconds of walking time. The total machining time is over 10 minutes, but the longest machining time is only 1 minute and 35 seconds. In a traditional machining center (a CNC machining center) might be designed to build to this component. However, the cycle time per part would jump from 2 minutes to over 10 minutes. This demonstrates the superiority of Lean manufacturing cells to conventional manufacturing. By adding or subtracting workers and redefining loops, the cycle time can be changed. However, it is immediately obvious that the machining time of 1 minute 35 seconds on Machine3 will limit the maximum cell output rate as currently configured.

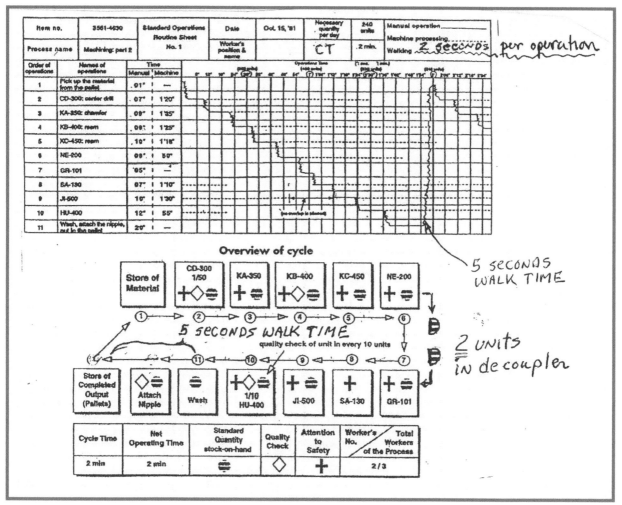

Figure 4.18
A Typical Completed SORS from a Toyota Supplier

4.17 Lean Manufacturing Cells with Manual Machining and Manual Assembly Operations

A U-shaped Lean cell can be entirely composed of automated machining operations, or contain a mixture of automated machining tasks and *manual* tasks such as assembly or manually operated machining operations. The latter is often found in interim cells. This will not change the basic operation of a cell, but might drastically affect cell cycle time and worker activities. Figure 4.19 shows a Lean cell with a combination of manual and automated operations.

Let MT_j designate an automated machining cycle at machine j, and HT_n designate a manual operation at machine k. Note that any one process step might have both automated and manual operations. The cell has 8 different workstations; 5 are automated machines and 3 are manual. The calculation of cell cycle time proceeds exactly as before, but all manual times in a worker's loop are added to the worker task times and walking times. Using only one worker and no decouplers, the cell cycle time is calculated as follows.

Work sequence	Name of operation	Time			Operations time (seconds)																	
		Walk	Manual	machine	6	12	18	24	30	36	42	48	54	60	66	72	78	84	90	96	102	108	114
1																							
2																							
3																							
4																							
5																							
6																							
7																							
8																							
				0	6	12	18	24	30	36	42	48	54	60	66	72	78	84	90	96	102	108	114

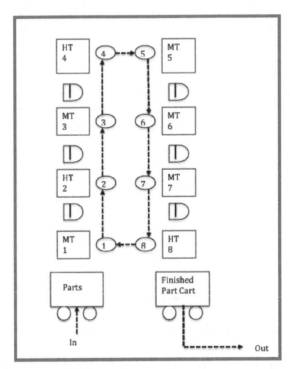

Operation	Unload/Load time (sec)	Machine or Process time (sec)	Manual time (sec)	Walk time (sec)
①	3	42	—	3
②	-	—	17	3
③	4	53	—	3
④	-	—	4	3
⑤	4	18	—	3
⑥	6	50	—	3
⑦	9	47	—	3
⑧	-	—	5	3
Total	26		26	24

Figure 4.19
Simple Standard Operations Routine Sheet

$$\text{Cell}_{Cycle\ Time} = (24\ \text{sec. walk time}) + (26\ \text{sec. manual operations}) + (26\ \text{sec. task times})$$

$$\text{Cell}_{Cycle\ Time} = 76\ \text{Seconds/Part}$$

The cell cycle time is greater than any processing time (MT).
The cell throughput time per part is given by:

$$\text{Cell}_{Throughput\ Time} = (76\ \text{sec})*(6) = 456\ \text{Seconds}$$

Notice here that the TPT is calculated based on 5 SCA machines, not 8 operations as the operator carries the work through operations 2, 4 and 8 while operations 1, 3, 5, 6 and 7 require that the part be left in the machine for one loop. Thus 6 loops are needed to get a part through the cell from input to output.

4.18 Design for Customer Satisfaction

Product and system designs are done using two fundamental goals: (1) Design for manufacturing and (2) Design for customer satisfaction. Note that both design goals are focused to the customer. Design for manufacturing is aimed at the *internal customer* and design for customer satisfaction is aimed at the *external customer*. The first design goal is to *minimize non-value added activities* and the second to *maximize value- added activities*. The first design activity is directed to lower production costs, while the second activity is directed to maximizing profits (sales). In a real sense these two design activities are tightly coupled, since the external customer demands the highest quality product at the lowest cost. A major problem that has developed over the last 20 years is the separation of the design engineer from the manufacturing engineer. The use of CAD/CAM, virtual reality design tools and other computerized techniques has created what we will call *Functional Foxholes* and *Islands of Automation.* This has created tremendous problems for floor manufacturing. It is an all too common practice for the design engineer to *lob* a new product design over the fence and simply say *make it.* This is a sure recipe for manufacturing cost escalation and customer dissatisfaction. A good example is as simple as replacing an automobile starter. This should be simple, but American cars or trucks sometimes require the removal of the fan belt, the cooling fan and a morass of other things just to get to the starter, so the task might take hours. At least one clever European car manufacturer has moved the starter to the top of the engine block. Replacement can be done in 15 minutes or less.

One activity that must be done is to execute *benchmarking* of the market competition. Although not covered in this text, the use of market surveys and the *House of Quality* is emphatically recommended. It would be a good exercise for the student to go to the internet and write a short essay on each topic.

It is a serious mistake to ignore the customer and the world around us in the design function. The world that surrounds us is more and more beginning to influence and even dictate the way manufacturing will be done in the next 20 years. Product design is now being influenced by more than just placing machines and moving product through the plant. Environmental engineering, Safety engineering and Green engineering are now playing a major role in the product design function, and any major change in traditional plant layout factors will in turn affect the way parts and products are manufactured. Another major factor which will begin to influence how products are being designed and manufactured is how products will be retired and raw materials are disposed. For example, one day when an automobile is ready for the junk yard, there will be no junk yard. Every piece of the automobile will be recovered, reused or recycled. The next factory revolution will be the mandate to consider all of these areas when building commercial products. The flexibility of Lean systems will be ideally suited to adapt and retrofit to meet new social and legislative constraints. In the Factory of the Future, or more appropriately in a *Factory with a Future* (Black), there will be fewer workers than machines or processes. Hence, the Lean production system will provide each worker with a natural environment for job enrichment. Greater job enrichment will result in a satisfied worker who will produce a higher quality product.

4.19 The End Game: Benefits of Converting to Lean Manufacturing and Interim Cells

The process of applying Lean principles, eliminating non-value added activities, and simplifying production avoids and mitigates many unknown risks and makes automation easier. Conversion to a linked-cell lean manufacturing system can result in huge savings over a 3-5 year period of time. Many companies who have begun and completed this journey report significant reductions in WIP, setup costs, cycle time and throughput time along with higher quality. Lean thinking and Lean reorganization also has other benefits. It simplifies management functions, eliminates many traditional production planning and control mechanisms, and paves the way for automation. Conversion and progression from job shops, flow shops and mass production systems to manned linked cell, Lean systems must be accomplished in logical, economically justified steps…. each stage of conversion building upon the previous stage. Even in early conversion activities, *interim cells* offer many advantages, including:

- Significant WIP reduction
- Greater worker productivity and improved moral
- Balanced production rates
- A reduction in material handling costs and movements
- No expediting or changing priorities
- Setup and changeovers simplified and less time consuming
- Better quality and less rework/scrap
- Smoother, faster flow of products (shorter TPT)
- Less bottleneck problems
- Reduction of variability
- Greater production flexibility
- Cheaper automation costs

4.20 Interim Cells to Mature Lean Cells

We have not discussed the advantages (and sometimes necessity) for a Lean supply chain, but a major part of migrating from *interim Lean cells* to permanent Lean cells is leaning out all subcontractors and activities in the supply chain. Lean production facilities usually change over to a supply chain with sole source suppliers. These vendors are also encouraged and sometimes forced to convert their manufacturing facilities to Lean production. Here is where mature vendors become unique, developing specialized equipment and tools to support *Just-in-Time* delivery at minimum cost. Because the main production lines cannot tolerate defects or rejects, sole source vendors are forced to supply defect free products to the end user. A main requirement for migrating from interim Lean cells to mature Lean cells is to fully integrate supply chain vendors into the Lean factory, quickly followed by Lean marketing, warehousing and distribution. At this point, the *Lean Enterprise* begins to emerge. See Chapter 18. There are many other steps needed to upgrade interim cells to mature Lean cells. Figure 4.20 shows how a particular interim cell was upgraded into a more functional Lean cell.

The interim cell with two operators is changed into a cell with only one operator. The new cell has a much smaller footprint than the old cell, and has been equipped with more automation. Certainly not comprehensive, but the following tasks are almost always completed to accomplish conversion; not necessarily in this order.

- Complete the full implementation of U-shaped Lean cells with a much smaller footprint. Put machines as close together as possible.
- Convert every machine in the cell to single-cycle automatics to support automatic machining cycles and one-piece flow.
- Implement Total Quality Assurance and Total Preventative Maintenance procedures using cell workers

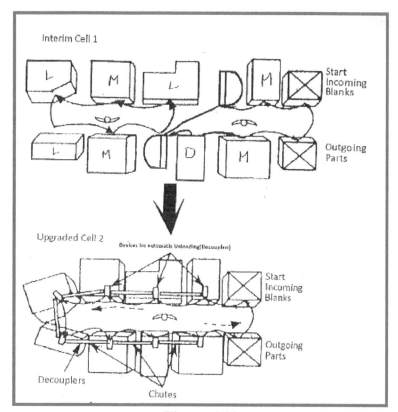

Figure 4.20
Conversion of an Interim Lean Cell to a Mature Lean Cell

- Complete setup time reductions, add Poka-yoke devices and walk away switches.
- Automate machine loading and unloading using chutes and decouplers as much as possible. The worker will do the rest.
- Refine and upgrade decouplers

139

- Add sensors and monitoring devices to constantly check for tool wear, vibration, misalignments, etc.
- Design flexibility into each cell to support Mixed Model production.

4.21 Poka yoke Devices

Poka-yoke (ポカヨケ) is a Japanese term that means *mistake-proofing*. A Poka-yoke is any mechanism in a lean manufacturing process that helps an equipment operator avoid (*yokeru*) making a mistake (*poka*). Its purpose is to eliminate product defects by preventing, correcting, or drawing attention to human errors as they occur (Wikipedia) .The concept was pioneered and the term first used by Shigeo Shingo as part of the Toyota Production System. It was originally described as *Baka-yoke*, but as this means i*diot-proofing*, the name was later changed to *Poka-yoke*.

Shigeo Shingo recognized three types of Poka yokes for detecting and preventing errors in a mass production system (Wikipedia).

1. The *contact* method identifies product defects by testing the product's shape, size, color, or other physical attributes.
2. The *fixed-value* (or *constant number*) method alerts the operator if a certain number of movements are not made.
3. The *motion-step* (or *sequence*) method determines whether the prescribed steps of the process have been correctly followed.

    Either the operator is alerted when a mistake is about to be made, or the Poka-yoke device actually prevents the mistake from being made. In Shingo's lexicon, the former implementation would be called a *warning Poka-yoke*, while the latter would be referred to as a *control Poka-yoke*. Shingo argued that errors are inevitable in any manufacturing process, but that if appropriate Poka-yokes are designed and implemented then mistakes can either be completely eliminated or caught quickly; hence preventing defects. Poka-yokes proactively eliminate defects at the source, as opposed to using reactive Control Chart or inspection procedures. Figure 4.21 illustrates a Poka-yoke device for a drill press to prevent drilling a hole too deep (Black).

4.22 Ergonomic and Human Factors Considerations in Lean Cell Design

The following Ergonomic and Human Factors features are representative of what might take place in a superior cell design.

- Regardless of whether workers are seated or standing, design a workplace to avoid stress, fatigue and unusual tasks that might be physically harmful.
- Eliminate any loop/worker/machine assignments that cross other workers cells or workstations. Mark walking paths.
- Provide good lighting at all machines to support visual inspection
- Follow good ergonomic design rules that dictate how much can be safely lifted, angles of extension and span or reach
- Safety in the workplace is the most important design factor to consider
- Design walk-away switches to be easy to use and easy to access
- Encourage worker interactions. Periodically hold brainstorming sessions to discuss problems and problem solutions.

- Provide comfortable, ergonomically designed shoes for all workers; cell workers walk a lot on concrete floors
- Task a trained Industrial Engineer to establish standard times and standard operating procedures for all operations.
- Task a trained Manufacturing engineer with developing customized jigs, fixtures and material orientation devices to speed production and save time

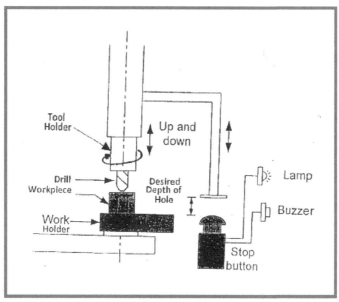

Figure 4.21
An Example of a Poka yoke Device to Control Hole Depth in Drilling

4.23 Summary and Conclusions

This Chapter has addressed a wide range of general topics associated with U-shaped Lean cell design, and provided an overview of cell design considerations and how Lean cells play a major role in Lean manufacturing systems. We have attempted to unify Lean thinking and show why Lean manufacturing is superior to traditional Mass manufacturing systems. The particular focus of this Chapter has been to discuss the design and operation of U-shaped Lean manufacturing cells, but we have also touched upon both the supply chain and subcontractor roles. Lean cells can be configured for both manual assembly operations or automated machining operations or both. Creating worker loops with specific workstation requirements provides a way to provide cell flexibility, and to respond to changes in both product demand and types of products. The maturation of Lean cell design over the past 30 years has destroyed the myth that Lean manufacturing only works in production systems with steady demand and single part production. Another common myth is that Lean systems cannot be designed and built for job shop operations… nothing could be farther from the truth. The same design principles can be used in both single and Mixed Model production systems. Leveling and balancing production are two key issues which largely influence how well a Lean system performs. Both work-in-process and production rates are closely tied to simple Kanban control mechanisms.

To summarize, the following fundamental goals must be met to realize the full benefits of Lean cell-based manufacturing systems.

- All setup and changeover times must be reduced to a minimum
- Zero defects are mandatory along with zero rework
- Takt time and production rates for all operational components must dictate production and be determined by final assembly and intermediate subassembly requirements
- All final assembly lines must be converted to Mixed model production lines
- All work must be balanced and leveled to smooth production
- All component and subassemblies should be converted to U-shaped Lean cells
- Everyone in the company must be trained in Lean thinking, and realize their role in the transformation
- Every cell worker must be cross trained to work any machine in the cell
- Every cell worker must be trained to be a quality inspector and perform simple preventative maintenance tasks
- All subcontractors in the supply chain must be converted to Lean subcontractors. Each must strictly adhere to just-in-time supply strategies
- Material handling carriers must be standardized by area of usage and equipped to carry Kanban control cards

The motivation and reasoning behind the conversion of all component and subassembly Mass production lines into Lean cells are summarized as follows.

- U-shaped Lean cells improve productivity and flexibility
- Workers are more motivated and focused in Lean cells
- Lean cells provide the mechanism to rapidly adjust to changes in product demand and model types
- Cells encourage teamwork and allow operators to assist one another if needed
- Cells promote higher quality products
- Cells are by design self-balancing
- Cells are designed with a WIP cap
- Cells cannot overproduce: Production and output WIP is tightly controlled by Kanban signals
- Lean cells promote one-piece-flow and eliminate large lot and batch production
- Production problems and machine malfunctions are more easily detected
- The cost of tooling can be significantly reduced by lowering key machining parameters such as speed and feed. Lower cost tools can sometimes be used under these conditions.
- Cell cycle times are independent of component machining times
- Lean cells significantly reduce the floor space (footprint) required
- Historical reports from companies who have successfully converted to Lean manufacturing reveal huge benefits.
  - 30% to 40% fewer operators do the same amount of work
  - There is 80% to 90% less work-in-progress
  - There is a 50% to 90% reduction in rejects and rework. This significantly improves as interim cells become mature Lean cells.
  - A 20% to 30% reduction in floor space is required

- The cost per unit of production is reduced from 20% tom 30 %. This enables a company to maintain the same profit margin while investing more money in R&D and technological innovation.

It is the premise of this book that if any company decides to begin the long journey to Lean manufacturing, that a new breed of Manufacturing, Mechanical and Industrial Engineer is needed to lead the transition. We call this new engineer a *Lean Engineer*.

Finally we exit this Chapter with a word of caution. The transition from traditional Mass manufacturing to Lean manufacturing is not instantaneous, nor is it simple. A meaningful transition will usually take from 3-5 years, but the journey never ends. Changing market forces and the need to introduce new or better products demand that Lean transition and Lean thinking never ends. Also be aware that once the transition is started, things will usually get worse before they get better. There will be resistance to change at all levels, but the leadership must stay the course. As cellular production kicks in, Kanban control replaces traditional scheduling and order release; and the work force is retrained… there will be significant gains in cost reduction, responsiveness and versatility.

Review Questions

1. In designing single-piece flow cells, what will be the consequence of all the hidden variations that existed in the job shop design?
2. How will you prepare the workforce to deal with the conversion to the lean shop?
3. What can be done before cell formation that will minimize the variation in cycle time?
4. Who has the best ideas about how to improve the operations in the cells?
5. Who should be tasked with sustaining the maintenance of the standard work instructions?
6. What is standard work?
7. Why is standard work necessary?
8. What is Takt time versus CT for the cells?
9. If a machine's processing time exceeds cycle time in a manufacturing cell, what alternatives might the manufacturing engineer consider to ensure that a cell meets the necessary cycle time?
10. What are the advantages of single piece flow in a manufacturing cell?
11. What are the machine design changes for interim cells (for example a walkaway switch)?
12. What is the key role of the worker in the cell?
13. Assume that the Takt time is 60 seconds for a cell, but a finishing process has a cycle time of 65 seconds, and there is no way known to reduce that time.
    a) What would you do, ideally and why?
    b) If a new finishing process machines costs $5,000, what would you do?
    c) If a new machine costs $50,000, what would you do?
    d) How are the parts pulled through a cell?
    e) What is an example of mistake-proofing devices in common practice?
    f) Why is mistake proofing so important in a cell's operation?
    g) Assume that you are designing a machine cell for a product that has an expected life of one year. What would you automate? What would you not automate?
    h) Why do cells use single-cycle automatic machines?

Problems

1. The manufacturing Cell shown in Figure 4.A has 5 single-cycle automatic machines and manual operations. Assume the walk time between all stations is two seconds.
2. What is the operation routine sheet, cell cycle time for one worker?
3. Develop a standard operation routine sheet for one required cycle time period for one operator.
4. Can the Takt time be met? If not, then why not?
5. Which operation would you have to improve before adding a second worker?
6. Assume certain improvements and provide a workable cell design for two workers.
7. Figure 4.18 Shows the SORS and layout of a lean manufacturing cell.
   a) How is the cycle time of 2 minutes calculated?
   b) What is the TPT assuming the part stays in the decouplers between steps 6 and 7 for 2 cycles?
   c) What is the cycle time for two operators?
   d) What can you do to improve the cell?
8. The cell design showing in Figure 4-B has eight stations with 5 SCAs located at Operation Number 1, 2, 4, 6 and 7
   a)   Determine the cycle time for 1 operator,
   b)   2 operators (as shown)
   c)   3 and 4 operators
   d)   Plot output per hour versus number of workers (on horizontal) and comment on your result.
   e)   What is the TPT for 1 operator versus 2 operators?
9. A simple SORS showing the relationship between the operator and the single-cycle machines tools in the manufacturing cells shown in Figure 4-C. Notice the cell has a transfer line.
   a)   Calculate the cycle time for 1 worker (no decouplers)
   b)   Calculate the throughput time for one operator
   c)   Calculate the cycle time for 2 operators (show the best distribution of the two workers in the cell)
   d)   Calculate the TPT for 2 operators considering two decouplers and the transfer line.
10. Manufacturing cells are used to produce parts one at a time using standing, walking workers. The interim manufacturing cell shown in Figure 4-D has seven machines and can be manned by one or two operators. These operators are multifunctional and multiprocess. Each of the machines is at least single cycle automatics; that is, they can complete machining cycle automatically once it has been initiated by the operator (using a walkaway switch). The machining times for the processes are in seconds shown on the machines beginning with the saw and going CW around the cell to the grinder. The final inspection machine is also automatic and takes about 15 seconds to load in a part, depending on which component from the part family is being inspected. The time for the inspection process is 15 seconds. The machining times given above are for part A from the part family. Machining times (MTs) for some operations are somewhat smaller for parts B, C and D. The operator (or operators) take about 10 seconds at each machine to perform various manual operations like unloading the machined part, checking the part,

perhaps deburring the part, and loading the next part to be machined. The operators spend about 3 seconds walking from machine to machine. The aisle between the machines is about 4 feet wide.

Answer the following questions based on the cell design above and the information provided.

a) The grinder is a single-cycle automatic. How long does it take to complete its machining cycle? Why is it important?

b) For the lathe, the MT could be calculated for this equation $MT = L + All / (Ns + fr)$. Which term depends on cutting speed?

c) For the cell with one operator, what is the cycle time in minutes?

d) For the cell with one operator, what is the production rate?

e) For the cell with one operator, what is the TPT in minutes?

f) For the cell with two operators, what is the CT?

g) For the cell with two operators, what is the TPT?

h) Suppose the time on the grinder needs to be increased to improve surface finish. The new MT is 1 minute. What is the impact of this change on the TPT for the cell?

i) What is the impact of increasing the MT of the grinder to 1 minute on the TPT of the cell B – two operators?

j) What design change (s) would you make to the cell to increase the TPT for the cell?

Machine Tool	Operation or Process	Machining Time	Operator Manual time	Walk Time
Saw	Cut bar to length	20	6	3
Lathe 1	Rough turn	25	20	3
Lathe 2	Finish turn	25	10	3
Horizontal Mill	Mill Step	30	10	3
Vertical Mill 1	Mill slot	30	15	3
Vertical Mill 2	Drill and tap holes	20	10	3
Surface Grinder	Grind slot	45	10	3
Final inspect	Final Inspection	15	15	3

Table for Figure 4-D

11.	Complete the SORS for Figure 4.19

Thoughts
and Things.......

Chapter 5
Subassembly Cells

"The arrival of an empty container at a subassembly cell is the signal to make more"
Taiichi Ohno
"I ain't supposed to do nothing unless somebody tells me to"
General Motors line worker

5.1 Introduction

The Toyota Motor Company has risen to a position of world prominence in the automobile industry by redesigning the traditional Mass manufacturing system into a Linked-Cell Manufacturing System (L-CMS), or what is now known worldwide as *Lean Manufacturing*. In redesigning the mass production system final assembly is converted into a mixed model production line, and many subassembly operations into U-shaped lean production cells unless they are in sequence to or synchronized with final assembly. Final assembly is set up to produce to Takt time, and requires line balancing for any mix or volume of parts. Subassembly cells are required to have a cycle time slightly less than the Takt required for downstream operations. All machines in a lean cell are redesigned to operate on a basic Make-one, Check-one and Move-one-on (Mo-Co-Moo) basis. One or more workers are assigned to each cell and machining operations are assigned to each worker in such a way to satisfy Takt time requirements. This is called a *loop*. Decouplers are placed between each workstation and at loop completion points (*baton passing*) to make each machining operation and each worker loop independent of any other. The term decoupler is derived from the axiomatic-design rules developed by Nam Suh.

Concurrent engineering (CE) and design for manufacturing (DFM) methods have traditionally focused upon the relationship between product design and process design, but often have ignored system design (SD). Hence, a true integrated manufacturing system design has been an important but neglected aspect of CE and DFM. If a company would first focus on system design using Lean manufacturing system principles, the gap between CE, DFM and SD would be narrowed if not closed. Both need to fully integrate the fundamental role of humans using sound human factors and ergonomic designs. The result would be to position any manufacturing company to more successfully compete in a world market by adding flexibility, saving production costs and reducing time to market.

A manufacturing enterprise (ME) is a complex organization of supply chains, manufacturing lines, and final product delivery. However, the *manufacturing system* is the beating heart of any manufacturing enterprise. The manufacturing system has been defined as *a complex arrangement of physical elements characterized by measurable parameters* (Black, 1988). The physical elements are machines, people, material handling systems and information systems. Ohno asserted that the most important element in the system is skilled laborers (people). Honda refers to them as the internal customer, why? The different processing centers required to make a final product are linked to both upstream (input) and downstream customers (output) by simple Kanban, visual control mechanisms. Each processing center is therefore a customer (*internal customer*) and a supplier (*internal supplier*). At the same time, a final assembly line is producing products to satisfy the demands of *external customers*. In terms of a harmonious and successful system design, this is a key concept. *A manufacturing system must be designed to satisfy the needs of both internal and external customers.* The Lean system is designed for two purposes: (1) Eliminate all forms of waste, and (2) transform an extremely complex system into one that is simpler to operate and understand, while simultaneously satisfying the needs of these two customers.

Automated conveyor lines are considered by most modern Manufacturing engineers to be the key element in implementing a mass production system. Many leading manufacturers are now revisiting this traditional approach while implementing a Lean manufacturing strategy. Figure 5.1 shows an example of a typical linear conveyor line used to pace a subassembly line with 9 workstations, each having one operator. This line will require balancing.

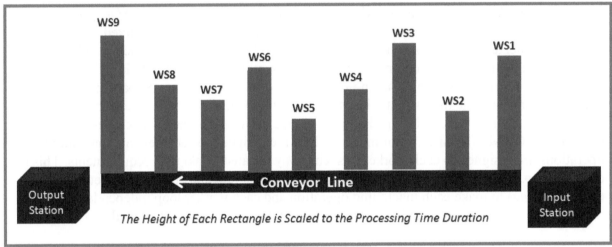

Figure 5.1
Mass Production Line with Conveyor Paced Material Handling

This conveyor line presents a set of problems which can typically be summarized by the following categories of waste (*Muda*).

- Underutilization of workforce due to the fact that line cycle time is bounded by the workstation with the longest processing time (WS9).
- Waste of time in reaching for the work piece on the conveyor and returning it to conveyor after task completion.
- Waste of inventory due to the holding of material between successive stations in buffers. In manned conveyor lines, keeping a buffer of parts at workstations is indispensable because although all the line workers try to follow the pace imposed by the automated conveyor, it is not possible to keep perfect balance in all workstations on the line.
- Waste of resource capacity during product model changeover. The longer a task time becomes when dividing the entire line into multiple workstations, the more time-consuming it is to deal with product model changeover readjusting and realigning tasks at each workstation to balance the entire line can require many technological and methodological changes, and can take a long time. The Mass production, conveyor paced design generally does not allow for handling multiple stations; so there is also waiting

148

time for workers operating partially-automated short-cycle processes. Manufacturing with conveyor lines have other types of latent problems such as:

- o Reliance on huge and expensive investments in facilities
- o Reliance on indirect and support personnel who generate to added value
- o Not fully utilizing the workers' intellectual capacity, and
- o No flexibility of the production facilities.

This last problem, the lack of *flexibility*, is really the main disadvantage of conveyorized Mass production lines.

- Product model design changes; introduction of new products; and changeover of work holding devices like jigs and fixtures are costly and time consuming.
- Reconfiguration of the design to respond to changes in demand is extremely difficult, requiring rebalancing of the entire line.
- Technological innovation which will result in shorter cycle times at each workstation will require line rebalance

The very nature of the line is subdivision into balanced workstations, using a sequenced, linear design. To follow these design principles, the conveyor line can bring other inherent disadvantages:

- The line only runs where all stations (and operations) are available and ready to operate
- The line can be ergonomically unsound with adoption of predominantly one-sided physical motions.
- Considering the conveyor line system from another point of view, larger lines are very problematic when production volumes decline. Given the huge investments needed for the installation of a conveyor line, management wants high utilization rates, and quite often these result in large work-in-process inventories and perhaps large final product inventories.

5.2 Converting Conveyor Lines to Lean Cells

An example of converting a conveyor fed serial production line into a U-shaped, Lean production cell (from Sekine) is shown in Figure 5.2. The original Mass production line used 10 workers. After conversion and redesign, the same product (a shaft and pinion gear) was made using only 4 workers (Cell design 1) and two workers (Cell design 2).The first cell design uses a cell staffing methodology called *rabbit chase*, which will be discussed later in this Chapter. The second cell design uses only two standing, walking workers who man sub cells (loops). The Lean cell with only two workers was made possible by modifying and upgrading the four presses, and the screw / name plating process. An *inspection station* was added that is shared between the two workers. Note that the *shaft BB* station was a short cycle operation that was also shared by both workers. The details of this conversion were made proprietary by the company, but the results of Lean conversion are impressive. These are examples of self-balancing subassembly (SBS) cells.

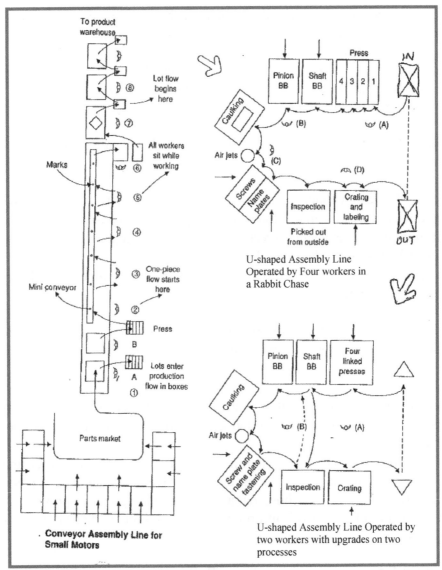

Figure 5.2
An Example of a Conveyorized Production Line That Was Converted to a Lean, U-Shaped Cell

5.3 The Toyota Production System (TPS)

The linked-cell manufacturing system (L-CMS) is firmly grounded on some fundamental principles based upon the concepts of producing only to (actual or forecasted) customer demand, eliminating all forms of waste (*Muda*), reducing Work-in Process (WIP) to a minimum, and scheduling production with a *pull* system. The commitment to continuous process improvement by systematic identification and elimination of waste through *Kaizen* events leads to superior quality levels. Many manufacturers in Japan's electronics industries relied on guidance or support provided by consultants who are expert in TPS to introduce cell production in their plants. The physical configuration of U-shaped cells, which are the elementary production units of the L-CMS, was conceived by Taiichi Ohno at Toyota. Beginning in the 1960s, U-shaped cells became the means to build a flexible manufacturing system, because the design and configuration assures a more rational materials flow than the linear conveyor lines, a Mass

150

production system or a Flow line. The job shop remains apart of small, individualized and customized components. Lean system design enables one-piece flow, reduces work-in-process inventory, and employs small work groups or individuals while exploiting workers' multi-functional capability. Lean, U-shaped cells were first adopted in machining and parts manufacturing processes by Ohno. Later under the L-CMS approach, subassembly cells and some component manufacturing have found application in the lines which feed final assembly; especially in the electrical and electronics industries. Figure 5.3 provides a second example of a traditional conveyor fed assembly line that is manufacturing small motors, redesigned into a cell with 2 workers who are using the rabbit chase method.

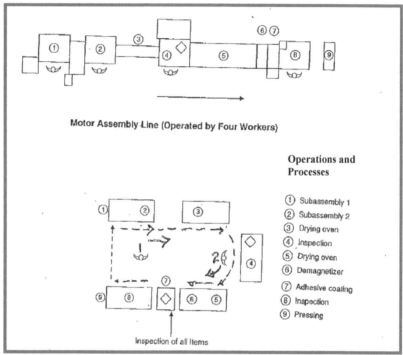

Figure 5.3
Converting a Conveyor Paced Assembly Line into a U-Shaped Cell
(Sekine, 1990 modified by Black)

The output can be changed by changing either the number of workers in a cell, or by reassigning machines to each worker. There is no need or requirement to equalize and balance the processing time cycles at individual stations. The operations in a Lean, U-shaped *assembly* cell are usually all manual so the operator must stay at the station until the task is completed. In a *machining* cell, the machining cycle time is separated from the workers loop time. The worker can start a machining operation and it will automatically run to completion. Of course, cells have a mixture of manual assembly operations and automated machining operations.

Converting from Mass assembly lines to Lean cells generally results in a marked improvement in both the productivity that cells can achieve, and other equally important measures of productivity such as lower cost, lower cycle times, higher output and less throughput time. Contrary to large lot production and long setup/changeover times reflected in traditional manufacturing, long setup/changeover times must be vigorously attacked and reduced so that the processes can be

151

changed over quickly from making one product to making another. This makes the manufacturing system *flexible*, and enables manufacturing cells to make piece parts and subassemblies for Mixed Model final assembly. Again, Lean cells are designed to manufacture specific groups or families of parts.

5.4 Subassembly Cell Design and Operation

Standard work is the basis for maintaining productivity, quality, and safety, and provides a consistent structure for performing the tasks in such a manner as to meet required Takt time, while simultaneously uncovering opportunities for making improvements in work procedures. There are three aspects in structuring standard work in a cell: (1) Takt time, (2) The sequence of required cell operations (3) and (minimizing) cell stock-on-hand.

5.5 Takt Time and Cell Operation

Takt time reflects *the pace* of sales in the marketplace. It is based upon a combination of forecasted plus actual demand. The Takt time required for final assembly dictates the cycle time for each cell. The working *sequence* is the sequence of operations that is the best way to perform the subassembly job. The *stock-on-hand* is the minimum number of parts needed in a cell to maintain a smooth flow of work.

Within a cell, a set of process steps performed by each worker, the set is selected to make each workers loop or sub cycle time equal to the required output Takt time as closely as possible. If the cycle time of each worker is equal to the required Takt time, this creates a perfectly balanced cell. Takt is the German word for a conductor's baton which is used to keep every musician in time. Takt time is calculated by dividing all available production time during any one day by the daily demand (DD).

$$\text{Takt Time} = \frac{\text{Production Time Available/Day}}{\text{Daily Demand}}$$

5.5.1 Takt Time (TT) Example

Suppose that over the next 6 months a certain item has confirmed and forecasted demand of 60,000 units. This is 10,000 units per month on the average. The normal work month is 20 days, two shifts per day with 30 minutes off each 8 hour shift for a meal. This results in a required production rate of 500 units per day, or 33.3 units per hour. The required Takt time 1.8 minutes per part.

Takt time is the *drumbeat* for a Lean cell. Every Lean Cell tries to produce to Takt time to meet the functional requirements of producing the right quantity at the right time. The L-CMS groups the operations or processes into U-shaped lines called *Lean cells*. The processes can be changed over rapidly so products can be turned out in greater variety in an almost customized fashion with no cost penalty for small production runs (small lots). The entire manufacturing system is designed such that the throughput time (TPT) is as short as possible. This requires a continuous redesigning of the manufacturing system (which is just another way to say *continuous improvement*) to shorten the TPT, the total time it takes each part to move through the plant.

5.6 Sequencing Cell Operations

Just as in Mass manufacturing, the sequence of operations is determined by the process routing sheet. The first operation to be performed in a cell proceeds the second, the second the third and so on until the entire sequence of operations is completed to build a component or subassembly. However, the order in which each machine is placed in a cell is different from the order that each cell worker follows as he/she proceeds through the loop of one or more machines in the loop. Each worker has an order in which each machine is visited, but that order does not need to coincide with the build sequence.

5.7 Minimizing Cell WIP: Stock-on-Hand (SOH)

One of the fundamental differences between a U-shaped Lean cell (LC) and a traditional Mass production system (MPS) is the way that cell WIP (Stock-on-Hand) is controlled and managed. In a LC, stock-on-hand can be found in a decoupler, in a machine or with the operator. SOH in decouplers is either between machines or at the point where two workers interface. In either place, there is usually no more than one unit. Hence, the amount of stock-on-hand in each U-shaped cell is tightly controlled and minimized. In a rabbit chase, it can be as low as only the number of workers in a cell.

In the L-CMS, the manufacturing cells and subassembly cells are often designed with cycle times slightly less than the TT of final assembly. This requires *balancing* the entire manufacturing system. Some subassembly lines and cells are *synchronized* with final assembly, so if final assembly stops (for a problem), they stop and then restart with final assembly. This balancing accomplished by a *linked-cell Kanban control system* will be described later. For now we will simply note that the linked-cell manufacturing system (L-CMS) synchronizes all manufacturing and subassembly cells to final assembly by means of Kanban inventory links. The Kanban links serve to carry production control information upstream to supply cells while delivering the inventory downstream to subassemblies and then to final assembly. The Kanban links represent the supply chain for the L-CMS and the management of inventory is almost automatic.

In subassembly cells, when an operator arrives at a machine or station, the operator usually remains at the station or machine and completes assigned processing duties before moving on to the next machine. Cells are typically manned by multifunctional operators trained to be able to perform many tasks and operate every process at every station in the cell. The machines in a cell are placed next to each other by a U-shaped or parallel row design to reduce walking times between machines and achieve one-piece flow within the cell. So, how are cells different from flow lines?

Redesigning a Mass production flow line into a U-shaped Lean cell using multiple workers automatically achieves a balance in task times by letting the workers sequentially visit a group of machines assigned to him/her in a walking cycle called a *loop*. This type of cell design is more flexible than the traditional Mass production line, and it is easier to change the output as needed by changing the number of workers in the cell than the number of machines or processes in the manufacturing sequence. We will see *it is not necessary to balance the cycle times at individual workstations* (no line-balancing) and the design leads to self-balancing by the operator.

From the perspective of facilities design, a linked-cell manufacturing system is fundamentally composed of multiple U-shaped cells. A cell is usually a connected set of machines, processes, or stations in which a single worker or multiple workers perform all the tasks needed to make a component or subassembly. The shape and size of designs vary greatly. Cell designs can be laid out in U-shapes, L-shaped lines, parallel lines or even more twisted lines. However, the U-shaped design is by far the most frequently adopted design for the L-CMS.

5.8 Operating Lean Cells

There are five basic methods for operating Lean cells. They are (1) *single worker*, (2) rabbit chase, (3) *Sub cells*, (4) *Bucket Brigade,* and the (5) *Toyota Sewing System*. None of these methods require that the individual workstation cycle times be balanced. We will describe these five methods within the context of a subassembly cell. In this case, the cycle time for each worker is composed of tasks at each machine or stations, walking times between in their loop, and the times to perform tasks at machine in the cell.

The superiority of a U-shaped Lean subassembly cell design over a linear flow line is the ease with which output can be increased or decreased. This characteristic is what gives the system its flexibility to ramp up and down to meet changes in demand. Adding or decreasing the number of workers changes the cell output rate in a linear fashion.

5.8.1 Single Worker

A lean cell managed and run by a single worker is simply assigning one worker to completely perform an assembly in a U-shaped cell over a series of sequential operations. All machines and stations are assumed to be manual.

5.8.2 Multiple workers

Rather than use only one worker, the cell operations can be divided into sub cells, each managed and executed by a single worker. We call these sub cells *loops*. The time it takes a worker to complete all operations in the loop is the walking time between stations plus the time it takes to complete all assigned duties in the loop, including loading and unloading machines and decouplers. One worker will load parts into the cell and remove parts from the cell. The maximum possible number of sub cells that can be created is one worker at each station. This also yields the fastest possible output rate for the cell. In this case, the cell output rate will be that of the bottleneck machine. The slowest output rate for the cell is when only one worker is used in the cell. Since the worker loop times are almost always different, blocking can occur at the point where each sub cell worker hands off a part to another sub cell worker (baton passing). There is usually almost no blocking with three or fewer workers in the cell. As the variation about the average processing time increases, blocking can occur more frequently. If blocking becomes a persistent problem, it can be temporarily eliminated by increasing the decoupler SOH. This is usually the preferred solution in an interim cell. The permanent solution is of course to balance all loop sub cell cycle times with Kaizen event(s).

5.8.2.1 Rabbit Chase

In Figure 5.4, two worker are executing all the cell operations using a *rabbit chase* (workers follow each other around the loop). The beauty of a rabbit chase is that it completely eliminates the need to line balance (make task times equal).

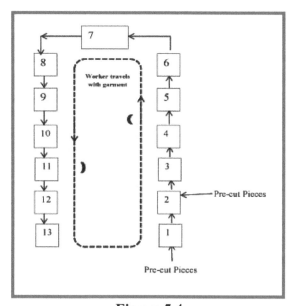

Figure 5.4
Design of a U-Shaped Assembly Cell with 13 Assembly Stations Operated by Two Workers in a Rabbit Chase

This is accomplished by using workers who are capable of performing every operation and running every process in the cell. Another significant advantage is that there is no need to maintain WIP in any decoupler between machines. In this example, two workers follow each other around in the cell, doing all the processes in sequence. This method is not usually used in the manufacturing cells. The need for precise line balancing for the entire subassembly cell is eliminated, and requires no decouplers because the work is always with the worker. After completing a garment at the last station, the operator returns immediately to Workstation1 and starts making another product. The rabbit chase method requires every worker to operate all the machines or processes, which may increase the cycle time, the possibility of making errors, and the training time for new workers.

The main disadvantage of this method is that the slowest worker dictates the cell's output. Blocking by the slowest operator usually occurs because all workers operate (assemble) at different rates. There is variation in their average assembly times for every task. Also, workers may be blocked by the workstation which has the longest average processing time. This is called a *bottleneck* workstation. A bottleneck workstation will increase idle time and may decrease the output rate of the cell.

5.8.2.2 *Bucket Brigade* (Bartholomew and Einstein)

The bucket brigade evolved from the basic philosophy of a rabbit chase. It is operationally the same at the TSS. The motivation was to resolve worker idle time and blocking. The operation of a bucket brigade is simple: Each worker carries a part from the first workstation in a cell to the last workstation in a cell, executing Mo-Co-Moo and any other assigned activities. The first worker will load apart to the first machine, execute duties, and move to the second machine in sequence. When the first worker finishes the cell cycle, he/she unloads the finished part from the cell, and then walks back upstream to take over the work of his predecessor, who walks back and takes over the work of his predecessor and so on, until after relinquishing his product, the first worker walks back to the first machine to begin a new product. If, in addition, workers are sequenced from slowest to fastest, then we call the system a *bucket brigade*. Bartholomew and Einstein have shown that as cell operation continues, each worker will spontaneously gravitate to the optimal division of work so that throughput is maximized.

Notice that workers must maintain their sequence: No passing is allowed and so it can happen that one worker is blocked by his successor, in which case we require that he simply wait until he can resume work, after his successor has moved out of the way. (This waiting is not necessarily bad because it is the means by which the workers migrate to their optimum locations). There are several potential benefits in using the bucket brigade cell staffing system.

- There is a reduced need for planning and management because bucket brigades make the flow line self-balancing.
- Production becomes more flexible and agile because bucket brigades 'tune' themselves, without time-motion studies or the other cumbrous endeavors of assembly-line balancing.
- Throughput is increased because bucket brigades spontaneously generate the optimal division of work.
- Secondary labor is reduced and quality improved because bucket brigades operate with the absolute minimal work-in-process.
- Training and coordination are simplified because it is easy for workers to know what to do next.

5.8.2.3 The Toyota Sewing System (TSS)

Toyota developed this method in the 1960s. The TSS uses multifunctional workers in a cell with typically three to five workers operating 10 to 15 processes or machines. TSS permits workers to share processes and pass work to one another, just as runners in a relay race pass the baton to one another at 10-meter sections of the track. The rabbit chase method is readily changed into the TSS. Thus, the TSS design has processes that are called *relay zones*. Between each workstation, a decoupler holds a certain number of parts, with one being the minimum number. The parts in the cell are called *stock on hand* (SOH).

Workers can travel either clockwise or counterclockwise through the cell, but the most efficient movement is counterclockwise. As long as a worker has a part to assemble in a succeeding workstation, that worker travels counterclockwise. When the worker is blocked, that worker puts the part in the decoupler between the two work stations, and travels clockwise until finding another unfinished garment to assemble. This unfinished garment may be either in a decoupler or at another workstation. An example of the Toyota Sewing System is shown in Figure 5.5, using the cell shown in Figure 5.4. The rabbit chase can be used to train workers for the TSS. The U-shaped layout for this apparel cell has sewing machines that are ergonomically identical. The cell can be operated by a variable number of workers, each cross-trained in all the different processes. Each of the cell's 13 sewing stations has a different mean processing time (Table 5.1). Thus, one of the unique aspects of the design is that processing times at the individual stations need not be balanced.

Table 5.1
Mean Processing Times for Sewing Cell

Workstation	Operation	Time (Seconds)
1	Sew Pockets	30
2	Attach pockets of legs	70
3	Connect legs	60
4	Sew legs	40
5	Sew leg opening	95
6	Turn inside out	5
7	Hem bottom legs	40
8	Attach elastic band	30
9	Stay	5
10	Buttonhole	10
11	Topstitch elastic	15
12	Inspect	20
13	Package garment	80

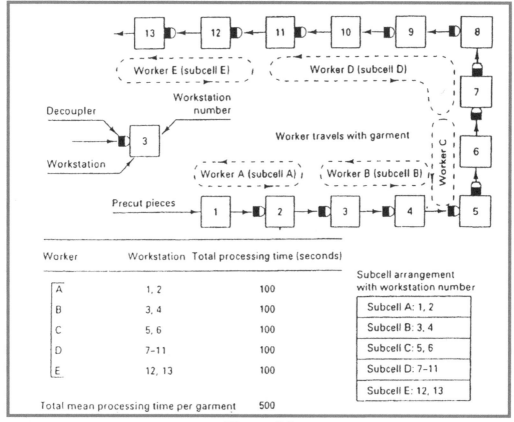

Figure 5.5
The Toyota Sewing System with 5 Workers Assigned to Multiple Stations

As an example of the Toyota Sewing System (TSS), let us look at using five-workers for the cell. Figure 5.5 shows the cell with decouplers and the loop assignments for each worker.

157

Every worker will travel in a counterclockwise flow pattern. Worker E completes a garment at Workstation13. With no more garments to assemble, worker E then walks clockwise. If worker E finds a decoupler that contains an unfinished garment, worker E travels counterclockwise assembling the garment. If worker E finds a workstation where a garment is being assembled, worker E takes over the job from that workstation's worker (D) and continues assembling the garment. Worker D then travels clockwise. If worker D finds a decoupler that contains an unfinished garment, worker D travels counterclockwise assembling the garment. If worker D finds another workstation where a garment is being assembled, worker D then takes over the job from that workstation's worker C assembling the garment. Worker C travels clockwise back toward worker B, and worker B loops between worker A and C. Worker A always starts a new garment. Workers can come and go at any time changing the output. With 5 workers, worker D would do operations 7 through 11, but sometimes would do 6 or 12. But, if one worker goes on break, the cell keeps going in the same manner with four workers. Over time, the workers develop work patterns with certain workers taking the entire responsibility for certain workstation, and other workers sharing the responsibility for some workstations. The best balance was achieved with 5 workers each averaging about 100 seconds of processing time per garment. Recall that if one worker was assigned to the cell, a garment was produced every 570 seconds; two workers using a rabbit chase averaged 285 seconds in steady state. Using the TSS, a garment can be produced every 100 seconds. One might suspect it would take $(570/5) = 114$ Seconds per part in steady state, but it took only 100 seconds. Why was the cycle time per part 14 seconds faster than expected? The answer is typical of what can be expected with TPS because in some circumstances a roving worker will share a machine cycle time with another worker. This advantage is difficult to discover without building and running a detailed simulation model. This is a clear case of why Lean modeling tools should be used to predict system performance in the design phase.

Note that in this example, the total processing times were perfectly balanced – each worker worked on each garment for 100 seconds. If the cell is feeding a final assembly, the cycle time can be adjusted to meet the demand of final assembly by altering the number of operators in the cell. Comparing the TSS staffing method with the rabbit chase method, the TSS was found to be more flexible because workers can help one another instead of being idle due to blocking.

Many companies in the apparel industry have adopted the Toyota Sewing System (TSS), which is also called the modular production system. TSS was developed by Toyota for making seat covers. TSS was known in the West in 1985 as Toyota's *standing up system* because workers work standing up. The features of TSS are similar to manned assembly cells. That is, all cells are U-shaped with workers making garments on a SPF basis. TSS has advantages compared with the tradition batch or bundle system, including less floor space, less creasing in garments, and a better working environment.

Each of these 5 techniques has staunch supporters, and each have been used in a wide variety of Lean systems. The bucket brigade seems to be particularly effective in order picking operations, and has not been widely applied to U-shaped Lean cells. The Toyota Sewing System evolved from the apparel industry, and has been used there with great success. The rabbit chase is said to be easy to install and learn, is self-balancing and has been reported to increase worker productivity and moral. The sub cell method is more widely used than any other method and we will use it in this book to illustrate cell design and operation. Of course, the single worker cell is a completely degenerate case of the multi-worker cell.

5.9 Linking Cells

Cells may be linked to each other in serial or parallel flows, and flow may be split into different branches or joined as shown in Figure 5.6. The linking of the cells may result in a large variety of cell designs which are given names like *spider line, spiral line, escargot line* or *heart line*.

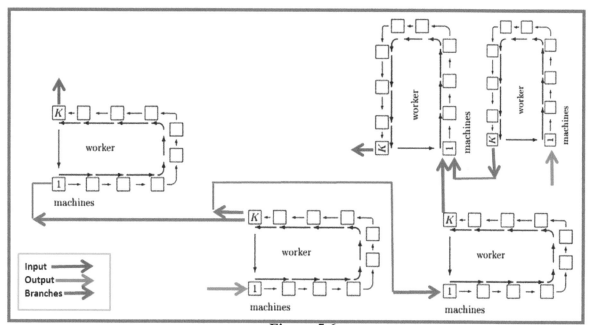

Figure 5.6
Different Ways that Cells are Connected or Linked

A principle that guides the design of the L-CMS is to make the cell as compact as possible by placing all machines in a U-shape, and narrowing the distance between different stations to minimize walking. Other motivations include the minimization of material handling, minimizing floor space (footprints), and minimizing total inventory both between machines and between worker loops.

In Lean production, conveyor lines are usually eliminated. Parts are transferred from one workstation to another by passing through small in-process inventory locations called *decouplers*. Large products can be moved to decouplers by overhead cranes or robot arms. Small parts may have either gravity chutes or robot transfer. A Kanban system is used to control the withdrawal of finished parts from the cells by another downstream cell or a final assembly line.

In Lean cells, the workers perform many more tasks than they would perform in a Mass production system. If the cell is an *assembly* cell, the number of components that have to be assembled within the cell can be very large. In a complex assembly cell, some parts might be pre-assembled in an area close to an interim storage area where parts to be used in assembly are picked from specific lot containers and organized on carts. This area is called *POUSAS* for *point of use storage and assembly,* and its purpose is to accommodate a limited transitory inventory used to cover short-term needs. Of course, these pre-cell assembly stations could be placed

159

directly into the cell itself, but by pre-assembling some components better control and specialization of labor can be accomplished.

In general, at each workstation there is one designated location used to place the POUSAS interim parts or subassemblies. These interim parts or subassemblies are delivered in standardized carts which contain a small number of assembled components. As in the cell, WIP is tightly controlled and carts are delivered using Kanban control. When a cycle of assembly tasks is finished in a station, the just-emptied cart is replaced by another cart containing the pre-kited set of parts to be assembled next. Empty containers are returned to inbound warehousing spaces called *stores* and exchanged with full containers. The arrival of the empty cart is the signal to make more. The stores are replenished by suppliers of components and modules as well. Material handling workers are assigned to control the inventory including pre-kiting tasks, and sustain the materials flow from stores to the cells.

POUSAS areas can hold components from another subassembly line, or parts used in an assembly that is purchased and supplied according to either a Kanban control mechanism, or stocked using an inventory replenishment model. For example, consider the assembly workstation shown in Figure 5.7. This workstation might be part of a U-shaped cell or an assembly line which has a worker at each station. The workstation in Figure 5.7 is part of a Lean cell which is building up a laptop computer. The station shown is installing a hard drive into a computer. As computers arrive at the workstation the worker will retrieve a hard drive from a storage area adjacent to his workstation. Note that there are groups of three hard drives stored on green, red and gold colored pallets. When the hard drives that are stored on the green pallets are all installed, the hard drives stored on the red pallets have moved to the front of the line. This is a signal to order more hard drives, from an upstream cell for hard drives. The reorder point is based upon a *Safety Stock Level* inventory control model. Once the order is placed upstream, it should be delivered just as the gold inventory is depleted. The entire system is controlled by a Kanban card system. It is visual, simple and effective; placing a WIP cap on the supply chain, so they represent a realization of the Toyota *5's system* of visual manufacturing.

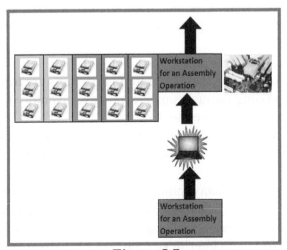

Figure 5.7
Example of POUSAS for an Assembly Station Making Hand-held Devices

5.10 Cell Performance Measures

Different system designs will result in different levels of measurable parameters. *Cell cycle time* (CT) is the most critical measurable parameter for evaluating cell performance. Closely coupled to cell cycle time is *throughput time (TPT)* and the amount of inventory in the cell. By systematically reducing TPT, a company can achieve world class status. In the lean manufacturing system, the manufacturing TPT is reduced by the systematic and gradual removal of inventory (cell WIP) as proved by the well-known Little's Law, which mathematically relates TPT and CT to product demand. Little's Law is stated as follows:

$$WIP = TPT * (1/CT) = TPT * \text{Production rate}$$

Since the cell production rate is fixed by customer demand, we see that TPT and WIP are directly related to each other. Inventory in the system is held in links (between the cells) in the system or in inventory storage locations called Supermarkets or Heijunka Boxes. Finished goods inventory temporarily holds finished products prior to shipment. All work-in-process inventories are moved by the demand of internal customers (other cells or final assembly). Movement of WIP is triggered and controlled using the Kanban system of production control. Kanban is discussed in detail in Chapters 12 and 15. The linked-cell manufacturing system (L-CMS) can have many cells linked to one another by Kanban inventory control.

The material in a Lean cell is usually referred to in the Lean world as *stock-on-hand* (SOH), and is not considered to be inventory or traditional WIP. However, the Operations Research area called *queuing theory* will neither recognize nor use this term. Little's Law uses the term WIP for any inventory in the system, whether it is in a cell, in storage, or in a process step, and it can be used to predict throughput time for any cell or the entire system given the demand rate. SOH material is either in the stations or in the decoupler elements. WIP in a cell (SOH) is strictly controlled and is a major factor in how a cell is operating. The SOH held between machines result in a *decoupled design*. The design of the factory and its physical elements must precede the design and manufacture of the product. Because the design includes people, ergonomic and human factors issues must also be considered. In the links between cells, the inventory is standardized, controlled, and minimized by never producing, storing or moving inventory unless required by internal customers. Final assembly triggers the entire system by either producing to final goods inventory or directly to customer demands. Kanban controlled links also provide production control information to the upstream processes and suppliers, telling them *what* to make, *when* to make it, and *how many* to make on a daily basis.

5.11 Staffing Lean Cells and Responding to Demand (Takt time)

In a Linked Cell Manufacturing System (L-CMS), all subassembly cells are designed to execute a particular cell cycle time to support the Takt time (TT) of final assembly. Cells with manual *assembly tasks* require the operator to remain at the station or machine and complete the process steps before moving on to the next station. Cells with *machining operations* usually contain machines which cycle unattended. Cells are typically manned by multifunctional operators trained to be able to perform multiple assembly tasks and to make sure that any machining tasks are executed correctly. The stations are placed adjacent to one another in a U-shaped or parallel row design to achieve one-piece flow within the cell. Let us look at an example of running a cell using four of the five different staffing methods previously described: (1) Single worker (2) Multiple workers or sub cells (3) Rabbit chase and (4) The Toyota Sewing System.

Cell staffing methods will be explained using an apparel SBS cell with walking workers and decouplers. A *decoupler* which holds only one part will be placed between adjacent workstations in the cell. Decouplers will also be used to separate worker loops. Decouplers also enable the cell to work on a single-piece flow (SPF). The ideal decoupler holds only one unit of stock-on-hand (SOH) and performs multiple input and output functions. Increasing decoupler capacity improves cell output as the processing time variation increases, but little improvement occurs after increasing the decoupler capacity beyond two. Of course, throughput time per part will almost double if two units are placed in each decoupler.

For this example, a pair of pants is being *manually* assembled, and there is a large variation in the processing times at each workstation. Apparel manufacturing is a very labor-intensive industry in which the operator has incentive to outperform a standard *piece rate,* allowing earning and productivity to exceed a predicted level. Most sewing machines are not automatic repeat-cycle processes.

5.12 Cell Description

The U-shaped layout for the cell is shown in Figure 5.4 and in Figure 5.8 with decouplers. The black numbers next to each machine are the average assembly times plus any other assigned duty. Note that the cell flow is CCW. Assume that walking times between all machines are all 5 seconds, and from machine13 to machine1 is 10 seconds. The sewing machines are ergonomically identical. The cell can be operated by one or more workers, each cross-trained in all the different processes. Each of the cell's 13 sewing stations has a different mean processing time. Recall that one of the unique aspects of U-shaped cell design is that the amount of time each worker spends in producing one completed part (piece of apparel) should be the same (balanced to Takt time), but processing times at the individual stations need not be balanced. This is a major departure from traditional Mass assembly lines.

5.12.1 Single Worker

The lowest production rate for this cell is when it is manned by a single worker. Since this is an assembly cell, the Cycle time is: $CT_{Worker} = CT_{Cell} = 570$ seconds. The output rate is 6.32 pieces/hour.

5.12.2 Two Workers in a Sub cell

Suppose that two workers are assigned to the cell. Can the workers find a feasible worker assignment such that the cell is balanced? By trial and error, the following feasible assignments were determined. Worker 2 will operate workstations 4-11, and worker 1 will operate workstations 1-3, 12 and 13. It can be verified that the time required for Worker1 to complete his loop is 290 Seconds and Worker2 will complete his loop in 280 seconds.

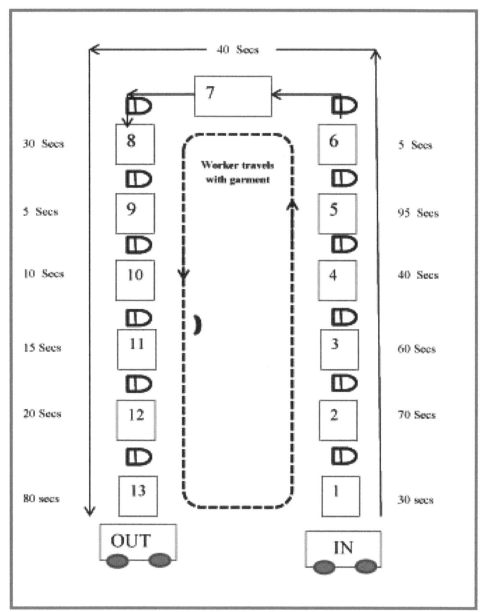

Figure 5.8
A Garment Subassembly Line

Hence, $CT_{Worker1} = 285$ Seconds, $CT_{Worker2} = 300$ Minutes and $CT_{Cell} = 290$ seconds. The cell output rate is now 12.4 pieces/hr. Note that this solution almost doubled the output of the one worker cell. Over time, the workers would share operations at the loop intersections, so the CTcell would approach 285 seconds and the output 12.64 pieces/hour.

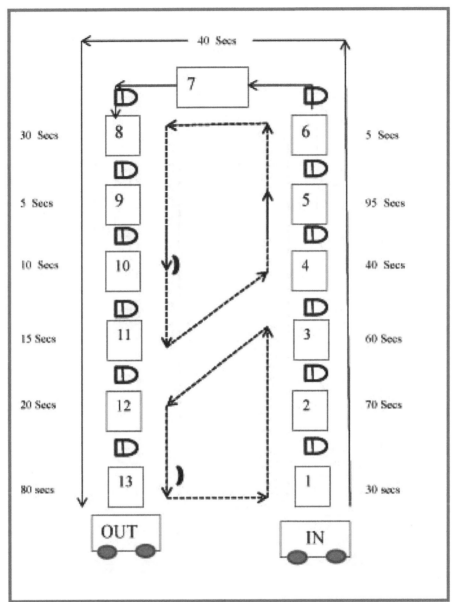

Fig 5.9
The Apparel cell with Two Workers in a Sub-cell Arrangement

5.12.3 Five Workers in a Sub-cell

Figure 5.10 shows the sub cell worker arrangement for five operators. The cycle times for workers C, D and E is 100 seconds plus walking times. Workers A and B are doing a rabbit chase with 25 seconds of walking time. The cell cycle times for each worker are as follows.

Cycle Time $_{WorkerA}$ = 112.5 Seconds
Cycle Time $_{WorkerB}$ = 112.5 Seconds
Cycle Time $_{WorkerC}$ = 110 Seconds
Cycle Time $_{WorkerD}$ = 110 Seconds
Cycle Time $_{WorkerE}$ = 125 Seconds
Cycle Time $_{Cell}$ = 125 Seconds
    Cell Output = 28.8 pieces/hour

164

The sub cell cycle times are not precisely balanced. Each worker is in charge of a sub cell's input and output, thus controlling the SOH within the cell. Regardless of the worker arrangement, if any worker's cycle time increases beyond the required Takt time for the cell, the cell's output rate cannot be met.

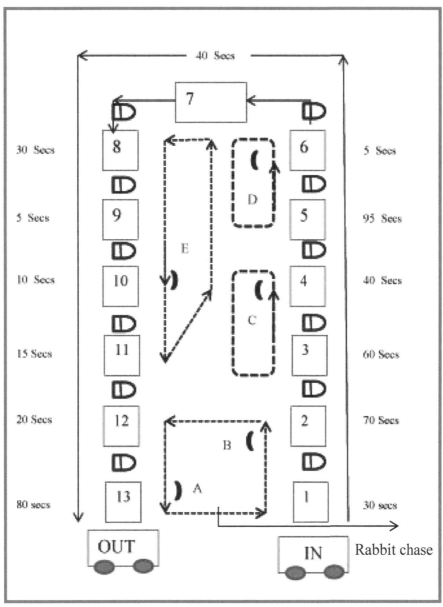

Figure 5.10
Five Workers Using a Sub-cell Arrangement in Apparel Assembly

Finally, the cycle times at each workstation used to determine loop times were considered times were considered to be an average time. In reality, these times are random variables with a mean and a variance. Even if every workstation had the same average cycle time, if these times had considerably different variances then both balance and cell cycle time requirement could be very hard to maintain. Increasing the decoupler capacity from one to two will tend to level out the effects of variance. Remember, however, that increasing the number of units in the decoupler

165

increases the throughput time for the cell. Perhaps the team will engage a Lean engineer and pursue a Kaizen event specifically directed at workstation variance.

5.12.4 Rabbit Chase

Consider a rabbit chase with two workers shown on Figure 5.4. The first worker will start at Machine1 and finish at time T=30. Walking to Machine2 he arrives at T=35, starts that cycle of 70 seconds, and finishes the Machine2 assembly at T=105. Meanwhile, Worker2 is delayed for 30 seconds and Starts a second part on Machine1 at T=30 and finishes at T=60. Walking to Machine2, he arrives at T=65. Worker2 is now blocked for another 40 seconds until Worker 1 finishes his cycle at Machine2 at T=105. The reader is encouraged to draw a timeline of both workers through one complete cell cycle. Clearly, Worker1 will produce a part every 570 seconds. The timeline will show that the second worker will not finish making the second part until T=675 seconds. Worker two will not be blocked again, and both workers will cycle through the cell in 570 seconds. The following sequence of part completion times will occur.

1st Part finishes at T=570 Seconds
2^{cd} Part finishes at T=675 Seconds
3rd Part finishes at T=1140 Seconds
4th Part finishes at T=1245 Seconds
5th Part finishes at T=1610 Seconds
6th Part finishes at T=1815 Seconds
7th Part finishes at T=2280 Seconds
8th Part finishes at T=2385 Seconds

The average completion time per part by the time the 8th part has been completed is 298 Seconds. Note that using two workers, one might expect the average cell cycle time to be (570/2) = 285 Seconds. The difference is due to the original non-stationary conditions where worker 2 started too soon. After 200 parts are produced, the average time per part is 287.5 Seconds. If the cell is allowed to run long enough, the average time per part will converge to 285 seconds as expected.

5.13 The Sub Cell Design Method (Multiple Workers in one U-shaped Cell)

The sub cell design is by far the most popular cell staffing method in use at this time. Because of its popularity, we will offer a few additional comments. The sub cell method, also called the *distributed method*, differs operationally from the TSS because it allows workers to cross the aisle, essentially separating the cell into sections or sub cells with approximately equal sub cell cycle times. See Figure 5.11 for a cell with 4 workers. We will sometimes call these sub cells *loops*. Most companies perform extensive analysis to determine how the cell's tasks should be distributed (or combined) into subgroups to get the sub cell time closest to the necessary cell Takt time. A correctly designed cell with decouplers between all the stations allows the operators to *self-balance* the cell and find the optimal task assignments for each number of workers. During the day, the number of workers in the cell might vary as workers go on break or lunch or to the bathroom, etc. The number of loops in the cell obviously depends on the number of workers. Each loop can be considered as a *sub cell* linked to every other sub cell and to downstream cell demands by the decouplers. The movement of parts between all cells and to final assembly is controlled by Kanban signals; thus forming a *pull* manufacturing system. A part (garment) is processed through these sub cells one at a time. The sub cells start making a garment only when the garment in the downstream decoupler has been removed by the next worker or withdrawn from his sub cell. This simple rule is all the production control needed in the cell. Like the rabbit chase and the TSS operations, the need to line balance individual stations is eliminated. Workers are standing, walking and multiprocessing as well as being multifunctional.

Cell output using the sub cell method will be approximately the same as either rabbit chase or the Toyota sewing system, showing the robustness of the sub cell method. Simulation studies have shown that sub cell cycle times decrease with increases in variability in processing times, but increasing decoupler capacity will level cell output. This is to be expected by the *Pay me now or pay me later* Law of Factory Physics. But improvements are very marginal when the decoupler capacity is more than two, even for large amounts of processing time variation. Figure 5.12 shows the L-CMS with the subassembly cell being fed by manufacturing cells (or other subassembly cells) while feeding final assembly. The subassembly cell can be staffed many different ways, but still generates about the same output per employee. Most companies

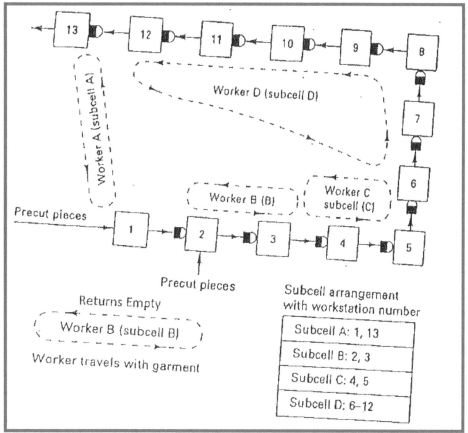

Figure 5.11
The Sub-cell design Allows Workers to Cross the Isle

find that letting the best employee (fastest) pace cell output is a good strategy. Most arrangements of multiple cells are not all as neatly laid out as that shown here, but are usually clustered so that the output from one cell can be assembled directly with the output from another cell. Figure 5.13 shows a design of multiple subassembly cells clustered together to make a product using one-piece flow. The primary difference between the *manufacturing cell* and the *assembly cell* is that the operations or the machines in the assembly cell are usually entirely manual. That is, the operator must stay at the station for the duration of the task.

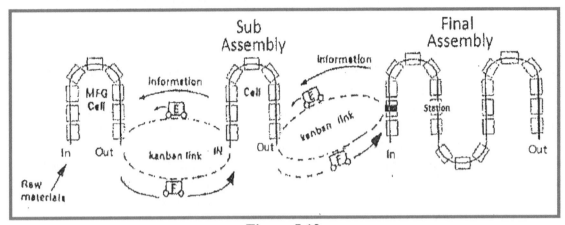

Figure 5.12
L-CMS Subassembly Cells Linked to Final Assembly by Kanban Links

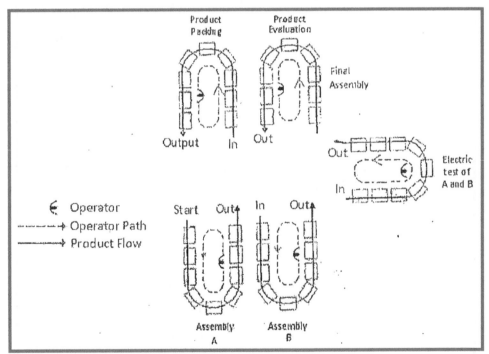

Figure 5.13
Example of a Cluster of 5 U-shaped Cells Working Together to Produce a Product

5.14 Flexibility and Cell Design

The hallmark of a properly designed linked-cell system is flexibility. Properly designed, there is a wide range of demand requirements which can be accommodated without radical design changes. A properly designed cell can handle single or multiple products, with either machining operations (automated), assembly operations (manual), or a combination of the two. The cell is balanced, by properly assigning machining and assembly operations to each worker in the cell

168

such that every worker works to Takt time for the cell; or as close as possible. The key to running a lean cell efficiently is its linkage to the upstream and downstream customers, and its ability to reconfigure worker assignments (not the cell) to rapidly and efficiently respond to changes in product mix or demand. Cell linkage to both upstream and downstream customers will be by Kanban or any other appropriate *pull mechanism*. Kanban methods will be discussed in Chapters 13 and 15. Aside from a single cell operator, the rabbit chase is the easiest to implement. In the rabbit chase, one or more workers are assigned to the cell, and each worker will sequentially start at the load station with a single part. Each worker will process a single part through the cell from load to unload executing MO-CO-MOO and walking between stations at a normal pace of work. There are no decouplers. Once the first worker has finished at Worktation1, the second worker will begin; again, the station times do not need to be balanced. When more than one worker is active in the cell there may be some periodic blockage since workers rarely work at exactly the same pace or rate at each individual workstation. This can be minimized by starting the fastest workers first, and following the operational rules previously described. Another feature of a U-shaped cell is that advice and experience are shared among workers. Eventually, a uniform methodology will emerge for each workstation. The *best* method is documented and posted, made visual, and becomes part of standard work. It is also good practice to have each worker discuss what they are doing at daily or weekly meetings to promote continuous improvement. Rabbit chase applications have suggested that workers are often more alert, compatible and uniformly efficient.

Although rabbit chase has been used for years in the TPS, most American manufacturers tend to use the sub cell method, in which two or more workers are assigned multiple stations or machines to work in independent loops. The cell cycle time is the maximum loop cycle time of any of the workers. If possible, the cycle time of each loop should be the same (balanced) and meet cell Takt time. However, this is sometimes difficult to achieve and some imbalance should be expected. If the differential is large enough, extra loop tasks can be assigned to idle workers.

Another cell operating methodology has been proposed by Bartholomew, who derived an operational methodology called the *Bucket Brigade*. The Bucket Brigade was developed for order picking operations, but it can also be used in Lean cells: It operates as follows. Suppose we assign 4 workers to the U-shaped cell. Each worker will start product into the system, one following the other in sequence. The first worker proceeds at his/her own pace through all of the workstations until a part is completed. At that point, the first worker returns to the worker following him or her and takes over the part that is currently being processed. The preempted worker hands his/her part off and then retreats to the worker following him/her. This preemption occurs until the 4th worker is forced to return to the start station. A new part is started into the system and the pattern repeats itself. So, the Bucket Brigade is essentially the same as the TSS. The TSS strategy is interesting in that after only a few parts are completed, the *handoff points* become fairly consistent and each worker settles into a span of workstations that fall within his or her capabilities. Handoff points can occur within the station or between the stations where the part is left in the decoupler. While the number of workstations that eventually wind up assigned to each worker may vary from worker to worker, the amount of work content by each worker tends to stabilize and equal out. In other words, the procedure is *self-balancing*. Based upon reported applications of this strategy, the worker/station assignments tend to rapidly stabilize and react quickly to changes in methodologies or demand. The minimum number of workers required is the number required to meet cycle time as based on the cell Takt time. Note that all these strategies avoid *crossing paths* and eliminate the need to enumerate or determine all of the possible loop assignments in the cell. The bucket brigade is not widely understood or implemented by American manufacturers, but it could possibly be the easiest and most effective strategy.

Suppose the assembly cell has two workers and eight stations as shown in Figure 5.14. One worker could perform the first, second, seventh, and eighth operations by walking from one station to the next. The other worker does the third, fourth, fifth, and sixth. Two decouplers then link the two workers. Only the total time it takes for the workers to work their loops must be balanced, and the number of workstations assigned to each worker is immaterial as long as each worker spends the same amount of time in his loop. Thus, one worker could do two tasks and the other six tasks if that will make the loop times balance. This methodology eliminates line balancing in the normal sense. The foreman and workers can usually decide on their own how balance can be achieved. However, in even the best cases there may be some imbalance. In this case if there are 8 stations in the cell, then anywhere between 1-8 workers can be assigned loops of work. If the loop times are designated as L_i, then the cell cycle time (CT) is given by: CT=Max $\{L_i\}$ i =1, 2...# cell workers.

Alternately, the two workers could simply follow each other around the cell and both perform each assembly step in sequence. This practice (the rabbit chase) also eliminates the need to do precise line balancing for the entire cell or the partial loops. Thus, the design of the cell with walking worker(s) eliminates the need for traditional line balancing within the cell no matter which method is used. Having the worker mobile as well as multifunctional is a key element in the design of U-shaped lean cells. The walking worker is a critical element in the design of manned cells, making the cell flexible and better ergonomically for the operators who experience far less cumulative trauma injuries. We have indicated that a worker can walk around his/her assigned loop in either a clockwise or counterclockwise direction. The difference is minor.

Workers can use lights mounted above the work stations to signal when they need assistance (yellow); when they may be delayed (red); or if work is proceeding as usual (green). When a problem is severe enough to halt the flow, the worker turns on the red light. This stops the line and all the workers. Other cell workers will respond to the trouble and help to resolve the problem. When the problem is resolved, the green light is turned on again and each worker begins his/her work again. Problems in the cell and how they were resolved can be documented and posted on a cell display board. At the end of the day, these problems are discussed with a Lean Engineer and improvements are suggested by the workers.

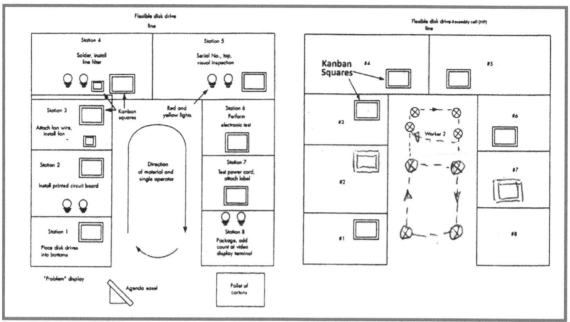

Figure 5.14
Assembly Cells operated by 2 Workers with Decouplers called Kanban Squares.

If the cell was operated at peak capacity (all 8 stations with an operator), the cell then behaves as if it was a Mass assembly line. Unless all eight workstations take approximately the same amount of time, workstations should be combined to require the same amount of time (on the average). Any imbalance will cause idle time for each worker except for one, who is working 100% of the time. That worker is at the *bottleneck* workstation. That is, traditional *line balancing* is required. Extensive redesign of the station, its hardware, and methods may be required to achieve a good balance. This greatly reduces the flexibility of the cell and is a difficult task. If workstations cannot be balanced, the cell output rate will be the same as the output rate of the *bottleneck process*.

Any assembly, machining or a combination cell can be operated by only one worker to meet the daily demand or multiple workers when higher cell output is required. This is *flexibility* in output. The big difference between manufacturing and assembly cells is that the machines in manufacturing cells are usually single-cycle automatics able to complete the process cycle untended, unless it is a simple manual operation or a process like seam welding. In the assembly cells, all the operations are usually manual or less automated, so the operator cannot let the process run untended.

5.15 Summary and Conclusions
The lean factory is based on a different design for the manufacturing system on which the sources of variation in time are minimized and delays in the system are systematically removed. In the linked-cell manufacturing system, manufacturing and subassembly cells are all linked together with a Kanban pull system for material and information control. Downstream processes dictate upstream production rates and Kanban links control inter-cell movement of in process inventory. The linked-cell strategy simplifies the manufacturing system, integrates critical

control functions, and makes it easier to implement new production innovations (automation, robotization, and computerization). This is the strategy.

- When implementing lean manufacturing, start at the top with the CEO/President. Systems level changes require strong leadership.
- Perfection is not optional. Defects can be prevented. The goal is zero defects. When something goes wrong, fix it right away.
- When asked to choose between cost and quality, take both by using one-piece-flow.
- Maintenance is not an option. The squeaky wheel gets oiled or replaced. Zero machine breakdowns and zero tool failures is the goal.
- Continuous improvement requires continuously changing and improving the manufacturing system design.
- The lower the level of inventory in the system, the faster the material moves through the system and the shorter will be the throughput time (Little's Law).
- Everyone who works in the factory should understand how the factory works; i.e., how it functions operationally. Therefore, the L-CMS is simple in concept and design.
- In lean manufacturing, the supply chain is a design decision; i.e., the production control function is built into the design and functions automatically.

Remember to not confuse the terms for cycle time with throughput time (TPT). The former is the reciprocal of the rate of production (PR) so $CT = 1/PR$ and is called Takt time when it refers to final assembly. CT's around one minute are typical in lean production of cars. The TPT refers to the in-process time and includes all the waiting and delay as well as the time in-process when value is added to the parts.

Many think that the best way for manufacturing companies to compete is to automate. In a nutshell, the traditional concept is to achieve required production rates through computerization and automation, while trying to computerize, robotize, or automate the complex mass manufacturing system. This is called *Computer-Integrated Manufacturing* or *CIM*. Companies know how to do this when there is little or no variety in the products, but variety and small lots is now a fact of manufacturing life. In traditional Mass production, highly automated, serial production lines must process in large lots to amortize expensive machines and minimize the impact of complicated and time consuming setup and changeover times. This is known as *economy of scale*.

Lean manufacturing takes a different approach. We call it *economy of scope*. First, simplify the system design and integrate the manufacturing system with the critical control functions, then computerize and automate at individual workstations. The development of self-balancing subassembly cells is the key step in integrating and automating the manufacturing system. One piece flow is a design goal within all U-shaped, Lean manufacturing cells, using small transfer lot sizes between cells. While costs of these strategic approaches are difficult to obtain, the early evidence suggests that the lean cell approach is significantly cheaper than the CIM approach.

Another part of lean manufacturing is *continuous improvement* (CI). Continuous improvement requires the continuous redesign of the manufacturing system. This is a fact of life for lean manufacturing companies, usually employing *Kaizen events*.

There has been a lot of talk about *technology transfer*. In lean manufacturing, technology transfer happens when the lean company shares its knowledge and experience in lean manufacturing with its vendors on a one-to-one basis. The lean company cannot afford to have multiple vendors supplying the same components or subassemblies, so lean manufacturers focus on sole sourcing each component or subassembly. For lean automobile manufacturers, the final assembly plant may have only 100 or so suppliers, with each supplier becoming a lean JIT vendor to the company. The strategy of single sourcing is in keeping with the proprietary aspects of lean production. Only one vendor knows how the cell is able to process racks with different teeth angles on a one-piece-flow basis as described in the case study of the rack bar cell in Chapter 6. The designs of the machine tools used in the manufacturing cells are never published. In the future, the number of vendors supplying an individual company will decrease. The subassemblies will become more complete as the vendors take on more responsibility for the on-time delivery of a larger portion of the interim and final products. The final assembly line will be much shorter, and the entire facility will be much smaller, even leaner.

The other hot topic today is supply chain management (SCM). In the L-CMS, the supply chain is linked (managed) by the Kanban system. This Kanban system is easy to operate and understand since it is simple, automatic and functions in real time. Because the manufacturing system has been completely redesigned, it is simple to understand and operate. That is, everyone who works in the system understands how the system works to produce superior quality goods at low cost in a timely flexible way.

It is our opinion that Lean manufacturing is here to stay. What company would not want to embrace Lean thinking that has as its primary goal waste (Muda) elimination? The magic formula for successful lean companies is to treat external customers like guests and internal customers with respect. Perhaps more radical but with the potential to transform an average company into a world class manufacturer are the methodologies of Lean manufacturing. These transformations will be led by a new breed of engineer which we call the *Lean Engineer*.

Review Questions

1. What are some of the pros and cons of conveyors – that is, using moving assembly lines for subassemblies?
2. What is involved in line balancing a moving assembly line?
3. What are the methods for staffing a subassembly cells?
4. What is the difference between the TSS and bucket brigade for manning a cell?
5. In Figure 5.10, why are workers A and B doing a rabbit chase?
6. What is the difference between the TSS and the sub cell methods for manning a cell?
7. How could you overcome a slow worker in the rabbit chase method?
8. Why is a slow worker (or new worker) not a problem in the TSS or sub cell methods for U-Shaped cells?
9. After the lean engineer and the workers have designed a cell, how would you determine how many workers are needed in the cell to meet the daily demand?
10. Show how the sewing cell might be manned by 3 workers in a sub cell arrangement. Se Figure 5.11 for 4 workers in the cell.

Problem (Sekine)

1. Calculate the Throughput Time for the conveyor line in the following diagram and compare it with the cell design
2. How could you improve the cell design to further reduce the number of operators?

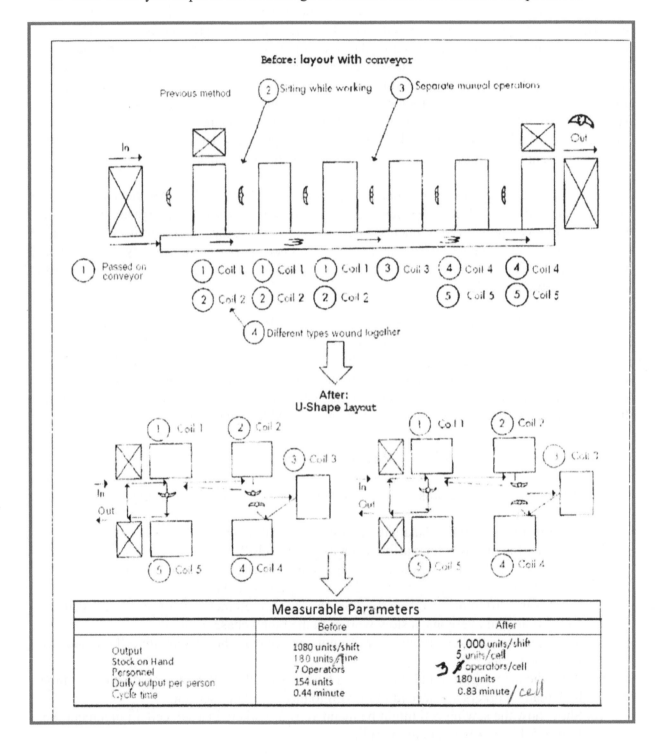

Measurable Parameters		
	Before	After
Output	1080 units/shift	1,000 units/shift
Stock on Hand	180 units/line	5 units/cell
Personnel	7 Operators	3 operators/cell
Daily output per person	154 units	180 units
Cycle time	0.44 minute	0.83 minute/cell

Chapter 6
A Plant Trip (Lean Shop vs the Job Shop for a Rack Bar)
Contributions by David Cochran

Seek and you shall Find...Knock and It Will be Opened to You   King James Bible

6.1 Introduction

In the early eighties, Dick Schonberger wrote a book called *Japanese Manufacturing Techniques* which sold over 100,000 copies and set Americans to thinking about JIT/TQC as the way to make low cost, superior quality goods in a new manufacturing system which had short cycle times and low work-in-process. In 1990, *The Machine that Changed the World* was published and we learned more about a new manufacturing system called the Toyota Production System (TPS). Soon everyone began calling this new machine the *Lean Production System*.

In *The Machine That Changed the World*, Womack, Roos and Jones compared lean manufacturing assembly plants (Toyota) with American mass production assembly plants. A typical automobile final assembly plant contains about 30% of the manufacturing and assembly work actually required to make a car, while about 70% of the manufacturing of the car takes place at the tier 1 and tier 2 suppliers. Therefore, understanding lean manufacturing not only involves a study of final assembly but also requires a visit to tier 1 tier 2 suppliers in order to understand the entire manufacturing system. There must be something very different about a Lean supply chain that renders a competitive advantage. So let us take a plant trip to see two vendors which supply the same component; a steering gear. One vendor supplies Toyota Camry and the other supplies American made General Motors vehicles. The two suppliers will be compared at the component level, where lean systems are building components using lean manufacturing cells, and the typical mass production system is using a traditional manufacturing system.

6.2 Description of the Study

Lean manufacturing second generation cells are different from interim manufacturing cells (Chapter 4) in the following ways. The interim cells, designed with machine tools that were originally designed for use in the job shop, are actually the prototypes for permanent lean cells. In a true lean manufacturing system, manufacturing equipment must be designed, built, tested and implemented into U-shaped cells. This means that the machines, tooling (work holders, cutting tools, jigs and fixtures), and material handling devices (part loading, unloading and decouplers) are all special made for the product family being made in the cells. Simple, reliable equipment that can be easily maintained should be specified. In general, flexible and dedicated equipment can be built in-house better than it can be purchased and modified for the needs of the cell. Many companies understand that it is not a good strategy to simply buy manufacturing process technologies available to other companies and then expect to make an exceptional product using the same technology as a competitor. When process technology is purchased from outside vendors, any unique aspects will be quickly lost. The lean company must carry out research and development on specialized manufacturing technologies as well as customized manufacturing systems in order to produce effective and cost efficient products. Properly designed Lean manufacturing systems also exhibit a very important characteristic previously discussed...*flexibility*. An effective, cost efficient manufacturing system makes research and development (R&D) in manufacturing process technology pay off.

This Chapter describes and analyzes two different types of system designs for the purpose of providing an understanding and a qualitative feel for the inherent differences between mass and lean production system designs within the automotive industry. The term *design* is used euphemistically. There is no indication that any unifying system design approach was used in the Mass manufacturing plant. Two factories are compared; both make the same product. The design of the assembly systems to make the final product is contrasted, as well as the manufacturing environments in which component parts are machined and assembled.

A production job-shop using a batch and queue system is used to manufacture parts in the Mass production system. The Lean system uses manufacturing cells to manufacture the same part. The Lean system uses machinery that was specifically designed for the lean, linked-cell manufacturing approach. The Mass production system uses standard mass production equipment and methodologies. The equipment in the lean cell is designed to meet specified production rates or Takt time. Traditional Mass manufacturing systems exhibit what is called a *push* production control strategy while a Lean manufacturing system basically uses a Kanban controlled *pull* system.

Figure 6.1 presents the model which Toyota uses in its production system design. It is very difficult to understand the Toyota Production System (TPS) design by looking at this diagram. By looking at plants that have physically implemented a lean production system, it is much easier to understand the fundamental principles of the TPS design.

One design concept that must be emphasized before looking at the plant design comparison is the idea of *balanced production*. A Lean manufacturing cell is designed to be manned by one or more workers, who have each been assigned a set of machines to operate in a specific order, which we will call a *loop*. If each worker in a multiple worker cell spends the same amount of time working in his/her loop, this is called a perfectly balanced cell. The goal of all Lean cell designs is to have a balanced cell, and each loop time is identical to the required Takt time. The ideal cell cycle time, or Takt time, reflects customer demand and determines the pace of each cell. Therefore, a balanced production *system* is one in which each cell runs at the pace of the customer demand or

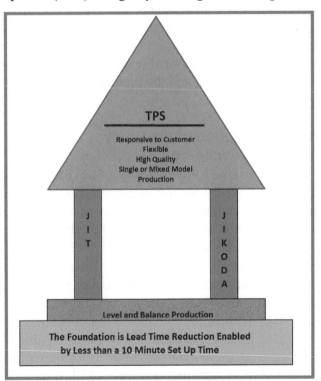

Figure 6.1
The Toyota Production System

the demand set by one or more downstream cells which enables the system to produce according to customer Takt time. Every machine in each cell is designed to run according to the design rules presented in Chapters 3 and 4, and should never overproduce. Therefore, the machines designed for a Linked-Cell Manufacturing System (L-CMS) are very different than machines designed for mass production systems, which typically run as close as possible to 100% utilization. It follows that a cellular based manufacturing system is designed and operated significantly different than traditional Mass manufacturing systems.

A typical mass production system plant design is illustrated on the left side of Figure 6.2. This is not a balanced system design. So why is this system not balanced? This design uses high-speed subassembly lines which produce parts with a cycle time of ten to twelve seconds. This single line feeds five different vehicle assembly plants. Each of these vehicle assembly plants has a cycle time on the order of fifty-five to sixty seconds. Since the products made by the subassembly lines are made at a ten to twelve second intervals, the system is not balanced by our previous definition. Furthermore, the machining of the component part that goes into the steering gear assembly is done in a departmental or large production job shop. This is shown in Figure 6.3. All the turning machines are grouped in the turning department to turn the part; all the broaches are in the broach department; all grinders are in the grinding department; and so forth. This is the typical *Functional* layout of a production job-shop.

This modernized job shop production layout has a designed capacity to meet the demand of three high-speed assembly lines. It was designed to produce four thousand two hundred units per shift, which equates to the demand from the assembly lines in the three other production facilities.

In order to produce to *customer* Takt time or at the pace of customer demand, the first question to ask is *who is the customer of the machining department*? Subassembly is the customer. Machining should produce at the speed of this one subassembly line, which is really the key design objective. To accomplish this goal, the machining department should be reconfigured into U-shaped cells that run at the pace of the subassembly. The second question is: *Is assembly balanced with its customer*? Apparently, planning and engineering aggregated the demand from five vehicle assembly plants into one program requirement to produce 1.2 million units per year for the five assembly plants. Therefore, the engineers built one large production job shop based on the demand for the program; they did the math and decided that twelve seconds must be the required production time per unit. The final result is shown in Figure 6.3.

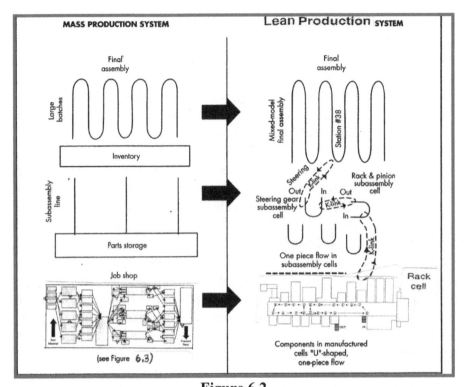

Figure 6.2
In the lean Plant on the Right, the Components Made in the Rack Cell are Combined into Rack and Pinion Gears which are Assembled into a Steering Gear

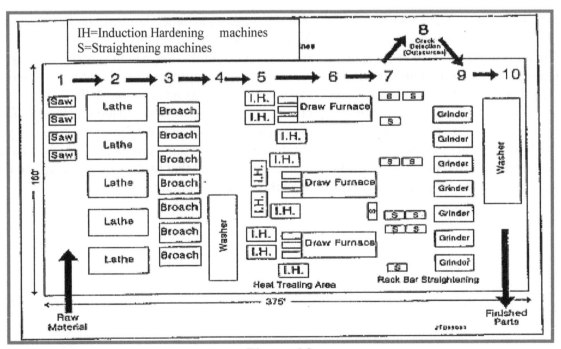

Figure 6.3
One Large Production Job shop

This plant believed they had converted to lean manufacturing because they had calculated a Takt time for the assembly line of twelve seconds. However, as previously noted, this is not balanced production. If the entire production facility was truly balanced, instead of one high-speed subassembly job shop feeding three vehicle assembly plants, there would be three subassembly lines, each feeding one vehicle assembly plant, as show in Figure 6.4. This approach makes it easier to match the production of one cell or one assembly line to its customer.

This Mass production Job Shop is to manufacture racks. The induction hardening furnaces (IH) for quench and temper and the drawing furnaces run large batches of parts. After heat treating, the rack bars are straightened (S); sent to inspect for cracks (B); and then returned to a grinding operation.

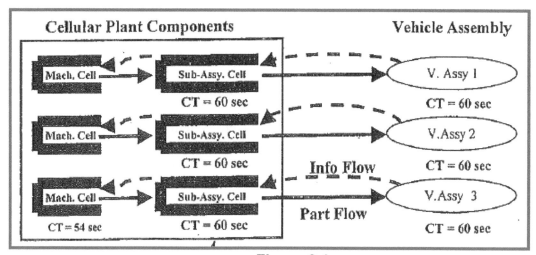

Figure 6.4
The Linked-Cell Manufacturing System uses a Balanced System Design with all the Parts Operating on One Piece Flow

Balanced production with the linked-cell system certainly has a major impact on inventory as well, because the inventory can be focused on supplying only one customer with each line, as opposed to supplying five customers with one line. In the mass production line, while a high-speed assembly line is producing for one vehicle assembly plant, the other four plants are drawing parts from inventory in order to be supplied. Recall what Toyota said in 1948; they wanted to produce to actual demand. *Pulling from inventory to supply four plants while producing product as if there was only one consumer plant is not producing to actual demand.* Finally, it should be recognized that if the one, multipurpose machine shuts down, the entire system shuts down.

In contrast, the lean, linked-cell manufacturing system shown in Figures 6.2 and 6.4 was designed when the first vehicle assembly plant was built. The production cell at the lean plant was built to feed the vehicle assembly plant and its Takt time. Likewise, a machining cell was designed and built to run at the pace required by its own internal customer. Balanced production is a key design attribute of the linked-cell system design. The system is balanced because the Takt time throughout the system is based on the final customer demand not aggregated demand

179

from multiple customers. A *pull* manufacturing cell was designed and implemented instead of a traditional batch and queue system.

There are several benefits as a result of each steering gear cell being linked to a single vehicle assembly plant. First, production can be performed to actual customer demand instead of building ahead of demand: The system can then be designed for minimum WIP. By the use of Little's Law for a fixed demand, if WIP is at a minimum, then throughput time will be also.

The second strategic benefit of setting up a Lean linked-cell system is the ability to manage product variety and to control system complexity. This type of manufacturing system is able to produce many different product types in very small run sizes within a single cell. One of the important design requirements in accommodating product variety is the reduction of changeover time. Of course, from the business point of view in this plant, when another customer is added and production demand increases, a second linked-cell system is added. This tier 1 supplier is, in fact, a North American plant that supplies many Japanese transplants in the United States. The benefit of replicating the system is that it is known to work and can also be improved. Each time the cell is replicated; more knowledge is gained in order to improve the cells further.

Another advantage of a properly designed Lean Linked Cell System (L-CMS) is inherent flexibility. For example, if a vehicle assembly plant is tooled up to make Camry automobile components and then it is discovered that no one likes a Camry, this system seems to carry very high risk. However, if the assembly plant is set up to make five different models that people like as well as the Camry, then the level of risk is actually minimized by having the platform variety.

To be fair, it should be pointed out that there are possible disadvantages with this system as well. The most obvious disadvantage of this system seems to be the increase in investment. With five manufacturing cells instead of one higher-speed machine tool, there are now five machines instead of one. The conventional wisdom is that one machine which can be configured to replace five machines will prove superior to five machines doing the same job. Another relevant question is: *Does it cost more to implement the lean, linked-cell manufacturing systems versus traditional mass manufacturing?* To answer this question, performance measurements related to investment, direct labor, throughput time, inventory, quality, and warranty must be examined and the total cost of each system calculated; not just purchase cost. Finally, it should be recognized that if the one, multipurpose machine shuts down, the entire system shuts down.

Figure 6.5 is a linked-cell system design. By calculating the customer Takt time, the system can be paced to produce no more or no less than is actually required. The Mixed model *final assembly* Takt time cascades all the way back through the manufacturing system and the supply chain, which is called the *value stream*. Everything runs to meet Takt time. The following equation shows how Takt time is calculated.

$$\text{Takt Time} = \frac{\text{Available Daily Production Time}}{\text{Average Daily Demand (DD)}}$$

Takt time is in units of time per unit, a cycle time. The inverse of Takt time is the required production rate.

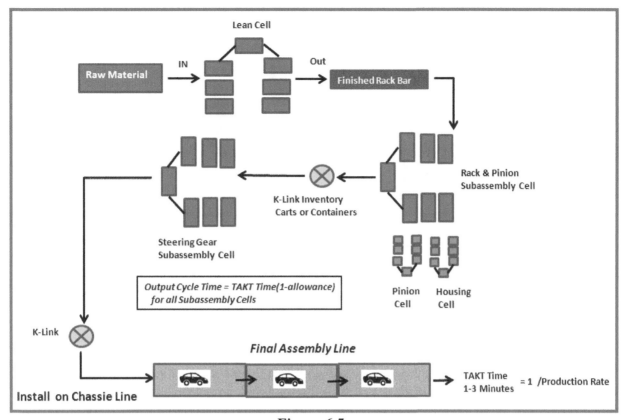

Figure 6.5
The Lean Production System has a Linked-Cell System Design which Operates Visually with Built-In Inventory Control and Workload Balancing

6.3 The Traditional Mass Production System

In a traditional Mass production system, subassemblies are produced on high-speed subassembly lines such as those shown in Figure 6.6. These lines run continuously at a very fast rate: In this case at a ten to twelve second cycle time. They are long, straight, and often conveyor linked. One of the authors of this book once observed a line which was 1.5 miles long! The line in this example was approximately is 310 feet long (about one football field). *What are the people doing in the mass assembly line?* Some are working continuously, and some may not be working at all unless the line is carefully balanced. They do not move from their stations because a conveyor moves the part (steering gears) on pallets from operation to operation. Some operations are automated, and some are manual assembly operations. The operators have only ten to twelve seconds to execute their work cycle. This is a typical case of division of labor and labor specialization. Most manual assembly operations are short term, repetitive tasks. This does not promote teamwork because the workers are treated like machines, and the work is completely dehumanized and boring. Operators are happy when one operation shuts down temporarily because then they can catch a break. If this is a union company, the senior people on the line have the easiest jobs because senior union people get their choice of jobs, and the new people are assigned the most difficult jobs with the most work content. It is interesting that many modern Mass production lines never learned a thing from Henry Ford. Management wonders why labor force turnover is high, and why there are frequent strikes.

181

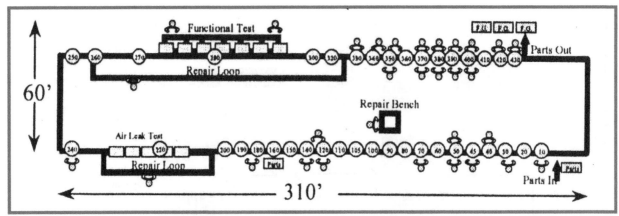

Figure 6.6
A High-Speed Subassembly Line Layout with a 12-Second Cycle Time

There is another phenomenon here. In order to increase break time, some operators pre-build their assemblies and then leave the line and have longer breaks. It is clear that there is no motivation for any harmony of production. The people have been treated as chimneys or islands. The reason one operator can temporarily overproduce and pre-build is because that workstation may have only six or seven seconds of work content, which results from the difficulty of traditional line balancing . The goal of line balancing is to have twelve seconds of manual time at each station, but due to many different reasons, that is nearly impossible to achieve. The theoretical solution to a typical Mass production line balancing problem is very difficult. General solutions are almost always based upon heuristic procedures (Chapter 11).

Note that the *test area* is completely isolated from the people producing the part. This is common practice in most Mass production facilities where the QC function is a staff, service organization. In this particular QC operation, there are eight testers working in parallel. There is also a feedback loop called a *rework and repair* loop. The only alternative to a rework and repair loop is to throw the defects away as scrap. This is institutionalized waste, and has four devastating effects on Takt time production: (1) Rework is costly, disruptive, requires extra material handling and disrupts line balance. (2) Rework inflates Takt time (3) Rework introduces extra WIP, and: (4) Variance in the system is tremendously increased. Reworking parts is often accepted by management as a way of life. Management also often believes that it is easier to accept defects and simply put up with them rather than expend time and effort of to fix the problem. The typical solution is to overproduce to meet final product demands. For example, if a particular line experiences a 10% scrap rate; it is not unusual to start around 112 parts to get 100 parts out. What a horrible policy decision! This is in stark contrast to Lean thinking and Lean manufacturing, which cannot tolerate either rework or scrapped parts. Rework and repair are strictly *Muda* and non-value added activities. It will be shown in the cellular approach that when the Takt time is slowed down to sixty seconds to match the customer demand, many different benefits can be reaped, provided that rework and repair loops are eliminated.

It should be clear that traditional Mass production lines present many problems. *First*, people are isolated at stations. *Second*, rework and repair loops must be built to deal with defective products. *Third*, heavy investment in trafficking and flow logic in the line is necessary. *Fourth*, some machines must run in parallel to absorb long loading and unloading times. *Fifth*, because the Takt or cycle time of the line is so short, long loading and unloading times and complicated set ups can destroy planned production rates. *Sixth*, the fact that the workers are isolated leads to a lack of teamwork, boredom and fails to promote little innovative ideas. If this system were to be improved, improvements would most likely be operational improvements because the basic floor plan is locked into place. Moving a machine would likely require a line shutdown, which makes it very difficult and costly to execute continuous improvement strategies. It is very important that a production system be designed so that it is easy to implement technological improvements or human factors/ergonomic modifications. This is part of *designed flexibility*. We will now discuss a Lean subassembly cell with a sixty-second cycle time. This cell is shown in Figure 6.7.

6.4 A Lean Subassembly Cell

The first thing to notice about this design is that it has a much smaller footprint. Also, it is a semi-closed loop; not a completely closed loop. *What are the people doing in this system?* They work at more than one station except for one person, who is putting on the bellows. Everyone can see and help each other. If operator C has a problem in this system, operator D will simply walk to the problematic station to help or even do the work because everyone in this system is completely cross-functionally trained. In fact, people sometimes rotate work loops on a two-hour basis. A cycle time around thirty to sixty seconds is optimal in most cases. When cycle times are short, very little manual work can be done, the workers are dehumanized and much automation is necessary. Conversely, with a much longer cycle time requirement, there are negative effects because there is only so much a person can remember. For example, Volvo assembly used one person to assemble one entire car. The operators spent a lot of time looking through manuals to complete the assembly. In contrast, the assembly line of Henry Ford used people who never moved, and whose work was optimized in short cycle tasks. This was boring, repetitive work and so there were high labor turnover rates.

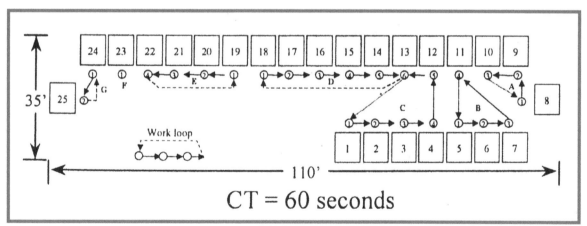

Figure 6.7
A Lean Subassembly Layout with 7 workers and No Conveyors

This single Lean cell does one-fifth the production of the Mass subassembly line so there are five of these cells. However, fewer people are needed in this assembly system overall. If one person does not show up for work, the work loop configuration can often be changed to run at roughly the same rate. Since all potential cell workers have been cross trained to work each machine, little if any productivity is lost by assigning another worker to the vacated work loop. The work environment in this new system is very much team-oriented. There is often a highly skilled team leader who works on this line and acts as a reliever. If one person has to leave, is vacant or ill; the team leader will take over. In a typical Mass production line, a foreman or supervisor sits in the office and watches people.

In a Lean cell, if there is a defect, everyone on the line immediately knows there is a problem. Since everyone is a Quality control inspector and can do every other person's job, the problems are usually resolved quickly, and rework and repair does not become institutionalized. One of the privileges of any cell worker is that if a potential problem is detected, each worker is empowered to shut down the entire cell without fear of retribution. This is a part of *Total Quality Management* (TQM) in Lean manufacturing. Consequently, this lean plant has only about five defective units per day whereas the mass plant can have as much as 10-15 percent defective parts per day. In effect, quality control and inspection have been integrated into the line. As the Lean facility matures, a realizable and prime goal of the Lean Engineer is to eliminate all defects and rework by using *Poka-yoke* strategies.

All the Quality control tests, rotational torque verification, load, and inner ball joint torque tests are integrated into the line. Operators walk past the test machines and clearly see the test results. Testing is not buried in some isolated area. Testing can be run in series because the cycle time is longer, which means the loading and unloading times do not have a great impact on cycle time. In order to produce the same volume in the new plant as in the old, there must be several of these cells so some direct labor might need to be added. However, since smaller machines are used in the cells, investment is approximately the same or less in the new plant. Also, in the original high-speed subassembly lines, the machines are completely automated or completely manual. The machines in the lean plant are single-cycle automatics. In other words, these machine tools may require the operator to load a part or do something related to the machine's operation, and then the machine does the processing automatically. Thus, the worker has been separated from the machine. The operator can unload a part; load a part; start the machine; quickly inspect the part; and then walk to the next machine in his/her assigned *loop* of machines.

6.5 Mass Plant vs Lean Plant Machining Areas

Let us continue our plant trip and explain the difference between lean and mass production at the component level. First, we will visit an automobile assembly plant where a car is being final assembled. To set the scene, we will go to one station where the steering gear is being installed. The station is shown in Figure 6.8. The car spends one minute at this station on the chassis line (one of about 450 steps) where the steering gear is installed by an operator.

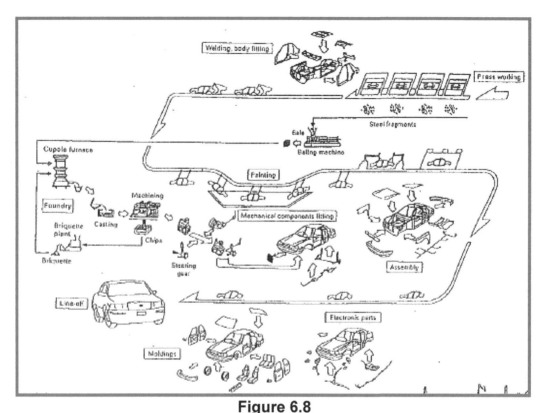

Figure 6.8
The steering Gear is installed at a Station on the Chassis Line in this Automobile Assembly Plant, After it has been Received from the Supplier and Brought to the Assembly Line.

The steering gear assembly is shown in Figure 6.9. It connects the steering column to the front axle. The operator is able to do this task for all the variations of the steering gear that go into this vehicle, including right and left hand drive. On the final assembly line for the Toyota Camry, the steering gears are made in Tennessee and delivered daily to an assembly plant in Lexington, Kentucky in boxes containing between 10-20 gears. They are delivered to the point of use on the assembly line at the first of each hour. All material movement and quantities are controlled by a Kanban system. The steering gear plant supplies the gears on a daily rate equal to what the assembly plant consumes. Toyota is making 1,000 Camry, 500 Sienna, and 300 Avalon's per day; for a total of 1,800 vehicles per day. This is the inverse of the average Takt time per vehicle.

The steering gear contains a rack and pinion subassembly which is the critical element in the steering gear. Let us look at the rack and see how it is made for the GM vehicles versus how it is made for the Toyota Camry. The component is a rack or rack bar which requires precision machining, heat treating, bar straightening, grinding and inspection.

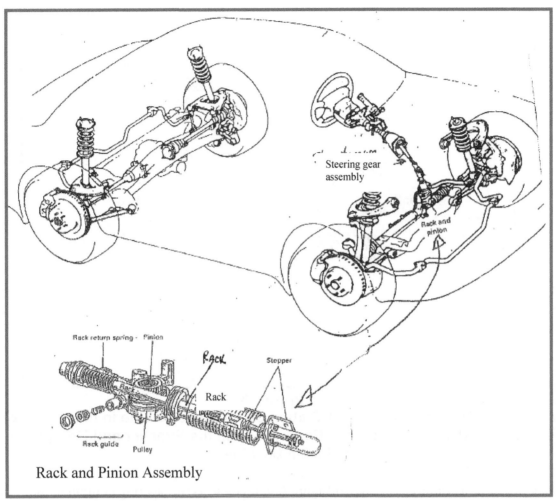

Figure 6.9
The Steering Wheel is Connected Through the Firewall to the Front Wheels via the Steering Gear which Includes the Rack and Pinion Assembly (left).

First we will visit the mass production supplier that makes all the steering gears for this assembly plant. We refer back to Figure 6.2. On the left side of Figure 6.2, a job shop is shown where rack bars are made. This job shop is shown in Figure 6.3. Steel bar stock is sawed to length and moved by containers to the five CNC lathes where all the drilling, turning, grooving, facing, and threading operations are performed. The bars are then moved in large batches to seven large broaching machines which cut the teeth in the rack. Next, the bar is heat treated in large batches in a large furnace using a staged quench and temper process. The bar is then shipped out for magnetic particle inspection to detect any cracks. When it returns from the inspection process, it is straightened and then finish- ground. The bar is then ready for subassembly into the rack and pinion gear.

This rack and pinion assembly shop for GM takes up about 2 acres. The throughput time is typically days. This is an example of one place that a traditional component manufacturing shop still exists. It is often called the *production job shop*. It is a modern day version of the old job shop where large volumes of goods are produced in batches of 50-200 pieces. This design still groups processes functionally, but in this case the walls have been removed. The materials move from left to right through the two acres of plant floor. While it was difficult to obtain an accurate count, the work-in-process (WIP) in this area was counted to be around 95,000 rack bars at any point in time. Using Little's' Law for this portion of the mass production system, the throughput time (TPT) was about 8 days and the production rate 8 parts per minute. The facility runs a three-shift operation 24 hours a day with 21 hours of actual production time. Thus: TPT x PR = WIP; 8x21x60x8 = 80, 640 parts as estimated by Little's Law. *Why the difference*? This is another issue, but experience has shown that if there is a major difference in actual WIP count and that predicted by Little's Law; the company has hidden WIP or unaccounted WIP somewhere in the system. This is a major reason to use Little's Law in process flow analysis. In this system, it is not possible for the operators to move a part from machine to machine (or operation) one component at a time.

6.6 The Lean Shop

We will now visit TRW/Koyo, the Japanese plant which makes the steering gears for Toyota Camry and many other cars and truck companies. All component and subassembly lines have been redesigned into a L-CMS. The connection or linkage of the subassembly and component manufacturing cells shown in Figure 6.5 is by Kanban cards. Let us examine how the Lean rack assembly makes the rack bar. Figures 6.10 shows the design and staffing of the rack bar cell. This cell is making the rack for the Camry, Sienna and Avalon. Almost all of the processing in the cell is done by machine tools that were not made by the traditional machine tool builders. These machines are custom built for the cell. Each machine performs one processing step on each bar. The bar stock is moved from machine to machine by operators through *decouplers equipped* with Poka-yokes to check what was done on the previous step. In some cases, the operator may be checking the part as part of his MO-CO-MOO methodology. Machine tools in the cell are simple, but are designed for automated, single machining cycles; very repeatable, reliable, and replaceable with narrow footprints to reduce the walking time from machine to machine. This is in sharp contrast to the machines made by American machine tool builders used in the General Motors rack bar production line which are great examples of the *super machine* … costly and complex with many capabilities which the user pays for but may never use.

The lean manufacturing system has some unique characteristics that are embedded in the manufacturing and assembly cells. The cells are manned by multifunctional workers. This means that they can perform individual tasks such as handling the material and operating the equipment for every machine in the cell. Assigned tasks include quality control and inspection to prevent defects from occurring; machine tool maintenance, setup time reduction and problem solving are also part of their assigned duties. The cells are usually U-shaped or rectangular, and each cell is designed so the worker can easily step across the aisle and work on machines on the opposite side. The concept is called *separation of machine's work and man's work*. The output of the cell can be varied by changing the number of workers and redefining worker *loops*.

Implementation and subsequent improvement of this system requires hard/systematic work. The Lean Engineer, trained and knowledgeable in all the tools and methodologies contained in this book is an ideal candidate to lead the Lean team. The design of the cells and how the system operates continues to evolve (change) and improve over time. *Continuous improvement* is forced through the gradual removal of inventory from the Kanban links between the cells. Most of the equipment in any Lean cell operates unattended (completes the cycle initiated by the operator), and has devices built-in to prevent defects from occurring (Poka yoke devices). The cells operate on a Make-One, Check-One, Move-One-On (MO-CO-MOO) strategy. Within the cell there are an exact number of parts that are either in the machines or in the decouplers between

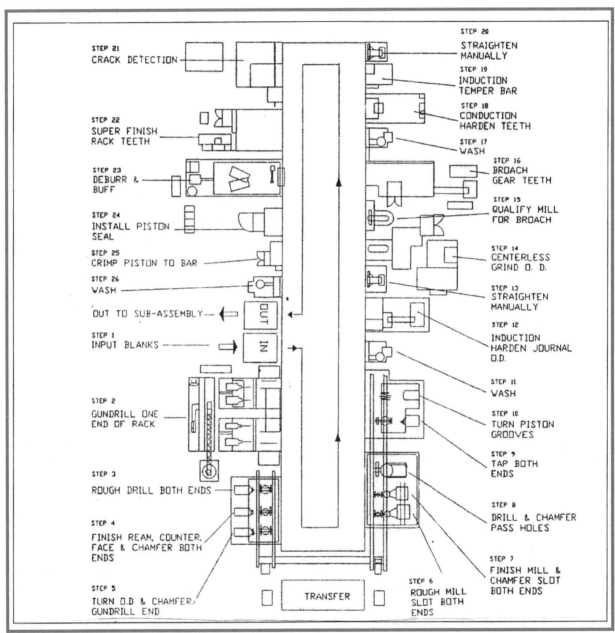

Figure 6.10
Layout of a Rack Bar Manufacturing Cell at a Toyota Tier1 Supplier

the machines. This provides a *WIP cap* on each cell. It should be noted the lean point of view often says there is no inventory within the cell and hence the often confusing term *Zero Inventory* has evolved. The material within the cell is sometimes called *stock-on-hand*. When a part is finished and exits the cell, another part can be started into the cell. The true secret of lean production is that manufacturing cells operate on a *one piece flow* philosophy. Very little has been published about these cells, but these lean manufacturing cells are the heart of the Toyota Production System (TPS) as hundreds of components going into the cars are made this way. This is why you need a trained Lean engineer to implement the lean production system.

6.7 How Lean Cells Operate

Understanding the TPS is to understand how a lean manufacturing cell works. Manufacturing cells are the proprietary element in lean production. However, the lean manufacturing cell shown in Figure 6.11 is very typical. This is a fairly large cell capable of high throughput rates. In addition to numerous machining operations like drilling and tapping, gear teeth milling, deep hole drilling, grinding and broaching, the rack requires heat treating, inspection and mechanical straightening. All of this processing (required to produce a finished rack, ready for subassembly) is in the cell. Comparing the areas required for Figure 6.3 and Figure 6.10, where both manufacturing systems are making a rack bar for a rack and pinion steering gear, each lean cell takes up about 300 square feet.

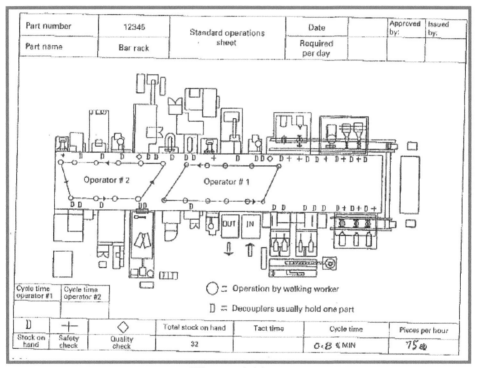

Figure 6.11
A Typical Lean Cell Design

This lean rack cell can make nine different types of racks for the Camry, Sienna and Avalon. The changeover at any individual machine occurs with *one touch* at the time the operator unloads the previous part from the machine (Shingo, 1985). Many machines are equipped with poke-yoke devices that can prevent the processes or the operator from making mistakes.

This cell is designed a bit differently from most Lean cells in that the work arrives and departs from the middle of the cell, but the cell still has a U-shape. The operations at the start (the right end of the cell) are the same for all the bars in the family. Therefore, automatic transfer devices (small robots and mechanical arms and levers) can be used to move the part from machine to machine. For the transfer line portion of the cell (steps 2 through 10) the times for the machining processes (deep hole drilling) are longer than the cycle time for the cell $M_{Drill} > CT$. The cell completes one finished rack bar every 48 seconds so CT = .8 min/rack. One rack is started through the cell every 48 seconds and the deep hole drilling and tapping processes are divided into stages so that on the average, the machining times at each stage are less than 48 seconds. In this area, the machines have automatic repeat cycle capability with automatic transfer devices moving the parts from step 2 to step 10 because all the parts in the family use identical processes in these areas. In total, the rack moves through 26 steps or operations, most of which are machining performed by single cycle-automatic machines. However, there are some manual operations in the cell other than loading/unloading. These include step 13 and step 20 to manually straighten the bar (which can warp after heat treating); step 21 to inspect the bar; and steps 24 and 25 to assemble parts onto the bar. The rack bar manufacturing cell is typically run by two operators. These workers move from machine to machine in *counterclockwise* (CCW) loops as shown in Figure 6.11.

Each operator completes a loop in about 48 seconds. Operator1 typically has 10 stations in his assigned loop of machines, and Operator2 is assigned 11 workstations. Most of the steps involve unloading a machine, loading another part into the machine, checking the part just unloaded and dropping the part into the decoupler between machines. The stock-on-hand (SOH) is 32 units, and each unit is either in a decoupler or on a machine. Each operator performs a manual bar straightening step after heat treating. The decouplers also can be designed to perform inspections for part quality or necessary process delays while the part heats up, cools down, or cures. The SOH is tightly controlled and kept as small as possible; since each part in a decoupler directly increases the throughput time per part. After completing all the tasks at a particular machine, the operator(s) walk to the next machine; hitting a *walk away* switch for the machine just left behind.

Sometimes a decoupler performs an automated inspection of the part, but usually they only serve to hold parts from one process to the next. Sometimes the decoupler performs a secondary operation like deburring or degaussing the bar to remove residual magnetic fields. The bar is made from steel and can become magnetized, causing small chips to adhere to the bar and perhaps cause the bar to be misaligned in the subsequent process.

One operator usually controls both the input and output of the cell, and this is by design. This also keeps the SOH quantity constant and keeps the cell working to the Takt time determined by the final or subassembly lines it is feeding. Operator 1 *loads* the centerless grinder, then moves across the aisle to unload the operation called *deburr and buff*. Operator2 *unloads* the centerless grinder and then loads a part to the next machine. Both move in a counterclockwise loop from right to left.

At the interface between the two operators, either operator can perform all the operations. The choice of which worker to use depends upon which worker is pacing cell output, or more specifically which worker is not. We call this the *Baton passing zone*. This cell can be run with one, two, or three operators.

6.8 The Rules for Lean Manufacturing Cell Design

1. Each machine or process or operation in the cell is ergonomically designed for a worker coming to the machine from the right and leaving to the left.
2. The workers move from to left because of the sequence of *Mo-Co-Moo* activities required at each machine.
3. The machines are designed to have walk-away switches that the worker hits when leaving the machine. The doors close and the machine begins the processing cycle untended. The machine completes the machining or processing cycle in machining time MT. The machine is at least a single cycle automatic.
4. The machines are arranged in the sequence of operations needed to process the part.
5. All the processes needed to make the part are in the cell and have a Machining Time (MT) less than the Cycle Time (CT). This often requires the development of some rather unique process technologies by the Lean engineer.
6. The width of an aisle in the cell is narrow to reduce walking times. This cell has an isle about 4 feet in width so that workers can quickly move across the aisle.
7. Between processes, decouplers hold one part and maintain the stock-on-hand. The decouplers are designed to connect the part flow between machines, reorient the part, deburr the part or even degauss the part. The decoupler can also hold the part for heating or cooling or curing or drying. Like the machine tools and the tooling, the decouplers are custom designed to hold each type of part in the family.
8. The cell is equipped with many Poka-yoke devices performing QC inspections. These devices can be in the decouplers or in the work holding device of the next machine.

6.9 Ergonomics of Lean Manufacturing Cells

Ergonomics deals with the *mental, physical and social requirements of a workstation, task or workplace; and how the workplace is designed (or modified) to accommodate human limitations.* For example, are the machines in the cell designed to a fixed height for all parts which must be lifted, or is the height adjustable? After all, workers can be either male or female and are of different heights. Are transfer devices designed for slide-on/slide-off? Are automatic steps equipped with interrupt signaling to help the worker monitor the process? When the job is defined as primarily loading or unloading, ergonomic concerns with bending over when placing parts in machines and operating work holding devices must be addressed to avoid back problems. In these systems, human intervention in detecting and correcting cell malfunctions will mitigate or avoid machine failure, and thus increase production efficiency and avoid costly disruptions. The design of machines for maintainability and diagnostics is critical. The original designer of the cell should try to incorporate ergonomic issues in the initial design phase rather than trying to come back later to fix problems. Properly designed cells should eliminate any Cumulative Trauma Syndrome (CTS) problems because the physical and mental demands on a cell worker vary from machine to machine.

What are some of the ergonomic issues that Toyota has identified?
- Posture
- Force
- Duration
- Reach/distance/weight
- Repetition.

It follows that the first step is defining *standard work*.

- Create or verify existing standard work
- Break-down standard work into actions required
- Use only actions which are necessary
- Include only actions performed every cycle
- Set time and motion standards for each manual activity

Consider *posture: What are some issues to be dealt with by a well-trained Lean or Industrial engineer?*

- Which parts of the body are out of neutral posture?
- How severely out of neutral or normal based on range of motion (ROM)?
- Numerically rank fatigue for each manual work element from 0 to 5. This is usually part of a *Rapid Upper Limb Assessment Program* (RULA)

Another *Methods* issue is to measure the amount of *force* required to execute a task. Things to consider include:

- Grip
- Sudden impact
- Constant pressure
- Torque
- Excessive manipulations or "indexing" actions
- Workplace environmental conditions
- Numeric risk values from 0 to 5 are assigned
- All tasks should be evaluated and studied in terms of *reach/distance/weight*
- Anthropometric differences
  - Measure reach from body's center of gravity (CG)
  - Measure vertical distance from floor or platform
  - Measure weight as part is processed
  - Numeric risk values from 0 to 5 are assigned
- Measure the *duration* of each task
  - Pace at which job is performed
  - Human process time and machine cycle time
  - Time at beginning, middle and end of shit
  - Experienced, mid-level and new employees
- Duration as a multiplier to other risk factors
  - Assigned risk x (process time/10)
  - Provides total risk of a job
- Develop the operator loops and balance with cycle time
- Assess the *repetition* of the tasks to avoid mental fatigue and boredom
- Number of actions per body part
- Number of actions in given amount of time (speed)
  - More constant the action, greater the risk
- No numeric value is assigned
  - Used with assigned risk to schedule rotations
- Ergonomic *machine design* issues include:
  - Uniform loading heights
  - Location of walk away start switches

Minimum reach (bending) into machine
Eliminating machine obstructions in aisle
Providing access to machine from the back
Position of load/unload decoupler (when part requires a two-hand load)
Part weight, size, shape, burrs, finish, etc.

In both manufacturing and assembly cells where operators manually control the operation of the machines, it is important that all the man/machine interfaces are ergonomically identical. Sewing machines in a cell are a good example (Black and Schroer, 1993). To the operator, all the machines should feel the same in terms of operation and control. Table 6.1 lists some ergonomic advantages of a Lean Cell. Ergonomics in manufacturing cells is the subject of ongoing research in cell design. For a complete treatment of this area see Neibel.

Ergonomics – Lean Production Manufacturing Cell Ergonomic Advantages

- Less Muscular Fatigue
  - CMS standing, walking worker
  - Venous pooling
  - Better circulation
  - Lower heart rates
- Less CTD Hazards
  - Non-repetitive movements
  - Forces exerted
  - Better posture
- Better Man/Machine Interface
  - Adjustable height work station
  - Controls/Displays compatibility
  - Scientific designed tools
  - Self aligning
  - EZ access
  - Self inspection
- Better Interface With System
  - Streamlined parts flow
  - More frequent moves
  - Minimum twisting
  - Minimum lot size

Table 6.1
Some Ergonomic Advantages of a Lean Cell

6.10 Advantages of Home-Built Equipment

There are some unique advantages in using single-cycle, uniquely modified machines in Lean cells. See Table 6.2.

•	Flexibility - Rapid tooling changeover - Rapid modification for new products - LTFC (less-than-full-capacity) design
•	Build exactly what you need
•	Maintainability, Reliability, Durability built into the machine tools
•	Single Cycle Automatics – with process delay if needed
•	Poka-yokes and decoupler in materials handling
•	Easy to load, unload & operate (walkaway switch and fail-safe operations
•	MO-CO-MOO operation
•	Economical to the annual build quantity
•	MT < CT with CT ≅ TT (1 – allowance)

•	Advantages of home-built equipment
	- Only company with this unique process technology
	- Other companies cannot get access to the technology from machine tool vendors
	- Cells can be designed and built with flexibility in mind so that it can adapt to changes in product design (both existing products and new products) and customer demand
	- Equipment is designed for (any or all of these)
	    standing, walking workers (right height, narrow footprint)
	    easy load/unload with CCW operation
	    easy maintenance, easy to clean, easy to repair
	    easy to use (operate)
	    easy to rearrange (wheels, quick-connect utilities)
	    easy to change tools (automatic tool changers)
	    easy to change workholders (rapid setup) leading to rapid changeover for existing products reliability and durability
	The equipment (and methods) are designed to prevent accidents (a fail-safe design) and to not produce defective parts

Table 6.2
Advantages in Designing Your Own Equipment

1.   *Flexibility* – (process and tooling adaptable to many types of products). Flexibility requires rapid changeover of jigs, fixtures, and tooling for existing products and rapid modification for new designs. All processes have excess capacity. They can run faster if they need to, but they are designed for less-than-full capacity operation.

2.   *Building exactly what you need.* There are three aspects to this. *First*, you are not paying for unused capability or options. *Second*, the machine can have unique capabilities that your competitors do not have and cannot get access to through equipment vendors. When you purchase equipment from vendors, you may be paying for capability your competition can get for free (from the vendor). For example, in the lean cell described in this example, a broaching machine for producing the gear teeth on a rack bar is needed. The angle these teeth make with the bar varies for different types of racks. Gear teeth in racks are made by broaching but the job shop broaching machines are not acceptable for the cell because of their large footprint and long changeover times. So from this cell, a unique machine tool for broaching has been designed and built. The process is proprietary. Your local machine tool vendor does not sell these machines because America's machine tool builders don't make machine tools for lean manufacturing cells! *Third*, the equipment should allow the operator to comfortably stand or be seated at the machine. Equipment should be the appropriate height to allow the operators to easily perform assigned tasks. Maintainability/reliability/durability is all built-in and monitored features. Equipment should be easy to maintain (oil, clean, changeover, replace worn parts, standardized screws). Many of the cells at vendors like TRW/Koyo are near clones of one other. The vendor company,

being sole-source, has the volume and the expertise to get business from many companies making essentially the same components or subassemblies for many OEMs. The vendors build a manufacturing cell for each OEM. Some of the equipment can be interchanged from one cell to another in emergencies.

3.   *Designing and building machines, material handling devices, decouplers and tooling for the needs of the cell and the system.* Machines are typically single cycle automatics but may have capacity for process delay. An example of process delay would be an induction heat treatment (IHT) process that takes four minutes in a cell with a one minute CT. The IHT process has the capacity for four units, each getting four minutes of heat treating but outputs a unit every minute.

4.   *Designing/modifying machines to process single units, not batches.* Small footprint, low-cost equipment is the best. MT (machining or processing time) should be modified so that it is less than the cycle time (the time in which one unit must be produced). Equipment processing speed should be set in view of the CT, such that MT >CT for all machines in the cell. The MT is usually related to the machine parameters selected. This approach often permits the reduction of the cutting speed, thereby increasing the tool life and reducing downtime for tool changes. This approach also reduces equipment stoppages, lengthens the life of the equipment and may improve quality.

5.   *Designing equipment to prevent accidents* (fail-safe).

6.   *Equipment should be designed to be easy to operate, load and unload.* Toyota Ergonomic and Industrial Engineers recommend unloading with the left hand, loading with the right hand, and walking through their assigned loop from right to left.

7.   *Machining or processing time (MT) at each machine must be modified so that it is less than the cycle time for the cell* (the time in which one unit must be produced). This may require increasing capacity at one or more workstations. Remember that one unique way to create capacity is to realize that since each machine will not have 100% utilization. Hence, the machining parameters (speed, feed, depth of cut) can usually be adjusted. Since tool failure depends mainly on cutting speed (V), decreasing V increases the tool life and reduces cutting tool replacements and machine tool maintenance. The same is true for the tool feed rate. For example, suppose the MT is 30 seconds for the CT = 1 minute. Both the cutting speed and the cutting feed can be reduced, thereby increasing the tool life and reducing downtime for tool changes. This approach also reduces equipment stoppages, lengthens the life of the equipment, and will usually improve quality.

8.   *Equipment should have self-inspection devices* (such as sensors, Poka yokes) to promote *autonomation.* Autonomation is not the same thing as automation. Autonomation is the autonomous control of quantity (don't overproduce) and quality (no defects). Autonomation should complement and not replace the worker. Hence, it is sometimes called automation with a human touch.

9.   *Equipment should be movable.* Machines are sometimes equipped with casters or wheels, flexible pipes, and flexible wiring. There are no fixed conveyor lines.

10.   *Equipment should be self-cleaning.* Equipment disposes of its own chips and trash usually to the rear of the machine where they can easily be picked up.

11.   Finally, *equipment should be profitable at any production volume.* Equipment that needs millions of units per year to be profitable (Schonberger calls them *super machines*) should be avoided because once production volume even slightly exceeds the maximum capacity that the first super machine can handle, it will be necessary to purchase another super machine. Most super machines cannot be profitable until it approaches almost 100% utilization. A good

example of this philosophy is found in the bar straightening machines using Mass production. In the Mass manufacturing job shop, the bar straightening machines are totally automatic and cost around $1 million per machine and take about 30 seconds to 1 minute to straighten a bar (estimated by engineers at General Motors). For the lean cell, the bar straighteners are manually operated, take only seconds to straighten a bar and cost around $100,000 per machine. In other words, some functions are better performed by a combination of humans and machines using Lean cells. Table 5.2 list some advantages of *home-built* equipment.

6.11 Measurable Parameters (Mass vs Lean)

In the traditional mass production system, the accounting system usually insists that reducing direct labor is the foremost approach to reducing production cost. The mass plant puts a high value on reducing direct labor and increasing machine utilization. Instead of focusing solely on labor reduction and machine utilization, the lean plant concentrates on a system design to simultaneously achieve the many lean manufacturing system goals.

The data in Table 6.3-a compares a subassembly cell at a lean plant to a subassembly line at a mass production plant. The results indicate that the shop at the mass plant had a lower direct labor rate, but requires more inventory, has a longer throughput time, was more expensive to build, and produces more defects.

The data in Table 6.3-b compares the square footage for rack machining in a lean manufacturing cell to typical machining areas required at a typical mass plant. Again, the mass plant has a lower direct labor rate, but requires more floor space, significantly higher inventory, longer throughput time, and the parts cost more and have more defects. So the mass plant succeeded in achieving a lower ratio in direct labor in both machining and labor, but if scrap rates and overtime are considered, both plants produced about the same amount of parts per labor hour. *So why all the fuss?* Lean manufacturing reduces WIP, increases the output rates and decreases throughput times. Lean manufacturing provides more flexibility and uses cells that are self-balancing. Inventory turns are significantly improved. All of these advantages not only save money in the long run, but all work to promote faster time to market, improve quality and increase customer satisfaction.

While visiting both plants, some intangible differences also became evident. The attitudes of workers at the two plants were noticeably different. The push to reduce direct labor at the mass plant causes workers to be wary of the goals set by management. It also tended to promote overproduction. The plant engineers seemed reluctant to seek suggestions from the operators on the plant floor and tended to minimize the importance of contributions from the internal customers. They seemed to be sometimes disconnected from chronic problems and shop floor workers. On the other hand, at the lean plant a very close relationship between the Lean engineers and the production workers had developed. Workers appreciated having their cell's engineer nearby and frequently provided suggestions for improvements. Likewise, the engineers respected the workers, appreciated their suggestions, and often implemented the proposed solutions or improvements.

Table 6.3-a
Assembly Measurable
(Source: Cochran and Dobbs, 1999)

Assembly measurables for design comparison (source: Cochran and Dobbs, 1999)	Lean	Mass
Floor area	1	1.1
In-cell inventory	1	2.8
Throughput time	1	1.6
Capital investment	1	1.3
Direct workers	1	0.70
Parts/labor-hour (w/overtime)	1	0.99
Line returns	1	1.2
Warranty claims	1	9.2

Table 6.3-b
Machining Measurable
(Source: Cochran and Dobbs, 1999)

Machining measurables for design comparisons (source: Cochran and Dobbs, 1999)	LEAN	MASS
Floor area	1	1.7
In-cell inventory	1	97
Throughput time	1	117
Capital investment	1	1.2
Direct workers	1	0.86
Goods parts/labor-hour (w/overtime)	1	1.0
Internal scrap	1	5.4

6.12 Manufacturing Process Technology: Lean vs Mass Production

All the machine tools in the Lean plant visited were custom built and most had proprietary modifications Figure 6.12 and Figure 6.13 show the difference for one process; *broaching*. In the mass plant, the broach has a large footprint, is not moveable and has a very long changeover time. The design of the broaching machine used to machine a steering gear rack bar is typical for this type of job shop. Here is how it works. First, the operator loads the machine manually by clamping the uncut rack bar into a fixture. Next, the operator initiates the machine cycle. When the broach tool finishes its downward stroke, the operator unloads the rack from the fixture, and the broach tool returns upward. In this particular case, the vertical broach tool, and hence the stroke of the machine, is 90" long. In order to allow the operator to load and unload the machine from the floor level, the machine is set into a 7' trench in the floor.

Notice how much larger the broach is in the mass plant. The broaching operation is totally immovable and takes 45 minutes to changeover. In the lean cell broaching occurs at Step 16. The broach in the lean cell is called a *rotary broach* and only requires the flip of a switch to change it from left to right hand rack teeth angles. Here is how the rotary broach works.

A machine designed specifically for machining gear teeth on steering gear rack bars is called the *rotary broach machine*. In this design, the operator loads the rack bar into the fixture at the front of the machine. The operator initiates the machine cycle by flipping a walk-away switch. The broaching process consists of three distinct phases. *First,* the machine pushes the rack toward the rear of the machine and past the roughing broach tool. *Second*, when the part reaches the rear of the machine, the broach turret indexes 90 degrees to the finishing broach tool. *Third*, the part is pushed back to the front of the machine, past the finishing broach. Once the cycle is complete, the clamps then release the rack and it rolls to the front of the machine. The rotary broach tool can be indexed to do either left hand or right hand gear teeth with the flip of a switch. The lean broach runs with a cycle time of 30 seconds.

The racks pass horizontally across the broaching tool and the tool is subdivided into two sections, a roughing side and a finishing side. The mass broach is set up for high volume production. The cycle time is approximately seven seconds in one continuous pass. It completes

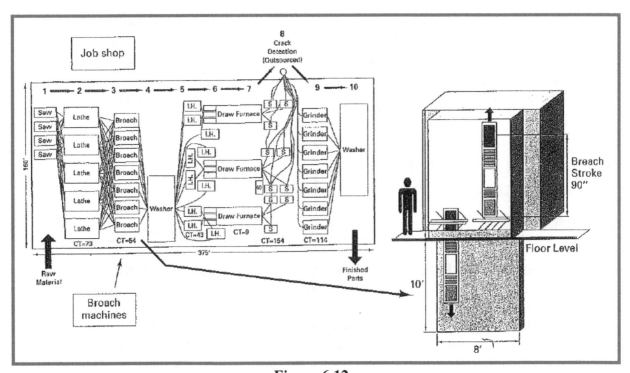

Figure 6.12
This Broach for the Job Shop has a Large Footprint and Long Changeover Times.

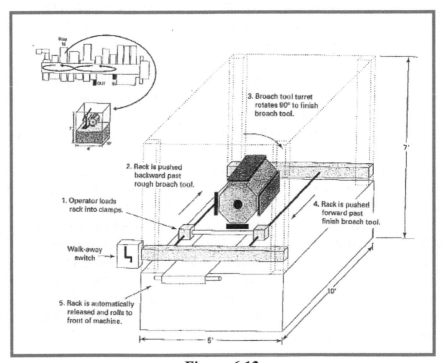

Figure 6.13
A Broaching Machine Tool Designed for the Rack Manufacturing Cell. It has a Narrow Footprint and can be Rapidly Changed Over.

the entire cutting and finishing of one rack in one shot. Notice what was done to the factory floor in order to enable the operator to load the rack at a comfortable level.

The rack remains stationary, and the broaching tool travels vertically so they cut a pit in the floor measuring eight feet wide by ten feet deep. This represents a large investment in facilities to install these broaches. In order to changeover this broaching tool, it takes about an hour with the help of an overhead crane after removing eight to ten bolts with an *Allen* wrench. The new tool must be *qualified* to insure that the parts are good after taking a trial cut. The lean broach is permanently set up for left hand and right hand racks with the roughing and finishing for left hand types on one side, and those for right hand types on the opposite side. Therefore, changeover is as simple as flipping a switch. This is an example of zero changeover time.

In contrast, the mass machine is designed for maximum speed, whereas the lean machine is designed to meet or exceed Takt time. A key point is the difference in labor costs. For the mass machine, labor cost is reduced by increasing the speed of the machine and forcing higher machine utilization. For the lean machine, labor cost can be reduced by increasing the capacity of each worker by changing the number of operations assigned to each person. This is only possible through separating the worker from the machine. That is the *second pillar* in the Toyota model. There is *Just-In-Time* on one side and *Jidoka* on the other.

In the mass plant, after broaching the racks are induction hardened and then fed through a drawing furnace; they are then moved to one of the straighteners by a series of conveyors. This is called a *Turtle layout* because a high-speed machine sends parts to many slower machines by conveyors. Typically, in the mass plant, one person is assigned to unloading three straighteners. There is automatic loading, but manual unloading. The straightener machine design is shown schematically in Figure 6.14.

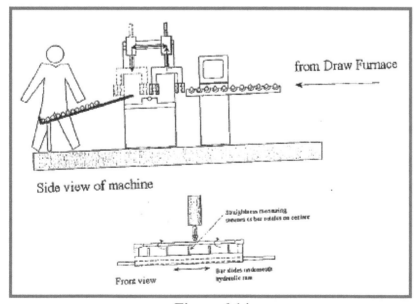

Figure 6.14
Mass Plant Automatic Straightener Design

As just described, parts come from the draw furnace and are loaded into the automatic straightener. The operator unloads parts from the straightener. The straightener picks up the circularity of the part in five locations and finds the high spot by looking at the run-out at each of those five locations. Then, the machine applies force at the high point on the bar and in multiple other locations along the bar to straighten the rack. The machine cycle time is highly variable and highly stochastic because it is difficult to predict how many times the machine will need to apply pressure to straighten the rack. The machine cycle time varies from thirty seconds to 120 seconds. Interestingly, there is a manual backup at the automatic straightening operation so if all else fails the racks can be straightened manually.

The straightening operation in the lean, cellular plant is shown in Figure 6.15. This is a manual machine with a run-out indicator. In this machine, the operator places the rack between two V-shaped blocks, spins the rack, finds the high point of run-out, and actuates a ram to apply force to the bar to straighten it. This machine design assumes that the rack is bowed, not S-shaped as in the mass machine. This different assumption about the characteristic of the part allows a simplified machine design.

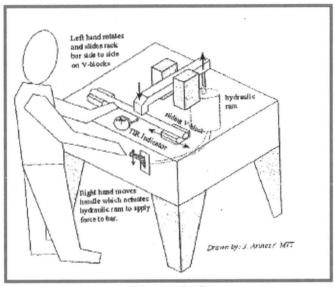

Figure 6.15
Lean Plant Manual Straightener Design

Let us apply typical Mass production cost based upon labor cost per unit to this operation. For direct labor per unit produced, *which machine is less expensive from the direct labor viewpoint, the manual machine or the automatic machine?* For the automatic machine, the operator unloads three straighteners and puts the racks into a large container called a *gon*. The worker carefully places the racks into the *gon* unless he/she gets in a hurry, in which case the operator throws or drops the hardened racks. This can sometimes cause quality problems. From an operations perspective, the automatic machine has less direct labor because one person unloads three straighteners. Since a part is processed roughly every sixty seconds, the operator has an approximate twenty second cycle time, or twenty seconds of direct labor per piece. In the manual machine, the rack may take twenty five to forty seconds to straighten. The investment savings is justified based upon the elimination of investment in the automatic machine, which is in turn based on the elimination of direct labor time. One would probably never put a manual machine in the mass plant based on direct labor reduction because it is less expensive to use the automatic machine when thinking in terms of the operation. If total cost is compared, the manual machine enabled the operator to do more than just straightening. That is where the key to the labor savings occurred. They invested $20,000 in each manual machine versus $800,000 in an automatic machine to produce an equivalent number of parts. This is indicated in Table 6.4. Therefore, most of the capital investment difference between the lean plant and mass plant went into straightening.

The lean plant, in order to simplify its process, also found a different way of heat treating the product to control the thermal distortion of the rack. There is a fundamental difference in heat treating, induction tempering (Step 12) and conduction hardening (Step 18). These two operations replace the traditional quench and temper processes in the job shop, where the heat treating is done in large batches with long processing times (1.5 hours).

Comparison of Straightening Operation

- **Plant 1 machine**
  - Produce Rate: 643 parts/shift (need only one machine in that cell to meet the specified demand rate)
  - Cost: $20,000

- **Plant 2 Machine**
  - Production Rate: 381 parts/shift (need 11 machines to produce the demand of 4200 parts/shift)
  - Cost: $510,000/machine

- **Normalizing with respect to Production Rate:**
  - Normalizing Factor = 1.69
  - Cost of machine at Plant 2 in order to make as many parts as Plant 1 machine = $862,000

Table 6.4
Comparison of Straightening As-is and To-be

The induction tempering in the lean cell is done in line with a cycle time of about 54 seconds. This is shown in Figure 6.16.

6.13 Mass Vs Lean Capacity

The mass production job shop is producing rack bars for 12 different vehicles (different models) with a WIP of around 95,000 parts and an 8 day throughput time, so it is producing around 12,000 parts per day. Note that WIP = TPT x PR. For the lean production system, the Camry is being produced in Lexington at a rate of 1.500 per day, so the Camry rack bars are also being produced at a rate of about 1,500 parts per day (one rack for one rack and pinion per car) with a cycle time of around 48 seconds and a throughput time of about 31 minutes. Figure 6.17 shows a capacity comparison of the mass system versus the lean system. In terms of capacity, it would take an estimated 8 lean manufacturing cells to produce the same annual volume as the large production job shop on the right.

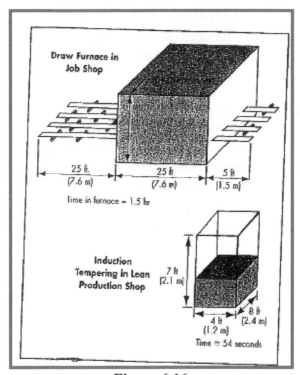

Figure 6.16
Batching in the Draw Furnace versus One Piece Flow in the Induction Tempering Process (Cochran, 1999).

At the lean plant, there are about 14 cells for rack bars; one cell for each Japanese automobile company. A single manufacturing cell is dedicated to each company's specific design, but each cell has similar processing technology. So, the machines for the Mitsubishi rack cell are all interchangeable (with some modifications) with the Camry cell in the event a machine tool in the Camry rack cell has a breakdown, or if there is a major change in the supply contract. If there is a problem with the rack bar, the engineers from Toyota would usually call upon a Lean engineer from the Camry rack bar cell, and the source of the problem is usually quickly resolved. The Lean cell design also isolates processes which need to be redesigned. Notice that crack detection (Step 8) in the job shop required outsourcing. This process was incorporated directly into the cell in Step 21. Between Steps 21 and 22 is a decoupler which transports the part and degausses the parts before it goes into the super finishing process.

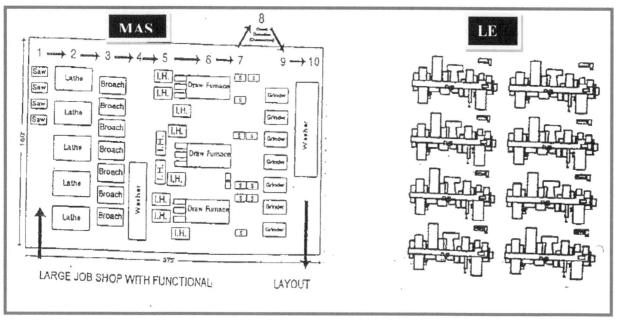

Figure 6.17
Capacity Requirements: Mass versus Lean (Cochran, 1999)

6.14 Summary

Table 6.6 shows the *General Motors (GM) Global Manufacturing System Division Guide*. This was recently published as part of GM's lean initiative.

This is a very large laundry list but is missing the key concepts of mixed model final assembly for leveling the demand on both the supply chain and at final assembly. However, U-shaped manufacturing and subassembly (for one-piece flow with zero defects and flexibility), setup reduction (for small lot sizes) and poke-yoke systems for defect prevention are described. It is not clear at this time how well GM and other U.S. companies will do in converting to lean manufacturing. However, if it is true that the future is already here then we can say the next factory revolution is already going on in the world today. This is certainly true if that factory revolution is converting traditional Mass manufacturing plants to Lean thinking and Lean Manufacturing strategies. In fact, a recent survey revealed that 70%-80% of all American companies are seriously looking at a Lean transformation, or are already underway. The second factory design revolution was based upon the pioneering efforts of Henry Ford and spurred on by the manufacture of weapons of war in WWI. The mass production factory met the demand for equipment and weapons for WWII. It should not be surprising that the lean factory design revolution is based upon the need of any company to now produce at low cost and fast delivery to compete in a worldwide market. This revolution just assisted by the technology developed to design an airplane (specifically a Boeing 747). Boeing has one of the most active Lean development organizations in the United States. The computer system that Boeing used to design their newest airplane, the 777, was used by the airborne laser team to combine the work of 22 design teams in eleven states. Both the product design and the manufacturing system design were all *simulated* in the computer using *virtual reality* methodologies. This same technology can be used to design Lean manufacturing systems and products simultaneously.

People Involved	Built in Quality	Continuous Improvement
• Vision, Values & Cultural Priorities	• Product Quality Standards	• Problem Solving
• Mission	• Manufacturing Process Validation	• Business Plan Development
• Health & Safety	• In-Process Control & Verification	• Andon Concepts
• Qualified People	• Quality Feedback	• Lean Design of Facilities & Equipment
• Team Concept	• Quality System Management	• Tooling & Layout
• People Involvement	Short Lead Time	• Early Manufacturing & Design
• Open Communication	• Simple Process Flow	• Integration (Design for Manufacturing)
• Shop Floor Management	• Small Lot Packaging	• Design for Assembly
Standardization	• Standardized Transportation	• Total Maintenance System
• Workplace Organization	• Fixed Period Ordering System	• Continuous Improvement Process
• Management by TAKT Time	• Schedule Shipping & Receiving	
• Standardized Work	• Temporary Material Storage	
• Visual Management	• Pull Systems	
•	• Level Vehicle Orders	
•	• Supply Chain Management	

Table 6.5
General Motors (GM) Global Manufacturing System Division Guide.

The products can be large subassemblies that are assembled on a final assembly line anywhere in the world. The second driving force behind a mad rush to convert to Lean is the success that Toyota has had in pioneering Lean concepts. American automobile manufacturers are starting to recognize the benefits of both Lean thinking and Lean manufacturing. Unfortunately, the conversion will be costly and slow in the U.S. automotive industries because of the high capital investment, time, and energy expended in developing traditional Mass production systems. Nevertheless, progress is being made. This will mean that the final assembly line for cars will be shorter (leaner) and possibly have more workstations to produce to Takt time rather than pure forecasts. There will be more fully integrated suppliers into the final assembly MMP line, with more dedicated subassembly cells. The vendors will not only be responsible for the lean manufacture and delivery of many components and subassemblies, but might also supply the labor to perform the installation on the customers' assembly line. There are some car assembly plants doing this in Brazil, which has become the test bed for future lean manufacturing system designs. In particular, Ford motor Company has just built possibly the most advanced automotive assembly line of any American Automobile on the Amazon River in Brazil.

Computationally, supercomputers will permit the simulation of the assembly of the entire product (the car) including the simulation of each workstation (with ergonomic subroutines for good workplace design). Similarly, the work cells with their walking workers which produce the components for the subassembly cells can be simulated in detail.

However, for the *Fourth Factory Revolution* the enabling technology will not be the supercomputer, but the realization that a factory built on total automation is too expensive, too inflexible and too complex. However, supercomputers will play a large role in interim cell and factory designs and in *Design for Manufacturing*. CAD/CAM and 3D designs with virtual reality will not just redesign products, but also the processes and machines in the entire manufacturing system. The MSD will be linked component and subassembly U-shaped Lean cells and Mixed model final assembly. Such systems are either in partial use today at Chrysler, Boeing, Electric Boat and a few other large select companies because the future is already here; it is just not everywhere yet.

Review Questions

1. How do interim manufacturing cells differ from true lean manufacturing cells?
2. Why are manufacturing cells not discussed in books like The Machine That Changed The World?
3. Why is the cycle time for the rack cell 48 seconds when the final assembly line is running at 57 seconds?
4. What are some of the reasons for sole sourcing the supplier for subassemblies like the steering gear?
5. How are cells ergonomically superior to job shops?
6. What is standard work for the lean manufacturing cell?
7. In the TPS model shown in Figure 6-1, what are JIT and Jidoka all about?
8. Why are the machine tools in the lean manufacturing cell home-made?
9. Explain the differences between line balancing and a balanced system.

Thoughts
and Things.......

Chapter 7
Balancing and Leveling Production

"The slower but consistent tortoise causes less waste and is much more desirable than the speedy hare that races ahead and then stops occasionally to doze. The Toyota Production System (TPS) can be realized only when all the workers become tortoises"　　　*Taiichi Ohno, 1988*

7.1　Introduction

A fundamental requirement of smooth, synchronized Lean Manufacturing is to balance, smooth and level production requirements. In order to implement the Lean practices of Just-In-Time, pull production, flexible manufacturing cells and mixed model manufacturing. It is important to create and maintain even, consistent flow of work-in-process, including raw materials. Production requirements need to be allocated to Workstations in fixed order release quantities and then synchronized by using small batch sizes. It is also critical to minimize setup and changeover times, and sequence mixed model production to create smooth, uninterrupted flow of products through an assembly or manufacturing sequence. We will categorically state that a balanced and leveled production system is a necessary prerequisite to Lean Transformation. This fundamental principle was recognized very early in the Toyota Production System (TPS).

"In general, when you try to apply TPS, the first thing you have to do is to even out or level production. That is the responsibility primarily of production control or production management. Leveling the production schedule may require some front-loading of SOME order release quantities, or postponing SOME shipments, and you may have to ask SOME customers to wait for a short period of time. Once the production level is more or less the same or constant every month, you will be able to apply the principles of pull systems and balance the assembly lines. But if production varies from day to day, there is no sense in trying to apply other (Lean Engineering) systems because you simply cannot establish standardized work under such circumstances"
Fujjo Cho, President, Toyota Motor Corporation

There are a couple of remarkable things to be noted about Mr. Cho's comments. *First*, this is not a Manufacturing Engineer or a Production Foreman making these comments.....this is the President of the company. Just as Deming observed and stated concerning Total Quality Management, there can be no real transformation unless higher management has completely bought into the program and supports a paradigm shift in manufacturing philosophies. *Second*, it completely dismantles the piecemeal approach to implementing Lean Manufacturing by starting with Kaizen Events and haphazardly implementing Pull Production. If Lean Manufacturing is going to be implemented, it MUST start at the top and require the involvement of every company employee. It MUST be preceded by a total and complete examination of the Manufacturing Enterprise, and it MUST be accompanied by analysis and modeling exercises to help determine the correct place to start. Piecemeal implementation may sometimes help, but our observations are that unless coupled to overall system objectives, it will only hinder the journey. It is our opinion that Lean Implementation must be preceded by a comprehensive Systems Analysis, and then followed by a roadmap and plan of integrated change. The primary focus of most companies on *muda, Kaizen Events* and *Kaizen Blitzes* is both appealing and intuitive, but sometimes extremely short sighted. It is relatively easy to *recognize* waste, but it is more difficult but correct path is to aggressively pursue waste reduction by executing *root cause analysis* and killing the source of the problem…not the symptom of the problem.

The subject of this Chapter is to address the difficult problems of balancing and smoothing workflow to support a Lean transformation. In the limited space available, we can do little more than to emphasize the importance and magnitude of this problem, and to demonstrate representative solution techniques. The process of leveling and smoothing product flow will be called *Heijunka*, and we will broaden the use of this term to include single or multiple product flows, mixed model production and the visual procedures of *Andon*. Historically, it is probably fair to say that production control, accounting functions and even shop floor foremen have been opposed to the Heijunka practices of mixed model sequencing and small lot sizes. This fact stems from the persistent use of MRP *push* order release systems; the accounting philosophy that large lot sizes are needed to amortize equipment procurement costs; and the manufacturing philosophy that long production runs spread setup and changeover times across many production units. This *strategy* is also consistent with the erroneous concept that material handling should follow two primary rules; move things in large quantities that usually match production runs, and move product when something is full and not when it is needed. Heijunka can properly be defined as *creating level or constant order release to every Workstation on the factory floor such that volume and product mix combine to create a lean and consistent flow of work*. **Lean flow** primarily means the avoidance of large lot sizes, while **consistency** means that every work Workstation will be working on the correct product, in the correct quantity, in the correct sequence, at the correct point in time. Large lot sizes contradict and defeat this logic. The Toyota Production System developed ways to effectively move and process individual units or small lot sizes. These practices were largely enabled and preceded by *SMED*. SMED is directed to minimize setup and changeover times and support the philosophy of Just-In-Time. The term Heijunka relates to two separate but interrelated principles. The first is to level production by *tasks,* the second is to level production by *volume*. Successful Heijunka has as its goal the coordinated movement of work-in-process from order release to order completion, such that all materials flow in a uniform fashion and arrive at final assembly or final production in the right *quantities*, at the right *time* and in the right *order*. Expediting and priority scheduling are to be avoided at all times. There are two key concepts associated with this definition of Heijunka which we will address in this Chapter. The first is *balance* and the second is *synchronization*. Balancing workload implies equalizing the amount of time spent at every Workstation, consistent with demand or Takt time. Synchronization is determining the correct product mix delivered to each Workstation such that variance is minimized and production resources equally utilized. The key issues are to **avoid** (1) Congesting and slugging the system with large lot sizes; (2) starving the system due to poor order release practices, and poorly sized transport lot sizes; (3) Workstation imbalance and (4) whiplashing the system due to poor order mix policies in Mixed Model Production (MMP). It should be clear that if these basic operating policies are accomplished, Just-in-time manufacturing and pull production are just a step away.

7.2 Balancing Workload

Balancing workload in either single model or multiple model production lines is a relatively old Industrial Engineering problem. Modern Lean Engineering production practices have caused this problem to reemerge with renewed importance. Traditionally, dedicated assembly lines and

Focused Factories were designed and implemented as older factories were being rebuilt. Large, cumbersome machines were placed in a common work area, and little thought was given to moving them around to different uses or locations. All of that changed with the advent of flexible and versatile manufacturing, particularly within the Lean Manufacturing context. Modern production management is driving toward flexibility to handle not only multiple products but multiple tasks. In any case, if the product mix, product demand or product design is changed; the production system usually needs to be re-sequenced, re-balanced and often re-configured. Two similar manufacturing paradigms have emerged. The first is the serial production line, which is really not new at all; dating back to the classic work of Henry Ford. The second is the U-shaped Manufacturing Cell or the U-shaped production cell. These are discussed in great detail in Chapters 2-6. In either case, any dedicated arrangement is now usually referred to as a *Focused Factory* (Burbidge), or as a Lean Manufacturing Cell (Black and Hunter). Each are composed of Workstations dedicated to specific products and tasks. The objective of Heijunka is to evenly distribute both the volume of work and the mix of work evenly across the production line. There are two major design decisions to be made, and they are not independent. The first is *how many Workstations are needed to satisfy production requirements (demand)*, and the second is *what set of tasks are to be allocated to each of the Workstations*? To address these issues, we must first decompose the production process into a linked set of individual tasks, and then allocate these tasks to separate Workstations. For each task, it is necessary to have a standard time. Up to this point, we have used the mean, variance and/or the Squared Coefficient of variation to describe each processing step. In this Chapter we will only be concerned with the average time to complete a task or processing step. There is current research directed to stochastic task times, but algorithms are currently not widely available to address this complexity. Hence, for our purposes (and in most industrial settings) we will only use an average time for each task.

Having specified a set of required tasks, and the corresponding time required to execute each task, we will now need to determine the precedence relationship between each of the tasks. Precedence relationships specify which task or set of tasks must be completed before other tasks can be started. An analogous problem is the set of precedence relations which define a PERT/CPM network. Having specified the set of required tasks, task times and precedence relationships between individual tasks; we are now ready to determine how many Workstations will be required and how much workload (tasks) to assign to each Workstation. The basic problem is that both customer demand and the amount of available working time will determine the limit on how much work/tasks can be assigned to any one Workstation. Assume that the planning horizon is one week of 5 days working 8 hours per day, and that a demand rate of 100 parts per week must be made and delivered. For this example, one part must be completed every 24 minutes. In this book we call this the required *Cycle Time* for each part. It is not the same as the Throughput time. The inverse of Cycle Time using this definition is the *Production Rate* (parts/min). Of course, any time unit or basic definition can be used. In this case, it is obvious that if multiple Workstations, assemblies or subassemblies are used to produce a single part, each of these areas cannot take more than an *Average* time of 24 minutes between part completions. This time is called the Takt time. This upper bound on time between part completions must be strictly enforced for single part production lines. Notice that in Mixed Model Production lines some parts may take a little longer and some a little less time, but *average* production must strictly adhere to Takt time. Of course, we recognize that there should be a minimum amount of spread between the minimum and maximum Cycle Times, and so we see that our old enemy *Variance* is once more very important and must be minimized.

7.3 Balancing Single Product Production Lines

Figure 7.1 shows a typical production system for a single product. Each Workstation will be assigned a fixed number of Tasks to execute. The number of Tasks assigned to each Workstation is constrained by both precedence relationships and by the required Takt time. The primary goal is to assign K production tasks to a series of N Workstations such that the total workload is evenly distributed across all Workstations.

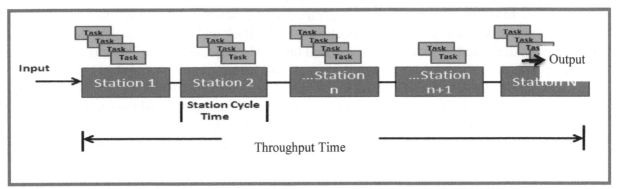

Figure 7.1
A Typical N Workstation Production Line

The required Throughput Time to complete all tasks assigned to the line is the total time required to execute all tasks. The required time between part completions must be strictly less than or equal to the *Takt time,* and is given by $Takt = \frac{Total\ Production\ Time\ Available}{Total\ Demand}$ over a specified planning period. Takt is more of a goal than an absolute, because it is rarely the case that tasks can be assigned to workstations such that each Workstation has identical throughput times due to variability in task times. An ideal balance for a production line is one where all workstations take exactly the same amount of time to execute the set of assigned tasks, and that time is as close as possible to the required Takt time. This primary goal is usually difficult to achieve because the assignment of tasks to workstations is dictated by the following conditions.

(1) Tasks must be assigned to each Workstation in a specific order that cannot be violated without design change. The order is defined by a set of *Precedence Constraints* which dictate that certain tasks must be completed before other tasks can be started.

(2) Workstation assignments should be done in such a way as to level or smooth workload. The goal is to meet Takt time.

(3) The bottleneck workstation *average* cycle time cannot exceed Takt time

(4) Variance in processing times cannot usually be totally eliminated. Since Tasks are assigned to workstations to meet Takt time, the variance of individual part cycle times must be Minimized

Assume that a production line consists of N Workstations in series such as those shown in Figure 7.1, and that there is a set of K Tasks which must be completed in this work sequence. The

n^{th} Workstation is to be assigned a set of K Tasks that will individually take an average of t_k k=1, 2.... K time units to complete. We assume that the sum of all task times allocated to a single Workstation k cannot exceed a specified Takt time and is given by tt_k. The following relationship must hold for any line assignment.

$$\sum_{n=1}^{N} tt_n = \sum_{k=1}^{K} t_k$$

The maximum workstation Takt time that can be allowed is calculated as the available production time per planning period (P_p) divided by the demand rate per planning period (D_p). In addition, the amount of work assigned to each workstation cannot violate any precedence constraints. To achieve workstation balance (Takt time -tt_n) at each workstation should be minimized. Note that the allowable Takt time is a constant across all workstations. In other words, we can define the problem as follows.

$$\text{Minimize } \sum_{n=1}^{N}(TAKT \ time - tt_n) \tag{7.1}$$
$$\text{Subject to precedence constraints and}$$
$$tt_n \le \text{Takt Time}$$

Note that the minimum of Equation 7.1 can be achieved by minimizing workstation idle time or by minimizing the number of workstations used, subject to the precedence and Takt time constraints. This is true because:

$$\text{Min } \{ \sum_{n=1}^{N}(TAKT \ time - tt_n)\} = \text{Min } \{N^*\text{Takt } time + \sum_{n=1}^{N} tt_n\} \tag{7.2}$$
Since $\sum_{n=1}^{N} tt_n$ is just a constant,
$$\text{Min } \{N^*\text{Takt } time\} = \text{Takt } time^* \text{ Min } \{N\} \tag{7.3}$$

Hence, one can choose to assign tasks to a series of N workstations while attempting to minimize idle time with respect to Takt time, or try to assign the set of tasks to a specified number of workstations without exceeding Takt time at any individual workstation. The exact solution to either problem is very difficult, particularly for industrial problems where the number of tasks and Workstations are large. For this reason heuristic solution procedures have been proposed by a number of researchers. We will generate an initial, feasible solution to this problem by using two representative techniques; (1) A simple Assignment Rule called *The Largest Eligible Time Rule* (LETR), and (2) an effective procedure called the *Ranked Positional Weighting Technique* (RPWT). We will then discuss heuristic based techniques to improve on the initial solution. The RPWT is widely used in industry, and lends itself well to microcomputer.

7.4 A Single Product Balancing Problem

Specialized Products, Inc. is under contract to Black's Sweat Shop to produce 400 Widgets per month. Black's shop works two 8 hours a day shifts, 5 days per week with 30 minutes off each shift for lunch. The first thing is to create a 'Precedence Diagram" which shows the precedence structure between tasks. The precedence diagram for our Widgets is shown in Figure 7.2.

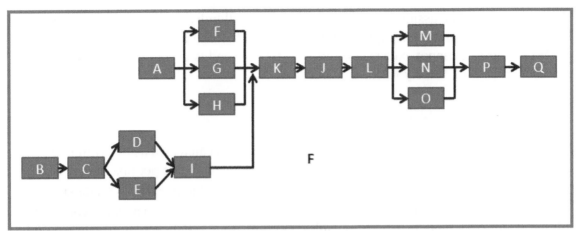

Figure 7.2
Precedence Diagram for Black's Sweat Shop

Black's available working time is $(8 \frac{hrs}{shift}*2 \frac{shifts}{day}) - (2\frac{shifts}{day}*0.5 \frac{hrs}{shift}) = 15 \frac{hrs}{day}$.

Daily product demand is $\{ \frac{400 \, Widgets/mo}{20 \, Days/mo} \} = 20$ Parts/Day, so Takt time is $\{\frac{15 \, Hrs/Day}{20 Parts/Day}\} = 0.75$

Hrs/Part or = 45 Minutes/Part. This is the *drumbeat* of the production line. In other words, assigned task times to each Workstation cannot exceed 45 minutes or 20 units per month cannot be produced on time. Table 7.1 shows the average task times to be assigned to each station. Figure 7.2 shows the precedence relationships.

Table 7.1
Black's Sweat Shop Precedence Tasks Times (Average)

Task ID	Task Time (Min)	Task ID	Task Time (Min)	Task ID	Task Time (Min)
A	21	G	15	M	21
B	36	H	22	N	36
C	15	I	25	O	12
D	25	J	20	P	19
E	14	K	12	Q	44
F	41	L	18		

The total task times are T=396 min. Since the maximum workstation cycle time is 45 min, the minimum number of workstations required is C=$\frac{396}{45}$ = [8.8] = 9, where the brackets indicate that any fractional number should be rounded up to the next integer. This is a lower bound, and does not imply that a balance can be found with only 9 workstations due to the differences in individual task times and the constraints imposed at each workstation. We are now ready to assign tasks to workstations. We will first use the Largest Eligible Time Rule (LETR).

7.4.1 The Largest Eligible Time Rule (LETR)
The LETR rule proceeds as follows. Let "j" represent the Workstation being considered.

(1) Construct a list of all tasks, task times and precedence constraints. Set j=1.

(2) Identify all tasks that can *immediately be* started (no precedence constraints) and place those tasks in a list **L**. Order list L from the task with the longest processing time to the task with the shortest processing time.

(3) Open Workstation j for assignment of tasks. The available unassigned time is initially the Takt time. From the ordered list L, choose the task with the LARGEST processing time that can be allocated to Workstation j without violating the unassigned time. If the first eligible task is too long in duration, try the second task in list L. Proceed until no task can be allocated or all tasks have been assigned. (By construction, at least one task can always be assigned). If the entire list of tasks has been assigned, terminate. Otherwise; empty set L, open Workstation j+1 and Repeat step (2).

7.4.2 The Assignment Procedure

Set j=1. Open Workstation 1. The set L is initially L = {A, B}. Unassigned time is Takt time=45. Since Task B has the largest processing time, assign to Workstation 1. Remove Task B from further consideration, and add Task C to set L. Set L is now L = {A, C}. The remaining unassigned time for Workstation 1 is [Takt time-t_B] = [45-36] = 9. The required processing times for both Task A and Task C exceed the time available at Workstation 1. Close Workstation 1. Set j=j+1=2.

Open Workstation 2. Set L is still L = {A, C}. Set the remaining unassigned time at Workstation 2 to Takt time. Choose Task A for assignment to Workstation 2 since it has the largest processing time and does not exceed the available Workstation time. Remove Task A from further consideration; add Tasks F, G, and H to list L. List L is now L={C, F, G, H}. The remaining unassigned time is [Takt-t_A] = [45-21] = 24. The largest task time which can now be assigned is Task H. Assign Task H to Workstation 2, and purge Task H from the list L. Note that Task F was passed over because it's processing time was too large for the unassigned time remaining at Workstation 2. List L is now L = {C, F, G}, and the remaining unassigned time is now [45-21-22] = 2. No further assignments can be made at Workstation 2. Close Workstation 2.

Set j=3. Open Workstation 3, and set L= {C, F, G}. Set the remaining unassigned time at Workstation 3 to Takt time. Proceed with this same methodology until all task assignments have been made. The final solution is shown in Figure 7.3.

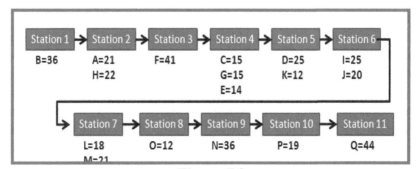

Figure 7.3
Black's Sweat Shop Initial Task Assignments

The task assignments result in a line with 11 workstations. The assignments all satisfy Takt time and no precedence relations are violated. However, *looking at the Workstation Utilizations we find that the* **line efficiency** *is:*

$$\text{Line Efficiency} = \left(\frac{\sum_{n=1}^{K} t_n}{Number\ of\ Stations * TAKT\ Time} \right) = \left[\frac{396}{(11*45)} \right] * 100 = 80\%.$$

Another performance measure often used is *Workstation Efficiency*, which is defined as the ratio of individual Workstation usage to Takt time. For example, Workstation 11 has an efficiency rating of $E_{11} = \frac{44}{45} = 0.98$. In contrast, Workstations 8 and 10 have efficiency ratings of only 0.27 and 0.42, respectively. The worker at Workstation 6 is busy 100% of the time, while the worker at Workstation 8 is busy only 27% of the time. The Balance is poor.

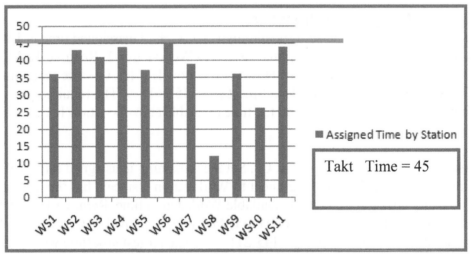

Figure 7.4
Station Time Assignments vs Takt Time

The Longest Time Heuristic just presented generates a feasible solution, but it is rarely optimal or even near optimal. In almost every case, the initial solution can be improved by moving around feasible task assignments. For example, at Workstation 7 we could we could assign Task 1 to be worked first and Task O to be worked second, resulting in a Workstation cycle time of 30 minutes. Workstation 8 can be assigned Task N with a Workstation cycle time of 36 minutes, and Workstation 9 assigned Tasks M and P with a Workstation cycle time of 40 minutes. Workstation 10 and Workstation 11 remain unchanged. The new assigned Tasks and Task times by workstation are shown below. The balance is now much better, with every workstation being utilized between 80% and 100%, with the exception of Workstation 7 which is 67%. However, this could be good since note that Workstation 6 is running at 100% utilization and is the bottleneck workstation. Since Workstation 7 is right next to Workstation 6, if Workstation 6 happens to get behind, Workstation 7 might help out Workstation 6. This, of course, emphasizes the need to cross train workers at all Workstations, and promote friendly relationships. Finally, observe that there is one less workstation in this allocation scheme (10) than in the initial solution (11).

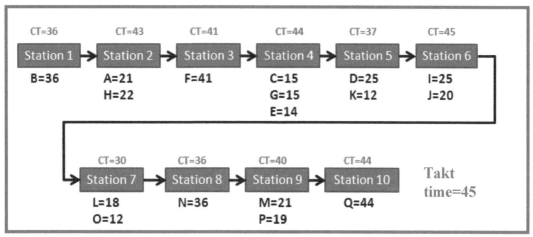

Figure 7.5
Modified Task assignments

7.4.3 The Ranked Positional Weighting Technique (RPWT)

It is intuitively clear that if a good initial solution can be generated, it can be used without further modification, or it can be used as a feasible starting solution for further refinement. The previous example was only given to present the basic principles of Line Balance. The second rule which we will use is very popular, and has been shown to usually dominate the LETR rule. This rule is called the *Ranked Positional Weighting Technique* (RPWT). The basic procedure is the same as before.

(1) Identify all tasks, task times and precedence constraints. Set j=1

(2) For each task, calculate the SUM of that particular task time and ALL OTHER task times for those tasks which cannot be started until that particular task is finished. This is the Ranked Positional Weight (RPW) for that task. Open up Workstation j.

(3) Identify all tasks that can be immediately started (no precedence constraints) and place those tasks in a list "L". Order list L from the task with the largest RPT to the task with the Smallest RPT.

(4) Open Workstation j for assignment of tasks. The available unassigned time is initially the Takt time. From the ordered list L, choose the task with the LARGEST RPT that can be allocated to Workstation j without violating the unassigned time. If the first eligible task is too long in duration, try the second task in list L. Proceed until no task can be allocated or all tasks have been assigned. (By construction, at least one task can always be assigned). If the entire list of tasks has been assigned, terminate. Otherwise; empty set L, open Workstation j+1 and repeat step (3).

7.4.4 Manufacturing a CD/VHS Combo Player Unit

Consider a simplified manufacturing line for a standard VHS/CD player unit. The basic process consists of cutting a case and bottom out of a piece of sheet metal, drilling holes, cutting out user interface slots, and installing mounting brackets, various controller cards, and screwing the unit together. The assembly flow chart is shown in Figure 7.6

215

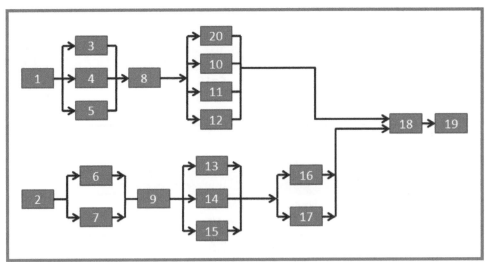

Figure 7.6
A CD/VHS Combo Player Unit

The tasks, task times and positional weights are shown in Table 7.2.

Table 7.2
RPT Example Data

Task ID	Task Time (Sec)	RPW	Task ID	Task Time (Sec)	RPW	Task ID	Task Time (Sec)	RPW	Task ID	Task Time (Sec)	RPW
1	50	331	7	42	290	13	30	143	19	15	15
2	40	363	8	20	186	14	25	138	20	42	85
3	40	226	9	45	248	15	35	148			
4	25	211	10	36	79	16	30	73			
5	30	216	11	25	68	17	40	83			
6	32	280	12	20	63	18	28	43			

The available working time is $(7.5 \frac{hrs}{day})(60 \frac{min}{hr})(60 \frac{sec}{min}) = 450 \frac{min}{day} = 27,000\frac{sec}{day}$, and the daily demand is $325 \frac{unit}{day}$. The total time required to assemble a unit is 650 sec. Hence, the Takt time at any one workstation is Takt $= \frac{27,000}{325} = 83$ seconds per unit. This is the calculated Takt time, which is the drumbeat of the production line. In other words, no more than 83 seconds of work can be assigned to any one workstation or 325 units per day cannot be produced. The total task times are T=650 min. Since the maximum Workstation cycle time is 83 seconds and the minimum number of workstations required is C=$\frac{650}{83}$ = [7.8] = 8. We are now ready to assign tasks to workstations. Open Workstation 1 and create the list L. In this case, L= {1, 2}. Since Task 2 has the highest RPW and its task time is less than the available remaining time, assign this task to Workstation 1. The remaining time available at Workstation 1 is [83-40] = 43. Remove Task 1 from the L list, and add Workstations 6 and 7. L= {1, 6, 7}. Task 1 is the next

216

preferred task to assign, but requires 50 seconds. Remove this task from L. Now L= {6, 7}. Both tasks can be assigned, but Task 7 has the highest RPW (290). Assign Task 7, and then remove Task 7 from the L list. The L list is now L= {1, 6}. The remaining time available at Workstation 1 is [83-40-42] = 1. The L list is now L= {1, 6}, but both require too much time to be assigned to Workstation 1. Close Workstation 1, clear the list L, Open Workstation 2 and repopulate L with L= {1,6} . Since Task 1 has the largest RPW, we assign it to Workstation 2. Remove Task 1 from the L list, and add tasks 3, 4 and 5. The L list is now L= {6, 3, 4, 5}. The remaining unassigned time at Workstation 2 is [83-50] = 33. Of these four eligible tasks, Task 6 is preferred with RPT=280. The required Task 6 time is 32 minutes. Assign Task 6 to Workstation 2, purge Task 6 from the L list, and add Task 9. Now L= (3, 4, 5, 9}. The remaining unassigned time at Workstation 2 is now [83-50-32] = 1. No tasks can be assigned. Open Workstation 3, reconstitute the L list, and proceed as before to assign tasks to Workstations until all tasks are assigned to some Workstation. The results are shown in Figure 7.7 and the line balance in Figure 7.8

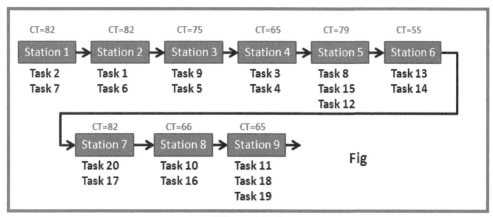

Figure 7.7
Combo Unit Initial Task Assignments

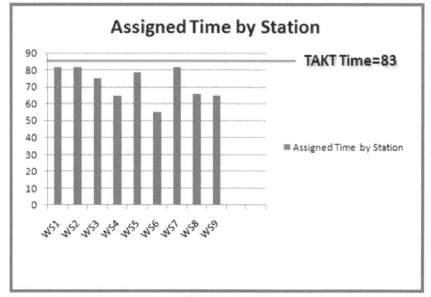

Figure 7.8
Station Times

The solution is fairly well balanced, with Balance Efficiency = ($\frac{650-103}{650}$)*100 = 84.2%. However, this balance wastes 15.8 % of capacity. Workstation 6 is only utilized 65.5%, and Workstations 1, 2 and 7 are working 97.7% of the time. Workstations 1, 2 and 7 are all bottleneck workstations with Workstation CT=82. This allows for some slippage since Takt=83, but the margin is slim.

Again, improvements can be made starting with this solution. The reader can verify that the following steps are feasible.

(1) Move Task 6 at Workstation 2 to Workstation 1.
(2) Move Task 7 at Workstation 1 to Workstation 3.
(3) Move Task 5 at Workstation 3 to Workstation 2.
(4) Move Task 17 to Workstation 6 and Task 16 to Workstation 7.

The new assignments are shown in Figure 7.9

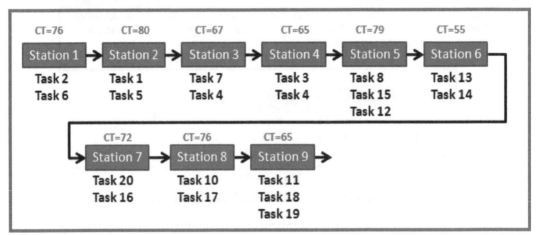

Figure 7.9
Final Combo Task Assignments

The results are significantly better. The maximum Workstation Cycle Time is now 80 at Workstation 2. The line balance is good except for the idle time at Workstation 6. The Line Balance Efficiency is now = $[\frac{(650-104)}{650}]$*100 = 84%.

7.5 Mixed Model Production

It is not uncommon to find Focused Factories and dedicated Production Lines in almost any type of manufacturing or production facility. Whether or not facilities can be dedicated to a single product rest upon many business factors; including volume of production, market share, inflexible production lines that are difficult to set up and the degree of automation desired. Dedicated lines often result in Batch production or long production runs, and we have already discussed how this inflates both WIP and throughput times. Modern Lean Engineering encourages flexible, versatile and heterogeneous production capabilities. As already noted, this necessitates minimum set up and changeover times, a cross-trained workforce, flexible and versatile production equipment and Just in Time material handling. Conceptually, the Balance of a Mixed Model Production Line (MMPL) is no harder than Single Model Production Lines (SMPL), but in practice it is more difficult to control, sequence and tie multiple production lines together. We will now demonstrate how to balance MMPL's, and then point out how to improve the initial balance as we did for SMPL's.

7.5.1 Bottleneck Processes and Pacing Parameters

Inherent in all Balancing models is the principle of *Workstation Capacity*. It is always assumed that adequate capacity is available to process the offered workload. This is tantamount to saying that all processing Workstations have or will have enough processing capacity to meet the required demand. This is also assumed in Line Balancing for Lean production facilities, but it is enforced in a different way. When balancing an assembly line for either single or multiple products, the available production time over a specified planning period is divided by the demand rate(s) to calculate what we call Takt time. Every Workstation produces according to this Takt time. In a typical push system, final assembly is dictated by the rate at which subassemblies or work in progress arrives to final assembly. In a Lean Pull system, the output rate is dictated by what we will call the *Final Assembly Schedule* or FAS. Pull systems will be discussed in detail in Chapter 13. The fundamental assumptions underlying a FAS driven system is that each workstation has enough processing capacity to meet the FAS schedule; there are only minor deviations from the planned processing Workstation times; quality is high (no rework); equipment has 100% availability; and the product resources required arrive just in time, just at the right place and just in the right quantities. However, we have observed that line throughput is strictly dependent upon the slowest Workstation in the line, and we have called that the *Bottleneck Process*. The bottleneck process can be anywhere in the production line, and in actual practice the bottleneck process will periodically change positions in the line due to variations in demand and product mix. This phenomenon is called the *Shifting Bottleneck Problem*. Loading a production line being constrained by a bottleneck process is inherent in assembly line balancing.

In a Lean product line, the cycle time at any process step, including the bottleneck process, cannot exceed Takt time. Takt time is the *pacing parameter*, and line output cannot be more than the rate at which the bottleneck process produces. This is called the *Bottleneck Rate*. Note that the converse is *NOT* true. Goldratt calls this the *drumbeat* of the production line. For multiple product lines, keeping the production line and all of its individual Workstations in sync with the required Takt time is difficult to enforce. Note that without some form of production control, every assembly line, subassembly line and manufacturing cell would have its own drumbeat. In an ideal *synchronized* Lean production facility, every component production cell, subassembly cell and assembly line is controlled by the required Takt time. In an ideal pull system, the *only* production component that is not driven by subsequent operations is the final assembly or shipping line, which sets the Takt time for the *entire system*. Ideally, the final assembly line will be the Bottleneck operation for the entire manufacturing facility, and every other manufacturing component follows the drumbeat of the FAS.

For single product production facilities, the manufacturing times at every workstation or process are simply the time to process each individual item, and Takt time is determined by the demand for that single product. However, in multiproduct production, each product type will have its own processing time and its own demand schedule. This is called *Mixed Model Production* (MMP) and necessitates the concepts of *average production time* and *average demand*.

The usual approach is to *aggregate* and calculate a *weighted* average service times using the appropriate demand rate for each product type to get a composite service time. This procedure will be used to balance multiproduct Lean production lines. This general family of problems is usually referred to as balancing *Mixed Model Production* (MMP) lines, and has two primary goals; (1) Balance the production line by assigning mixed model production requirements to each Workstation without exceeding an average Takt time; subject to any precedence constraints and (2) Establishing a Mixed Model Production Sequence (MMPS) which will minimize batching and provide smooth, uninterrupted product flow to each manufacturing operation.

7.5.2 Balancing Mixed Model Production Lines: Calculating a Weighted Average

We will demonstrate how to calculate a weighted average of processing times at each workstation by using a highly simplified example. The basic approach is to use a Weighted Average to determine the average Takt time and workstation processing time requirements across all products. In other words, we will balance to an average processing strategy. In this case, it is acceptable to have any one product exceed the Takt time requirement at any one workstation, as long as the *average* workstation cycle time across all products does not exceed Takt time. To demonstrate the basic procedure, we follow an example first published by Nicholas (18) in his excellent textbook *Competitive Manufacturing Management*. Consider the production of four different products with the demand, cycle time and processing time requirements shown in Table 7.3.

Table 7.3

A Four Product MMP Example

Product	Demand (per week)	Task 1 Time	Task 2 Time	Task 3 Time	Task 4 Time
A	500	1.40	0.60	***	1.20
B	250	1.55	2.20	1.75	3.20
C	300	2.60	***	3.22	2.70
D	200	1.35	2.75	1.25	1.00
Totals	1250	6.90	5.55	6.22	8.10
Weighted Average		1.710	1.12	1.323	1.93
	(Units)	(Min)	(Min)	(Min)	(Min)

We start by calculating the proportion of total production allocated to each of the four products. Define $p_k = (\frac{D_k}{\sum_{j=1}^{K} D_j})$, where D_k = demand for product k=1, 2.... K products. Hence, $p_1 =$ $\frac{500}{1250} = 0.40$ $p_2 = \frac{250}{1250} = 0.20$ $p_3 = \frac{300}{1250} = 0.240$ and $p_4 = \frac{200}{1250} = 0.160$. The weighted average production time for Task 1 is given by (1.4*0.40) + (1.55*0.20) + (2.60*0.240) + (1.35*0.160) = 1.710 minutes. Assume the number of production minutes available each week = 3300 minutes. Hence:

$$\text{Takt} = \frac{3300\ Min/Wk}{1250\ Units/Wk} = 2.64 \text{ min/unit}$$

The Takt time for this MMP is given by 2.64 Min/Unit. This implies that one unit must be produced on the average every 2.64 Minutes, or the demand cannot be met. Note that this not imply that the time required to process each individual part must be less than or equal to 2.64 minutes. Some parts may exceed this number, and some parts will be less than this number.

The only precedence constraints that will be enforced are that Task1 must precede Task2 and that Task2 must precede Task3 for all products. In general and more realistic problems with multiple products, each product will have routes with both common and unique Workstations. The goal is to assign product dependent tasks to a set of workstations such that Takt time of 2.64 min/part is not violated. We will use *the Longest Processing Rule* simply for demonstration purposes. The average Takt times are: Task 1=1.710, Task 2=1.12, Task 3=1.323 and Task 4=1.93. Hence, the Assignments would be Task 1 to Workstation 1, Tasks 2 and 3 to Workstation 2 and Task 4 to Workstation 3. Three workstations are required. The result is shown in Figure 7.10.

For now, we will simply note that there are two basic procedures which can be used to balance a line. The *first* is to Minimize the number of Workstations required given a set of products, associated tasks, precedence relationships and Workstation cycle time (Takt). The *second* is to Minimize Takt time given a fixed number of Workstations. Both objectives can be addressed by a slight modification of either the Longest Time Rule or the Ranked Positional Weighting Technique previously presented. Both rules will provide a feasible solution, which will almost always require improvement using some heuristic.

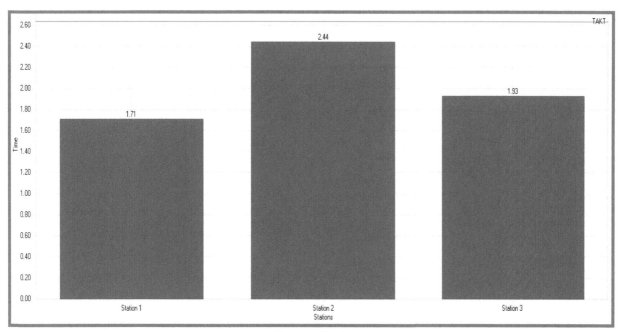

Figure 7.10
Initial Line Balance: LPT rule

We would like to point out that although some optimal solution procedures have been reported in the technical literature, any solution technique used to solve real problems will likely be based upon a heuristic procedure. That means that a two phase process will probably be needed to refine and rebalance an initial, feasible solution. This being said, the goal and fundamental issues of

221

MMLB which we will discuss are true regardless of the solution technique used. We will address the basic problem, and then conclude with a discussion of the impact of additional real world constraints and how they can be incorporated.

The solution procedure will be based upon the RWPT heuristic presented in Section 7.4.3, and an initial balance will be conducted using *average* production requirements. Tasks will then be assigned such that the number of Workstations will be minimized subject to Takt time and predecessor constraints. After the assignments are made, we will suggest a *Mixed Model Sequence* which will smooth and level product flow.

7.6 A Mixed Model Line Balancing Example

To demonstrate Mixed Model Line Balancing (MMLB) we will assume that three different products (A, B and C) are to be manufactured. There are 17 tasks that must be accomplished, each with a product specific task time. Not every task is required for all of the products A, B and C. Table 7.4 shows the processing time requirements for each product by task, and the Weighted Average for each task.

Table 7.4
MMLB Processing Times

Part	T1	T2	T3	T4	T5	T6	T7	T8	T9	T10	T11	T12	T13	T14	T15	T16	T17
A	8	8	16	*	*	8	8	*	*	10	16	*	*	*	10	12	10
B	10	6	*	*	*	12	*	10	16	12	*	*	14	12	*	8	12
C	*	*	*	6	18	12	18	*	*	10	*	8	*	16	*	6	10
Wt Avg.	6.5	5.5	8	1.5	4.5	10	8.5	2.5	4	11.5	8	2	3.5	7	5	9.5	10.5

An asterisk indicates that a specific task is not required for that particular product. The weekly demands for products A, B and C are A=20 units, B= 10 units and C= 10 units. The assembly line is expected to work 960 minutes per day. Hence, the Takt time for this line is Takt = $\left(\frac{960\frac{Min}{Day}}{40 Day}\right)$ = 12 Min/Unit. In other words, a part needs to be produced on the average every 12 Minutes to meet demand. This is the drumbeat of the production process. The objective is to assign tasks to assembly workstations to meet Takt requirements. The product mix is 50% product A, 25% product B and 25% product C. The strategy that we will use is to once again use the Ranked Positional Weighting Technique (RPWT) to generate an initial line balance, and then improve that initial solution. For Mixed Model Production, the RPWT will use the *average weighted* Workstation time using the logic presented in Section 7.5.1. The weighted service time will use the product demand percentages and the product dependent task times given in Table 7.4. The Weighted Average task times (WATT) are shown in the last row. For example, Task 6 WATT is given as 10. This is calculated as 0.50(8.0) + 0.25(12.0) + 0.25(12.0) = 10. The RPWT will now be applied as before to the set of *average* task times. The required tasks are numbered 1-17. The red letters of the alphabet (A, B or C) next to each task correspond to the products which must be processed through that task. For example, only product B requires Task 8, while

222

Task 10 is required of all products. The precedence diagram is shown in Figure 7.11. Below each task box in Figure 7.11 is the Ranked Positional Weight for each task. For example, the Ranked Positional Weight for Task 12 is given by (10.50+9.50+7.0+2.0) = 29.0. Of course, the time to execute each task by product is product dependent as shown in Table 7.4

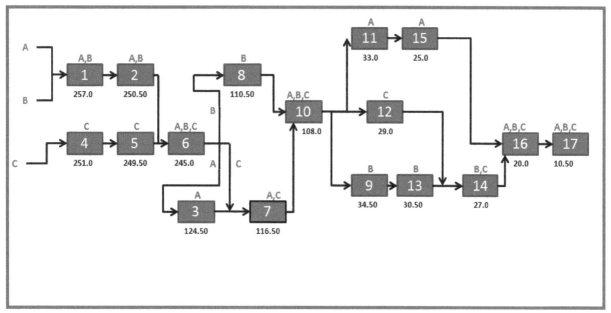

Figure 7.11
RPWT Results

The solution procedure is as follows.

Open up Workstation 1 for task assignments. The initial list L of eligible assignments (those with no precedence constraints) is L= {1, 4}. Tasks 1 has the largest positional weight (257), so remove Task 1 from the list L and assign it to Workstation 1. Workstation 1 now has [12-6.5] =5.5 min of unallocated time. List L now becomes L= {4, 2}. Since Task 4 has the largest positional weight (251) and only requires an average of 1.5 minutes, allocate Task 4 to Workstation 1 and Purge Task 4 from the L list. Calculate the remaining available processing time [12-6.5-1.5] = 4 Min. Add Task 5 to the L list so that L= {2, 5}. Task 2 now has the largest positional weight (250.5), but the average time required for Task 2 (5.5 Min) exceeds the remaining available time (4 Min). Pass Task 2, and consider Task 5. Task 5 requires an average of 4.5 Min of processing time, which also exceeds the available remaining time at Workstation 1. Hence, close Workstation 1; Open Workstation 2 and reconstruct L = {2, 5}. Task 2 has the largest positional weight (250.5) and can be assigned. Allocate Task 2 to Workstation 2. Calculate the remaining available time at Workstation 2 [12-5.5] =6.5. No task can be added to the L list, so that L= {5}. Task 5 requires only 4.5 minutes of processing time, and 6.5 is available, so allocate Task 5 to Workstation 2; Calculate the remaining available processing time [12-5.5-4.5]= 2. The L list is now L= {6}. Since the processing time for task 6 (10.0) is more than the remaining available time, Close Workstation 2; Open Workstation 3 for assignments; set L= {6}. Proceed to allocate Tasks to Workstations until all tasks are assigned to some Workstation, such that precedence constraints and Takt time are not violated. The final Line Balance requires 11 Workstations and is shown in Figure 7.12. The Workstation assignments are

223

shown below the task box and the Workstation Cycle Time above the task box. The Bottleneck Workstation is Workstation 7 (CT=12). The Manufacturing Cycle Time is given by (12*11) = 132 time units. The total line idle time is given by 24 Minutes, and the line Efficiency is given by [$(\frac{132-12}{132})$*100] = 90.91 %.

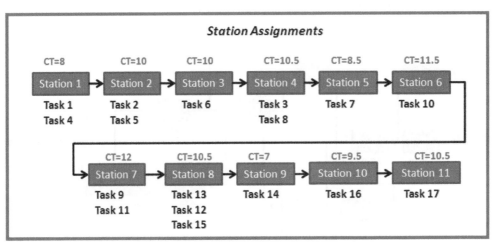

Figure 7.12
Final Station Assignments

A *Balance Graph* illustrating the required processing times for all four products and the average required time is shown in Figure 7.13. Note that some of the *individual* task times at each Workstation exceed Takt time. This is allowed, but the *Average Workstation Time* must not exceed Takt=12 Minutes.

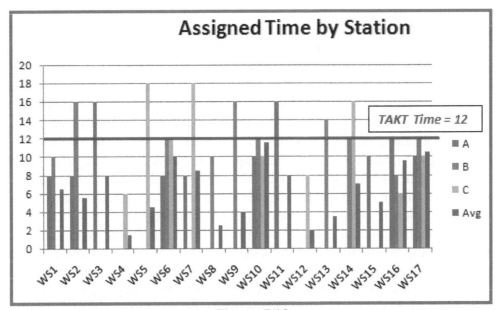

Figure 7.13
A Balance Graph

224

It might be interesting to ask the following question. If there was only room for 8 workstations, what would be the minimum Takt time to support this scenario? In this case, the procedure would be to increase Takt time until the minimum number of workstations required was equal to eight. The correct way to do this would be to determine a balance for a specified Takt time =13, 14, 15…..until eight workstations are all that is required. Alternatively, one could set the Takt time to some upper bound of, say, [2*Takt] and successively reduce the upper bound until a balance is found for eight workstations.

7.7 Maintaining Balance

The Lean Literature is full of philosophy as to how Balanced and Level production can be established, but there is not much on how it can be adjusted and maintained. The most recent line of thinking is to adapt flexible and versatile manufacturing capabilities such that machines and resources can be reconfigured and changed as production requirements and customer demands change. We have seen one or two companies with sophisticated production lines that can accomplish this, but this is generally very difficult to sustain. Another problem which we have not addressed is the use of *Key Workstations* that are environmentally sensitive and cannot be moved. These are *called Monumental Resources* by Dr. David Sly. It is true that line reconfiguration and relocation might be required as new products are introduced, but the realistic challenge is to adapt to normal fluxuations in demand and seasonal variations without disrupting ongoing operations. Experience has shown that with a capable, well trained workforce if demand, and hence Takt time, changes by only 10-15% over a short period of time; that this change can usually be absorbed. In any case, if demand changes over an extended period of time or demand spikes occur in excess of 10-15%, rebalancing will usually be required. In addition, when any demand change occurs, it will always be necessary to change the product mix to level production.

Assuming that Takt time needs to be maintained to keep from slipping pre-set delivery requirements, there are two fundamental ways to meet increased and changing demand. The *first* is to simply increase the amount of time allocated to production, and the *second* is to increase machining capacity. Increasing the amount of time allocated to production is perhaps the most common and easiest solution, and the common management response is to simply schedule overtime. If capacity is to be increased, this requires purchasing faster machines or adding parallel processing capabilities to the bottleneck workstations. In either case, some inventory buffering might be required to offset the fact that concurrent production rarely exists at the same time that overtime is being worked.

Overtime is often a good (short term) solution over a finite planning horizon, provided that key resources and Workstations are not being used continuously over a 24 hour day. Extra or partial shifts can often be scheduled. While management does not like overtime, most workers like it to earn extra pay. Changing the production capacity or rate of production is usually more difficult. This requires adding new or modified machines or executing dedicated Kaizen events to increase capacity. Reducing set up times, changeover time, providing quick mount jigs and fixtures and design for production can all create capacity. As we will demonstrate in Chapter 12, minimizing batch sizes, streamlining material handling, increasing product quality and implementing a good off-line TPM program all effectively create more operating capacity. All of these solutions require dedicated Lean Engineering work and will not happen quickly. However, there are two other realistic short term solutions: Adding temporary parallel capacity by subcontracting work or reassigning experienced workers to take advantage of their superior skills on some tasks. It must be quickly pointed out that if cycle time requirements at bottleneck processes are not reduced, nothing will help.

Consider the following operational scenario. Products A, B, C and D are scheduled for production. The manufacturing line operates two 7.5 hour shifts per day six days a week. Hence, there is currently $5400 \frac{Min}{Wk}$ available for production. Using the required production rates for each product, the Takt time is given by $Takt = (\frac{5400}{600}) = 9.0$. The product demands, Workstation Cycle Time, Product Mix and Takt time requirement is shown in Table 7.5

Table 7.5
Creating Excess Capacity

Product	Demand (Unit/Wk)	Required Workstation Cycle Time (Min)	Product Mix
A	200	27.0	0.330
B	200	27.0	0.330
C	100	54.0	0.167
D	100	54.0	0.167
	Total	Takt Time Requirement	Total
	600	9.00	1.00

Management would like to set up a single three-Workstation production line to produce all four products. The concept is to execute all Task 1 work for products A, B, C and D at Workstation1; all Task 2 work at Workstation 2; and all Task 3 work at Workstation 3. Each product requires all three tasks in the order 1, 2, 3. The required processing times for each product at Workstation 1, and the Weighted Average Processing Time for all four products are shown in Table 7.6. For example, the product mix demands that 33% of all products passing through Workstation 1 is Product A. The time per part is 7.50 min, so the Weighted time is (0.33)*(7.5) = 2.5 minute/part.

Table 7.6
Weighted Average Processing Times

Products	Product Mix	Normal Time Required Task 1	Weighted Time Task 1	Normal Time Required Task 2	Weighted Time Task 2	Normal Time Required Task 3	Weighted Time Task 3
A	0.330	7.50	2.50	8.70	2.90	6.30	2.10
B	0.330	9.60	3.20	8.40	2.80	7.50	2.50
C	0.167	8.20	1.37	9.70	1.62	6.00	1.00
D	0.167	5.40	0.90	11.82	1.97	7.40	1.23
Average Time Required:		7.97		9.79		6.83	

226

We observe that with Takt = 9.0, the proposed production line cannot meet demand requirements, since the Weighted Average Workstation (WAW) time at Workstation 2 (9.79) exceeds the required Takt time = 9.0 Recall that individual product task times can exceed the required Takt as long as the *average time* is not exceeded. If Takt time cannot be met, a typical management strategy would be to either increase available working time or increase the capacity to produce. Although the product demand profiles and the standard times at Workstation 2 are fixed, working overtime will temporarily fix this problem, but if this profile is expected to continue for some time, a more permanent solution may be preferred. Increasing the capacity to produce is the long run solution, but this will require capital acquisitions or some focused Lean Engineering work at Workstation 2. There is another solution which would provide both a short term and long term remedy: assign workers to tasks so that the fastest Task 2 workers (possibly more experienced workers) are assigned to Task 2. Faster workers at Workstation 2 essentially reduce the task time(s), effectively creating more capacity.

7.7.1 A Worker Assignment Strategy

Before we proceed, it might be beneficial to appreciate and understand what a *Standard Time* means. A standard time is the average time required to execute a fundamental operation or task by a qualified, experienced person under normal working conditions. Time standards can be pre-engineered standards or set by observation. If set by observation, the standard should be set using an average worker under normal operating conditions that has been fully trained to execute the task for which the standard is being set. A concept which we will define relative to a time standard is what is called *Normal Time*. Operators and working personnel assigned to specific jobs which work at the established time for that job are said to be operating at normal time or at 100%. Less trained and experienced operators may operate at some other percentage of normal time, say 110%. Experienced and seasoned veterans may take less time than the normal time, say 90%. A person who exceeds normal time on the average is typically rated between 100%-120%, while a person working less than normal time usually works between 80% -100%. One must be cautious in using these ratings. For example, if a task normal time is 50 minutes, and a worker is rated on that task at 110%, that worker will on the average complete that task in 55 minutes. A typical task time that is also sometimes adjusted with allowances, such as fatigue factors. We will not consider any predefined adjustments to normal time. The issue here is one of how to use standard times and worker task ratings to effectively increase Workstation *capacity* at no cost. Capacity can be gained by properly allocating skilled workers to bottleneck tasks, and we will use this example to illustrate the procedure. We recommend that every Lean Manufacturer or would-be-Lean Manufacturer maintain a worker/task rating database and follow this procedure to help in responding to short term capacity problems.

In this example, assume that such a data base exists, and for Worker 1 the Standard Time Ratings have been set by task and by product as shown in Table 7.7. For illustrative purposes, we will assume that this worker has been cross-trained and rated on all tasks and all products. This restriction will be relaxed in the next example.

Table 7.7
Worker 1 Standard Time Ratings

Products	Task 1	Task 2	Task 3
A	1.00%	88%	110%
B	95%	90%	120%
C	110%	100%	90%
D	90%	110%	85%

Using these standard time ratings, it is now possible to calculate the average times which one might expect to occur if Worker 1 is assigned to a particular task associated with a particular product. These times would be an adjustment of those times shown in Table 7.6 by the Worker 1 standard time ratings shown in Table 7.7. We will refer to these (adjusted) times as the *Task Execution Rating Matrix for Worker 1 (TERM1)*. The results are shown in Table 7.8.

Table 7.8
The TERM1 Matrix for Worker 1

Products	Task 1	Task 2	Task 3
A	2.50	2.552	2.310
B	3.04	2.520	3.00
C	1.507	1.620	0.90
D	0.810	2.167	0.680
TERM1	7.857	8.860	6.890

For example, the weighted normal production time listed in Table 7.6 for Product B, Task 1 is $(0.33)*(9.6)= 3.20$ minutes/part. From Table 7.8, worker 1 is rated at 95% on Task 1, Product B. Hence, that task can be completed on the average in $(3.20)*(0.95) = 3.04$ time units. All other entries in Table 6.9 are determined in the same way. The last line of Table 7.9 is the most important one to this discussion. If Worker 1 is assigned to Workstation 1, the average time to complete one set of products (A, B, C & D) is 7.97 time units. More importantly, the Cycle time for a set of Task 2 units (executed at Workstation2) is 8.860 time units, which is below the required Takt time of 9.0. In the simplest case, if workers 2 and 3 work all task times at the Normal rate, then assign worker 1 to Workstation 2, worker 2 to Workstation 1 and worker 3 to Workstation 3. This will result in an average sojourn time of 7.47 at Workstation 1; 8.860 at Workstation 2; and 6.83 at Workstation 3. This will result in a production line which can now satisfy Takt time requirements.

In general, all three workers will be rated differently when working certain tasks attached to certain products, and not all workers will be cross trained or qualified to work all tasks. We will now show haw a simple Linear Programming procedure called the *Assignment Problem* can be used to find a feasible strategy to satisfy Takt time, provided such a solution exists. Following the procedure just discussed, suppose that all three workers are task/product rated and the TERM matrix in Table 7.9(a) is available from the Industrial Engineering Methods & Standards group.

Table 7.9(a)
A Composite TERM Matrix for all Three Workers

	Task 1	Task 2	Task 3
W1	7.857	8.86	6.89
W2	****	7.21	8.10
W3	8.52	9.81	7.25

Notice that worker 2 has not been trained to do Task 1. Also note that Worker 3 cannot execute Task 2 for all products without exceeding Takt time. The assignment problem that we wish to solve is to assign M workers to N Workstations to Minimize the total time required to process all products at each Workstation. We can formulate the Assignment Problem as follows.

$$\text{Minimize } z = \sum_{i=1}^{M} \sum_{j=1}^{N} c_{ij} x_{ij}$$

$$\text{Subject to: } \sum_{i=1}^{M} x_{ij} = 1 \quad i=1,2....N$$

$$\sum_{j=1}^{N} x_{ij} = 1 \quad j=1,2....M$$

$$x_{ij} = 0 \text{ or } 1 \quad \text{all I, all j}$$

The c_{ij} values are the ith row and j^{th} column entry in Table 7.9. The x_{ij} values are the solution variables. If Worker i is assigned to Task j, then x_{ij}=1. If not assigned, then x_{ij}=0. This formulation assures that only one worker will be assigned to each Workstation and that the objective function will be minimized. If a Worker cannot be assigned to a task, then add a large number to the largest c_{ij} in the Term matrix and enter that number into the corresponding cell. Note that M=N (required). The Assignment problem can be solved in a matrix form in its own space. The rules can be found in any Operations Research Textbook, such as Phillips, Ravindran and Solberg.

Step 1: Construct a 3x3 matrix from entries in the Composite TERM Matrix. Note that from Table 7.9 (a), Worker 2 cannot be assigned to Task 1. Hence, the matrix entry in Row 2 and Col 1 will be arbitrarily assigned a large number (10.0 + 9.81) = 19.81 (See table 7.9 (b)).

Table 7.9 (b)

	Task 1	Task 2	Task 3
W1	7.857	8.86	6.89
W2	19.81	7.21	8.10
W3	8.52	9.81	7.25

Step 2: Examine each row and column. (a) Subtract the ***minimum*** number in each row from every other number in that row. Next (b) from this set of new numbers, subtract the ***minimum*** number in each column from every entry in that column. Step 2 will always result in a number of cells that have zero entries (See table 7.9 (c) and table 7.9 (d))

Table 7.9 (c) (d)

	Task1	Task2	Task3
Worker1	0.967	1.97	0.0
Worker2	12.60	0.0	0.89
Worker3	1.27	2.56	0.0
Worker1	0.0	1.97	0.0
Worker2	11.33	0.0	0.89
Worker3	0.0	2.56	0.0

(c) applies to the first three rows; (d) applies to the last three rows.

Step 3: Since we are trying to assign a single worker to each Workstation in such a way as to minimize the total assignment *cost (*time*)*: We now try to make an assignment using matrix (d) which will result in *zero cost*. To do this, there must be only one assignment in each row and each column. By inspection, there are two solutions which exist for this problem. The time required at each Workstation must be the appropriate Task times recorded in Table 7.9 (d).

Solution 1: Assign Worker 1 to Task 1, Worker 2 to Task 2 and Worker 3 to Task 3
Solution 2: Assign Worker 3 to Task 1, Worker 2 to Task 2 and Worker 1 to Task 3

(1) Solution 1 will yield a Workstation 1 sojourn time of 7.857, a Workstation 2 sojourn time of 7.21 and a Workstation 3 sojourn time of 7.25.

(2) Solution 2 will yield a Workstation 1 sojourn time of 8.52, a Workstation 2 sojourn time of 7.21 and a Workstation 3 sojourn time of 6.89

Both solutions have resulted in all Workstations operating below the Takt time requirement of Takt = 9.0, without any extra investment or cost and without working overtime. The modeling approach is the Assignment problem and if a feasible solution cannot be immediately obtained from Steps 2 and 3, the solution procedure may require a number of extra steps other than those shown here. This methodology can also be used with more workers than machines, and will always find a solution. The reader is referred to any one of many Operations Research textbooks on the market, including Phillips, Ravindran and Solberg (20). Once Task and Worker assignments have been determined which will satisfy Takt time, it is necessary to determine a production sequence which will result in a smooth flow of products and level production downstream.

7.7.2 Sequencing the Products

Now that the task allocations and worker assignments have been determined which satisfy the Takt time requirement, it is now necessary to turn our attention to the production sequence which will be used at each workstation. A common production sequence which is sometimes used is to simply release product using production lot sizes as the demand quantities for each product. This is called a *no lot splitting policy* or *maintaining order integrity*. For this example, the production requirements are A=200, B=200, C=100 and D=100. A possible production sequence is shown in Figure 7.14.

This is not a good production sequence (although it might minimize set-up/changeover times), since Product D cannot be sent downstream until all of the previous A, B and C products have been completed. Of course, one could still prevent starvation of downstream processes if enough make-to-stock is stored as in-process WIP in front of downstream processing stations.

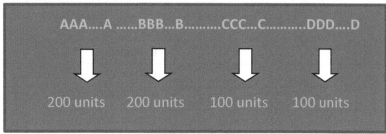

Figure 7.14
A Possible Production Sequence

However, this is not an option to the Lean Engineer. Such practices should be avoided, since they cause excessive WIP and long throughput times. The correct Lean Engineering plan is to use a Mixed Model Sequence (MMS) which continuously produce a product mix and levels downstream production. One possible sequence is as follows. A=10, B=10, C=5 and D=5.....repeated 20 times. The astute reader will recognize that while there are many possible MMP sequences, each sequence will require a set of associated set-up and changeover (SCT) times. For example, the first production sequence (A=200, B=200, C=100 and D=100) will require four different SCT....A to B, B to C, C to D and D back to A; assuming that this demand sequence will repeat itself. For our given demand rates, the first production sequence will require a total of (4*20)=80 different SCT .If any one setup and changeover time took between 5 and 10 minutes, between 400 minutes and 800 minutes of production time would be lost. This is the reason that Ohno put so much emphasis on SMED in the Toyota Production System. To properly frame the problem and propose a solution procedure, we need to recognize that set-up and changeover times are usually sequence dependent. For a four product sequence, there are exactly (4*3*2*1) = 24 different sequences that are possible. It is also possible that set-up and changeover times could be Workstation dependent. An optimal basic production sequence is one that will minimize total SCT. To simplify the presentation, we will assume that SCT is not Workstation dependent, although this is not a limiting assumption. For numerical purposes, assume that Table 7.10 contains the set-up and changeover times for products A, B, C and D in minutes.

Table 7.10
Product Dependent SCT for Four Products (Minutes)

	A	B	C	D
A	0	6	10	14
B	4	0	3	9
C	6	1	0	8
D	4	3	6	0

The Table 7.11 lists the 24 different sequencing options for four products, and the total SCT time for each permutation.

Table 7.11
Four Product Permutations

Sequence	Time Required	Sequence	Time Required
A-B-C-D	21	C-A-B-D	27
A-B-D-C	27	C-A-D-B	26
A-C-B-D	24	C-B-A-D	25
A-C-D-B	25	C-B-D-A	24
A-D-B-C	26	C-D-B-A	25
A-D-C-B	25	C-D-A-B	21
B-A-C-D	25	D-A-B-C	22
B-A-D-C	25	D-A-C-B	23
B-C-A-D	26	D-B-A-C	25
B-C-D-A	21	D-B-C-A	26
B-D-A-C	24	D-C-A-B	27
B-D-C-A	27	D-C-B-A	25

For example, the sequence B-D-C-A yields a value of (9+6+6+6) = 27. There are three alternate optimum solutions: A-B-C-D=21, B-C-D-A=21 and C-D-A-B=21. We now need to note that since the product demand is A=200, B=200, C=100 and D=100 the three optimal solutions must follow the following ratio of lot size sequences: (A, B, C, D) = (2:2:1:1). It should be clear that, for example, one of the optimal lot size sequences is to run (B=200, C=100, D=100 and A=200). This is one of three solutions that minimizes SCT=21 minutes. However, we have already noted that such a solution is not a good Mixed Model Sequence. Here we can note that any subset of this solution that maintains production lot size ratios of (A, B, C, D) = (2:2:1:1) will also be optimal for that particular subset solution. For example, the Mixed Model Solution of (B=10, C=5, D=5 and A=10)….repeated 20 times is optimal with respect to that particular production lot size and Mixed Model Sequence. The only possible set of such solutions for this example is shown in Table 7.12

The question of which solution is best is dependent upon many factors, but it is generally true that the preferred solution would be one with minimum lot size. However, recall that in order to satisfy customer demand, Takt time cannot be violated. That means that any feasible solution is one in which the sum of processing plus SCT time cannot exceed Takt time. For this example, the total weighted average processing time for each option was calculated in Table 7.6. Recall that the SCT times are the same at each Workstation.

Table 7.12
Subsets of All Optimal Solutions

ID	Solution (2:1:1:2)				Reps	Time(Min)
	B	C	D	A		
1	200	100	100	200	1	21
2	100	50	50	100	2	42
3	50	25	25	50	4	84
4	20	10	10	20	10	210
5	10	5	5	10	20	420
6	4	2	2	4	50	1050
7	2	1	1	2	100	2100

We can now combine the required Processing times and the Set-up/Changeover times to get a total time required at each Workstation. These results are shown in Table 7.13
There are several interesting conclusions that can be drawn from Table 7.13.

Table 7.13
Total Average Sojourn Time

WS	PT (Hr)	Process Time + SCT1 (Hr)	Process Time + SCT2 (Hr)	Process Time + SCT3 (Hr)	Process Time + SCT4 (Hr)	Process Time + SCT5 (Hr)	Process Time + SCT6 (Hr)	Process Time + SCT7 (Hr)
1	7.857	8.207	8.557	9.257	11.357	14.857	25.357	52.857
2	7.210	7.560	7.910	8.610	10.710	14.210	24.710	52.210
3	7.250	7.600	7.950	8.650	10.750	14.250	24.750	52.250

(1) Since Takt is 9.0 hours, the only feasible solutions are those shown in the blue shaded cells. The best MMP solution is to run production lot sizes of B=100, C=50, D=50 and A=100, in that order, repeated 4 times. The total time (Production + SCT) at each Workstation is Workstation 1 = 8.557 hours, Workstation 2 = 7.910 hours and Workstation 3 = 7.950 hours. The system Throughput Time is TPT=23.607 hours, and the system Output Rate=one unit every 8.557 hours.
 (2) There is an opportunity to use production lot sizes of B=50, C=25, D=25 and A=50 if a SMED- Kaizen study is directed at the set-up and changeover times at Workstation 1. A reduction of (Process time +SCT) = 9.257 to below 9.0 would allow these production lot sizes.
 (3) Table 7.13 provides targets for SMED studies which could allow for other Mixed Model production sequences. (4) Recall that there were three alternative optimal solutions using different Production Sequences. A complete investigation of optimal operating policies would require the construction of two more tables similar to Table 7.13, which reflect different changeover times. It is possible that there is a better solution than that which we have obtained. The reader will be asked in a homework problem to see if this is true or not. Finally, if the SCT times are actually Workstation dependent, the above procedure is still valid but will require more Workstation specific calculations and Tables. The entire procedure is easily programmed in EXCEL or any programming language.

7.7.3 Computational Difficulties

The sequencing solution presented in Section 7.7.2 is based upon total enumeration of the different permutations (ordering) of N products. The example we used required the examination of N=4 products with 24 different permutations. For N=5, this would require 120 permutations and for N=6 it would involve 680. Since most dedicated production lines or *U-Shaped Manufacturing cells* are dedicated to a small number of products, modern computers are capable of enumerating a fairly large number of products in a short period of time. But what if N is large, say 25-50 products? In this case a *Monte Carlo Procedure* is suggested. The procedure is to randomly generate a large number of possible sequences and evaluate the time required for each successive sequence, retaining the pair-wise least time feasible solution sets. Assume that a million solutions are randomly generated and keep only those solutions which meet a specified Takt time. Depending upon the value of N, the feasible solutions can be rank-ordered. Limited computational experiments by the authors have yielded interesting results. For N=52, the best solution obtained for several test problems was within 5%-10% of the global optimum solution. In any case, such a solution is certainly better than a guess and serves as a platform for improvement.

7.8 Set-up and Changeover Times Revisited: General Observations

For multiple product production lines, there are several batching policies that could be used to satisfy demand, each of which will directly impact the amount of production time lost to set-up and change-overs. In Lean Manufacturing, if N products consisting of $n_1, n_2, n_3 \ldots n_N$ units are to be produced, and if a Kaizen event directed to SMED (Setup Time Reduction) has resulted in minimum setup and changeover times, then any product sequence can be used. Line rebalance is unnecessary in this case since the RWPT balancing heuristic previously used is based upon an average of all parts produced. For example, suppose that there are N=3 parts that are to be produced with demand n_A=500, n_B=200 and n_C=100. The required production mix is in the ratios 5:2:1. One way to do this is to produce 500 units of part A, 200 units of part B and 100 units of part C. Each production machine must be set up 3 times in a full production run. This production sequence has in fact historically been specified by production scheduling and accounting because the setup times for A, B, and C can be amortized over the maximum production sequence. However, we clearly demonstrate in Chapter 13 that this sort of batching causes variance components to significantly increase, with a corresponding increase in WIP and Throughput Time.....*the pay me now or pay me later syndrome*. Architects of the Toyota Production System (TPS) recognized this phenomenon fairly early, and spent a great deal of time and effort to minimize setup and changeover times. This effort was known as SMED. Further, emphasis was placed on using simple multipurpose machines which were designed to minimize setup time. Production and design engineers joined forces to design for minimum changeover. The results are well documented and often spectacular, reducing setup and changeover times in some cases from hours to minutes. As a result, Toyota was able to operate and balance their lines with almost any mixed model production sequence which could then be modified at any time to reflect changes in demand.

For this example, the Lean Engineer would study the downstream effects of Mixed Model Sequencing, and might prefer to execute the sequence AABACBAA three hundred times. Even with SMED exercises, this strategy may not be feasible without dedicated production work cells, multiple servers or both. In any case, the success of MMP greatly depends upon the amount of effort put into SMED, KAIZEN and TQM. Finally, recall that if t_p time units of available working time are consumed by setup and changeovers, the net effect is a significant and avoidable loss in production capacity....a fact often overlooked or ignored by manufacturing managers. For long term changes parallel or dedicated production lines may need to be introduced. To illustrate several principles, consider the weekly production requirements shown in Table 7.14. Available working time is 960 minutes per day.

Table 7.14
Weekly Production Requirements

Product	Weekly Demand	Percent of Production	Required Cycle Time
A	400	53.3%	2.4 min
B	250	33.3%	3.84 Min
C	100	13.4%	9.6 Min
Totals	750	100%	1.28 Min

Since the product demands are 400 A's, 250 B's and 100 C's per week, a possible mixed model production sequence is AAAABBCCBBBAAAA, repeated 50 times. Now suppose that the demand for product C increases to 200 units per week. The weekly production profile for a 960 min workday is shown in Table 7.15. The required cycle time is 1.13 min or 85 parts per day.

Table 7.15
Weekly Profile

Product	Weekly Demand	Percent of Production	Required Cycle Time
A	400	47.1%	2.4 min
B	250	29.4%	3.84 Min
C	200	23.5%	4.8 Min
Totals	850	100%	1.13 Min

Note that the original required cycle time decreases from an average of 1.28 min per part to 1.13 min per part. This will require a working day of 1088 min or 2.13 hours of overtime each day. This perhaps the best alternative, although the production sequence will change and in-process inventory will increase. For example, a product mix of 8 A's, 5 B's and 4 C's repeated 50 times will now satisfy demand. As usual, the way in which this production mix is sequenced will depend upon a Kaizen/SMED Blitz to reduce set up times as much as possible. Note that if the initial mixed model production sequence had been used prior to this change in demand, the sequence AAAABBCCCCBBAAAA can now be implemented with minimum disruption, and will not change the number of setups required.

Now suppose that management is not willing to schedule overtime (a common phenomenon). In that case, the current amount of production time per day will not change, and other ways must be found to accommodate the new production requirement for product C. It is clear that the production cycle time for product C must be roughly cut in half. Some time might be gained by labor reallocation, but the decrease in product C cycle time might be too much to realize. If business is booming and production requirements for all products are expected to increase, parallel processing might provide a solution. If the increase in demand for product C is expected to continue for some time and then possibly decrease to the previous level, subcontracting might provide a reasonable solution. In any case, it will probably be necessary to restructure line balance, examine the amount of in-process inventory levels to smooth production and possibly use a combination of alternatives.

7.9 Responding to Demand Changes: General Observations

There is only one thing which is certain to remain unchanged in production planning, and that is the fact that demand will constantly change. The general strategy of MRP is to try and forecast or predict future demand, and try to smooth production by averaging order release quantities across a specified planning horizon. As discussed elsewhere in this text, even MRP II systems with capacity planning modules will release work to the shop floor based upon a set of lead times and estimated processing times, often without knowledge of current operating conditions. A properly executed Lean order release strategy always depends upon production status, since it is driven by a Pull philosophy which virtually eliminates scheduling based upon lead times. In many computerized order release systems, serious imbalances in production are covered up by long lead time quotations to the customer, and usually result in expensive and inefficient make to stock or make to order inventory policies to compensate for fluxuating customer requirements. In all cases, Work in Process is always inflated. Conversely, the proper Lean Strategy is to institute pull strategies which dynamically and directly respond to customer demand by initiating upstream production only when downstream WIP or production requires more parts. This strategy prevents WIP buildup and places a *WIP Cap* on all inventory in the system. This is particularly evident in U-Shaped production and assembly lines which are paced by labor and not production capacity. Short term imbalances can usually be absorbed using the tools and techniques discussed previously, but beyond that point radical line changes may be required. In all cases, it is likely that the production mixed model sequence will need to be adjusted and modified to balance production and maintain level flow.

7.10 Solving Short Term Imbalance in Production Requirements

We simply mention three viable alternatives with selected comments.

Parallel Redundancy

A typical way to do this is by temporarily assigning another machine(s) to the bottleneck workstation. This is sometimes called *parallel redundancy*. There are several reasons why parallel redundancy might be the best option.

(1) If the parallel machine is a flexible machining platform, it might be shared between two production lines provided that set-up and changeover times are minimized. This is particularly important when dealing with bottleneck processes. Of course, some advantages might be offset by corresponding increases in variance components.

(2) Parallel machines rarely if ever fail at the same time. Parallel redundancy protects against line stoppage, and removes the need for excessive in-process inventory to protect against unreliable equipment. Of course, this in no way eliminates the need for a good preventative maintenance program.

(3) Introduction of temporary or periodic changes in line requirements can often be difficult or even impossible at bottleneck Workstations without parallel available capacity.

(4) Production line workers often work more efficiently and make fewer mistakes when not constantly pressured to maintain Takt time. The issue here is not to allow for worker idle time, but to try and avoid work induced variance components due to excessive demand requirements and pressure to maintain Takt time.

(5) Since parallel redundancy almost always results in available capacity, the need for working overtime and frequent line rebalance is often mitigated

Working Overtime

Another common solution in dealing with temporary production imbalance and changing demand is to work overtime. For the example in Section 7.4.3 one can verify that by working one hour per day overtime at Workstation 2, the Takt time of 9.0 can be achieved. However, since Workstation 3 works only 7.5 hours per day, there will be an increase of in-process inventory at Workstation 3 during this overtime period, and will need to be worked off the next day. This would be true in any case, but particularly for a one shift operation.

Subcontracting or Outsourcing

If product demand is too great to absorb over a long period of time, an alternative strategy to increase production might be to outsource or subcontract either assemblies or subassemblies. In fact, many companies are moving to this strategy to take advantage of subcontractor efficiency and specialized knowledge, avoid capital investment costs to procure additional scarce resources and to enforce just-in-time strategies on assembly lines. One word of caution: If too much technology and expertise is exported to a second party, it is easy to lose the understanding of how to satisfy technical requirements in complex assemblies/subassemblies. This loss of intellectual capacity could prove disastrous if the subcontractors pull out.

7.11 Summary

In this Chapter we have considered the need to balance and smooth production. Several methods were presented that can be used to accomplish these goals. Both single model and mixed model production scenarios were presented and demonstrated in a series of numerical examples.

Review Question

1. Work-in-process (WIP) has been called the *root of all evil* in manufacturing systems. Is this true? Explain your answer. Discuss the meaning of this term and the role of WIP.
2. What is *root cause analysis* and how is it related to *Kaizen events*?
3. What is *Heijunka* ?
4. Discuss (a) *process leveling* and *process balancing.*
5. What is Andon? Use the internet to provide examples of Andon.
6. Why have large production lot sizes been used historically?
7. What is consistency in Lean Manufacturing?
8. Define SMED. Why Is SMED so important to Lean Manufacturing strategies? Use the internet to find SMED success stories and report three of these.
9. Explain the difference in *balancing, synchronization and leveling.*

10. What is a Focused Factory? Under what conditions would a Focused Factory be appropriate? How would one determine which products would be produced?

11. Precisely describe the term Takt time. What is the relation between Takt time and output rate? If a Manufacturing System has multiple Subassembly cells or work cells dedicated to producing a finished product, would they all follow the same Takt time?

12. Discuss the impact of variance on Line Balancing.

13. The following data has been collected for a sequenced production system.

Task ID	Task Time (Min)	Task ID	Task Tim21e (Min25)	Task ID	Task Time (Min)
1	22	7	16	13	24
2	34	8	25	14	32
3	10	9	22	15	15
4	18	10	18	16	23
5	25	11	15	17	39
6	36	12	19	18	17

The following precedence relationships must be maintained.

Task ID	Predecessor Tasks	Task ID	Predecessor Tasks	Task ID	Predecessor Tasks
1	None	7	1	13	12
2	None	8	1	14	12
3	2	9	4,5	15	12
4	3	10	6,7,8	16	13,14,15
5	3	11	10	17	13,14,15
6	1	12	11	18	17

The production line operates 16 hours per day, 6 days per week. There is a 15 minute rest time each shift, and a 30 minute lunch break. (1) Draw the precedence diagram (2) Calculate the required Takt time. (3) Calculate the minimum number of required workstations. (4) Assign tasks to workstations using the Largest Eligible Processing Time Rule. (5) Calculate Workstation Efficiency and Line Efficiency. Is your final solution a good line balancing solution? Improve the initial solution by visual inspection. (6) Compare the improved solution to the initial solution.

14. Rework Problem 13.0 using the Ranked Positional Weighting Technique (RPWT). (1) Compare the initial RPWT solution to the final solution obtained in Problem 13.0. (2) Improve upon the RPTP initial solution. (3) Would the RWPT solution technique always be superior to the Largest Eligible Processing Time Rule? Why or why not?

15. The following Task – Time data has been collected for a proposed manufacturing assembly line.

Task ID	Task Time (Min)	Task ID	Task Time (Min)
1	12	7	8
2	5	8	6
3	6	9	2
4	4	10	5
5	4	11	6
6	13	12	8

The following precedence relationships must be maintained.

Task ID	Predecessor Tasks	Task ID	Predecessor Tasks
1	None	7	3,4
2	1	8	7
3	2	9	5
4	2	10	9,6
5	2	11	8,10
6	2	12	11

If Takt time has been set at 16 minutes; (1) Draw the precedence diagram (2) Calculate the required Takt time. (3) Calculate the minimum number of required workstations. (4) Assign tasks to workstations using the Largest Eligible Processing Time Rule. (5) Calculate Workstation Efficiency and Line Efficiency. Is your final solution a good line balancing solution? Improve the initial solution by visual inspection. (6) Compare the improved solution to the initial solution.

16. The following Task – Time data has been collected for a proposed manufacturing assembly line. (Gaither)

Task ID	Task Time (Min)	Task ID	Task Time (Min)	Task ID	Task Time (Min)
1	6	7	7	13	7
2	3	8	4	14	3
3	9	9	9	15	5
4	5	10	8	16	8
5	8	11	11		
6	10	12	12		

The following precedence relationships must be maintained.

Task ID	Predecessor Tasks	Task ID	Predecessor Tasks	Task ID	Predecessor Tasks
1	None	7	4	13	11
2	1	8	4	14	10,12
3	2	9	5,6	15	13,14
4	2	10	6,7	16	15
5	3	11	8		
6	2	12	9		

If Takt time has been set at 16 minutes; (1) Draw the precedence diagram (2) Calculate the required Takt time. (3) Calculate the minimum number of required workstations. (4) Assign tasks to workstations using the Largest Eligible Processing Time Rule. (5) Calculate Workstation Efficiency and Line Efficiency. Is your final solution a good line balancing solution? Improve the initial solution by visual inspection. (6) Compare the improved solution to the initial solution.

17. In Line Balancing problems with a fixed Takt time, if any one Task is greater than the Takt time, a common procedure is to either work overtime to meet Takt time requirements, or to add one or more machines to reduce the average task time per unit. Both Gaither and Chase have proposed a Line Balancing algorithm which automatically calculates the number of machines required as part of the heuristic solution procedure. This method is called the *Incremental Utilization Heuristic* or the IUH. This heuristic is appropriate when one or more required task times cannot meet Takt time. The IUH proceeds as follows. Assume for simplicity that there is a single start task. Create the first Workstation with that task using a single machine. If the task time is greater than Takt time, add the required number of machines to that Workstation to make Workstation utilization less than one, add tasks to that workstation in order of precedence one at a time continuing to add machine capacity until Workstation utilization is equal to 100% or **Utilization Decreases.** When this happens, close out this Workstation; start a new one; and repeat the process until all tasks are assigned to workstations. Using the following set of data, use the IUH heuristic to determine line balance, and how many Workstations and Machines are required. Calculate both machine and line efficiency. Takt time is 0.10 Minutes. What is the minimum number of machines required.

Task ID	Task Time (Min)	Task ID	Task Time (Min)	Task ID	Task Time (Min)
1	0.18	7	0.38	13	0.48
2	0.11	8	0.42	14	0.30
3	0.32	9	0.30	15	0.39
4	0.45	10	0.18		
5	0.51	11	0.36		
6	0.55	12	0.42		

The following precedence relationships must be maintained.

Task ID	Predecessor Tasks	Task ID	Predecessor Tasks	Task ID	Predecessor Tasks
1	None	7	6	13	11,12
2	1	8	7	14	13
3	1	9	8	15	14
4	1	10	9		
5	2,3,4	11	10		
6	5	12	10		

18. If *a single machine Workstation has a processing time which is greater than required Takt time, the processing time per unit can be reduced by 50 % by adding another identical machine.* Under what conditions will this statement be true? Under what conditions will it be false?

19. Three products (A, B and C) are to be sequenced through a single machine. If the processing times are A=5 min, B= 7 min and C=8 minutes and the set-up and change over times are as given below in minutes, determine an optimal sequence for A, B and C which minimizes total Throughput time.

Product	A	B	C
A	0	12.5	16.2
B	9.5	0	10.7
C	8.2	12.3	0

20. Using a hand simulation, determine the throughput time to make one set of products using a single machine if A=6 min, B=4 min and C=3 min. if the sequence is A then B then C. How much time can be saved by adding a second identical parallel server? What sequence would minimize throughput time with one server? Two servers

21. A Mixed Model Production Line makes Products A, B and C for the demands A=50 Units/week, B= 25 Units/week and C= 25 Units/Week involving 12 different machining tasks. The time required per task (in minutes) by product type, and the precedence relationships are shown below. For example, all parts go through Tasks T1, T2 and T3 and tasks T1, T2 and T3 must all be completed before task T4 can be started. Assume that the line runs three 8 hour shifts per day. (1) Calculate the Takt time for this line. (2) Calculate the weighted average processing time per task. (3) Use the Largest Positional Weight Technique to assign tasks to Workstations. (4) Show the final Workstation assignments (5) Calculate line efficiency (5) draw a Balance graph for the final solution.

Mixed Model Production

Part	T1	T2	T3	T4	T5	T6	T7	T8	T9	T10	T11	T12
A	10	8	*	*	11	12	10	11	*	11	*	9
B	12	11	*	11	9	9	13	14	*	10	8	12
C	8	12	7	10	12	14	12	10	12	9	13	10

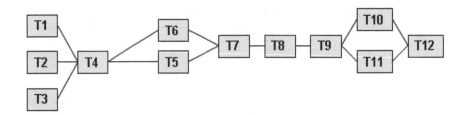

22. Assume that the production time requirements (Minutes/part) for four products (A, B, C, D, and E) assigned to three Workstations is as follows.

Products	Normal Time @ Workstation 1	Normal Time @ Workstation 2	Normal Time @ Workstation 3
A	8.25	8.75	7.76
B	7.22	7.56	9.78
C	9.10	9.22	6.98
D	7.81	6.97	7.22
E	5.32	7.23	8.55

The mix of products to be produced are A=10%, B=20%, C=40% and D=30%. (1) What is the minimum Takt time that can be achieved with this data? (2) The three workers normally assigned to each Workstation are given a one month training course to increase their efficiency. The following Table describes the new worker ratings at each Workstation by Product.

Worker 1

Products	Rating: WS1	Rating: WS2	Rating: WS3
A	0.98	0.91	0.95
B	0.92	0.88	0.85
C	0.79	0.98	0.88
D	1.00	0.91	0.90
E	0.88	1.0	0.89

For example, Worker 1 took 8.25 minutes per part on Product A at Workstation 1 before training, but can now accomplish this same task in (8.25 * 0.98) = 7.61 Min per part. (1) Calculate the new Product/Workstation standard times for each worker (2) determine the new minimum Takt time which can now be attained assigning Worker 1 to Workstation 1, Worker 2 to Workstation 2 and Worker 3 to Workstation 3. (3) Assume that each Worker will be assigned to a Workstation by using the Assignment algorithm presented in Section 7.7.1. What is the best Takt time which can be achieved and where will each worker be assigned?

23. Assume that a process needs to produce A=50, B=25, C=50 and D=10 units of product. The Product dependent Set- up and Changeover times in hours for these four products are as follows.

	A	B	C	D
A	0	1.67	0.87	1.26
B	1.23	0	0.76	1.23
C	1.00	0.89	0	2.21
D	1.55	1.22	1.10	0

(1) List all of the product changeover sequences that are possible and calculate the total time consumed by each. (2) What is the optimal product mix which can be used? (3) How many different subset optimal solutions can be identified? (3) Calculate the Total Average Time required for each of the subset optimal solutions. (4) Recommend a product mix and production plan that minimizes the total time required.

24. Reconsider the Line balance obtained for Black's Sweat shop. Show that a better solution can be obtained by using a lean, U-shaped cell.

Thoughts
   and Things……

Chapter 8
Lean MSP Implementation Methodology

If you keep doing the same bad things over and over again, how can you expect a different outcome?
Mark DeLuzio, Danaher

8.1 Introduction

In designing the TPS, the Toyota Motor Company redesigned the final assembly production system. They changed the final assembly lines into a Mixed Model final assembly line to level the demand on their suppliers. They converted the linear subassembly lines into U-shaped subassembly cells. They redesigned the job shop into a lean shop composed of manufacturing cells.

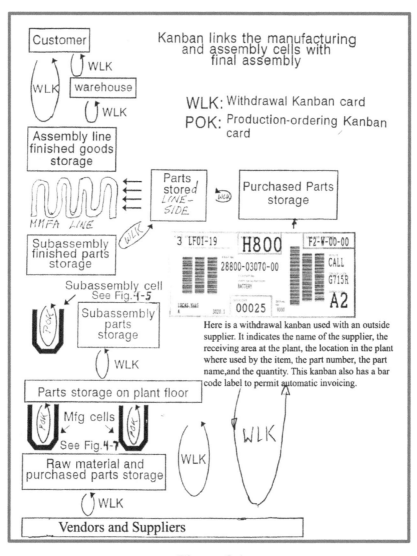

Figure 8.1
A Lean Linked Cell Manufacturing System

Final assembly operates with a Takt time based on a daily demand and designed cells balanced to produce that daily demand. The new linked-cell system, shown in Figure 8.1 can operate on a single-piece flow basis. Many think of lean production conversions as a collection of lean tools or building blocks as shown in Figure 8.2. But clearly all these tools and techniques

KAIZEN FOR CONTINUOUS IMPROVEMENT			
PULL / Kanban	MANUFACTURING, MACHINING AND ASSEMBLY CELLS		TPM
JIDOKA / QUALITY / 6 SIGMA	POUS	SMED	VISUAL CONTROL
STANDARD WORK	POKA – YOKES DECOUPLERS	A3 REPORTS	TEAMS
5S SYSTEM	PROCESS MAPPING VSM	JUST IN TIME Takt TIME	MMFA /

Figure 8.2
Typical Display of Lean Tools. Which One Do You Do First?

cannot all operate independently of one another. This Chapter provides a proven, integrated methodology for implementing lean manufacturing. In Chapter 3 we learned that the design of a lean manufacturing system can be based on axiomatic design rules. These rules are operationally true but you cannot derive them - you just evolve them out of the practice i.e., how the system works.

The final assembly (FA) line operates with a Takt time based on the daily demand for the goods being made in the factory. The daily demand is based on the monthly demand which is based on the annual demand. The leveled daily demand is what permits all the part-producing elements to be balanced to produce the same AMOUNT every day in whatever mix the downstream customer requires. Takt time is easy to calculate but it requires lots of engineering to execute.

Final assembly has to be converted to a Mixed Model final assembly that makes any of the goods in any order. The inverse of the Takt Time is the production rate for final assembly. The Takt Time sets the pulse or production rate for the factory. Making FA mixed-model provided a leveled or smoothed demand for goods for the rest of the factory and all of the supply chain (Chapter 7).

All the goods producing aspects of the system operate on a MO-CO-MOO basis. This is also called one-piece flow or a single piece flow. Therefore, the subassembly and component part manufacturing factories must be totally redesigned into U-shaped manufacturing and subassembly cells to also achieve one piece flow with flexibility. Final assembly is already doing one-piece flow, but now all the subassemblies and all the component part manufacturing are also performed on a Make One -Check One Move One On (MO-CO-MOO) basis.

In this Chapter we will describe manufacturing cells as having machine tools that have single-cycle capability while subassembly cells will have only manual tasks. In reality, most cells are a combination of these two designs. See Chapters 4, 5 and 6 for an example.

The design of the manufacturing cells is based on a design rule wherein the machining time for any part in any machine in the cell is less than the necessary cycle time (CT). The CT is based on the Takt time, being slightly less to provide a margin of safety for the suppliers to final assembly.

For the lean system, the production and inventory control system is a pull system known as Kanban. The subassembly and manufacturing cells are connected (linked) to final assembly with Kanban links which withdraw material from the subassembly and component suppliers as needed by final assembly and give production orders to all the suppliers automatically. This rule governs the maximum inventory in any link. The maximum inventory is equal to the daily demand (DD) times the lead (L) time plus a safety stock. The lead time accounts for all the waiting and transportation time, all the delay time, processing time etc. in the links connecting the subassemblies and component manufacturing to final assembly.

Design rules help us design a linked-cell manufacturing system. Naturally it still must be operated by people using automation where it makes sense to meet the quality (perfect) and quantity (don't overproduce) needs of the system. The result of these system design rules is a robust system than can make complex goods but is simple to operate. Therefore everyone who works in the system understands how the system works, but the system can make low-cost, superior quality complex products with the minimum of daily information.

8.2 How Manufacturing System Designs Have Evolved

The linked-cell manufacturing system (L-CMS) design is an outgrowth of the previous two manufacturing system designs, the job shop and the flow shop. The job shop as a manufacturing system design evolved during the 1800s. These early factories replaced craft or cottage manufacturing when it became necessary to have powered machines. A functional design evolved because of the *method* needed to drive or power the machines. That is, water power. The job shop design became known to the historians as the American Armory System. People from all over the world came to New England to see this system, and it was duplicated around the industrial world.

In the early 1900s, the first vestiges of the flow shop began to emerge. Flow line manufacturing began for small items and culminated with the moving assembly line at the Ford Motor Company for automobile assembly. Just as in the 1800s, the world again came to see how this system worked, and this new design methodology was spread around the world.

Over time, a hybrid system -a mixture of job shop and flow shop- evolved. It is called mass production. This system permitted companies to manufacture large volumes of identical products at low unit costs. The job shop produced components in large lot sizes according to the economic order quantity strategies. The subassembly lines fed components to storage systems and to final assembly. The mass production system produced the goods that helped America win WWII and become a world leader in manufacturing. The physical design was the result of achieving certain function requirements like low cost. By the 1980s, the production control was operated by computer software

(MRP → ERP) and has been termed a push system.

After WWII, the Toyota Motor Company, led by the genius of a manufacturing engineer named Taiichi Ohno, developed a new manufacturing system design known initially as the Toyota Production System (TPS) and later the just-in-time/total quality control (JIT/TQC) system or world class manufacturing (WCM) system. In 1990, it was *finally* given a name that would become universal, *lean production.* What was different about this system compared to mass production? It was its *design!*

First, in lean production, final assembly is converted to Mixed Model so that the demand for subassemblies and components from the suppliers is the same every day. This is called leveling or smoothing production. Second, the subassembly lines are converted into U-shaped cells, often eliminating conveyor lines. Third, the job shop is converted to a lean shop (U-shaped cells). The cells operate on a one-piece flow basis like final assembly. The *subassembly* and *manufacturing cells are linked* to final assembly by Kanban to form an integrated inventory and production control system. This is now called a pull system. The result is low-cost (high-efficiency), superior quality (no defects), and on-time delivery of unique products from a flexible system. The new design operates in ways the old mass production system could not. The final assembly lines were designed (flexibly) to handle mixes of models so that quantity of all the components pulled into the final product was the same day after day. The cells are designed to also handle changes in volume and mix as well as changes in design. So when we say the lean system is flexible, we mean the system can readily adapt to changes in customer demand (volume and mix) and change in product design (new products or changes to the existing product). Flexibility in the cells is achieved by greatly reducing the setup and changeover times.

The TPS has integrated control functions which means that the system level control functions (for quality control, inventory control, production control, and machine tool reliability) are designed into (integrated) the manufacturing system. The lean production system is designed to produce superior quality products. Toyota believed in companywide quality control and taught it to everyone, from the company president down to every production workers. They were able to evolve a system that could give customers products of high reliability. This was accomplished through the design of the manufacturing system where single-piece flow methodology was used for every part and every subassembly and final assembly.

The system is designed for efficiency through the elimination of waste everywhere. The efficient use of space, people, and machine tools was a management philosophy that integrates the lean system into their business. The new design also led to reductions in variations in part sizes, lot sizes, and processing times.

Uniqueness and creativity on the part of engineers and workers lead to new and improved processes, reduced equipment failures, reduced changeover/setup time and continuously improved the design of the workplace (methods improvement). More significantly, the placement of the operations and processes in the manufacturing and assembly cells requires unique processing solutions to bring the processing times well under the Takt time. While many of these processes are well guarded company secrets, some examples are given in Chapter 5 on lean cell design.

8.3 Manufacturing Philosophy Must Change

Redesigning the manufacturing and production system requires a change of philosophy within the company. Employee involvement and teamwork are rooted in the idea that no one employee is better than another. Everyone is an associate. There are no private offices. There is no executive lunchroom. There are no preferred parking places (except for the associate of the month). The system in which management tells workers what to do and how to do it must be changed. However, this change requires that management, led by the CEO, has the courage to shift some of the decision-making power from management to the people on the factory floor. This big change is psychological and requires convincing the workers on the factory floor that nothing is more important than what they think and how they feel about the manufacturing system. For this change to work, the operators must have input to the new manufacturing system design. The way the work is done must be clearly defined by the workers. It must be *their cell design.* Their achievements must be recognized. The TPS is all about improving the capabilities of individuals.

Toyota developed a new and different factory design that was flexible, delivered quality products on time at the lowest possible cost, and did so day after day. They educated their work force and placed their best engineering talent on the production floor rather than in the design room. Those who say that Toyota is not creative, inventive or ingenious have simply looked in the wrong places for evidence. However, much of the unique processing technology was hidden in the manufacturing cells held by the sole-source vendors and suppliers to the main plant (the final assembly plant). This is one of the reasons why lean manufacturers use sole or single source suppliers. The proprietary process technology is captured and held in one place. They build their own equipment and they don't write about it in books and journals. It is difficult for those outside the company to get access to the proprietary process technology.

The TPS is a revolutionary successor to the American Armory System (the job shop operated by the Taylor's Scientific Management methods) and the Ford System but adoptions have been slow in coming. Why? The TPS is difficult to implement successfully without adopting a management philosophy that integrates the lean system into their business strategy. The company must recognize that every function in the enterprise is going to change. Lean is really the maniacal pursuit of the elimination of waste from every business process. See Chapter 18 for more discussion on the enterprise.

Integration of the production system functions into the manufacturing system required commitment from top-level management and communication with everyone, particularly manufacturing. Total employee (and union) involvement is absolutely necessary. The union leadership for the production workers will raise barriers to lean production if they feel this is just another program designed to make them work harder or eliminate jobs. In reality, it is those in middle management that have the most to lose in systems-level changes as their jobs get integrated into the manufacturing system.

Before starting a lean conversion, everyone from the production workers (the internal customer) to the president must be educated in lean production methodology and philosophy so everyone understands how lean is different from mass. Physical simulation of cells versus job shops is very helpful in accomplishing this step. The entire company will be involved in the change. The top people must be totally committed to the change, set an example, and be active leaders, and in fact, system designers. Management leaders must understand that lean MSD will lead to decisions that are opposite to current accounting practices that they have used to manage their mass production system. The selection of a few measurable parameters that track the system level change is critical. Everyone must be committed to the elimination of waste as the lean system is put in place. Placing stop cords on final assembly and converting lines to one-piece flow puts a huge burden on the internal customer. The operators must be empowered to design the cells; implement quality control, perform routine maintenance, execute production control, monitor inventory, execute process improvements and reduce setup time at each machine. The company must spread the success and reward the teams and share the gains with those who contributed. Many companies feel that bonus payments are the way to go to reward people. The reward structure of middle management must be changed to support the system design.

The TPS was invented and implemented by Ohno at Toyota over many years. Many books and papers have been written to describe the system. These authors describe all manner of characteristics of the TPS using a whole new set of terms and techniques, now referred to as Lean Tools or Lean Production Tools (Table 8.1 and Chapter 17). However, those authors failed to describe how the changeover from mass to lean should take place. What is the order in which these tools should be used?

Table 8.1
Another Example of a Lean Production laundry list

Another Example of a Lean Manufacturing Laundry List
• Value stream mapping
• Cellular Manufacturing
• JIT/TQM
• Teams
• Rapid Setup (SMED)
• Kanban (pull) for production/inventory control
• Process Mapping
• Leveling, Balancing, Sequencing
• 5-S and 5 Why Methods or Jidoka
• Autonomation
• Poka-yoke and decouplers
• Elimination of 7 Wastes
• Total Productive Maintenance (TPM)
• One-Piece Flow (Single-Piece flow)
• Standardized work
• Visual management
• Self and Serial Inspection
• Takt Time (cycle time for final assembly)
• Line side storage
• Kaizen events

The basic implementation strategy outlined in this Chapter is the amalgamation of the methods used by any companies to successfully implement lean manufacturing. These companies were not suppliers to Toyota and were not privy to the Toyota Supplier training and philosophy.

- Design the L-CMS for flexibility
  - Make final assembly Mixed Model
  - Design U-shaped subassembly cells for SPF
  - Design U-shaped manufacturing cells (interim) for SPF
  - Implement SMED for rapid changeovers -fast setups
- Integrate the critical control functions.
  - Quality control (zero defects through defect prevention)
  - Machine tool/equipment control/reliability(zero breakdowns)
  - Production control (right time, right quantity, right place) by pull -Kanban
- Autonomation (true lean cells)
  - -Autonomous control (computers, automation, robotics) of quality and quantity in the cells and systems (homemade machines)
  - Design the enterprise around the L-CMS

This design strategy is broken down into critical steps or methodology. See Figure 8.3. Notice that *autonomation,* part of lean cells design, the autonomous control of quality and quantity, comes very late in the methodology, after functional integration has been achieved. The system must be redesigned, simplified, and integrated before computers; robotics and automation are applied as part of last or concurrent engineering step. The last step is really the first step of an effort to restructure the rest of the enterprise, improving the cell designs (making them true lean cells) using machine tools and processes designed in-house for single-piece flow.

The order of the design steps outlined in Figure 8.3 is important. Many companies have implemented Kanban (pull) or made significant reductions in inventory levels before the necessary system and cell design steps were implemented. Such implementations often result in failure. Even worse, many companies have tried to implement costly computer integrated and MRP strategies without first implementing lean production resulting in computerization of an inefficient and complex manufacturing system.

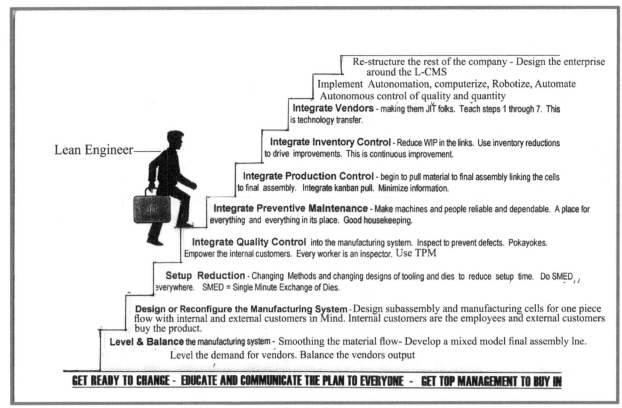

Re-structure the rest of the company - Design the enterprise around the L-CMS

Implement Autonomation, computerize, Robotize, Automate Autonomous control of quality and quantity

Integrate Vendors - making them JIT folks. Teach steps 1 through 7. This is technology transfer.

Integrate Inventory Control - Reduce WIP in the links. Use inventory reductions to drive improvements. This is continuous improvement.

Integrate Production Control - begin to pull material to final assembly linking the cells to final assembly. Integrate kanban pull. Minimize information.

Integrate Preventive Maintenance - Make machines and people reliable and dependable. A place for everything and everything in its place. Good housekeeping.

Integrate Quality Control into the manufacturing system. Inspect to prevent defects. Pokayokes. Empower the internal customers. Every worker is an inspector. Use TPM

Setup Reduction - Changing Methods and changing designs of tooling and dies to reduce setup time. Do SMED,, everywhere. SMED = Single Minute Exchange of Dies.

Design or Reconfigure the Manufacturing System - Design subassembly and manufacturing cells for one piece flow with internal and external customers in Mind. Internal customers are the employees and external customers buy the product.

Level & Balance the manufacturing system - Smoothing the material flow- Develop a mixed model final assembly lne. Level the demand for vendors. Balance the vendors output

Lean Engineer

GET READY TO CHANGE - EDUCATE AND COMMUNICATE THE PLAN TO EVERYONE - GET TOP MANAGEMENT TO BUY IN

Figure 8.3
Lean Production Implementation Methodology or Strategy

After completing the lean steps, the company will have a new look on the factory floor. If you took a picture from high above, it might look like Figure 8.4 kind of jumbled in some areas as the cells are arranged for good part movement and access by operators who may be working in more than one cell. Overall, it is a simple linked-cell system capable of making complex products -a system that everyone who works in the system understands how the system works or operated. The mass system is controlled by MRP computer software. Do you understand how MRP software keeps track of material moving through the factory (i.e. Production and inventory control) and makes sure nobody runs out of supplies? Could you fix the MRP software if something went wrong? Here is some detail on the steps to lean implementation.

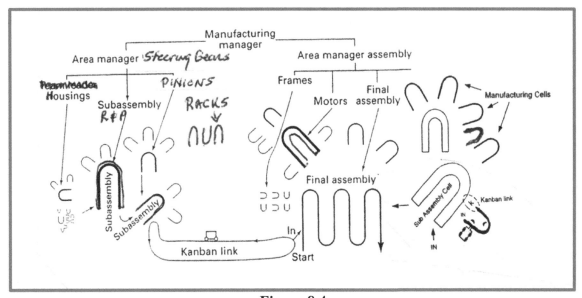

Figure 8.4
An Overview of a Linked-cell Manufacturing System with Subassembly and Manufacturing Cells Connected to Final Assembly by Kanban Links

8.4 *STEP 1*: Leveling the Manufacturing System

The lean Mixed Model production system depends upon l*eveling the manufacturing system.* In order to eliminate variation or fluctuation in quantities in feeder processes, it is necessary to eliminate fluctuation in final assembly. Leveling means the demand for the subassemblies and components from the suppliers is made equal to the daily demand. Here is a single example to show the basic idea. First, we need to calculate the production rate (PR) and its reciprocal, Takt Time, which is based on the daily demand (DD)

$$DD = Daily\ demand\ for\ parts = \frac{monthly\ demand\ (forecast + customer\ orders)}{number\ of\ days\ in\ each\ month}$$

$$PR = \frac{average\ daily\ demand\ (parts)shift}{available\ \frac{hours}{shift}} = \frac{DD}{\frac{hrs}{day}} = \frac{parts}{minute};$$

$$Takt\ Time = \frac{1}{PR}\frac{min}{part}$$

Suppose that the forecast requires 480 vehicles per day and there are 480 production minutes available (60 minutes x 8 hours/day) and there are four (4) models of cars. See Table 8.2. Thus the Takt Time is equal to one minute, meaning every one minute a car rolls off the line. Suppose you have four different models of cars going down the assembly line. The sub processes that feed the two-door fastback are controlled by the Takt Time for this model. Thus every 2.4 minutes, a subassembly line making a deck for the left back model produces a deck, specific for the left back (correct color and options). Every 1.6 minutes the door line is assembling doors for the coupe lift and the lift back as they use the same size doors at a Takt Time of 4.8 and 2.4 min/car. The doors are mounted on the body and initially painted with the body, then removed

253

when the car exits painting, placed on an overhead conveyor which transports the doors to the door line which has the same mix of product and cycle time as the main body line. The doors are trimmed (i.e., filled with windows, door handles, etc.) in sequence and in synch with final assembly. If the final assembly line stops, so does the door line. Other subassemblies, like the seat assembly lines and other main subassembly lines for parts which are special to the car model are done this way. After trimming, the doors are then returned to the line and goes back on the same car body from which is was earlier removed. This is an example of *synchronizing* or *producing in sequence.*

Table 8.2
Example of Mixed Model Final Assembly Line that Determines the Cycle Time for Multiple Car Models

Quantity	Vehicle Mix for Line Model	Takt Time for Model	Production Minutes by Model	Sequence (24 Vehicles)
100	Two-door coupe	4.8	100	TDC,TDF,TDF,FDS,FDW
200	Two-door lift back	2.4	200	TDC,TDF,TDF,FDW,FDF
50	Four door sedan	9.1	50	TDC,TDF,TDF,TDS,TDW
30	Four-door wagon	3.85	130	TDC,TDF,TDF,FDW
	Vehicles/ 8 hour 480/Day		480 min/ 480 = 1 minute per vehicle	

Most components are simply *sequenced* to the line. Every car, regardless of model type, has an engine. So engines are produced at a rate of one every one minute (480 minutes/480 automobiles). Each engine needs four pistons. Therefore every 1.0 minutes, four pistons are produced. Parts and assemblies are produced in their minimum lot sizes and delivered to the next process, controlled by Kanban. Engines are often made elsewhere, shipped to the assembly plant and put in the right sequence for the cars coming down the line.

Balancing is making the output from the cells equal to the necessary (or daily) demand for parts downstream. Many parts or components are not made in sync with final assembly, only the daily quantity is the same. In summary, small lot sizes, made possible by setup reducing within the cells, single-unit conveyance within the cells, and standardized cycle times are the keys to having a leveled manufacturing system. Over time, one strives to make the cycle time in the cells about equal to the Takt Time for final assembly but at the outset, matching the daily demand is sufficient. Ultimately, every part, subassembly or assembly operation, has the same number of specified minutes as the final assembly line.

However, changing final assembly into a Mixed Model final assembly (MMFA) is perhaps the most difficult part of instituting a lean manufacturing system. This aspect of manufacturing requires a concerted effort from all people involved in the process. This means everyone from the floor worker and engineers to upper management as well as the vendors and customers have to give input into the system in order for it to work properly. Everyone working together as a team can help ensure that materials arrive as needed, daily production of all models is leveled to meet customer demand, as well as the overall quality of the models. The basic idea behind this system is to be able to meet customer demand no matter how erratic and unpredictable the customer orders arrive. This principle can be achieved through the design and scheduling of the final assembly area to ensure a smooth production rate.

The main component of Mixed Model final assembly is its quick changeovers which allows for many different models to be assembled in one day rather than producing one model for several days and then changing to the next model which can take several hours to convert the tools and dies. This process includes the elimination of adjustments, which are the most time consuming elements of a changeover, giving flexibility to the equipment that allows for different models to be produced by a station or machine in a single day.

The smaller batch sizes mean that more models can be produced in a single day. It also helps out with leveling because models with larger Takt Times can be made in smaller batches and can be balanced with larger batches of models with smaller Takt times. Also this smaller batch size takes advantage of the quick changeover times that exist through implementing rapid tool and die exchanges.

MMFA needs a strong Preventive Maintenance to help reduce downtime. When a part of final assembly is down, the entire assembly goes down due to the nature of the pull system. This program will also help reduce variability in parts due to machines running in the same fashion without need for any adjustment. The lack of downtime means the assembly time with downtime built in can be reduced and turned into production time.

Production planning deals with the final assembly schedule. Leveling is the process of planning and executing an even production schedule, to try to set the daily production for each model based on demand. The necessary overall Takt Time is set in order to synchronize the rate of production with the rate of consumption.

The design phase sets the stage for being able to specifically cater to external customer needs and their specifications. This design allows for changes in the model to meet the changing needs of the customer or consumer. This freedom in marketing and design allows the product to satisfy more customers thus, increasing sales.

There are several important aspects of the scheduling process. In order for Mixed Model final assembly to be accomplished, a scheduling system must be in place. This scheduling system ensures daily, monthly, and yearly customer demand will be met along with meeting the demand for material flow into final assembly. These scheduling processes discussed in Chapter 7 have been developed so that the system can support itself once started.

The scheduling process needs to be flexible so the schedule and the design can be changed rapidly to meet changes in customer orders or specifications. Daily and final assembly schedules are distributed. The scheduling department and the purchasing department need this information to complete their jobs. However, when it comes to the factory floor personnel, only the first station in the final assembly line needs to be informed (leaving other employees to focus on things such as quality). This is because final assembly is using Kanban to pull parts from the subassembly cells and feeding the assembly line. Without this scheduling system the system would collapse due to the significantly reduced inventory levels as well as the demands of Mixed Model output on a daily basis. Even when properly instituted, the system can experience problems produced by faulty design and scheduling of the system.

8.5 *STEP 2*: Restructuring Subassembly and the Job Shop

While the final assembly area is being changed into MMFA, the subassembly (flow) lines (that typically use conveyors) are reconfigured into U-shaped cells to make these systems operate on a single-piece flow basis. The operations in an assembly cell are usually all manual so the operator must stay at a station until the task is completed, but the operators can move from station to station to perform multiple operations. Of course, the long setup times typical in many flow lines must be vigorously attacked and reduced so that the cell can be changed over quickly from one product to making another. This makes them flexible and compatible with the cells designed to make piece parts and with the other subassembly lines and final assembly lines.

The cells are designed to manufacture specific groups or families of parts. In the U-shaped layout in Figure 8.5 workers cover multiple operations. Notice that some workers share operations. *The need to line balance (make task times equal) the flow line has been eliminated.* This is accomplished by the design and by using standing, walking workers who are capable of performing multiple operations. For the cell, the cycle time is 3.5 minutes (to produce 130 units per day).

In L-CMS, the lean shop replaces the job shop. Restructuring the job shop is critical to reorganizing the basic manufacturing system into manufacturing cells that fabricate families of parts. This prepares the way for systematically creating a linked cell system for one-piece movement of parts within cells and for small-lot movement between cells. Creating cells that can produce to the daily demand is also critical to designing a manufacturing system in which production control, inventory control, and machine tool maintenance are functionally integrated.

8.5.1 Designing Cells

Conversion of the job shop system into a lean shop that is part of the linked-cell system is a design task. Just as design engineers now try to design products which are simpler and easier to manufacture, lean engineers try to design a factory which is simpler to operate.

Many companies *design* their first cell using group technology (GT) techniques (Figure 8.6). In GT, similar components are grouped into families of parts made by a common set of

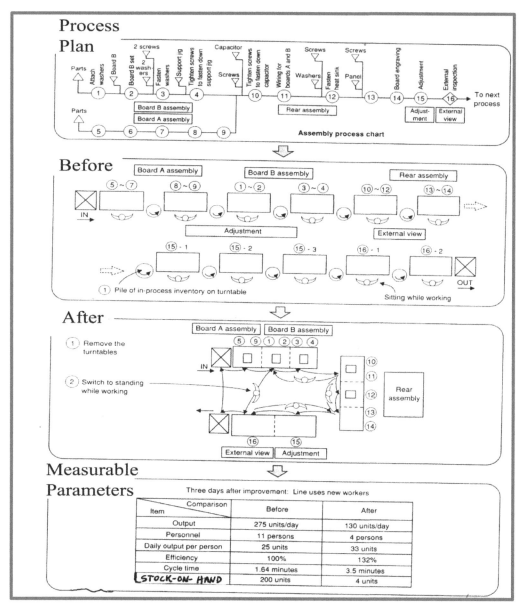

Figure 8.5
The Redesign of a Power Supply Assembly Area into a Subassembly Cell (Sekine).

processes. The operations in the cell can include metal forming or metal cutting operations, heat treating, inspection, assembly, and even grinding and super finishing, making all the variations of a component. See Chapter 17 for more discussions on GT.

8.6 Value Stream Mapping

Another very widely used technique for identifying the collection of processes for a manufacturing cell is called value stream mapping (VSM). VSM is sort of an advanced product flow diagram that documents the movement of a product group from raw material to finished part. A visual representation is drawn using VSM symbols that represent every process and activity in the flows, detailing both material and process flows and related information.

	Parts							
	1	2	3	4	5	6	7	8
A	1		1			1		1
B	1					1		1
C		1						1
D	1		1			1		1
E				1			1	
F		1			1			1
G					1			1
H	1		1			1		
I				1			1	
J		1					1	

(Machines — rows A–J)

Initial Matrix from PFA Analysis of Process versus Product

	Parts							
	1	6	3	8	5	2	7	4
B	1	1						
H	1	1	1					
D	1	1	1					
A	1	1	1	1				
G				1	1			
F				1	1	1		
C				1	1	1		
J						1	1	
E							1	1
I							1	1

(Machines — rows B, H, D, A, G, F, C, J, E, I) — with annotations: **Potential Cells**, **Exception**

Figure 8.6
Group Technology Develops a Matrix of Jobs (by number) and Machine Tools (by code letter) as Found in the Typical Job Shop. The Matrix can be Arranged to Yield Families of Parts and Associated Groups of Machines that Form a Cell.

The lean engineer should walk the process himself to get the steps and times correctly noted. This is called producing the current-state map. Often a kaizen event is used to determine this information and develop the plan for the future. This activity can prepare the way for a new cell design based on a value stream map showing the improved layout (called the future state map). Value stream mapping was introduced in 1998. (Rather and Shook, 1998)

The first step in any mapping activity is to identify a product family, a group of similar items that proceed through the same processes and equipment. This is the value stream, all the actions or steps required to bring a product or product family from raw materials to the customer. Mapping is greatly simplified, and the benefits of mapping are maximized, if careful thought is given at the very outset to the formation of products into families.

The next step, often using A3 analysis, is to determine what problems are being encountered on the company product from the standpoint of the customer and the company. For example, the customer may be demanding a price reduction or there may be chronic quality problems or there may be a need to increase output without adding more operators or making a significant investment in new equipment. An A3 report is a short form with a specific size and format, as shown in Figure 8.7. The report is used for problem solving where the problem is defined, analyzed, planned cures or fixes outlined, results of the activity and future plans are all captured on one 17 x 11 report. Figure 8.8 shows a completed A3 report prepared by the lean engineer. Whatever the problems, it's critical to have agreement with the customer on just what that problem is prior to the start of mapping. Otherwise, it's likely that mapping will fail to address the real issues. Or, equally likely, mapping will fail to spur any improvement in the process at all.

Once the problems with the current design have been identified, it's time to take a walk along the value stream and draw a map of the current state. This map looks at the process flow and the information flow. Ideally, mapping will be done with a team involving everyone connected to the value stream, so that complete agreement can be reached on the condition of the entire stream. In some cases, of course, this isn't practical. But lean engineers have learned over the years that if a much smaller team, or even one person, walks the value stream, it's critical to walk it from one end to the other, so that the whole value stream is captured. Assigning small teams or individuals to walk different segments of the stream usually leads to inaccurate maps that aren't trusted even by the team that draws them.

Figure 8.7
Blank A3 Form

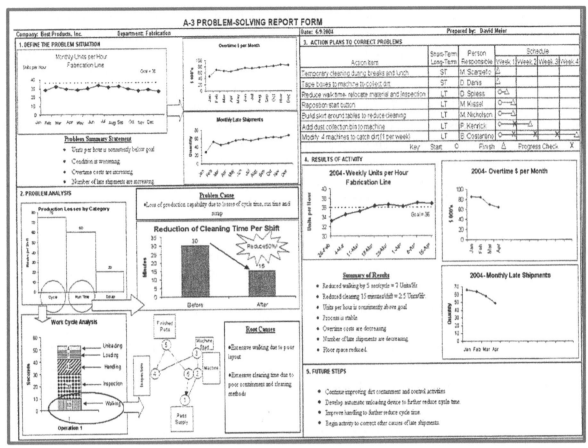

Figure 8.8
Completed A3 Form

Getting the current state correctly mapped is critical, because the performance problems in the value stream are the direct result of the way it is configured and managed. Improvement can only be based on accurate mapping of the current state. In practice, operators and managers involved in the current process are usually working very hard to make higher-level managers happy. Thus, it is natural for them to explain away problems that are observed while mapping as abnormalities not worth worrying about. It is very important to record real data on how the system really works, not how it is supposed to work on those days when nothing goes wrong and when no customer changes an order at the last minute or when a supplier makes a mistake.

Once the problem in the current-state value stream is understood, and a team is organized to walk the value stream, there is still a key question: where to start?

Since the first step was to identify a product family, experienced lean engineers usually start at the customer end of the value stream and work upstream to the point where the problem occurs. It's important to map the value stream through both the downstream and upstream departments, developing a standard method and a common language, because many members of the organization will need to conduct value stream mapping over time.

The objective in drawing the map is to identify each significant element required to create the desired value. These are carefully written down, along with information about the performance of each action.

260

Specifically we want to know whether each process step has:

- **Value**. Does it actually create value from the standpoint of the customer? The simplest measure of the value of a step is to ask if the customer would be less satisfied with the product if this step could be left out. For example, leaving out the step of painting a car would be a problem for practically all customers for motor vehicles. They think that paint adds to the value of the product. But leaving out all the rework and touch-up required to get a good paint finish wouldn't bother any customer. These latter activities are waste and need to be steadily reduced.

- **Capability or process capability**. Does the process achieve a good quality result is achieved every time. This is the core concern of the quality movement, and the starting point for many six sigma projects.

- **Reliability and Availability**. Will the process be ready and able to operate when it is needed? This is the core concern of Total Productive Maintenance. In typical operations, many processes can't produce a quality result all the time and are down a significant fraction of the time. Lean engineering often combines the issue of capability with the issue of reliability to estimate the stability of a process.

- **Capacity**. Does the process have the capacity is in place to respond to the changes in customer orders as needed? Usually there is an excess in capacity. Most processes have more than the necessary capacity, and this is waste of a different type. This waste occurs because equipment designers still want to build large machines designed for lower cost per step at high production rates and volumes. However, from a lean engineer's point of view, the one thing that is certain is that market forecasts of demand are wrong. Such errors lead either to chronic overcapacity or intractable under capacity, when getting even a small amount of additional capacity requires purchase of another large machine. The lean engineer's approach is to "right size" equipment whenever possible to create the right balance for labor and equipment to meet demand.

- **Flexibility**. This requires the ability to add and subtract increments of manpower, (fully utilized) over a wide range of volumes. The ability to change a process over quickly and inexpensively from one member of a product family to another is called flexibility. Rapid changeover permits the production of very small batches with many benefits for the entire value stream. Flexibility has, of course, been a hallmark of the Toyota Production System.

8.7 VSM Process Steps

VSM and the similar lean value-stream tools exist within the discipline of industrial engineering. Previous process mapping tools focused solely on the processes whereas VSM focuses on both of material and information flow. Understanding production as an integrated material and information flow system is a major innovation of the Toyota Production System.

VSM provides a graphical representation of the material and information flows from raw material to the delivery of finished goods. The objective is to render in simple form, something that seems complex and confusing so that we can change the focus from narrow operation efficiency to system level optimization.

The major steps to value stream mapping process are shown in Figure 8.9. The arrows show a feedback/learning loop between the current state, future state and implementation plan. The value stream mapping process is not linear. One doesn't really finish a current state map, then finish a future state map and then shift to implementation. There is considerable overlap and feedback between these stages and it may be necessary to periodically go back and gather more data, as needed.

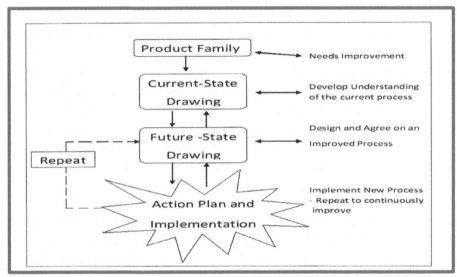

Figure 8.9
The VSM Methodology

VSM is a methodology to gain an understanding of the process but does not represent the entire set of tools to design optimum flow of product nor does it introduce any of the calculations needed to design manufacturing cells. VSM is simply a tool to assist in process design and thus the description of lean concepts is kept very simple Remember, VSM IS only a beginning step in improving material and information flows.

Draw the maps by hand. This forces the individual to go look, observe, and to try to really see what is going on at the process and system level, not just individual processes. Toyota Production System is about developing people and that requires people observing learning and continuously improving. This is why the current state maps are drawn by hand, with paper and pencil, as the system is walked. The looking, drawing and redrawing may seem like a lot of manual work, but it is in fact a Plan-Do-Check-Act (PDCA) learning cycle that keeps the LE focused on the flow and deepens understanding of the current production system.

With Practice, the LE can quickly and accurately draw dock-to-stock material and information flows but having a perfectly drawn current state map is not the point. What's important is learning to see what's going on. The person who best understands the material and information flows that the map represents is the person who drew the map. The map creation process, more than the map itself, helps the LE learn to visualize the process while developing eyes for flow and eyes for waste. Walking the process and drawing the maps helps the LE grasp the overall process from which position the LE can develop useful future-state value stream maps.

8.8 Determine Product Family
First determine the product family, then draw the current and future state maps and finally,

assemble the implementation plan and review. Lean engineers always start with the product or service. After all, the customer's only interest is the product, with value typically derived from cost, quality, delivery or service. To surface waste in the value streams, a map should be drawn for every product group and the first step is to identify the groups. In some companies this is simple, especially if you produce only a few items on a daily basis. However, if you have hundreds of products it obviously does not make sense to draw a value stream map for each one. Therefore, select the main component and limit the number of branches.

A Product Family Matrix can group many products into a few product families using a Group Technology method called product flow analysis (PFA) method. To create a product family matrix, list the machines down the left column, and the components across the top. An example is shown in Figure 8.6. In most cases, it won't be necessary that all the steps be listed. Then place a mark in each of the squares where the step applies to the product. Next, look for common process steps that apply to several products. These products can then be grouped into families. Usually the ones closer to the downstream customer are enough to be able to differentiate product families. Typically, downstream processes are used in the value stream to define product families. The processes need not be absolutely identical, because later, a cell may be designed wherein several products can pass through each process with some variations. In most commodity parts manufacturing companies, the product is truly differentiated to the customer's requirements in the downstream assembly steps. Upstream processes like stamping presses and machine tools may serve multiple product families, so it is difficult to define a product family based on these steps, as these steps may in fact serve most of the products.

In a custom parts organization, everything is produced to customer order and continuous flow is set up from the point of order release straight through to shipping. In these cases, the product family matrix may also include more upstream process, as these steps are closer to the point of order release. Commonly, in custom parts producers, the key processes that drive the pace of the operation are further upstream than in a commodity parts producer. In a custom parts organization, the key is to understand which of the orders can be grouped by the processes they follow. The product family matrix can still be used in this case, however there may be more upstream processes included.

The GT analysis may produce a surprising outcome. Some products thought to be totally different from each other may in fact travel through a similar set of processes. If this is the case, then they can be grouped into product families, even if they are not related from a marketing viewpoint.

Defining the product families by observing the common processes that they travel through is an essential step. However, don't let an extensive GT analysis hamper the efforts to perform the value stream mapping. Developing a product family matrix in of itself offers no value to the customers, so keep it simple. There always seem to be some exceptions, processes which don't seem to fit in any of the product families, but which need to be produced none the less. Use the Pareto Rule and focus on the 20% of the parts that make up 80% of the cost.

In summary, a product family is typically picked from the customer end of the value stream, based on products that pass through similar processes and through common equipment in the downstream processes.

8.9 Mapping the Current State: The Process Flow

Once a product family has been selected, the next step is to document the processes or steps from raw material to finished part, and draw a map using VSM symbols that represents processes and activities in the material and information flows. The map consists of three perspectives; physical material flow, information flow and the summary metric line, typically a timeline. A starter set of VSM symbols is summarized in Figure 8.10. Note that this is just a starter set. Others symbols relevant to a particular organization or process can be added.

For an example, a wiring harness manufacturer will be examined. The goal is to draw a current state map showing what is happening to this product on a daily basis. To get started, gather a cross functional team of managers and operators responsible for moving this product from start to finish. The team is usually six to eight people because large teams tend to be too unwieldy while smaller ones are too narrowly focused. Obviously, the team will talk to people involved in the process and can bring in subject matter experts if needed to the mapping exercise. A lean engineer may be needed to train the organization in lean tools & techniques, guide the development of the VSMs, facilitate the improvement effort and mentor people to think and act lean.

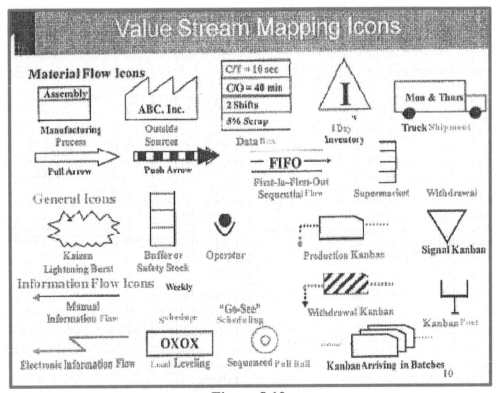

Figure 8.10
Summary of Icons for Value Stream Mapping

264

The mapping event is usually a one or two day event with team members dedicated solely to the team to avoid the distractions of their day-to-day jobs. There may need to be some preparation prior to starting, namely obtaining the customers' requirements and expectations, equipment uptime and availability. Remember that the map is not a layout of a plant but rather the path one product family follows through the plant.

Remember these two important points as the mapping event begins:
1. Do involve someone from top management, from start to finish.
2. Don't divide the value stream into segments and assign segments to different members of the group. Instead, have the whole team walk the entire stream together so everyone can see the whole process.

Most traditional process mapping efforts focused only on the movement of the material, but VSM also looks at the information flow. This is because information releases to the factory floor can cause all types of waste if not done properly. Figure 8.11 shows a map for a wire harness assembly process. The bottom portion of the map is for material flow proceeding from left to right, while the top half of the map is for information flow, proceeding from right to left.

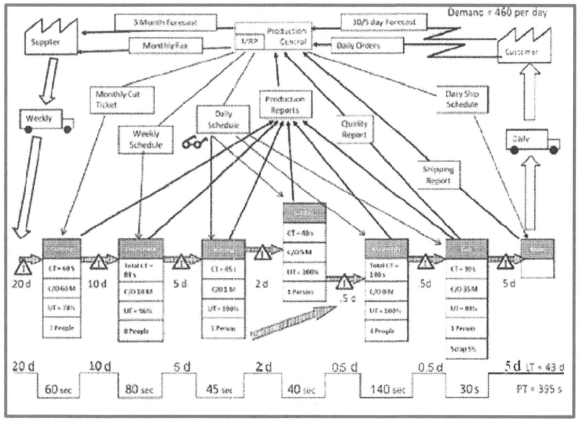

Figure 8.11
Current State Map – Wire Harness Assembly

The 9-step method, developed by Roth and Shook, for current state mapping for (a wire harness assembly process) is outlined as follows:

Step 1: Determine Customer Requirements for the product family, including the quantity required (units/day), delivery quantities and packing requirements.

Step 2: Quick Walk Through to Identify Main Processes in Order. Once a product family is selected, put yourself in the position of your customer and walk backwards from finished product back to raw materials. Try to capture the big picture the first few times that the process is walked from dock-to-dock and seek to understand the current state walking the process as a team.

Steps 3 & 4: Draw process boxes and fill in information gathered in the walk through and indicate where inventory was found

Step 3: Start at the end of the value stream on the factory floor and proceed upstream documenting the physical aspects of the process.

Step 4: Document the information or data on the processes and put it with process flows. Some typical process variables are shown in Table 8.3. The team must gather, validate and verify the data, including inventory amounts. Map upstream toward the raw material receiving dock. Branches can be added to the main value stream for every component being added to the product, but keep the maps simple.

Table 8.3
Process Data Shown

Process Data Shown
Cycle Time (CT) or Production Rate (PR)
Changeover Time (CO)
Uptime
Scrap Rate
Pack Out Quantity
Number of People
Available Time

Step 5 &6: Determine customer delivery requirements and identify what your suppliers are providing. In the mass system, the main customers send monthly and weekly forecasts via EDI to production control. Firm orders are sent daily by EDI. In turn, production control issues a 3 month forecast by FAX to suppliers for raw materials, while a weekly firm order is also faxed. The initial information is usually available from the production control department where (hopefully) someone can be found who understands how the MRP program works.

Step 7: Determine how the processes know what to make and when to make it (i.e., the Information Flow). In each process, the LE must know how the operator knows what to work on next. This is an important question because overproduction can be driven by faulty information. Usually, in the mass system, the build schedule is produced by the material requirements planning system (MRP). The shipping department gets a daily ship schedule from production control, based on the MRP program. Because the various operations have erratic availability and because the sales department often asks for a particular customer order to be expedited, the supervisor often goes to the operation to see at how much inventory is present and adjust the work schedule on the fly. We show this *go see* scheduling by drawing a set of glasses. Often companies use red tags on items to be expedited.

Step 8: Determine where material is being pushed and where it is being pulled. Since each process builds to a schedule, completed inventory is pushed to the next process, whether it needs it or not. Operators are rewarded by how fast they can produce parts in their individual processes. This inventory "push" is shown by crosshatching arrows between processes.

Step 9: Calculate summary metrics -Lead Time/Process Time

Once all the activities have been mapped and analyzed as to whether or not they create value, the elapsed time, from start to finish or the throughput time is determined and compared to total processing (value adding) time. Typically, more than 95 percent of the elapsed time is non-value added. Also, the current state map shows where inventory has collected all along the value stream, which dramatically impacts the throughput time, requires more floor space and containers and requires more handling of this inventory.

8.9.1 VSM Summary, Current State

Current state maps always trace the flow of materials toward the customer and the flow of information back from the customer, to create a closed loop of information and material. Always avoid having backward flow. If backflow in the process occurs such as rework, then include it. P This allows appropriate and accurate summary metrics to be calculated.

Avoid thinking that value stream mapping is in itself a goal. Drawing maps of all the value streams may lead to a better understanding of the whole process, but not necessarily to any measurable improvement. Results come from active implementation of an improved future state, a condition that goes beyond what is possible today. Value stream mapping can be a useful tool in this pursuit, yet it is still only a tool, not the actual improvement itself. Many maps are developed long before their time, in other words, before anyone is committed to making improvements to the process. Instead of mapping everything and expecting good things to happen, it may be more effective to start by improving one value-adding step in the process by doing a process capability study for example to improve quality. Then progressively migrate into more of the value stream and support functions as is necessary to be able to further improve the whole process and you will always be working on critical aspects of the process.

Don't wait until the current-state map is perfect since this will never happen. Get the map close enough to right to be a guide for action and start the future state map, which leads to process improvement. Avoid an overemphasis on counting inventory. Throughput time is a great metric. Reducing inventory accumulations will reduce the TPT from dock to dock. But don't let this become more important than understanding why the inventory is there. Inventory accumulations always occur because of problems in order release, machine downtime, variance in processing times and overutilization of machines. Always ask: *What is causing us to hold so much inventory here?* Inventory is always there for a reason. Go after the reasons.

8.10 Creating the Future State

The future state map represents the improved process. The future state map does not show the specific details, but presents a simple, visual, easy to understand, graphic picture of the target conditions. With a current state map, managers should be able to see the potential of lean thinking by asking if each activity creates value from the standpoint of the customer.

Waste, or *muda*, is non-value added activities which create no value for the customer but are currently performed to produce an acceptable product. Manufacturing waste can be categorized as defects, overproduction, waiting, not utilizing employees, transportation, inventory, motion and extra processing. The goal of the future state map is to show the plan to eliminate, reduce or simplify manufacturing wastes. Customers want to pay for a product that works perfectly, not for extra steps needed to create value when processes are incapable or out of control. In other words customers are indirectly paying for this waste.

The future state VSM will show how to eliminate waste while improving quality and customer response. Overall design goals include increasing awareness of the customers' requirements, improving flexibility, reducing throughput time by creating one-piece flow of connecting processes (see Chapters 3, 4 and 6 for details on cell design), implementing pull control and simplifying information flow. The VSM identifies processes that can be put together to form a cell but does not design the cell or indicate the changes that are needed to have a one piece flow manufacturing (to the processes) cell.

8.11 Value Stream Mapping Summary
- Helps you to visualize the total system rather than just focusing on single processes.
- Can be compressed to a single process or expanded to include other facilities, suppliers, and distribution facilities.
- Links material and information flows.
- Provides a common language for all participants.
- Develops a blueprint for a cell (identifies the processes).
- Ties together lean concepts and techniques.
- Can be used to identify waste and areas for improvement in any process.

Using the VSM, many companies develop a pilot cell, rearranging existing equipment so that everyone can see how a cell functions. It will require time and effort to train the operators, and they will need time to adjust to standing and walking. Simply select a product or group of products that seems most logical. The operators *must be involved* in designing the cell or they will not take ownership. The pilot cell will show everyone how SPF cells operate and why setup times on each machine must be reduced. Machines will not be utilized 100%. Machine utilization rate usually improves but may not be what it was in the functional system where *overproduction* is allowed. The objective in manned linked-cell manufacturing systems design is to utilize the people fully, enlarging and enriching jobs, allowing operators to become multifunctional. The operators learn to operate many different kinds of machines and perform many functions that include assembly, quality control, machine tool maintenance, setup reduction, and continuous improvement. (In unmanned robotic cells and flexible manufacturing systems, the utilization of the equipment is more important because the most flexible and smartest element in the cell, the operator, has been removed).

In manufacturing cells, standard work is the basis for maintaining productivity, quality, and safety at high levels, providing a consistent structure for performing the tasks to the designated Takt Time while uncovering opportunities for making improvement in work procedures. A standard operating routine sheet is used to describe the cell. As shown in Figure 8.12 there are three aspects to structuring standard work in a cell.
- Cycle time based on system Takt time
- Working sequence -sequence of operations

- Stock-on-hand
- Cycle time reflects the pace of sales in the marketplace or the daily demand. The working sequence is the sequence of operations that is the best way to perform a task. The stock-on-hand is the minimum number of work pieces needed in a cell to maintain a smooth flow of work.
- Standard work provides detailed, step-by-step guidelines for every job in the lean shop. Team leaders and operators determine the most efficient working sequence and make continuing improvements -kaizen -in that sequence. Kaizen thus begets new patterns of standard work.
- Cells have many features that make them unique and different from other manufacturing systems. Parts move from machine to machine *one at a time* within the cell. For material processing, *the machines are typically* capable of completing a machining cycle initiated by a worker. The U-shape puts the start and finish points of the cell next to each other. Every time the operator completes a walking trip around the cell a part is completed. The cell is designed so that this cycle time (CT) is equal to or slightly less than the Takt time (cycle time) for final assembly. This is referred to as the necessary cycle time. The machining time (MT) for each machine needs only to be less than the time it takes for the operator to complete the walking trip around the cell. Thus, the machining time can be altered without changing the production schedule. Conversely, and much more significantly, the production schedule can be altered without changing the machining times. Let's say that again in another way. The L-CMS design permits you to change the final assembly output without having to redo all the subassembly and component supplier schedules.

The cell is designed to make parts at a pace required by downstream processes and operations. There is no overproduction. Overproduction will result in the need to store parts, build huge automatic storage and retrieval systems, transport parts to storage, retrieve the parts when needed, keep track of the parts (paperwork), and so on. All this requires people and costs money but adds no value. In cells, the cycle time is readily changed by adding or subtracting operators. In manufacturing cells, there is no need to balance the MTs for the machines. This can be a very difficult task in automated flow lines (transfer lines). In cells, it is necessary only that no MT be greater than the required CT. The machining speeds and feeds can be relaxed to extend the tool life of the cutting tools and reduce the wear and tear on the machines as long as the MT does not equal or exceed the CT. The jigs and fixtures in the machines are designed (by the LE) to hold any part in the family so parts are easily loaded/unloaded, cannot be loaded incorrectly, and defective parts cannot be advanced.

Between the process steps *decoupler* elements are placed to provide flexibility, part transportation, inspection for defect prevention (Poka-yoke) and quality control, and process delay for the manufacturing cell. The decoupler element can inspect the part for a critical dimension and feedback adjustments to the machine to prevent the machine from making oversize parts (as the milling cutter wears). A process delay decoupler would delay the part movement to allow the part to cool down, heat up, cure or whatever is necessary for a period of time greater than the cycle time for the cell. Decouplers and flexible fixtures are vital parts of both manned and unmanned cells.

8.12 STEP 3: Rapid Exchange of Tooling and Dies

When cells are formed to make a family of parts, the problem of process and machine change over from one part to another must be addressed. Therefore, everyone on the plant floor must be taught how to reduce setup time using SMED (Single-Minute Exchange of Die) techniques. SMED is a four stage methodology developed by Shigeo reduces tooling and die exchanges times, that is, reduce setup times. See **Figure 8.13.** A setup reduction team trains the production workers and foremen in the SMED process and demonstrates the methodology on a project, usually the plant's worst setup problem. Reducing setup time is critical to reducing lot size and the idea for SMED is to make the setup process fast and easily done by the operators as part of their daily routine.

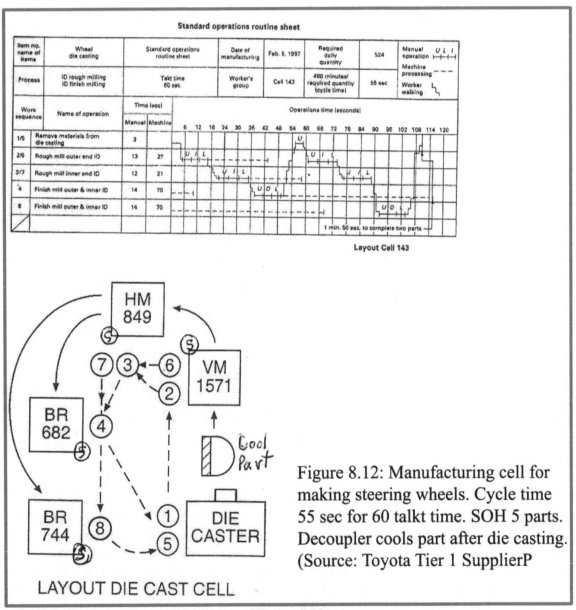

Figure 8.12: Manufacturing cell for making steering wheels. Cycle time 55 sec for 60 talkt time. SOH 5 parts. Decoupler cools part after die casting. (Source: Toyota Tier 1 SupplierP

Figure 8.12
Manufacturing Cell and SORS for Making Steering Wheels. Decoupler Cools Part after Die Casting. (Source: Toyota Tier 1 supplier)

The lean production approach to manufacturing demands that small lots be run. This is impossible to do if machine setups take hours to accomplish. Setup times can be reduced, resulting in reduced lot sizes and throughput times while improving the flexibility of the system. Successful setup reduction is easily achieved when approached from a methods engineering perspective. Much of the initial work in this area has been done by time and motion studies (MTM) while applying Shingo's SMED rules for rapid exchange of tooling. Setup time reduction occurs in four stages. The initial stage is to determine what currently is being currently being done for setup in the operation. The present setup operation is usually videotaped and everyone concerned meets and reviews the tape to determine the setup's elemental steps. The next stage is to separate all setup activities on the list into two categories, *internal* and *external.* Internal elements can be done only when the machine is not operating, while external elements can be done while the machine is in operation. Just this elemental division will usually shorten the setup time by 50%. Stages three and four focus on reducing internal time. The key is for workers to learn how to reduce setup times by applying the simple SMED principles and techniques. If a company must wait for the setup reduction team to examine every process, a lean manufacturing system will never be achieved.

The similarity in shape and processes needed in the family of parts allows setup time to be reduced or even eliminated. Initially setup times should be less than 10 minutes. As the cell matures, the setup times are continually reduced. Reducing the setup time until it is equal to or less than the cycle time (1-2 minutes) for the cell is usually quite easily accomplished. This will permit a significant initial reduction in lot size. The next goal is get setup times down to around 10-15 seconds, what is commonly called one-touch exchange of dies (OTED).

In the 3rd and 4th stages of SMED, it may be necessary to invest capital to drive the setup times below one minute. Automatic positioning of work holders, intermediate jigs and fixtures, and duplicate work holders represent the typical kinds of hardware needed. The result is that in many cases long setup times can be reduced to less than 15 seconds in relatively short order (Chapter 17).

When the setup time is down to less than the time needed to load, unload, inspect, deburr and so on at the machine, the operators can quickly change the machines over from one component to the next. Now what happens when the cell is changed over from part A to part B? The setup operation flows through the cell one cycle at a time as the cell is changed over from part A to part B. After each setup, at each machine, defect-free products should be made right from the start. The first B part out of the cell will be good. Ultimately, the ideal condition would be to eliminate setup between different parts. This is called no touch exchange of dies (NOTED). In summary, the savings in setup times are used to decrease the lot size and increase the frequency at which the lots are produced. The smaller the lot, the lower will be the inventory in the links; resulting in shorter throughput times, and improving quality.

8.13 STEP 4: Integrate Quality Control

A *multiprocess* worker can run more than one kind of process. A *multifunctional* worker can carry out tasks than only operating processes or machines. Such a multifunctional worker is also an inspector who understands process capability, quality control, and process improvement. In lean production, every worker has the responsibility and the authority to make the product right the first time, every time, and has the authority to stop the operation when something goes wrong. This line stop authority for workers is critical to the successful operation of the factory. The integration of quality control into the manufacturing system markedly reduces defects while eliminating inspectors. Cells provide the natural environment for the integration of quality control. The fundamental idea is to inspect to prevent the defect from occurring, never allowing a defective product leave the manufacturing cell and get into the system.

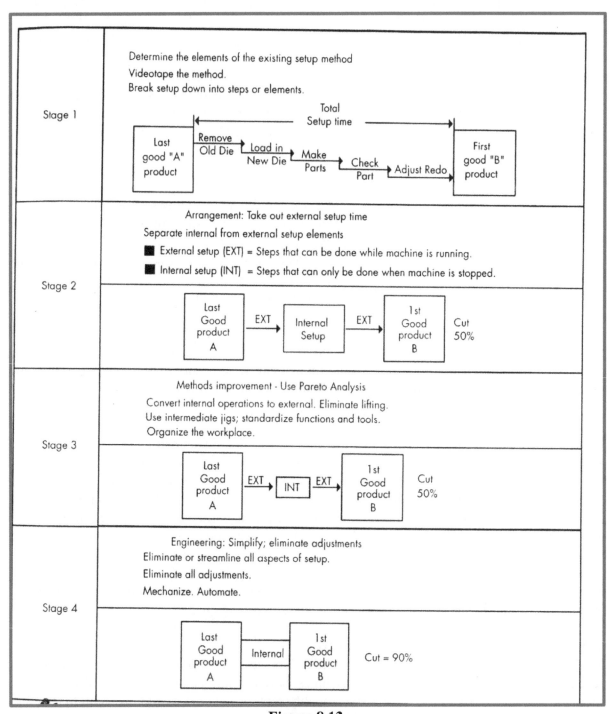

Figure 8.13
The Four Conceptual Stages of Single-minute Exchange of Dies (Shingo 1989)

272

When every worker is responsible for quality and able to use the seven basic tools of quality control (detailed in Chapter 14), the number of inspectors on the plant floor is markedly reduced. Every MfE, LE or IE should know these tools for QC and should be able to apply them daily to the factory. More likely, these are the people who will be called on to teach the quality tools to the operators. In Chapter 14, Six Sigma Methodology will also be outlined. LEs should know all these techniques, but are not usually taught to the operators. Products that fail to conform to specifications are immediately uncovered because they are checked or used immediately.

In lean manufacturing every worker is an inspector responsible for quality. What happens when a worker on the assembly line stops the line if he/she finds something wrong? Everyone's attention is focused on the problem holding up production. Problems get solved fast and permanently so the line is not stopped again. *Inspecting to prevent the defect from* occurring rather than *inspecting to* fine the defect after it has occurred becomes the mode of operation in the L-CMS, where devices to prevent defects (called Poka-yokes) are liberally implemented into the processes. Ultimately, the concept of *autonomation* evolves (see step 9) which means to automate to prevent the occurrence of defects based on lessons learned from the operation of the cell. Literally, the autonomation means automation with a human touch.

Now here lies the fundamental difference between lean production systems and other manufacturing systems. Through redesign, cells produce parts one at a time, just like assembly lines. This is called single-piece flow. Within the cells, the internal customers operate on a "make one, check one, move on" (MO, CO, MOO) basis. The operator is checking what the previous process produced to assure it is correct 100 percent of the time. On the assembly lines, pull cords are installed to stop the lines if anything goes wrong. For instance, if workers find a defective part, or if they cannot keep up with production, or if production is going too fast according to the quantity needed for the day, or if a safety hazard is found, then they are obligated to pull the cord and stop the line. Once the problem is located, it is rectified immediately. Meanwhile, the other workers on that line that are now stopped, work on other required tasks such as maintaining equipment, changing tools, sweeping the floor, or practicing setups; but the line does not move until the problem is solved.

For manual work on assembly lines, a system for tracking defective work is called *Andon. A* Andon is actually an electric light board that hangs high above the conveyor assembly lines so that everyone can see it. When everything is going according to plan, the board's lights are green. But, when a worker on the line needs help, he can turn on a yellow light. Nearby, (multifunctional) workers who have finished their tasks within the allotted cycle time move to assist the worker having a problem (called mutual assistance). If the problem cannot be solved within the cycle time, a red light comes on and the line stops automatically until the problem is solved. Music usually plays to let everyone know there is a problem on the line. In most cases, the red lights go off within ten seconds. If the line stops, for restarting, a green light comes on, with all the synchronized processes beginning together. The name for this system is Yo-i-don, which literally means *ready, set, go*. Such systems are built on teamwork and a cooperative spirit among the workers, fostered by a management philosophy based on harmony and trust.

Of course there are more ways to control quality than the *seven (7) tools*, but the seven are fundamental to the integrated process because these tools are used by the operators. Mapping and analyzing the process to understand its behavior is the key to finding out what went wrong. By designing the system and all the processes into a single-piece flow methodology, the causes of the defects (the effects) are quickly isolated and identified, thus cures can be readily implemented.

The L-CMS strives for continuous improvement in the processing to reduce the variability (or spread in measurements about the process mean). See Chapter 12. Thus reducing the process variability improves its process capability as shown in Figure 8.14. This figure depicts what is at the heart of six sigma projects. The process capability (accuracy and precision of a process) is improved by significant reduction in the variability in process, as measured by the standard deviation, known as sigma in the statistical world. Of course, the name six sigma is really a misnomer and should be 12 sigma -i.e.- 12 standard deviations between the upper and lower specification limits. In lean manufacturing, the continuous improvement aspects are driven by the systematic removal of inventory; see step 7 for further discussion.

So now we understand the key to achieving superior quality in lean manufacturing lies in isolating process steps in a serial fashion in the manufacturing cells, subassembly cells and final assembly. Multiple processes for the same step can increase the process variability. The outcome from each step is checked before the component is moved to the next step. The operators have complete feedback of each step in the processing. This feedback involves all of the operator's senses, hearing, sight, smell, touch, and even taste, making them superior to machines. They are on the front line for problem identification and resolution.

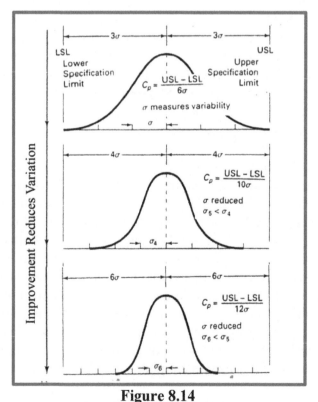

Figure 8.14
To Achieve 6 sigma Capability from 3 Sigma Capability, the Variance of the Process Must be Greatly Reduced

274

8.14 STEP 5: Make Output Predictable: Integrate Preventive Maintenance

Making machines operate reliably begins with the installation of an integrated preventive maintenance program, giving workers the training and tools to maintain their equipment properly. The excess processing capacity obtained by reducing setup time allows operators to reduce the equipment speeds or feeds and to run processes at less than full capacity. Reducing pressure on workers and processes to produce at full speed or maximum capacity fosters a drive to produce perfect quality.

Producing to Takt time, at the pace of customer demand, in a balanced system (step 1) defines the appropriate pace of production for the workers and the processes. Prior to lean, companies thought they could lower cost by running machines as fast as possible, all of the time. Lean recognizes that such strategies result in overproduction, machine failures and poor quality. The multifunctional operators are trained to perform routine machine tool maintenance. Adding lubricants (oiling the machine), checking for wear and tear, replacing damaged nuts and bolts, routinely changing and tightening belts and bolts, and listening for telltale noises that signify impending failures are that can do wonders for machine tool reliability while giving the worker a sense of pride of ownership. The maintenance department must instruct the workers on how to do these tasks and assist workers in preparing routine check lists for process maintenance. The workers are also responsible for keeping their areas of the factory clean and neat. Thus additional functions that are integrated into lean manufacturing system are maintenance and housekeeping. Originally Toyota developed a 4S strategy and over the years, companies who adopted it expanded it to 5S and some to 6S (safety). The Five S helps to identify problem areas and waste. However, lean production depends on everyone's active involvement. Thus, every member of the factory must follow the Five S principles before results are notices and sustained on a daily basis. See Chapter 17 for additional discussion of 5S tools for housekeeping.

Naturally, the processes still need attention from the experts in the maintenance department, just as an aircraft is taken out of service periodically for engine overhaul and maintenance. One alternative is to two 9-hour shifts (8 work hours plus 30 minutes for lunch plus two 15-minute breaks) separated by two 3-hour time blocks for machine maintenance, tooling changes, restocking, long setups, overtime, early time, and other functions. This is called the 8-4-8-4 scheme. The main advantage that equipment has over people is that processes can decrease variability, but processes must be made reliable and dependable. Smaller machines are simpler and easier to maintain and therefore are more reliable. Small machines in multiple copies add to the flexibility of the overall system as well. The linked-cell system permits certain machines in the cells to be slowed down and therefore, like the long-distance runner, to run farther and easier without a breakdown. Many observers of the lean shop come away with the feeling that the machines are *babied*. In reality, they are being run at the pace required to meet the Takt time demand.

True lean producers build and modify much of their manufacturing process technology. This is where their processing becomes unique and proprietary. Homebuilt machine tools are equipped with vibration sensors, temperature monitors, coolant/lubricant sensors, and of course, process controls for single cycle automatic operation of the machines and walk away switches to start the machining cycles. The sensors provide early warning of problems or can shut off the machine if a problem is detected. Because the machines are inexpensive, multiple copies of machines are built for cells that make the similar products for different companies. The lean supplier for steering gears has a cell for making rack bars for rack and pinion steering gears. The Mitsubishi manufacturing cell for racks makes 6 different racks for their SUV. In the same plant, the Toyota cells makes rack bars for Camry, Avalon and Sienna and can make 10 different kinds of rack bars in the same cell (See Chapter 6). The two cells are similar in their design because the rack bars require a similar set of processes. In the event of a machine failure in the Mitsubishi cell, a machine from the Toyota cell can be borrowed, modified, and used during second shift in the Mitsubishi cell. Processing capacity and capability is replicated in proven increments. Because the increment (the cell) has an optimal design, this is an economic choice as well as having the security of dealing with a proven manufacturing process technology. Modifying the existing equipment shortens the time needed to bring new technology on stream. Manufacturing in multiple versions of small-capacity machines retains the expertise and allows the company to keep improving and mistake-proofing the process. In contrast to this approach is the typical job shop, where a new super machine (large, expensive, multiple operation) would be purchased and installed when product demand increases. That is, many companies try to increase capacity by buying new, untried manufacturing technology that may take months, even years, to program, debug and make reliable. See Chapter 16 for additional discussion on TPM.

8.15 STEP 6: Integrating Production Control

The function of the people who work in production control (PC) is to schedule the manufacturing system which means they determine *where* the raw materials, purchased parts, and subassemblies are to go, *when* they should go there, and *how many* should go at any point in time. In short, production control determines, where, when and how many. In the traditional (mass) production system, the PC function is labor intensive and many have tried to computerize it with software called Material Requirements Planning (MRP), MRP II, or ERP. Many companies' experience with these computerized control systems has been that of great expense, wasted time, disappointment and frustration. Why? Because these software packages were designed for planning, not control. The lean manufacturing approach redesigns the manufacturing system so that production control functions are integrated into system design. Steps 1 through 5 provide the physical stage to integrate production control functions. Production Control in a lean factory is typically achieved by the use of Kanban control.

Integration of production control is achieved by linking the cells, subassemblies and final assembly elements utilizing Kanban. See Figure 8.15. The layout of the manufacturing system will define paths that parts can take through the plant. By connecting the elements with Kanban links, the need for route sheets is eliminated because the job shop is eliminated. Here is the design rule for each Kanban link,

$$K = \frac{L * DD + SS}{a}$$

Where: K =the number of carts or containers in the link
           SS =the safety stock
           a is the number of parts in each container,
           DD is the daily demand
           L =lead time.

Lead time accounts for all the waiting, delay and transportation time for a cart to make a trip around the link. The subassemblies and component parts (i.e., the in-process inventory) move within the structure on clearly defined paths. All the cells, processes, subassemblies, and final assemblies are connected by the Kanban links, pulling material to final assembly. This is the integration of production control into the manufacturing system, forming an L-CMS.

Kanban is a visual control system that is only good for lean production with its linked cells and its namesakes; it is not good for the mass production. Linking the cells together provides control over the route that the parts must take (while doing away with the route sheet), controls the amount of material flowing between any two points, and provides information about when the parts will be needed.

In the dual card Kanban system, there are two kinds of Kanban: *withdrawal* (or conveyance) *Kanban* (WLK) and *production-ordering Kanban* (POK). The WLK shown in Figure 8.15 is the link connecting the output side of one cell with the input point of the next cell. This link is filled with carts or containers that hold parts in specific numbers. Every cart holds the same number of parts and has one WLK and one POK. If there are K carts, then:

Maximum inventory in a link= K (number of carts) x a (number of parts in each cart = Ka

The arrival of an empty cart at the manufacturing cell initiates the order (the POK) to make more parts to fill the cart. The WLK Kanban cards tell the material handler where to take the parts. Suppose the link has six carts and each cart holds 50 parts. The maximum inventory (WIP) is 300 parts. The same kind of link connects the subassembly cells to final assembly. All the other cells in an L-CMS are similarly connected by the pull system for production control. See Chapter 15 for more discussion on Kanban.

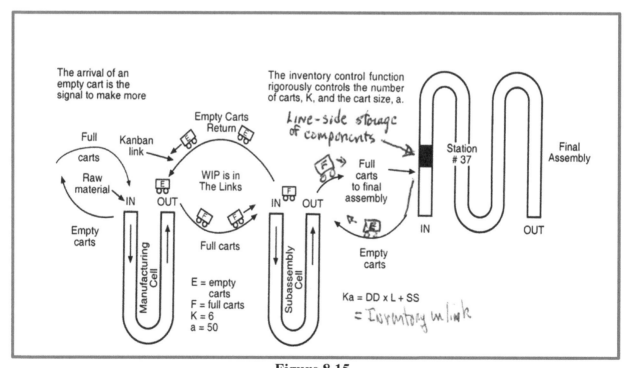

Figure 8.15
The Production Control Function in the L-CMS is Executed by the Kanban Links

8.16 STEP 7: Integrating Inventory Control

All of the material in a manufacturing system is considered inventory. There are three basic types -raw material, in-process or work-in-process (WIP), and finished goods. Step 7 involves the integration of control of the in-process inventory. In the L-CMS system, the work-in-process (WIP) inventory in the system is held in the links. The WIP inventory has been analogized to the water in a river, as shown in Figure 8.16. A high river level is equivalent to a high level of inventory in the system. The high river level covers the rocks in the riverbed and the boat can cruise safely. Rocks are equivalent to problems. Lower the level of the river (inventory) and the rocks (problems) are exposed. This analogy, developed by Taiichi Ohno and published by Shigeo Shingo, is quite accurate. In lean manufacturing, the problems receive immediate attention when exposed in the Lean system. When all the rocks are removed, the river can run very smoothly with very little water. However, if there is no water, the river has dried up. The notion of zero inventory is incorrect. While zero defects are a proper objective, zero inventory is not possible. (Within the cell, parts are already handled one at a time, just as they are in assembly lines. The material in the cell is called the stock-in-hand [SOH] so technically there is no inventory in the cells [hence zero inventory].)

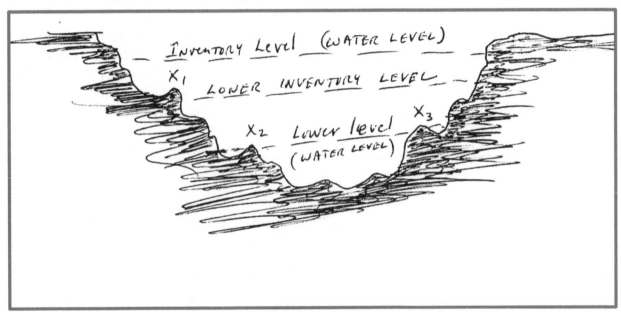

Figure 8:16
The Improvement Function Inherent in Kanban Exposes Problems by Gradually Lowering the Inventory Level. (Shingo, 1981).

The idea is to minimize the necessary WIP between the cells. This is how the inventory in a lean manufacturing system is controlled. The level of WIP between cells, subassembly, and final assembly is actually controlled by team leaders in the various departments. The control is integrated and performed at the point of use. Here is what team leaders do. Suppose that there are six carts in the link and that each cart holds 50 parts. The maximum inventory in this area is therefore 300 parts. The foreman goes to the stock area outside the cell and picks up the Kanban cards (one WLK, one POK), which puts one full cart of parts out of action. The (maximum) inventory level is now 250 parts. The foreman waits until a problem appears. When it appears, the foreman immediately restores the Kanban, which restores the inventory to its previous level.

278

The cause of the problem may or may not be identified but restoration of the inventory relaxed the condition until a solution can be enacted. Once the cause of the problem is identified and solved, the foreman repeats this procedure. If no other problems occur, the foreman then tries to drop the inventory to 200 parts. This procedure is repeated daily in the links all over the plant. After a few months, the foreman may be down to the three carts of 50 parts. Over the weekend the system will be restored to six carts between the two points, but this time each cart will hold only 25 parts. If everything works smoothly, with the reduced cart quantity, the foreman will soon remove a cart to see what happens. More than likely, some setup times will need to be reduced. In this way, the inventory in the linked-cell system is continually reduced, exposing problems. The problems are solved one by one. The teams work on solving the exposed problems including long setup times. The effect on the system of removing inventory is to continuously improve the TPT. This is how continuous improvement works in the lean production factory.

The minimum level of inventory that can be achieved is a function of many factors; the quality level, the probability of a machine breakdown, the length of the setups, the variability in the manual operation, the number of workers in the cell, parts shortages, the transportation distance, and so on. It appears that the minimum number of carts is three, and, of course, the minimum lot size is one. The significant point here is that inventory becomes a controllable independent variable rather than an uncontrollable variable dependent on the demands of the users of the manufacturing system for more inventory. More details on Kanban control are found in Chapters 4, 5, 13 and 15.

8.17 STEP 8: Integrate the Suppliers

In lean production, one tries to reduce the number of suppliers and have a single source for each purchased component or subassembly. Vendors and suppliers are educated and encouraged to develop their own lean production system for superior quality, low cost, and rapid on-time delivery. They must be able to deliver perfect parts downstream to their customers when needed and where needed and without incoming inspection. The linked-cell network ultimately should include every supplier. Suppliers become remote cells in the L-CMS.

In the traditional job shop environment, the purchasing department permits its vendors to make weekly/monthly/semiannual deliveries with long lead times-weeks and months are not uncommon. A large safety stock is kept just in case something goes wrong. Quantity variances are large and this, coupled with the normal late and early deliveries produce a vendor supply situation that hovers on chaotic. This situation leads to expediting, where people try to find the parts that are delaying assembly and get them moving (this is a total waste).

In mass production, as a hedge against vendor problems, multiple sources are developed. The suppliers' profit margins are cut thin and the supplier is unreliable and may be in jeopardy of bankruptcy. Meanwhile, the purchasing department may claim that pitting one supplier against another gives the company a competitive advantage and lower costs for parts.

The lean manufacturing system handles its vendor program very differently. Just-in-time purchasing is a program of continual long-term improvements. The buyer and vendor work together to reduce lead times, lot sizes, and inventory levels. Both the supplier and the customer become more competitive in the world marketplace because of this teaming relationship. In this environment, longer-term (18-24 months) flexible contracts are drawn up with three or four weeks delivery lead times at the outset. The buyer supplies updated forecasts every month that are good for 12 months, commits to long-term quality, and perhaps even promises to buyout any excess materials. Exact delivery is specified by midmonth for the next month. Frequent communication between the buyer and the vendor is the norm. The Kanban subsystem controls the material movement between the vendor and the buyer. The vendor is now considered a remote cell. Long-range forecasting for six months to a year is utilized. As soon as the buyer sees a change, the vendor is informed; this knowledge gives the vendor better visibility instead of a limited lead time view. The vendor has *build-schedule stability* and is not *jerked up and down* by the build schedule.

The buyer moves toward fewer suppliers, often going to local, sole sourcing. Frequent visits are made to the vendor by the buyer, who may supply engineering aid (quality, automation, setup reduction, packaging, and the like) to help the vendor become more knowledgeable on how to deliver, on time, the right quantity of parts that require no incoming inspection. This is true technology transfer. The vendors learn from about lean manufacturing from the customer. The buyer and seller must build a bond of trust, be willing to work together to solve problems, and share costs savings.

The advantages of single sourcing are that resources can be focused on selecting, developing, and monitoring one source instead of many. When tooling dollars are concentrated in one source, there is a savings in tooling dollars. The higher volume should lead to lower costs. The vendor is now more inclined to go the extra mile or exceed expectations for the buyer. The buyer and the vendor learn to trust one another and this may be a developmental process cultivated over time. An addition benefit to the single-source buyer is that quality is easier to monitor when there is a single source more consistent and variation is reduced.

Finally there is the aspect of proprietary processes. Toyota, Honda, and other users of lean manufacturing have published very little about their manufacturing and assembly cells. Why is that? This is where the unique processes exist that give companies like Toyota and Honda the edge in manufacturing strategy. The lean Engineers develop machine tools and the processes in-house rather than buy them from a machine tool vendor. It is a well-known fact that machine tool vendors cannot keep secrets. Their task is to sell processes and machines, so naturally machine tool vendors want to tell all prospective buyers of a new process they just developed for another company. On the other hand, lean manufacturers gain their competitive edge by developing unique process technology in-house and keeping it "locked up" in the cells. Table 8.4 summarizes the key points in managing a supplier plant. This list was obtained from the plant manager at a first tier supplier to Toyota Camry.

8.18   STEP 9: Autonomation

Autonomation is defined as the autonomous control of quality and quantity. Stop everything immediately when something goes wrong; control the quality at the source instead of using inspectors to find the problem that someone else may have generated. The workers in the lean factory inspect each other's work; this method is called successive checking. Taiichi Ohno, Toyota's former vice president of manufacturing, was convinced that Toyota had to raise its quality to superior levels in order to penetrate the world automotive market. He wanted every worker to be personally responsible for the quality of the piece part or product that they produced. This process is called source or self-checking and is part of the MO-CO-MOO methodology.

The need for automation simply reflects the gradual transition of the factory from manual to automated functions. Some people may think this is Computer Integrated Manufacturing (CIM) but lean engineering recognizes that people are the most important and flexible resource in the company and sees the computer as just another tool in the process but not the heart of the system. For example, inspection devices are placed in the machines (source inspection) or in devices (decouplers) between the machines, so the inspection is performed automatically. This generally reduces the cycle time in the cell. Remember, the idea is to prevent the defect from occurring rather than to inspect to find the defect after the part is made. Inspection by a machine instead of by a person may be faster, easier, and more repeatable but it may not be able to replace laying hands on the parts.

Autonomation means inspection becomes part of the production process and does not involve a separate location or person to perform it. Parts are 100 percent inspected by devices that either stop the process if a defect is found or correct the process before the defect can occur. The latter requires feedback to the process controller. Sensors are used to shut off a machine automatically when a problem arises preventing the damage to tooling or equipment. The machine may shut off automatically when the necessary parts have been made to prevent overproduction. This is part of inventory control.

Over time, the entire L-CMS is being continuously improved through changes in the system design. The interim manufacturing cells are gravitating toward lean manufacturing cells wherein many of the machine tools are custom built for the cells. This aspect of the L-CMS design strategy is discussed in Chapter 4 and Chapter 6.

8.19 STEP 10: Restructure the Production System

Once the factory (the manufacturing system) has been restructured into a linked-cell manufacturing system and the critical control functions well integrated, the company will find it expedient to restructure the rest of the company. The Danaher Production System was restructured, fueled by Danaher core values. The DPS engine drives the company through a never-ending cycle of change and improvement. Exceptional people develop outstanding plans and execute those using world-class tools to construct sustainable processes, resulting in superior performance. Superior performance and high expectations attract exceptional people, who continue the cycle. Guiding all efforts is a simple philosophy rooted in four customer-facing priorities: Quality, Delivery, Cost, and Innovation.

Many companies remove the functionality of the various departments and form teams, often along product lines. Concurrent engineering teams decrease the time needed to bring new products to market. This movement is gaining strength in many companies and the idea is to restructure the production system to be as waste - free as the manufacturing system.

Clearly, shifting from one type of manufacturing system design to another will affect product design, tool design and engineering, production planning (scheduling) and control, inventories and their control, purchasing, quality control and inspection, and, of course, the production worker, the foreman, the supervisors, the middle managers, and so on, right up to top management. Such a conversion cannot take place overnight and must be viewed as a *long-term transformation* from one type of *production system* to another. This kind of downsizing can be very traumatic for the business part of the company and usually has a negative impact on the morale of the company. This is why lean manufacturing is difficult to implement.

Table 8.4
Managing the Lean Production System Supplier (From a Plant Manager of a First Tier Supplier to Toyota)

The Lean Production System is the basic philosophy and concept used to guide production processes and environment. The LPS includes the linked-cell manufacturing system (cells linked by a Kanban pull system), the five S's, standard operation, the seven tools of QC, and other key organizational elements.	
Kanban pull system (Step 6)	The production processes that use a card system, standard container sizes and pull versus push production to accomplish just-in-time production.
Five S's (Seiri, Seiton, Seiketsu, Seis, Shitsuke) (Step 5)	The five S's are proper arrangement, orderliness, cleanliness, cleanup, and discipline.
Standard work in Manufacturing Cells (Step 2)	The production process used by technicians that combines people and process. The components of standard work include cycle time, work sequence, and standard stock-on-hand in the cells
Morning meeting	A daily meeting held for the purpose of sharing production and safety information, quite often by a quality circle.
Key Points: Process Sheets	The process sheets, which are visually posted at each workstation, detail the work sequence and most critical points for performing the tasks.
Tooling: Rapid Changeover and setup (Step 3)	The machine setup that takes place when an assembly line changes products
See seven tools of quality (Step 4)	The seven tools to quality are: A Pareto diagram, check sheets, histograms, cause-and-effect diagram, run charts for individuals, control charts for samples, and scatter diagrams.
Production behavior	Rules that include information on personal safety, safety equipment, clothing, restricted area, vehicle safety, and housekeeping.
Visual Management	Each line in the plant has a complete set of charts, graphs, or other devices, like Andons, for reporting the status and progress of the Area.

Therefore, everyone needs to recognize the need for the rest of the company to reorganize (get lean). This effort often begins with building *product realization teams* designed to bring new products to the marketplace faster. In the automotive industry, these are called *platform teams* and are an example of concurrent engineering. Platform teams are composed of people from design engineering, manufacturing, marketing, sales, finance, and so on. As the notion of team building spreads and the lean manufacturing system gets implemented, it is only natural that the production system will follow suit. Unfortunately, many companies are restructuring the business part of the company without having done the necessary steps 1 through 8 to get the manufacturing system lean and efficient. Downsizing the enterprise without simplifying and redesigning the manufacturing system can lead to difficult time for the enterprise. Other companies have tried to automate their way to productivity without first simplifying the system and integrating the control functions. Difficult times are the outcome.

The design of the manufacturing process technology must be done early in the product development process. The manufacturing process technology within the manufacturing cells must be part of (i.e., elements within) a well-designed integrated manufacturing system. Flexibility in the design of an integrated manufacturing system means it can readily accept new product designs. Flexibility in the process technology means the process can readily adapt to product design changes and be engineered to accept new products.

8.20 Why Lean Manufacturing Implementations Fail

There should always be a champion or take-charge person who gets behind the program and see it through. Lean manufacturing implementations will fail if the top management is not committed to it and if the company does not have a business plan and understand how a lean system will help them achieve those objectives. Many rooted in customer-facing priorities: like Quality, Delivery, Cost, and Innovation.

Problems in quality will prevent any real reduction in inventory and lead times. If the company cannot get to zero defects and zero breakdowns, they will not get to lean manufacturing. Lean exposes problems and is difficult to implement so the company must be ready to deal with problems in a fact based manner. Getting everyone to understand what "root cause" means takes time.

Watch out for management (managers) with hidden agendas. If the upper management already knows in advance what solution they want and uses the lean process direct the employees to their (management) desired ends, then the employees will tire of game playing, loose interest and even work to sabotage the new methods.

Middle management opposition will stop the lean manufacturing. The engineers, quality or MRP folks come to believe that the lean effort is a negative reflection on their expertise or competence and try to quietly sabotage the program. Their involvement in planning, data analysis, and status reviews must be assured. This problem also occurs with supervision if they are not fully brought into the effort.

System changes are inherently difficult to implement. Changing the design of the entire manufacturing system is a huge job. Changes on the factory floor will force changes for the entire enterprise. Companies spend freely for new manufacturing processes but not for redesigning the manufacturing systems. It is easier to justify new hardware for the old manufacturing system than to rearrange and modify the old hardware into a new manufacturing system (linked cells). Anyone with capital can buy the newest equipment but this is like creating another island of automation.

Converting to linked cells will free up additional capacity (setup time saved) and capital (funds not tied up in inventory), but such conversions are long term projects (lean is a journey) and will require expenditure of funds for equipment modifications and employee training in quality, maintenance of machine tools, setup reduction, problem solving and so forth.

The top management of the company may lose interest if they don't see quick results or get involved in some new fad which dissipates support for lean manufacturing. Generally, few of the companies have long range plans and programs that support concepts like lean manufacturing. Top management tends to manage for quarterly results and quick fixes. Therefore it is critical that the steps outlined here are followed for a successful implementation of lean manufacturing. Top management must participate in hands-on Kaizen events. They must do at least 10 to 12 hands-on Kaizen events in the first year. *Presidents, CEO's, and vice presidents* have to learn how to walk and talk *Gemba*. First it shows support, but second, their light bulbs are going to get turned on. The top management must participate in kaizen report-outs. All the Kaizen teams have report-outs on Friday afternoons. They have to participate in those. They cannot ignore them.

Decision making is choosing among the alternatives in the face of uncertainty. The fear of the unknown can lead to uncertainty. The greater the uncertainty, the more likely that the *do-nothing* alternatives will be selected leading to the sustaining of the status quo. In addition, many managers harbor faulty criteria for decision making. Decisions should be based on the ability of the company to compete (quality, reliability, unit cost, delivery time, flexibility for product or volume change) rather than on price alone. Getting the internal customer (production workers) involved in the decision making process is a significant change and critical to getting them committed to the new standardized work methods for lean manufacturing. Internal customers are pretty smart and logical especially when they see something that is going to make their life easier. Operators often say "if you go back to the way we used to do it, I'm quitting". Their life is chaos until you fix the system and they don't like to live in chaos every day.

8.21 Leadership
Real leadership is a rare commodity. To be a leader at any level from company president to middle manager the following qualities are required:
- **Vision:** To be a real leader, you must see beyond the day-to-day tasks and challenges. You must look beyond tomorrow and discern a world of possibilities and potential. You must learn to see what others do not or cannot and then be prepared to act on your vision.
- **Integrity:** A real leader must have integrity. Without this, real leadership is not possible. Nowadays, it seems that traits like integrity or honor or character are kind of quaint; a curious, old-fashioned notion.
- **Honor:** There are many people for whom personal integrity and honor are as important as life itself. For a real leader, personal virtues, self-reliance, self-control, honor, truthfulness, morality is absolute. These are the building blocks of character, of integrity and only on that foundation can true leadership be built.
- **Deep Conviction:** An additional quality necessary for leadership is deep conviction. True

leadership is a fire in the mind that transforms all who feel its warmth that transfixes all who see its shining light in the eyes of a man or woman. It is strength of purpose and belief in a cause that reaches out to others, touches their hearts, and makes them eager to follow this leader.

- **Self Confidence:** This is still another quality of leadership. Not the chest-thumping, strutting egotism we see and read about all the time. Rather, it is the quiet self-assurance that allows a leader to give others both real responsibility and real credit for success. The ability to stand in the shadow and let others receive attention and accolades. A leader is able to make decisions but then delegate and trust others to make things happen. This does not mean turning your back after making a decision and hoping for the best. It does mean trusting people at the same time you have a regular reporting mechanism, and are holding them accountable. The bottom line: a self-confident leader does not cast such a large shadow that no one else can grow.

- **Courage:** Leaders have to have courage: the courage to chart a new course; the courage to do what is right and not just what is popular; the courage to stand alone; the courage to act; the courage to speak the truth. In most business, government, and military training programs, there is great emphasis on team-building, on working together, on building consensus, on group dynamics. But, for everyone who would become a leader, the time comes when he or she must stand alone and say, this is wrong or I disagree with all of you, and because I have the responsibility, this is what we will do. That kind of action takes courage. Vision, integrity, deep conviction, and self-confidence are not enough to make a leader; a leader must have the courage to act, often against the will of the crowd. As President Ford once said, the greatest defeat of all would be to live without courage, for that would hardly be living at all.

- **Common Decency:** A final quality of real leadership is simply common decency: treating those around you decently and all your subordinates with fairness and respect. An acid test of leadership is how you treat everyone. A real leader, from the lowest rung of the ladder to the top, treats every person with respect and dignity. Use your authority over others for constructive purposes, to help them to watch out and care for them, to help them improve their skills and to advance, to ease their hardships whenever possible. All of this can be done without compromising discipline or mission or authority. Common decency builds respect and, in a democratic society respect is what prompts people to give their all for a leader, even at personal sacrifice.

8.22 Outsourcing

Lean engineering is not just about changing the MSD from a job shop to a lean shop. It is also about changing the way you think. The key is how leaders think about their business. That is what change is all about. Many companies are doing their strategic planning and are deciding to outsource their manufacturing, thinking that this is a good way to get rid of their batch-and-queue problems rather than having to go through the turmoil associated with a lean implementation. But this is not getting rid of anything. This is just handing over factory designs that were probably not lean anyway and they still have inherent problems within them. But you are adding a delta somewhere to the costs. Many companies think of manufacturing in terms of buying large increments of capacity (outsourcing). But if we think of lean in a machine design sense, increments of capacity that can be quickly changed over make you flexible. It can be easily adaptable to new product designs and can be easily moved within the plant so that an extra

10 percent of capacity can be added without any problem. The investment is small because you are not adding another $500,000 machine to add just 10 percent more capacity.

A lot of the companies that are making these outsourcing decisions have not thought that through from a machine design point of view. A superior competitive advantage can be built through machine design. Any company can go out and buy the same materials and equipment and compete for the same labor. But the lean company can build a competitive advantage with home built equipment, then the company has technology that others, especially their competition, do not have. Much of this technology is held in the lean manufacturing cells.

8.23 Summary

Implementing lean on the factory floor will impact the rest of the enterprise: human resources, accounting, financing, design engineering, purchasing, product development, sales and marketing will all be affected. Lean is extremely difficult to implement because every function within the enterprise is going to change. If it doesn't change, the company has not gone lean. For example, the traditional accounting systems are geared toward SEC and IRS statutory requirements. But *absorption accounting, for example, and purchase price variance drive the wrong behaviors. They are anti-lean. The finance manager will say they are good measures but they are not. So the accounting system has to change.* The HR people have to change. HR people need to understand that the work environment will be truly different in that people are doing a lot of different jobs and they're going to be multi-skilled. A new award system is needed so there is an incentive for people to learn new jobs. How the sales and marketing people take orders definitely impact the factory. If they take batch orders and allow their customers to order once per month to buy three month of supply, then the company never will be able to level the load on manufacturing. Level loading is almost impossible without the sales department changing how they incentivize the sales force, and how they manage their distributors and their customers. Every single function in the enterprise has to get on the same page. See Chapter 18 for additional Discussion.

Review Questions

1. What are the problems encountered by the company that are shown in the A3 analysis?
2. What kinds of things does Value Stream Mapping check for?
3. VSM is done for a product or product family. How do you define a "product family"?
4. Discuss the parameters Value, capability, capacity, flexibility and reliability of each step that we check from VSM?
5. Explain how capacity of machines is set in lean engineering verses how it is done in the mass production setting.
6. Outline the major steps in VSM methodology.
7. How can you group various products into families?
8. What are the important points that need to be followed from the start to finish of VSM to the end?
9. What are the steps involved in current state mapping?
10. What are the manufacturing wastes typically associated with lean production?
11. What measurable parameters are usually contained in the process data box?
12. What is the value of drawing the current VSM? Explain with an example.

13. What are the overall design goals of future state mapping?
14. Give a detailed definition of a Linked-Cell Manufacturing system. Be sure to include all the components of a manufacturing system.
15. List and define the lean tools of the new integrated manufacturing system design.
16. What are the major objectives of the Linked-Cell Manufacturing System design?
17. Describe the basis for maintaining these objectives?
18. Define the internal elements and external elements of a setup operation.
19. What are the advantages of integrated quality control?
20. What is the fundamental difference between how quality control is implemented in a lean system and how quality control is implemented in other systems?
21. What common element exists in implementing both integrated quality control and integrated preventive maintenance?
22. Describe the difference between leveling and balancing in the first two implementation steps.
23. How does the Kanban system accomplish pull manufacturing?
24. Describe how the Kanban system can be used to lower inventory levels.
25. A top-level manager at a manufacturing company just read a journal article on lean manufacturing and the Kanban system. Currently his manufacturing company does not operate on a lean system. However, the manager was so influenced by the article that he thinks his company can benefit from the ideas he read about, and he wants to take immediate action to implement a Kanban system. He asks you, the lean engineer, if the Kanban can be implemented successfully. What is your reply to him? DO you think the manager's current plant will be successful? If not, develop a successful plan of action, listing the steps required to achieve lean production. Be sure to warn the manager of the common reasons attempts at lean production fail.
26. Figure 8-A shows a manufacturing cell.
    a) What is the cycle time (show how it is calculated)?
    b) How much of the cycle time is walking time?
    c) Do you think the walking time for the two loops is the same?
    d) Why or why not?
    e) Why are there two BR machines?
27. Design Question: The interim manufacturing cell has seven machines and one operator
    a) How would you recommend improving the design shown in Figure 8.A?
    b) Calculate the throughput time (TPT) for one part going through the cell.
    c) What would be the problem with having two operators in the cell?

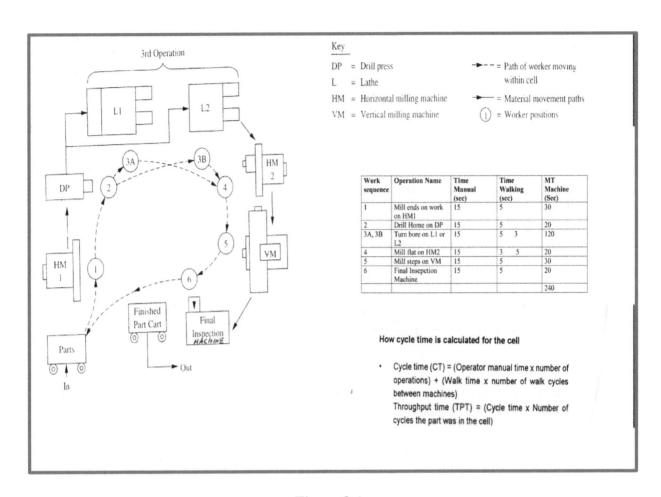

The figure includes the following key and table:

Key

DP = Drill press
L = Lathe
HM = Horizontal milling machine
VM = Vertical milling machine

▶--- = Path of worker moving within cell
▶—— = Material movement paths
① = Worker positions

Work sequence	Operation Name	Time Manual (sec)	Time Walking (sec)		MT Machine (Sec)
1	Mill ends on work on HM1	15	5		30
2	Drill Home on DP	15	5		20
3A, 3B	Turn bore on L1 or L2	15	5	3	120
4	Mill flat on HM2	15	3	5	20
5	Mill steps on VM	15	5		30
6	Final Insepction Machine	15	5		20
					240

How cycle time is calculated for the cell

- Cycle time (CT) = (Operator manual time x number of operations) + (Walk time x number of walk cycles between machines)
  Throughput time (TPT) = (Cycle time x Number of cycles the part was in the cell)

**Figure 8.A
An Interim Lean Cell**

Chapter 9
Fundamentals of Workstation Modeling

A credo for all Lean Engineers..........
"If you are waiting on me, you are backing up...and it is not easy to back up if you are moving forward."
East Texas Lingo

9.1 Introduction

In this Chapter we will address how to model manufacturing Workstations in a production facility. Our intent is to show how the fundamental constructs of Queuing theory can be used to accurately recover system performance measures. In this context a *system* will be defined as a single processing station with single or multiple servers. We will use the term *server* in a general context. A server can be a person, machine, material handling delay or any other processing component which causes a time delay. In Chapter 10, we extend the notion of a system to series configurations of servers which process only one part type, with either single or multiple processors. In Chapter 11, we present general modeling methodologies for multiproduct systems with single and multiple parallel processing components. The motivation for this Chapter is to gain insight into how to model single workstations, with single or multiple servers, and to explore what fundamental insight can be gained from modeling these basic systems. Before proceeding, it will be helpful to establish a common vocabulary.

Station or Workstation: The terms station and workstation will be used interchangeably to represent one or more identical servers (machines, people, computers, etc.) configured to complete one processing step in a series of processing steps leading to a finished product, an assembly or a subassembly. A station or a workstation typically includes a queue and one or more servers.

Server: A basic operational component of a manufacturing line which individually completes an assembly, operation, machining step, quality control inspection or any other basic unit of work. A server may be a machine, a person or an inanimate object which causes a delay. The granularity of work content and its definition is up to the modeler to define, and must be consistent with the basic definition of a Station or a Workstation. For generality and compatibility with most if not all Queuing reference sources, we will simply say that a single Workstation can have one or more identical servers (people, machines, computers, etc.).

Process Step: A typical factory will produce one or more finished products by transforming both raw material and paperwork into a finished product. The lowest cycle of work to be modeled is called a *Process Step*. A process Step can be an automated, semi-automated or manual operation which a modeler chooses not to subdivide for modeling purposes.

Process Routing Sheet: A process routing sheet is a routing document commonly used in a factory. A routing sheet lists of all of the processing steps required to produce a unit of finished product and usually also contains processing time specifications for each processing step, set up instructions, machine settings and other important information. Within any one processing step, the basic unit of work might be a finished product, a partially completed unit, a subassembly or an assembly depending upon where the modeler places the model boundary. Routing sheets are often referred to as travelers.

A Unit of Work: A unit of work is usually a piece of raw material which is transformed by a processing step at a Workstation to an intermediate or final product. However, the designated unit of work could be paperwork, physical raw material or a partially completed product (assembly or subassembly). This unit of work will usually be referred to simply as a *Part* unless specifically defined otherwise. Although it seems to be a paradox, a part can actually be anything moving through the system: a piece of raw material, a piece of paper, a person or even an abstract signal such as a Kanban card. The modeler must define the context. A part can change its definition from step to step as long as the conversion is properly identified and modeled. For example, one might designate a pallet of individual items as a single unit/part for material handling purposes, but if the pallet is unloaded one at a time for station processing, it might generate many individual *parts*. This phenomenon is part of what we call *Batch Moves*. The context and definition of a part is usually clear from problem definition. If parts are to be processed or moved in groups of size k, this unit will be called a batch or a *lot* to distinguish it from individual part movement.

Throughput Time (TPT): An important measure of performance is what the Lean Engineer would call *Throughput Time (TPT)*. Throughput time is defined as the total amount of time that a part or a defined unit of work takes to move from a designated point of entry to a defined point of exit. Throughput time can be measured at the individual Workstation level, a Manufacturing cell level containing many Workstations or at the factory level. *Throughput time* is usually called *Cycle Time* in the Operations Research, Production Control and Queuing technical literature. This definition is unique to the Lean community and must be carefully noted.

Cycle Time (CT): Another important system performance measure is called *Cycle Time*. In this book, Cycle time is defined to be the time which will elapse between part departures. Cycle Time must have context, such as the time between part completions at a single Workstation or time between part releases from a subassembly cell. In Lean engineering, Cycle Time is usually equated to or driven by *Takt* time. Cycle time is often defined as the inverse of the required production rate (PR) of an operational component or the customer demand rate (DR) from the Workstation, cell or factory. This definition is also unique to the Lean community. In practically every other literature source, Cycle Time is used in the same way as we have defined Throughput Time. In this context, Cycle time is often called the *time between departures*.

Takt Time (TT): As just noted, a special type of cycle time is called *Takt time*. Takt time is the minimum amount of cycle time per unit at any one workstation, assembly, manufacturing cell or assembly line required to satisfy customer demand. Takt time is determined by calculating the total amount of production time available over a fixed planning period, divided by the customer demand over that same period of time. The inverse of Takt time is often referred to as Cycle Time. Takt time is a term unique to the Lean community and can be interpreted as the inverse of the required output rate adjusted to available working time necessary to satisfy customer demand.

Machine Tool: A *machine tool* is a special type of server which completes a single processing step and is synonymous with a Station or Workstation server or individual processing unit. A machine tool might perform a single operation or multiple, sequential operations with a designated Throughput time. A typical tool might be a lathe, drill, mill, worker or any other physical entity at a Workstation. A special class of machine tools called *single cycle automatic machine tools* will be defined in Chapter 10 when we cover Manufacturing Cell Design and operation.

Operators: A human often executes manual processing, assembly, inspection and packaging steps on manufacturing lines. The time delay or throughput time for manual operations will be treated exactly the same as a machine delay. Manual or labor intensive steps in a routing sheet are simply another Station or Workstation modeling unit in our lexicon. How the work content is broken down by individual tasks must be consistent with the basic representation and operation of a workstation. The interaction between workers and machines was fully explored when we introduced-Shaped, Lean Manufacturing Cells in Chapters 3-4.

Work in Process (WIP): WIP will be defined as all of the units that are in the system being modeled. WIP is composed of two basic categories; WIP waiting in queue for the next available server, or a unit of work actually in the act of being served. Finished goods inventory will not be counted as WIP unless specifically indicated in a Factory Model. *System WIP is to be counted as the total number of units in the system from the time that work is released into the first server's queue, until it exits the last processing step. Workstation WIP includes the number of units that are waiting for service and are being served within the boundary of a Workstation model.*

System Performance Measures: A manufacturing model based upon Queuing theory might be used to generate many system performance measures. These measures will be demonstrated and defined through a series of examples in this Chapter. However, there are a set of performance measures which we will always want to calculate for a single workstation that are of particular interest to the Lean Engineer.

- The Expected number of parts waiting in queue at any one Station or Workstation
- The total number of parts in each Workstation. This is the expected number of units waiting in queue, plus the expected number of units in service.
- The average queue waiting time per part (Queue waiting time or queue throughput time)
- The average throughput time at any one Workstation or process step. The average throughput time is the expected amount of time spent waiting in queue, plus the expected amount of time spent in service per part.
- The average workload placed upon a workstation. This will be referred to as the *offered workload*. The offered workload will define the minimum number of servers/machines required to process the workload.
- Server Utilization. Server utilization is the proportion of time that any one service facility is busy. Server utilization will be greater than or equal to zero, and strictly less than one. Server utilization is defined as the offered workload divided by the number of servers.

9.2 Workstation Components and Characteristics

Figure 9.1 illustrates the components of a basic Workstation.

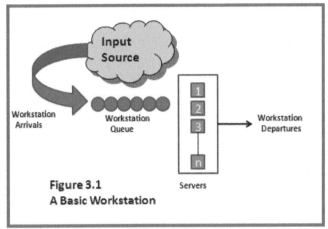

Figure 9.1
A Basic Workstation Configuration

Input Source Population: Arrivals to the Workstation can come from an outside source population that is either finite or continuous. In practice, an infinite source population implies a steady, uninterrupted source of parts to individual processing Workstations at the designated input rate.

Part Arrivals: Parts arrive to a Workstation according to some predetermined arrival distribution function. Arrivals are characterized by either a *Rate of Arrival* or by *Time between Arrivals* (TBA). A commonly assumed arrival rate distribution is the *Poisson Density function* given by:

$$p_x = \frac{\lambda^x e^{-\lambda}}{x!} \qquad x = 1, 2 \ldots \ldots \infty$$

$$\mu = E(x) = \lambda$$

$$\sigma^2 = Var(x) = \lambda$$

Where λ is the arrival rate to the Workstation.

If arrivals are characterized by time between arrivals (TBA), a commonly assumed distribution function for TBA is the Exponential Density Function given by:

$$f(x) = \theta e^{-\theta t} \qquad t \geq 0$$

$$\mu = E(x) = \frac{1}{\theta}$$

$$\sigma^2 = Var(x) = \frac{1}{\theta^2}$$

The characteristics of these two Probability Density functions and how they are related are covered in practically every Operations Research textbook and will not be repeated here. However, the following fundamental properties will be given without proof

Property 1: If arrivals to a Workstation follow a Poisson density function with parameter λ, then time between arrivals are distributed according to an Exponential distribution function with parameter $\theta = \frac{1}{\lambda}$. We will usually express λ in terms of arrival per hour, but any time unit could be used. Hence, θ is expressed as hours per arrival (Time between Arrivals). Please note that the Poisson is a discrete density function and the Exponential is continuous by construction.

Property 2: The Exponential density function has a unique property called the *forgetfulness property*. In simple terms, suppose that an arrival occurred at time T=6 time units, and that time between arrivals (TBA) follow an Exponential density function with mean TBA =4 Minutes. Now suppose that 10 minutes goes by without any arrival occurring. One would suspect that the next arrival is way overdue, and that one should occur at any moment. However, the expected time to the next arrival from time T_{now}= 14 time units is still 4 minutes!!. In effect, the Exponential process fails to recognize how much time time ΔT has elapsed following any arrival at time T. It simply *forgets* and projects the same expected time to the next arrival as E(t) = $\frac{1}{\theta}$. This property gives rise to what is called a *Markovian Process*, in which every time between arrivals is independent of the last arrival time. This is enormously important, since if an arrival (or a service) process is not independent (or cannot be assumed so), one must track past behavior.

Property 3: Every probability density function has three descriptive measures which we will be interested in defining. The first is the mean μ of the random variable being described, and the second is the Variance σ^2. The mean of a random variable (μ) is the point at which 50% of the density function is to the left and 50% to the right. The Variance (σ^2) is actually the second moment about the mean, and describes the spread of the random variables about the mean in terms of the distance squared. Clearly, σ^2 is greater than or equal to zero. σ^2 is referred to as a measure of *Variance,* and the square root of $\sigma^2 = \sigma$ as the *Standard deviation*. Both σ^2 and σ are relative measures of spread, relative to the mean μ. For example, if someone told you that the Standard Deviation of a random variable was one pound, this would not be much when describing the weights of all individuals over 250 lb., but would be disastrous if the random variable was the weight of a large bag of potato chips. A third measure of variability for a random variable is called the *Squared Coefficient of Variation,* which is the ratio of the Variance (σ^2) to the Mean (μ) squared.

$$CV^2 = \frac{\sigma^2}{\mu^2} .$$

Notice that in the above relationship, any two of the three parameters are sufficient to uniquely define the third. In this Chapter we will usually use μ and σ^2, but in more general station and manufacturing system models, we will use μ and CV^2. It is worth noting at this point that if a random variable (Time Between Arrivals, Service Times) follow an Exponential Density

Function, the CV^2 for those random variables will be $CV^2 = \frac{Var(x)}{E(X)^2} = \frac{\frac{1}{\theta^2}}{(\frac{1}{\theta})^2} = 1.0$

We will shortly see that in many modeling applications, it will not be necessary to establish what the exact form of the inter-arrival or service time density function might assume. The mean μ, and either σ^2 or CV^2 are all that is needed. Using the Exponential Density function as a "neutral point", more variance can be induced by simply increasing the value of CV^2, and less variance by decreasing CV^2. If a service time or time between arrivals has no variance, it is a constant and CV^2=0. Hopp and Spearman have suggested that the following values are characteristic of the Squared Coefficient of Variation values which might be observed in most manufacturing processes.

Low Variability (LV)	$CV^2 \leq 0.75$
Moderate Variability (MV)	$0.75 < CV^2 \leq 1.33$
High Variability(HV)	$CV^2 > 1.33$

Service Times: Because of the Markovian property, the Exponential Density function is often used to describe the distribution of service times in developing basic queuing methodologies. The properties of service times which follow an Exponential Density function are identical to those just described for Exponential time between arrivals. We will relax this limiting assumption later in this Chapter.

The Waiting line or Queue: The waiting space in front of a single or multiple server workstations is commonly called a *Queue*, being a French word which means *line*. Hence *Queuing Theory* is the study of waiting lines. In Lean Manufacturing, the study of Queues is of paramount importance. Taiichi Ohno, a chief architect of the Toyota Production System, called WIP (work in progress or a Workstation Queue) the *root of all evil*. Of course, as I tell my students WIP is not the root of anything; it is not the problem but a *symptom* of a problem. Much like when a sick person goes to the doctor, the doctor asks *what is your problem*? The response might be; *I have a terrific sore throat, I am coughing a lot and running a fever*. Of course, these things are not the problem at all, but symptoms of a problem. From those and other symptoms, a cure is found. That is how WIP should be understood. The notion of *Zero WIP* is an insidious myth, but excessive WIP can and does cause multiple problems. Excessive WIP is a direct signal to execute *root cause analysis* if the problem(s) are not immediately obvious. WIP will be our *guiding Light* to Lean Implementation and KAIZEN events. In general, the waiting line or queue can be of finite length or infinite length. Another parameter of the queue is the *queue discipline*, which can be First-in First – Out (FIFO), Last- in First- Out (LIFO), Random or a host of other priority rules. It is also true that in our presentation of Lean Manufacturing, we continually stress that *Variance* is *Public Enemy Number One*. A Factory Physics Rule (Hopp and Spearman) clearly states that if variance is not controlled and minimized, the system will pay for it in one of four ways: (1) Long Cycle Times (2) High WIP levels (3) Wasted Capacity or (4) Lost Throughput. These are our *Four Horsemen of Inefficiency*. These four performance measures represent manufacturing paradigms which cannot be tolerated in Lean Manufacturing. Variance in any form MUST be viciously attacked and eliminated on the journey called Lean Engineering. With this said, we will use a FIFO queue discipline and infinite queues in all of our models. A FIFO queue discipline should be stressed and adhered to unless unusual circumstances dictate otherwise. The common industrial practice of *expediting* and prioritization must be avoided at all cost. Such practices inherently and inevitably lead to degraded system performance. Infinite WIP will allow the modeler to immediately detect symptoms of the Four Horsemen, and will point the way to cause and effect analysis, followed closely by Kaizen event(s). Of course, one will argue that there are no infinite storage areas in real manufacturing systems. To this we reply that we have never seen work in progress simply disappear forever because of limited space. Excessive inventory *ALWAYS* goes somewhere. Unfortunately, off-line storage such as *Supermarkets* and ASRS systems often exist to cover up visible shop floor inventory and not to facilitate smooth, uninterrupted flow. These issues are discussed elsewhere in this textbook.

The Servers: As discussed previously, a Workstation or a *processing station* can contain one or more servers dedicated to a particular process step or processing function. A server can be a machine, a worker, a computer or any other limiting resource which provide processing capabilities. Basic models will be presented for both single and multiple server systems. The

294

most important characteristic of a server is the processing time required per part. Processing times can be both part and process-step dependent in multistep manufacturing processes. This Chapter will deal only with single step, single part type processing stations. Chapters 10 and 11 will extend these results to multistep, multiprocessing systems with both step and part dependent processing times. It is always necessary to know from direct observation, standard times, or design engineering the Mean and the Variance or Squared Coefficient of variation for the server processing time. In initially developing the concepts of single Workstation analysis, we will initially assume Exponential time between arrivals and Exponential service time distributions. This assumption will be relaxed in Sections 9.12-9.16.

9.3 The Server Utilization and the Offered Workload

As previously noted, a key issue is to determine the server workload and the minimum number of servers required to accommodate that workload. It is always necessary to determine the input of parts per unit time into the service facility. The total inflow will be represented as λ and usually designated as Parts/Hr for convenience. This is called the *Arrival Rate*. An equivalent measure is the time between arrivals which is simply $\frac{1}{\lambda}$ (Hrs/Arrival). Although it may be a misnomer, we will refer to processing capability as the product of the number of machines or workers available that are providing service (C) times the average service rate (μ) expressed in terms of ($\frac{Parts}{Hr}$) . The Offered Workload will be denoted by $\rho = \frac{\lambda}{\mu}$. Since the expected service time is $E[s] = \frac{1}{\mu}$, we will often use the following notation $\rho = \lambda * E[S]$. We use ρ to represent offered workload, since the value of ρ will dictate the number of machines or workers needed to satisfy external demand rate. ρ can be interpreted as the ratio of the mean arrival rate (external demand) to a single mean service rate (individual processing rate). The value of ρ can be any positive real number, but the ratio of $\frac{\lambda}{\mu} = \lambda * E[s]$ must be strictly less than one when an infinite queue is assumed, or WIP will build to infinity. To represent this requirement, we define "service facility efficiency" or "processing efficiency" as $u = \frac{\lambda}{c\mu}$ or $u = \frac{\lambda * s}{c}$. The value of C is the number of identical machines or workers required to render u<1. For example, suppose that parts arrive to a service facility at a rate of λ=5.5 units/hr, and the server is capable of producing at a rate of μ=6.1 parts/hr. Hence $\rho = \frac{5.5}{6.1} = 5.5(0.16393) = 0.90164$. In this case, c=1 and u=0.90164 since only one server is required. However, suppose that λ=5.5 and μ= 2.3. In that case, $\rho = \frac{5.5}{2.3} = 5.5(.4348) = 2.3913$. For these values of λ and μ, there will be three servers required (c = 3) and $u = \frac{5.5}{(3)2.3} = \frac{5.5(.4348)}{3} = 0.7971$. In the latter case, we would expect 2.393 servers out of 3 to be busy at any one point in time, and the service station utilization will be 0.7971. We cannot say exactly which machines or workers will be in service, only that 2.393 out of 3 will be busy on the average.

9.4 Kendall's Notation

As various queuing disciplines, time between arrival distributions and service time distributions are introduced, a wide variety of queuing models are possible. The assumptions can combine in so many ways that model identification becomes a necessary component of the modeling process. Fortunately, a standard format was developed long ago and is called the *Kendall's Notation*, after the individual who first proposed the scheme. The basic concept is to identify a particular model with a six-field notational scheme, with a particular model identified by the six fields separated by a semicolon and four slash marks as shown below.

Field1/Field2/Field3; Field4/Field5/Field6

Field 1 is an alphabetic code for the type of arrival process
Field2 is the same alphabetic code for the type of service time
The most commonly used codes are as follows.

   M for Exponential time between arrivals or service times
   D for constant time between arrivals or service times
   E_k for Erlang-k time between arrivals or service times
   G_I for general, but independently distributed time between arrivals
   G for general service times

Fields 3, 4 and 5 designate the number of servers (workers and/or machines), the capacity of the system (WIP) and the size of the input source population, respectively. *Field 6* is for the queue discipline. We will usually omit the last three terms based upon the typical but robust modeling assumptions of infinite source population, infinite WIP and FIFO queuing disciplines. For example, an M/M/1 model would represent Exponential service times Exponential time between arrivals, and a single server (Default conditions are assumed). For a second example, M/M/5 is the same model with C=5 identical servers.

Finally, a G/G/3 model would have a general (usually unknown) service time distribution and time between arrivals, with 3 servers. M/M/C models are often referred to as *Markovian Models*, a term which we will discuss later.

9.5 General Notation

The following notation will be used in Workstation modeling.

Table 9.1
General Notation

λ	**Workstation Input Rate**
μ	**Server Processing or service rate**
S	**Service or processing time per unit**
C	**Number of Workstation Servers**
ρ	**Workstation Offered Workload,**
U	**Workstation Utilization**
WQ	**Expected Waiting time in Queue per unit**
TPT	**Workstation Throughput Time**
CT	**Time between Workstation Departures**
LQ	**Expected Queue Length**
L	**Expected Number of units in the Workstation**

9.6 General Comments

Lean Engineering strategies and philosophies are both operational bodies of knowledge that vigorously attack the evils of long throughput times, excessive WIP and poor output rates. A partial list of strategies to mitigate if not eliminate these evils is shown in Table 9.2;

Table 9.2
Lean Strategies

Symptoms of poor Performance	Lean Strategies
Erratic arrival rates	Pull Systems, Batching Rules, Variance Reduction ,Kanban Cards, Group Technology, Supply Chain Analysis
Inconsistent Service or processing Rates	Use Less Batching, implement JIT philosophies, Examine Training, enforce worker motivation, implement at-the-source TQM & TPM, Eliminate variance
Long Move times	Coordinate movements, Pull systems, optimize batching, material handling planning, JIT Strategies, eliminate expediting, eliminate reprioritization
Unrealistic Lead Times	Base lead time on reality, not slop; use critical path planning, improve material handling, eliminate batching
Excessive Load & Unload Times	Kaizen SMED events, use standardized procedures, design Jigs & fixtures, automate load/unload functions
Long Throughput Times	Minimize variance-minimize variance-minimize variance! Avoid server utilization over 85% unless variance is low in both the arrival & service processes, Reduce WIP, Implement Pull, Enforce Line balance, Improve Quality & TPM practices
Excessive WIP	Improve capacity planning functions, eliminate high utilization, avoid prioritizing queues, stop expediting, investigate poor reliability/availability, avoid batching, implement pull strategies, investigate use of work cells with unit flow
Long Material Handling Times	Level production, balance lines, move to Cellular Mfg. strategies, automate wisely, avoid batching, Use Pull Strategies, use JIT, schedule to actual workstation demand.
Excessive Rework Items & Re-entrant Flows	Investigate poor quality control methods, eliminate work sampling and control charts as a control strategy, look for poor or outdated training, eliminate non-homogeneous raw material, relax unrealistic design specifications and avoid multistep machines

The main reason to develop manufacturing system and workstation models are to analyze system behavior, gain insight into system performance measures, identify bottleneck processes, experiment with remedial solution methodologies, reduce throughput time, improve cycle time, develop plans to reduce WIP, prioritize and identify Kaizen events and quantify the effects of variance.

9.7 Example 1

Consider a multipurpose production support Workstation which manufactures spare parts in a Machining cell. Request for manufactured items are frequent from several different sources, and for manufactured products which are similar but not identical. However, each part to be manufactured takes about the same amount of time which seems to follow an Exponential Density function. These characteristics indicate and justify a Poisson demand process and an Exponential density function to represent manufacturing work center demand and individual machine processing times. Parts arrive to a single waiting area (queue) and are processed one at a time using a FIFO waiting line discipline. The current estimated work load is $\lambda=5.6$ parts/hr and the average service/processing rate is $\mu=6$ parts/hr. First, we need to recognize that under these assumptions an M/M/1 queuing system is justified. Note that c=1 because $\rho = u = \dfrac{\lambda}{\mu} = \dfrac{5.6}{6} = 0.933$

What performance characteristics can be calculated knowing only the limited problem data? It turns out that quite a few performance measures are readily calculated from just λ , μ and c. The results shown in Table 9.3 are available from the Queuing literature.

All of these results are called *steady state* results, and are valid as long as the system is in a stable, non-transient pattern and $\rho = \dfrac{\lambda}{\mu} < 1$. These results can be derived in a variety of ways, including rate-in, rate-out equations and *Birth-Death* differential/difference equations (Phillips, Taha, Winston, Curry). The derivation(s) can be found in almost any Operations Research textbook, and will not be repeated here. For this M/M/1 problem, the results are shown in Table 9.4.

These performance measures should be informative and interesting. In fact, management was surprised by these basic results. A long standing operating policy was that each machine should be 90% utilized, and this machine is 93.3% utilized. However, the total throughput time per part (queue time + service time) is 2.4876 hours, with a standing average backlog of a whopping 13.067 parts waiting in queue. To mitigate these results, the predictive equations were examined. It is obvious that the demand stream could not be changed and might even increase as business got better. The only other parameter to adjust is the rate of service, or equivalently the average processing time. This was set by the Process Engineers and followed historical results predicted by an old work standard. However, a new Lean Engineer feels certain that he can increase the average service rate (decrease processing times) based on a LEAN Kaizen event.

The question that needs to be answered is, "What is the payoff for a dedicated Kaizen study?" He asks, "What service rate (μ) would be needed to reduce the System Throughput Time to 1.2 hour or less?" If this could be done, how would the server utilization and WIP be effected? To answer the first question, he needs to find the value of μ such that Throughput is TPT $\leq$ 1.2 hr. The value of μ is that which will satisfy the following inequality.

$$\text{TPT} = \frac{1}{\mu-\lambda} \leq 1.2 \text{ hr} \quad \text{or} \quad \frac{1}{\mu-5.60} \leq 1.2 \text{ hr}$$

This is easily solved to yield $\mu \geq 6.433$ parts/hr, or 0.1554 hr/part.
The system WIP is now 4.67 parts, which is a 66.5% reduction in WIP !! The Lean Engineer is quick to notice that the server utilization has now dropped to $\rho = 0.824$, which does not please

Table 9.3
M/M/1 Workstation Performance Measures and Notation

Performance Measure	Notation and Formula
Expected Arrival Rate	λ
Expected Time Between Arrivals	$\text{TBA} = \frac{1}{\lambda}$
Expected Service or processing rate	μ
Expected Service/processing time	$S = \frac{1}{\mu}$
Offered Workload	$\rho = \frac{\lambda}{\mu} = \lambda * S$
Number of Servers, machines or operators	C=1
Server Utilization	$u = \rho$
Expected Queue Length	$\text{LQ} = \frac{\lambda^2}{\mu(\mu-\lambda)}$
Expected Number in a Workstation service activity	$\rho = \frac{\lambda}{\mu}$
Expected WIP in the Workstation	$L = \frac{\lambda}{\mu - \lambda}$
Average Waiting Time in Queue	$\text{WQ} = \frac{\lambda}{\mu(\mu-\lambda)}$
Average Time in the Workstation or Throughput Time	$\text{TPT} = \frac{1}{\mu-\lambda}$
Probability of n parts in the Workstation	$p_n = (1-\frac{\lambda}{\mu})(\frac{\lambda}{\mu})^n \quad n \geq 0$
Probability the Workstation is Empty	$p_0 = 1 - (\frac{\lambda}{\mu})$

Table 9.4
Computational Results: Example 1

Performance Measure	Results
Expected Arrival Rate (λ)	5.60 parts/hr
Expected Time Between Arrivals ($\frac{1}{\lambda}$)	0.1786 hr/part
Expected Processing Rate (μ)	6.00 parts/hr
Expected Service Time (S)	0.1667 hrs/part
Offered Workload (ρ)	0.933
Number of Lathes required (c)	1
Server Utilization (u)	0.933
Expected Queue Length (LQ)	13.067 parts
Expected Number at the Lathe (ρ)	0.933 part
Expected WIP in the Workstation (L)	14.0 parts
Expected Waiting Time in Queue (WQ)	2.333 hrs
Expected Time in the Workstation (TPT)	2.50 hrs
Probability the Workstation is Empty (p_0)	0.067

the *bean counters*... (accounting). Notwithstanding, this level of server utilization certainly improves consumer response time, which should quell recent complaints regarding late delivery by the customer. Intrigued by this result, management now wants to know how the System Throughput Time and WIP behave in general as a function of the server utilization; ρ. It is intuitively obvious that as system WIP increases, the system throughput time will also increase. Note that $L = \frac{\lambda}{\mu-\lambda} = \frac{\rho}{(1-\rho)}$. Hence, the Lean Engineer produces a plot of ρ versus system WIP. The plot is shown in Figure 9.2.

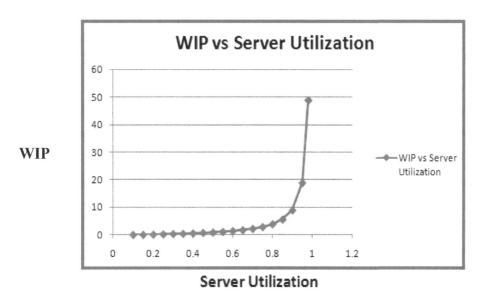

Figure 9.2
WIP vs. Server Utilization

This is a simple graph but illustrates a fundamental principle that every Lean Engineer should know. System WIP remains very low until server or process utilization exceeds about 80%. At that point it increases rapidly in a nonlinear fashion, increasing at a particularly alarming rate past about 95%. In fact, as predicted beforehand, system WIP will continually build and increase to infinity as utilization approaches 100%. This is intuitively obvious and results in infinite WIP when $\lambda=\mu$. But the Manufacturing Engineer with little experience in Lean Engineering principles will say, "This is ridiculous…I have seen many machines operating at near capacity, approaching 100% utilization at all times". We do not doubt this statement, but where is the paradox? The answer lies in our number one enemy…*Variance*. For the M/M/1 queue, both time between arrivals and service times are exponentially distributed, and the standard deviation is equal to the mean. This yields a squared coefficient of variation of $CV^2=1.0$ for both the arrival and service processes. In actual practice, it is not unusual for general service and arrival processes to have a $CV^2 \gg 1.0$ due to the degrading influence of variance. Clearly it is universally true that the Lean Engineer should constantly work to get this value as close to zero as possible. If the arrival rate and the service rate happen to be equal with no variance, then everyone is happy and the Manufacturing Engineer is correct. It should be noted that Figure 9.2 was constructed for the M/M/1 case, but if the "corrupting influence" of variance is present….and it almost always is…then the graph just shown will hold as a universal truth for any M/M/1, M/M/C or G/G/C system and one should not attempt to push server utilization past 85%-90% for these systems.. The Lean Engineering implication is that unless variance components cannot be aggressively attacked and reduced, the processing station or server must always run at reduced efficiency. In fact, 90% efficiency seems to be a good target but it could be higher depending upon the amount and type of variance components present. We will return to this discussion later in this Chapter.

Note finally that the expected number in queue represents the expected number of arriving parts or entities which must be stored while awaiting service. In actual practice, it is also necessary to estimate both the maximum number of units which might be expected and a probability of this level of storage required. These questions can be addressed by (1) Determining the Standard Deviation of the expected queue length and (2) calculating the maximum queue length expected equal to LQ plus $3\sqrt{Var(LQ)}$.

Feldman and Valdez-Florez have shown that the Variance of the expected number in queue for an M/M/1 service system is given by the following equation.

$$Var[LQ] = \frac{\rho^2(1+\rho-\rho^2)}{(1-\rho)^2} \qquad \text{Where:} \quad \rho = \frac{\lambda}{\mu}$$

Using Chebychev's theorem, an estimate of the maximum expected queue length is therefore given by:

$$max[\,LQ] = \{\frac{\rho^2}{(1-\rho)}\} + 3\{\sqrt{\frac{\rho^2(1+\rho-\rho^2)}{(1-\rho)^2}}\,\}$$

Note that this is only a rough estimate based upon $\mu_{LQ}+3\sqrt{\sigma_{LQ}^2}$. For this example,

$$max[\,LQ\,] = \{\frac{0.933^2}{(1-0.933)}\} + 3\{\sqrt{\frac{0.933(1+0.933-0.933^2)}{(1-.933)^2}}\,\} = 56.35 \text{ Parts}$$

It is also possible to calculate the probability that the number in the system will be of any length. The formula which can be used for an M/M/1 queue is as follows.

$$p_n = (\rho^n)(1-\rho) \qquad n=0, 1, 2... \infty$$

In this case, the reader can verify that the $p_r[n \geq 50] = 0.0296$
The general universal corrupting influence of variance and its basic causes will be studied in Chapter 12.

9.7.1 Performance Measures-Revisited

From Table 9.3 we observe that for an M/M/1 service system, system WIP $= \frac{\lambda}{\mu-\lambda} = \frac{\rho}{1-\rho}$. The expected time spent in the system is defined as the Throughput time (TPT), so that TPT $= \frac{1}{(\mu-\lambda)}$ $= \{\frac{1}{1-\rho}\}*s$, and the demand rate or required production rate (PR) is λ. Using these two performance measures, one can show that for the M/M/1 Workstation:

*WIP = Production Rate * Throughput Time = PR * TPT = λ*TPT*

This phenomenon was first investigated by John D. C. Little in the early 1960s. Little proved a remarkable and tremendously useful fact. He was able to show that the relationship between systems WIP, Production Rate and Throughput time will hold for ANY queuing system, and will be true for single station models, multiple station models, networks of Markovian queues and even networks of non-Markovian queues. This relationship is now known as *Little's Law*, and is the Manufacturing Systems equivalent to F=MA in Physics (Hopp and Spearman). Consider any Workstation with single or multiple servers with a demand rate $=\lambda$, service rate $=\mu$, and an infinite queue. A stable queuing system must have $\lambda<c\mu$ or system WIP will build to infinity. In this case, it should be noted that the output rate= input rate $=\lambda$. Recall the following notation.

L = Number of parts in the Workstation (WIP)
LQ= Number of parts in Queue
TPT= Throughput time in Workstation
WQ = Waiting time in Queue

λ = Demand or Input rate

The following set of relationships will be referred to as **Little's Laws** in this textbook.

Little's Laws at Steady state

$L = \lambda * TPT$
$LQ = \lambda * WQ$

Little's Laws will greatly simplify and increase our ability to calculate System Performance measures for an unlimited number of Workstation configurations. In this textbook, we will exploit these relationships by assuming that the Input Rate or Demand Rate λ is known or can be calculated. We will then usually calculate the expected waiting time in queue, WQ. Having these two values, the remaining Systems performance measures are readily determined since TPT=WQ+S.

9.8 Example 2

A tool crib has been proposed to support several manufacturing areas, and has been recommended as a way to improve the control and usage of specialized tools used in a manufacturing cell. A Manufacturing engineer has been asked to run a queuing analysis to predict tool crib behavior and recommend the number of tool crib attendants that would be needed to support production. For this example the tool crib is the Workstation and mechanics arriving to check out tools define the arrival rate. After some field investigations, it was determined that the average request rate would probably be 6 tools per hour. It is estimated that retrieving the tool and executing the necessary paperwork would take about 8 minutes. An assumption was made that the rate of requests follow a Poisson distribution function and the service times Exponential. The offered work load is $\rho = \lambda * S = 6[8/60] = 0.80$. Hence, only one tool crib attendant will be required as long as tools are requested at a rate of less than 7.50 requests per hour to keep $\rho < 1.0$ Keep in mind that as ρ increases beyond 0.85-0.90 , the waiting line, waiting time and system throughput time for tool service will grow at an increasing rate. At $\rho=1.0$, these performance measures will grow to infinity. For the base case, the Workstation (Tool Crib) Throughput Time is given by $TPT = \frac{1}{(\mu - \lambda)} = \frac{1}{7.5 - 6} = 0.667$ hrs and by Little's Law L = $\lambda *TPT = 6(0.667) = 4.02$ people. Because $\rho = \frac{\lambda}{\mu}$, it is predicted that $(4.02 - 0.80) = 3.22$ people are in line at any one time on the average. Again by Little's Law, the expected waiting time in line per customer is WQ= 0.533 hrs. It is also interesting to note that the probability that the tool room clerk will be idle is $p_0 = (1 - \rho) = 0.20$. It may seem very strange that more than 3 people are waiting for service on the average but the Tool Clerk is idle 20% of the time. To gain perspective of what is happening, recall that the mean and the variance of the Exponential service time distribution are the same. This creates an opportunity for an occasional very long service time. During this service period, several arrivals might arrive and form a long line. Of course, by the law of averages the Tool Crib attendant will eventually "catch up", and in fact the system will also be idle for possibly a long period of time. This erratic behavior and excessive waiting line is caused again by our old enemy **variance.**

It should be noted that this example was to illustrate the principles of queuing theory and not to promote "best practice". The use of a tool crib from the Lean Engineering perspective is a very bad idea. The reasons should be obvious. The tool crib promotes better control over operating resources, but causes interruption of work and results in lost operating capacity. Of course, these two results are both only observable symptoms of the real problem. Use of a central tool crib causes an increase in system variance way beyond the service facility itself. These additional corrupting influences cannot be quantified without a larger systems model, the objectives of Chapters 10 and 11. Any source of induced variance is an enemy of the Lean thinker and with proper identification, hard work and Lean thinking, the Lean Engineer can reduce if not mostly eliminate variance. Nevertheless, the correct Lean thinking is to provide each worker with his own individual tool kit which he is expected to manage properly. Large or expensive tools should be dealt with using Hyjunka or common "shadow boxes".

9.9 Multiple Server Workstations

To this point we have only presented modeling methodologies for single server M/M/1 Workstations. We will now introduce the queuing theoretical results needed to analyze multiple server Workstations and again demonstrate the universal application of Little's Laws. For now, we will restrict ourselves to M/M/c service systems. One notational change is to define $\rho=\frac{\lambda}{\mu}$ and $u=\frac{\rho}{c}$. Rho (ρ) will be referred to as the *offered workload* and **u** as the *server utilization*.

The Operations Research literature is full of derivations for the M/M/C service system. It is possible to recover in closed form the exact solutions corresponding to those previously presented for the M/M /1 system. Exact closed form solutions generally depend upon the value of p_0, which in turn can be calculated from closed form equations. Our ultimate objective in introducing the M/M/C model is to allow us to model more complex networks of G/G/C service stations. However, as we will see, this model and its extension have many interesting uses as a stand -alone analysis tool. The most useful result that we have found in the literature to analyze M/M/C service systems is a closed-form approximation for the expected waiting time in queue first proposed by Sakasegawa (1977). Both Curry &Feldman and Hopp & Spearman have both tested this approximation extensively and found it to be quite accurate.

$$ WQ(\ M/M/C)\ =[\frac{S*\ u^{\sqrt{2c+2}-1}}{c(1-u)}\ ] \qquad \text{where:} \qquad u=\frac{\lambda}{c\mu}=\frac{\lambda*S}{c}=\frac{\rho}{c} \qquad (9.1) $$

It is also instructive to note that Equation 9.1 can be written as follows.

$$ WQ(M/M/C)=[\frac{u^{\sqrt{2c+2}-2}}{c}]*(\frac{u}{1-u})*S \qquad \text{where:} \qquad u=\frac{\lambda}{c\mu}=\frac{\lambda*S}{c}=\frac{\rho}{c} $$

In this form, we recognize that $WQ(\ M/M/C)=[\frac{u^{\sqrt{2c+2}-2}}{c}]*WQ(M/M/1)$

For the expected time in the M/M/C system we can use

$$ TPT=\{\frac{S*\ u^{\sqrt{2c+2}-1}}{c(1-u)}\}+S \qquad \text{or} \qquad [\frac{u^{\sqrt{2c+2}-2}}{c}]*(\frac{u}{1-u})*S\ +S \qquad (9.2) $$

Note that the workstation offered workload is $\rho=\frac{\lambda}{\mu}$ and the server utilization is $u=\frac{\lambda}{c\mu}$. The use of Equations 9.1 and 9.2 only require λ, S, μ and the required number of servers (c) that are needed to make server utilization u < 1. Straightforward calculations using Little's Laws will provide all of the other performance measures.

9.10 Example 3

A turret lathe work cell is being planned to operate under a workload of $\lambda=16$ parts/hr and the lathe will be capable of processing parts at a rate of 5.9 parts/hr. Assume that the arrival rate and the service rate are Poisson and that there is infinite queue space available. What are the number of lathes required, system throughput time, WIP in the system and the expected time that each part waits for processing? The offered workload is $\rho = \frac{\lambda}{\mu} = \frac{16}{5.9} = 2.712$ Hence, three lathes will be required and this is an M/M/3 system. The server utilization with three servers is $u = \frac{\rho}{c} = 0.904$. The Waiting time in queue and the throughput times are:

$$WQ = \frac{S * u^{\sqrt{2c+2}-1}}{c(1-u)} = \frac{\left(\frac{1}{5.90}\right)0.904^{\sqrt{2(3)+2}-1}}{3(1-0.904)} = 0.4891 \text{ hrs.}$$

$$\text{TPT}= WQ + S = 0.4891 + 0.1695 = .6586 \text{ Hrs}$$

By Little's Laws
$$L = \lambda * \text{TPT} = (16)(0.6586) = 10.5369 \text{ parts}$$
$$LQ = \lambda * WQ = 7.825 \text{ parts}$$

It may be of some interest to know that exact solutions to the M/M/3 queuing system are available (Taha or Phillips, Ravindran & Solberg). The exact value of $WQ=0.4847$.

9.11 Example 4

A manufacturing flow shop has two assembly lines dedicated to producing replacement units for F16 fighter planes under a long term contract. The flow lines are balanced, and produce an average of 4.5 parts per day per line. Before moving to shipping, each unit of product requires a Quality Control (QC) check. QC is operated out of a central facility. Once a part is finished, it takes an average of 1/5 day to clear Quality Control. Each assembly line foreman is complaining about slow service (sound familiar?). To silence complaints and reduce Throughput time, a Manufacturing Engineer has recommended that QC provide a dedicated inspector to each line. Everyone thinks that this is a great idea except for a single Lean Engineer. He is not sure and requests a quick study be performed to compare options. First, consider the proposed solution. Assuming a Poisson rate of inspection and Exponential inspection times, each line produces parts at a rate of $\lambda=4.5 \frac{units}{day}$, and the typical QC response time is $s=1/5 \frac{day}{unit}$. The new (proposed) QC configuration will each operate as an M/M/1 service center with $\lambda=4.5$ units/day and $\mu=5$ units/day. The analysis is straightforward and yields the following performance measures.

Model 1: Two M/M/1 Systems

For *each* dedicated QC Workstation: ρ=u= 0.90, WQ=1.80 day/part, TPT= 2.0 days/part, L=9.0 parts and LQ=8.10 parts.

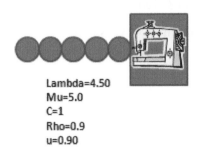

Lambda=4.50
Mu=5.0
C=1
Rho=0.9
u=0.90

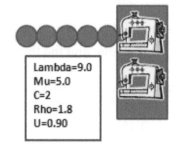

Lambda=9.0
Mu=5.0
C=2
Rho=1.8
U=0.90

Model 2: One M/M/2 System

The status quo model requires two QC inspectors to respond to demand, so both QC operations would employ the same number of QC inspectors. Since the demand streams are now combined, λ=9.0 and μ=5.0 for this model. Hence, ρ=1.80 and with two inspectors u= $(\frac{1.80}{2})$ = 0.90. The analysis requires use of the M/M/2 approximation model to calculate the average waiting time, and then application of Little's Laws. The results are as follows.

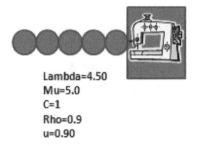

Lambda=4.50
Mu=5.0
C=1
Rho=0.9
u=0.90

WQ= 0.86 day/part TPT= 1.06 day/part L=9.53 parts and LQ =7.73 parts

The results are almost too hard to believe. The Manufacturing Engineer's solution to assign dedicated QC services to each flow shop performs worse than the single, pooled system. Of course, we have assumed that the travel time to a QC check is roughly the same, but this may or may not be a deciding factor. What is more remarkable is that this same conclusion is valid for *ANY* values of λ, μ and ρ, and extends to any number of required QC people (servers). This example is revealing since intuition would seem to lead one to the opposite conclusion. The result is valid even if the two lines generate unequal demand streams, and clearly point out the value of using Lean Modeling Methodologies to support rational decision making. The principle being modeled here is called *pooling of servers*, and the law which emerges is that it is always advantageous to pool services, all other things being equal. The insight behind this *law* is simple but universally true. When two separate service systems are used and no line switching is allowed, an arrival in one line can get caught behind someone experiencing a very long service time, even though the other server might in fact be idle. This cannot happen in *pooled systems*. This principle is frequently intuitively recognized in some Manufacturing and service systems. Most if not all airline reservation counters at an airport, Bank tellers and Post Offices all use pooled systems. A paradox may be noted that most ATM drive-in bank services and large grocery stores use parallel lines. In this case, there is less service efficiency assuming that line switching is not possible. But each of us has switched lines in a supermarket to avoid a long wait. When to do this is an interesting policy analysis. However, a parallel arrangement might help the problem of limited line lengths. This is undoubtedly the reason that Supermarkets do not use pooled checkout services. Of course, in service systems human behavior often plays a critical role. For example, a customer may prefer a particular bank teller or checkout station. Nevertheless, it can be categorically stated that strictly from the standpoint of the system performance measures and our modeling assumptions, it is always best to pool servers (Taha).

9.12 Example 5

To again demonstrate the value of even simple queuing models in Lean Thinking, we will consider the following scenario (Phillips, Ravindran & Solberg). In the Midwest wheat is the principle if not the sole source of income for many farmers. During harvest season, trucks loaded in the fields must drive to a grain elevator and be unloaded as quickly as possible so that they can repeat their mission. The unloading process involves paperwork, weighing the loads, unloading and a few other details. Farmers are very concerned about long system throughput times. Once the harvest is started, a heavy rainstorm or windstorm could prove disastrous. To keep the problem simple, let us suppose that the average inter-arrival time for trucks is 6.667 minutes and the average unloading time is 6 minutes. There is a single unloading station, and we will assume an M/M /1 model. One can verify that the average TPT= 1.0 hour per truck, which closely matches actual observed times. This is deemed to be too long, so at a meeting of the local Farmers Co-Op three solutions were proposed to reduce system throughput time.

1   Installing sideboard extensions to permit more hauling capacity would increase the interarrival time to 10 minutes. By making some minor changes at the unloading station, the average unloading time is expected to be only 4 Minutes. These modifications would cost an estimated $30,000 dollars.

2   Some farmers feel that although these changes might help, they are of the opinion that a new, modern weigh-in and unloading station should be built. Arriving trucks would join a single line at the unloading facility, but there would be two separate processing stations accessed on a first come-first in line basis. The cost of this new facility will be $400.000.

3   Another group of farmers feel that the current grain elevator needs to be duplicated on the other side of town. This proposal would roughly split arriving traffic in half, but would require acquisition of new land and right-of-way. Cost: $1,000,000.

4   At this point enter our Lean Engineer. He feels that the best alternative can be found by conducting a systems analysis. In terms of queuing models, the first proposal only involves changing the parameters of the M/M/1 model. The second proposal is clearly a M/M/2 model with a single waiting line. The third model will require two identical service facilities, each represented by an M/M/1 model. The Poisson arrival assumptions may possibly be justified, since arrivals are from several different places. The Exponential service time assumption may be questionable, but as we have seen it provides reasonable results with service time SCV's of 1. As with most modeling exercises, we are looking for comparative performance and not precise predictions. The three proposals and status quo modeling results are shown in Table 9.5

307

Table 9.5
Comparing Different System Configurations

Proposal	Model	Arrival Rate	Service Rate	System Throughput Time	Cost
Existing	M/M/1	9 per hr	10 per hr	60 Minutes	$0.0
1	M/M/1	6 per hr	15 per hr	6.667 Minutes	$30 k
2	M/M/2	9 per hr	10 per hr	7.72 Minutes	$400 k
3	M/M/1	4.50 per hr	10 per hr	10.91 Minutes	$1000 k

The results are startling. The least expensive proposal, and possibly the least preferred by the farmers, provided the greatest benefit. It is hard to imagine that these comparative results could be a'priori determined based upon intuition alone. The example clearly demonstrates what can happen if system changes are left to opinions and not modeling results. The human mind can often be fooled if complex interactions are part of the problem. From a pedagogical standpoint, the relative differences in results can almost exclusively be related to the offered workloads ρ and the server utilizations, u. The interested reader is encouraged to play with different combinations of these parameters.

9.13 General Non-Markovian Workstations

We have presented results for both the M/M/1 and the M/M/C service systems, both of which require the Markovian assumptions of Exponential service time(s) and Exponential time between arrivals. We will now address more general Workstation models which do not require these assumptions. We will first consider an M/G/1 model and then the more general G/G/1 and G/G/C models.

9.13.1 The M/G/1 Model

The M/G/1 model requires Poisson arrivals and an infinite queue, but can accommodate any continuous service time density function. A solution to the expected number in queue for the M/G/1 model was derived by Pollaczek and Khintchine and is found in the literature under the name *P-K Equation*. The result is given by Equation 9.3.

$$L_q(\text{M/G/1}) = \frac{(\frac{\lambda}{\mu})^2 + \lambda^2 \sigma_s^2}{2(1 - \frac{\lambda}{\mu})} \qquad (9.3)$$

It should be noted that Equation 9.3 provides an ***exact*** solution to the M/G/1 service system for the expected queue length. It will be convenient for us to derive another form of this equation which involves the squared coefficient of variation of the general service time distribution.

Denote $CV_s^2 = \frac{\sigma_s^2}{\mu^2} = \sigma_s^2 s^2$, adding the expected service time and invoking Little's Law (recall that the arrival stream is Poisson, so $CV_a^2 = 1.0$ and $u = \rho = \frac{\lambda}{\mu}$, we obtain Equation 9.4 for the expected time in queue.

$$WQ \, (M/G/1) = \frac{(1+CV_s^2)}{2} \{\frac{\rho}{1-\rho}\} *S \qquad\qquad (9.4)$$

We recognize that this result is related to the M/M/1 result.

$$WQ \, (M/G/1) = [\frac{(1+CV_s^2)}{2}] \{WQ(M/M/1)\} \qquad\qquad (9.5)$$

Equation 9.5 is an interesting and informative form of the M/G/1, WQ equation. This form of the P-K result shows that the Waiting Time in Queue for an M/M/1 Workstation is inflated by a factor of $\frac{(1+CV_s^2)}{2}$, which directly involves CV_s^2. As this term increases, the time in queue increases proportionately. Since the CV_s^2 has the variance of the service time in the numerator, it is clear that queue waiting time, and hence cycle time, increases as the variance increases. Here we conclusively show the *Degrading influence of Variance* on this model.

9.14 Example 6

Joe's Taco Shack is a one man operation that serves customers out of a small building with no permanent parking lot. Customers drive through and order out of their car window. There have been some complaints from the city concerning the line of cars that periodically develop. Joe has a Lean Engineering friend who is asked to analyze the operation and show using a queuing model how the service system behaves. Joe is only concerned about a peak period between 11:00am and 2:00pm. Our Lean Engineer collects some data and finds that cars arrive at an average rate of about 8.2 per hour and require about 5.1 minutes to fill their order. Because customers arrive at random from many places, the Lean Engineer feels that a Poisson arrival rate is justified. However, the service rate data does not follow any recognizable density function including the Exponential, but the Service time SCV can be calculated and turns out to be $CV_s^2=1.32$. Since Joe is the chief cook, bottle washer and server, the appropriate model is an M/G/1. The Pollaczek- Khintchine (P-K) equation provides an exact solution to this problem.

The offered workload is $\rho=\frac{\lambda}{\mu}=\frac{8.2}{11.765}=0.697$, so it appears that Joe is quite able to service the level of requests by himself. Using Equation 9.4, we obtain the Expected time any one customer waits in line for service.

$$WQ = \frac{(1+CV_s^2)}{2} \{\frac{\rho}{1-\rho}\} * S$$
$$WQ= \frac{(1+1.32)}{2} \{\frac{0.697}{1-0.697}\} (.085) = 13.61 \text{ min/car} = 0.227 \text{ hrs}$$

The expected time in the system per customer is:
$$TPT = WQ + S = 13.61 + 5.1 = 18.71 \text{ min/car} = .3118 \text{ hr/car}$$
The expected Queue length is:
$$LQ= \lambda * WQ = (8.20/60)(13.61) = 1.86 \text{ cars}$$

And $\quad L = LQ + \rho = 1.86 + 0.697 = 2.56 \text{ cars}$

Note that by Little's Law,

WQ $= LQ/\lambda=$ (8.20) (1.86) = 13.61 min/car as expected.

The expected waiting line is 1.86 cars, and there are 2.56 cars expected to be in the system at any one time. Joe can accommodate up to 5 cars in line without becoming a public hazard, so it appears that any complaints were unfounded except in extreme cases. The situation needs to be further examined by obtaining the probability distribution of n= 0, 1, 2 ....N cars in the system to further characterize the possible line lengths. A derivation and presentation of necessary results can be found in Gross and Harris. The reader will be asked to do this as a homework exercise. We may also be interested in examining the effects of increased arrival rates on system performance measures.

9.15 The G/G/1 Model

Notice that the first term in the numerator in Equation 9.5 is one, because the $CV_a^2=1$ for Exponential time between arrivals. This suggests that we could generalize the P-K equation to the G/G/1 case as follows.

$$\text{WQ (G/G/1)} =[\frac{(CV_a^2+CV_s^2)}{2}]\{WQ(\text{M/M/1})\}$$

or: $$\text{WQ (G/G/1)} =[\frac{(CV_a^2+CV_s^2)}{2}]\{\frac{\rho}{1-\rho}\}*\text{S} \qquad (9.6)$$

Witt has shown that this conjecture yields a good approximation for a wide range of problems, and that Equation 9.6 provides an accurate approximation for the expected time in queue for a single server system with general service time and general arrival time distributions. Note that Equation 9.6 clearly depicts a "double whammy" on the expected time in queue due to variance components, since the expected time in queue is inflated by both $CV_a^2 + CV_s^2$, which in turn depends upon σ_a^2 and σ_s^2. Equation 9.6 is often called the "Kingman Diffusion Approximation". For the remainder of this Chapter and in Chapters 10 and 11 we will use the following results.

$$WQ(\text{M/G/1}) = \{\frac{1+CV_s^2}{2}\}\{\frac{\rho}{1-\rho}\}*\text{S} \qquad (9.7)$$

$$TPT(\text{M/G/1}) = \{\frac{1+CV_s^2}{2}\}\{\frac{\rho}{1-\rho}\}*\text{S} +\text{S} \qquad (9.8)$$

$$WQ(\text{G/G/1}) = \{\frac{CV_a^2+CV_s^2)}{2}\}\{\frac{\rho}{1-\rho}\}*\text{S} \qquad (9.9)$$

$$TPT(\text{G/G/1}) = \{\frac{CV_a^2+CV_s^2}{2}\}\{\frac{\rho}{1-\rho}\}*\text{S} +\text{S} \qquad (9.10)$$

Where: $$CV_a^2 = \frac{\sigma_{TBA}^2}{E[TBA]^2}$$

and $$CV_s^2 = \frac{\sigma_s^2}{E[Service\ time]^2}$$

9.16 Example 7

A milling operation consists of a single, semi-automated milling machine. Industrial Engineering has compiled operating data on both the service times and the time between requests for milling individual parts.

Table 9.6
Operating Data

	Time Between Arrivals	Service Time
Mean	12 Minutes	10 Minutes
Standard Deviation	7 Minutes	12 Minutes

From this data, the $CV_A^2=0.34$ and $CV_s^2=1.44$. Since both CV_A^2 and CV_s^2 are not equal to 1.0, then it is clear that the Markovian assumption of Exponential service times and Exponential time between arrivals fails to hold, and so the M/M/C models cannot be used. The offered workload is $\rho = (\frac{\lambda}{\mu}) = (\frac{5}{6}) = 0.833$. Since $\rho < 1$, the use of a single mill is adequate to service the offered workload and u=0.833. The proper model is G/G/1. The corresponding system performance measures are

TPT=0.9083 hrs   L=4.5417 parts   WQ= 0.7417 parts   LQ=3.709 parts. The service facility is performing well, largely because the server efficiency is only u=0.833, which is just at the point where WIP starts to grow rapidly. If request for service increases beyond the current rate, a variance reduction exercise would certainly be a good idea to prevent a WIP explosion and a corresponding increase in throughput time.

9.17 The G/G/C Model

The last model which we will present is for a multiple server processing station with general time between arrivals and general service times. This is called the G/G/C Workstation model. There are several procedures which have been proposed to extend the G/G/1 Kingman approximation model to accommodate multiple servers. We will adapt and use a methodology demonstrated by Allen and Cuneen. Their approach is similar to the G/G/1 approximation procedure previously discussed. They propose use of the approximation given in Equation 9.11 for calculating the expected waiting time in queue.

$$WQ(\text{G/G/C}) = [\frac{(CV_a^2+CV_s^2)}{2}] \ \{ WQ(\text{M/M/C})\} \tag{9.11}$$

We will use the WQ (M/M/C) approximation procedure given by Equation 9.1

$$WQ(\text{M/M/C}) = \frac{S* u^{\sqrt{2c+2}-1}}{c(1-u)} \qquad \text{where:} \quad u = \frac{\lambda}{c\mu} = \frac{\lambda*S}{c} = \frac{\rho}{c} \tag{9.12}$$

Hence, using Equation 9.12 with Equation 9.11, the approximation procedure is given by Equation 9.13

$$WQ(\text{G/G/C}) = \{ \frac{CV_a^2+CV_s^2}{2}\} \ \{ \frac{u^{\sqrt{2c+2}-1}}{c(1-u)} \ \}*S \tag{9.13}$$
$$\text{Where:} \quad u = \frac{\lambda}{cu} = \frac{\lambda*S}{c} = \frac{\rho}{c}$$

9.18 Example 8

A Greenfield Factory design team has proposed a packaging and shipping cell for use by final product delivery. The cell handles a wide variety of individual products, but the packaging procedures have been engineered to deliver all products in a standard packaging configuration. The cell takes an average of 15.6 minutes to pack and ship an individual item. Products arrive about one every 5.5 minutes. It has been determined that the $CV_a^2 = 1.20$ and the $CV_s^2 = 1.35$ How many packers are required to pack and ship any one order in less than 20 minutes from when the order arrives to the packing and ship cell? The model needed to answer this question is the G/G/C . The cell workload is $\rho = \frac{10.91\ arv/hr}{3.846\ serv/hr} = 2.836$ Therefore, the minimum number of workers needed to service the offered load is 3. The service efficiency is given by $u = \frac{\rho}{c} = \frac{2.836}{3} = .9456$. Since the packing Station utilization is almost 0.95, we would expect that with only 3 packers, the throughput time in packing might be larger than desired. From Equation 9.13, the expected waiting time in queue is given by:

$$\text{WQ(G/G/C)} = \{ \frac{CV_a^2 + CV_s^2}{2} \} \{ \frac{S * u^{\sqrt{2c+2}-1}}{c(1-u)} \} = \{ \frac{1.20+1.35}{2} \} \{ \frac{0.9456^{\sqrt{2*3+2}-1}}{3(1-0.9456)} \} (0.260)$$

$$\text{WQ(G/G/C)} = 1.8328 \ \text{hrs/part}$$

The time spent in the Workstation (Throughput time) can be calculated as:

$$\text{TPT} = WQ\text{(G/G/C)} + S = 1.8328 + (15.6/60) = 2.093 \ \text{hrs /part} = 125.568 \ \text{min/part}$$

We can also calculate the expected waiting line using Little's Law.

$$\text{LQ} = \text{Arrival rate} * \text{WQ} = (10.91)(1.8328) = 19.96 \simeq 20 \ \text{parts}$$

From this analysis, the minimum number of packers required to satisfy design specifications must be greater than 3. Table 9.7 shows the time in the system as a function of the number of people working in pack and ship.

Table 9.7
Time in System vs. Workers

Number of People	Time in the Cell (Min)	Waiting Line (No. of Orders)	ρ	u
3	125.58	20	2.837	0.9456
4	23.736	1.478	2.837	0.7092
5	17.874	0.4137	2.837	0.5673

From these results, at least 5 pack and ship employees will be required to meet the maximum specified time in the cell. It might come as a surprise to the Design Engineers to find that that only 2.837 of the 5 workers are expected to busy at any one time, and that the utilization of personnel is only 56.73%. This is no surprise to the Lean Engineer who understands the fundamental principles of what Hopp and Spearman call "Factory Physics". The "smoking gun" is, of course, our old enemy, *variance*. Both the CV_a^2 and the CV_s^2 are over 1.0, but this is only indicative of the current cell operational procedures which currently exist. In general, this illustrates a fundamental principle of how all manufacturing systems operate. Inevitably, the factory will pay for variance in one of three ways: (1) Capacity (2) Time or (3) Inventory. (See Hopp and Spearman for a detailed discussion of this *Law*). In this case, since variance was not dealt with and *Throughput Time* needed to be reduced, which is directly tied to *WIP* or *Inventory*, the relief valve had to be *Capacity*. Indeed, this is the direct result which we observed to achieve our goal.....Server utilization was a poor 56.73% , which created operating capacity to respond to load. As Hopp and Spearman stress in their landmark text, *Factory Physics*, "know the laws, observe the fundamental principles and make money". We add to that principle, "Understand the Tools of Lean Engineering and use Queuing models for cause and effect analysis".

9.19 Summary

This Chapter has presented a set of fundamental Workstation models which are powerful Lean Analysis tools in their own right, but are also presented to serve as building blocks for more complex multi-server, multi-step factory performance models with multiple products being produced in the same facility. Several numerical examples were given to show the computational procedures, but hopefully to also point out the power and necessity of building and analyzing manufacturing performance models to aid the mental and decision making processes. Hopefully, all goals have been mostly accomplished in the short space available to spend on this subject. Other types of models abound in the literature for specific reasons. The body of knowledge presented in this Chapter is sufficient for us to move forward.

Review Questions

1. Discuss the difference between Workstation Offered Workload and Workstation Utilization. What are the design implications associated with these two measures?
2. A general manufacturing Workstation can have a "finite calling population" or an "infinite calling population". Explain what these two terms mean. Give two examples of each type of calling population.
3. List 5 ways that a Lean Engineer can : (1) reduce set up times (2) reduce work-in-process (3) improve workstation capacity.
4. The founders of the Toyota Production System called work in process the "root of all evil". Discuss why they would make that statement. Is WIP really the root of all evil?
5. Discuss the detrimental effects of poor quality.
6. What is the "Markovian assumption"?
7. Using the closed form solutions in Table 9.3, prove that the expected number in the system is the expected number in queue plus the server offered workload for an M/M/1 Workstation.
8. Using the closed form solutions in Table 9.3, prove that the amount of WIP in an M/M/1 Workstation is the product arrival rate to the Workstation multiplied by the expected time in the system.
9. A result frequently given for an M/M/1 queue is the expected waiting time of an arrival that is forced to wait. In other words, given that any one arrival must wait in queue, what is the expected waiting time? Derive this result.
10. If arrivals to a Workstation follow a Poisson distribution with $\lambda=5$ arrivals per hour, what is the probability that no arrivals will occur during any one hour? What is the number of arrivals that are expected to occur? What is the probability of 3 or more arrivals in any one hour?
11. Explain how Poisson arrivals to a Workstation are related to time between arrivals.
12. One of the properties of an Exponential distribution is that it is said to be "memoryless".

This implies that the time between two successive arrivals is not influenced at all by how much time has elapsed since the previous arrival. Let the Exponential random variable " t " be the time between successive arrivals, the time " T" be the time of the last arrival, and Δt be the time which has elapsed since the last arrival. The "memoryless property" implies that the following relation is true.

$p_r\{\, t > T + \Delta t \mid t > \Delta t \,\} = p_r\{\, t > T \,\}$. Prove this property of the Exponential Density.

Hint: $p_r\{\, t > x \,\} = e^{-\theta x}$, where $f(x) = \theta e^{-\theta x}$ $x \geq 0$

13. If the average arrival rate to a Workstation is Poisson with $\lambda = 5$ arv/hr, write down the probability density function that would describe the arrival stream, and the probability density function that would describe the interarrival times. What is the variance of the arrival rate and the mean and variance of the time between arrivals? What is the Squared Coefficient of Variation for arrivals and the Squared Coefficient of Variation for time between arrivals?

14. If the average customer arrival rate to Don's Donut Emporium is 10 customers per hour, what is the probability that 12 or more customers will arrive in the next hour? What is the probability that the time between arrivals of the next two customers is 8 minutes? It has been 24.5 minutes since the last customer arrived. What is the probability that a customer will arrive in the next 5 minutes?

15. In Don's Donut Emporium, every third customer buys a dozen donuts for $4.50 per dozen. Every other customer buys on the average, two donuts for $0.85. Every other customer buys a cup of coffee for $1.25. During the first hour the store is open, how much will total sales be?

16. Prove that the Squared coefficient of Variation for the Exponential Density function is one.

17. Approximately one person in South America is killed every day by a poisonous snake. If the distribution function of the number of deaths per day is Poisson, how many people would be expected to die in one 365 year day?

18. Pedro's Taco Place is a one man drive-through operation that serves arriving customers on a first come, first served basis. Customers arrive at an average rate of 8 per hour, and place an order at the drive-in window. The average time to take and fill an order is 10.5 minutes. What is the offered workload to Pedro's Taco Place? What percentage of time is Pedro idle? What is the average number of cars in the system? How many cars are waiting to place an order on the average?

19. Write down Little's Law for calculating the expected WIP in a system from the average arrival rate and the expected time in the system. Discuss the logic behind Little's Law. Does it make logical sense? Explain.

20. The longest that a fresh tomato can usually stay in a grocery store before it starts to rot is about 3 days. If a grocery store stocks 500 tomatoes every day , how many customers per day would be required to sell all the tomatoes stocked without losing any to rotting?

21. Harry's Hamburger Joint serves about 40 customers per day between 11:00am and 1:30 pm lunch hour. During any one lunch hour, there are on the average 12 customers eating. How long does each customer spend in the joint?

22. Consider the M/M/1 queuing model. Suppose that λ is doubled. How is the expected number in queue affected? Suppose that μ is halved. How is the expected WIP affected? Suppose both double simultaneously. How is WIP and the total throughput time affected?

23. The Industrial Engineering Department at Texas A&M University is trying to decide the minimum cost lease option between a slow copy machine and a fast copy machine. The

average salary for a copy clerk is $9.50 per hour. A slow copy machine can be leased for $4.00 per hour, and a fast copy machine for $6.50 per hour. On the average, there are 5 manuscripts per hour that needs to be copied. The slow machine can copy one manuscript every 11.5 minutes on the average, and the fast machine one every 10.5 minutes on the average. Assume that the number of jobs per hour follows a Poisson density function, and copy time an Exponential. Recommend which machine should be purchased and why. Is cost the deciding factor?

24. The Pollaczek-Khintchine formula is used to analyze what type of queuing Workstation? What can you say about the accuracy of system performance measures using the P-K formula?

25. Derive the result for an M/G/1 queue for expected waiting time in queue.

26. A manufacturing workstation receives work at an average input rate of 8.2 parts per hour, and the time to process each part has a mean of $\mu=9$ parts per hour with a 110 parts per hour. The input rate is Poisson distributed. Calculate (1) the offered workload (2) the Workstation utilization (3) the expected queue length (4) the expected time in queue and (5) the Workstation throughput time.

27. If a manufacturing Workstation is characterized as a G/G/1 Workstation, what are the inflationary effects of non-Markovian behavior on system performance measures?

28. Construct a general x axis-y axis plot of the M/G/1 expected time in queue and the expected system WIP. What can you deduce from this plot?

29. After further analysis, it was determined that the input rate to the Workstation in problem 9.24 had an average input rate of 8.2 parts per hour and a Squared Coefficient of Variation of 1.33. Compare the performance measures in Problem 9.24 with those in this new scenario. Compare both of these to a pure M/M/1 Workstation.

30. A Lean Engineer is asked to perform a quick workstation analysis at the Steady Manufacturing Emporium. On a visit to the plant site, the Lean consultant visits the workstation and is able to get the following information from the Workstation operator. The most likely time between part arrivals is 12.6 minutes, with the shortest time equal to 9.1 minutes and the longest time equal to 15.3 minutes. The most likely service time per part is 10.2 minutes, and if everything goes right service time could be as low as 8.5 minutes. If things go wrong, it could take up to 13.8 minutes per part. Dr Phillips suggests that the Lean Engineer use a Triangular Density function to analyze this system. The density function is given by:

$$f(t) = \frac{2(t-a)}{(m-a)(b-a)} \qquad a \leq t \leq m$$
$$= \frac{2(b-t)}{(b-m)(b-a)} \qquad m \leq t \leq b$$

$$E(t) = \frac{(a+m+b)}{3} \qquad Var(t) = \frac{(a^2+m^2+b^2-ma-ab-mb)}{18}$$

Calculate (1) the offered workload (2) the Workstation utilization (3) the expected queue length (4) the expected time in queue and (5) the Workstation throughput time.

Thoughts
  and Things…….

Chapter 10
Single Product Factory Flow Models

"I'll be glad to improve myself, he said...But I don't know how to go about it.
What shall I do? "   From the Saggy, Baggy Elephant
K. B. Jackson@1947

10.1   Introduction

In Chapter 9 we introduced the basic M/M/1 model to study single server systems with Poisson input and Exponential service times. This basic model was then expanded to an M/M/C multiple server model. The M/G/1 Pollaczek-Khintchine model for Poisson inputs and general service time distribution for a single server was presented, and then analysis capabilities were extended to G/G/C multiple server models using approximation procedures. The models used in Chapter 9 provide the basic "building blocks" for more complex multi Workstation, serial production line models. Manufacturing system queuing models generally fall into two major classifications (1) Serial production systems and (2) General production models involving routing sheets, reentrant flows and feedback loops. These two general model classifications can be further broadly characterized as single product flow models or multiple product flow models. Finally, one needs to consider whether or not batching, probabilistic branching, downtimes, poor quality and a host of other perturbations need to be included. In this Chapter we will concentrate on single product models and serial flows. Probabilistic branching will be presented for a class of models called Jackson Networks. Modeling the behavior of serial production systems is important to Lean Implementation, since dedicated Flow Shops, U-Shaped manufacturing cells, and Assembly Lines are all basic variations of serial production systems. Probabilistic branching and multiple Workstation visits by the same product will be addressed, but within the context of product flow described by "routing sheets". Routing sheets are commonly used to specify the manufacturing sequence, Workstations visited and various manufacturing specifications. We will assume that processing times at each Workstation are described by the expected processing time and the service time squared coefficient of variation. Product arrival rate is determined by customer demand. Before modeling general single product flow networks, we will first motivate basic understanding by looking at simple serial production systems with constant processing times.

10.2   A Serial Production Line with Constant Processing Times

The system which we will first study is shown in Figure 10.1 and is composed of four processing Workstations in series. Let the expected processing time at each server be equal with no variance. Each Workstation produces a single part every two hours, and has only one server. We wish to study the throughput time of this system and the system output rate as a function of the Work-In Process (WIP) allowed in the entire system.

Figure 10.1
A Typical Serial Production Line

Define the following terminology.

Work in process (WIP) is the total amount of product that is either in queue waiting for service or in a service facility anywhere in the multi -Workstation serial production line.

Throughput Time (TPT) is the amount of time that a single item/part/product spends in the system from when it is *released* to the first service Workstation to when it *finishes* processing at the last service Workstation. Note carefully that this is a Lean definition and is not typically used in the queuing or Operations Research literature.

Production Rate (PR) is the amount of product that is released from the system per unit time. There is clearly some relationship between the amount of WIP allowed in the system, system production rate and throughput time. We will study this relationship by *specifying* the maximum amount of WIP allowed in the system at any one time, and observe the production rate and throughput time of the system. Entry into the system will be controlled as follows. Items can be released into the system as long as WIP is less than or equal to the specified maximum allowed. Once the maximum level is reached, an item will not be allowed (released) into the system unless one item departs from the system. Such a system is called a *closed loop system*, and is representative of what we refer to as a *pull system with limited WIP*. This is also a form of what is called a CONWIP system in the technical literature.

The system shown in Figure 10.1 is called a *Balanced production line* since every processing Workstation produces at the same rate; otherwise it is an *Unbalanced Production Line*. For any balanced or unbalanced production line, we define the *Bottleneck Rate* as the maximum rate of production that can be achieved by the line to be the slowest production rate for any one machine or Workstation in the line. Call this rate r_b. In general, the bottleneck rate r_b is defined to be the *rate of production at the machine / Workstation having the highest long term utilization* (Spearman). This is typically the Workstation with the minimum average output rate. Define the *Raw processing Time* to be the minimum possible throughput time in the system, which is simply the sum of the individual processing times with no waiting. Call this T_0. Intuitively, there will always be a system WIP level which will yield a minimum throughput time and a maximum output rate. Call this WIP level the *Critical WIP* or WIP_c. We will show that critical WIP is equal to $WIP_c = r_b * T_0$. For this example, every Workstation has the same processing time equal to 2 hours/part, or an output rate = 0.50 parts/hr. This is also the system output rate. The raw processing time is $T_0 = 8$ hrs. Hence, $WIP_c = r_b * T_0 = 4$. Note that the critical WIP is equal to the number of service Workstations. This will always be true for a balanced production line. To study the relationship between WIP, output rate and throughput time; we will set the maximum WIP allowed in the system, let the system settle into steady-state, and observe the corresponding throughput and cycle time. To begin, let system WIP be only one unit. At time T=0, the unit is sitting in front of Workstation 1 and will immediately be released. At time T=2, this unit is just finishing processing at Workstation 1. After time T=8, the part leaves the system and immediately starts over again. For WIP=1, the system throughput time is 8 Hours and the system output rate is 1 unit in 8 hours or (1/8) unit/hr.

Now let the maximum WIP allowed in the system be two units. Hence, at time T=0, the two units are sitting in front of Workstation 1 and are ready to be released. At time T=2, the first unit has just finishing processing at Workstation 1 and the second unit has waited two hours.

Immediately following T=2, the first unit proceeds to Workstation 2, and the second unit begins processing at Workstation 1. After time T>2, no unit ever waits again. Figure 2 provides a state diagram of how after time T=12, the system cycles in a predictable and repeating pattern at steady-state. Steady state throughput begins at time T=4 time units and repeats after T=12 time units. In other words, the system reaches *steady state behavior* at T=12 time units and always repeats the same flow patterns. . Between T=4 and T= 12 time units, there are two parts which exit the system. Hence, the system output rate is two parts in 8 hours or 0.25 Parts/Hr. System throughput time for each part is 8 Hours.

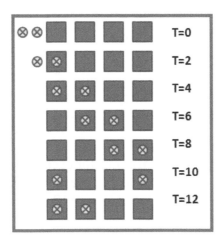

It should be apparent that when WIP=3, the second unit waits 2 hours and the third unit waits 4 hours, but after time T=14, the system starts to cycle with each part experiencing a throughput time of 8 hours. All 3 parts are produced every 8 hours, so the output rate is (3/8) parts/hr. Consider WIP=4 parts.

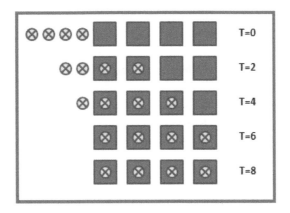

From time T=8, the system is continually full. At T=16, the system cycles continuously with 4 parts at steady state. The throughput time is 8 Hours/Part, and the output rate is 4 parts in 8 hours or 0.50 parts/hr. Now, it should be clear that if WIP=5, the fifth part will have to wait 2 time units before it can enter the system, and after that time every part starting over has to wait 2 time units. The system throughput time is 10. The output rate is 5 units every 10 hours, or 0.50 units/hr. Table 10.1 summarizes system behavior for WIP =1, 2....8 parts allowed in the system.

Table 10.1
WIP vs Throughput vs Output Rate

WIP (Parts)	Throughput Time (Hrs)	Output Rate (Parts/Hr)
1	8	1/8
2	8	2/8
3	8	3/8
4	8	4/8
5	10	4/8
6	12	4/8
7	14	4/8
8	16	4/8

The Table shows several interesting results

(1) Output Rate increases by 1/8 unit per hour for every unit increase in WIP until the rate reaches 0.50 parts/hour, at a WIP level of four units, then it remains constant.

(2) Throughput Time is constant at 8 hours until WIP reaches 4, then increases by two time units for every added unit of WIP

(3) The WIP level which results in a *Minimum* Cycle Time and *Maximum* Throughput rate is WIP=4. Note that this is the number of Workstations in the line. Recall that the WIP level which results in maximum output and minimum throughput time is called the *Critical WIP*, and is given by $WIP_c = r_b * T_0 = (1/2)8 = 4$.

Finally, note that the product of output rate and throughput time is always WIP. What does all of this mean? Recall in Chapter 9 we introduced an important relationship called 'Little's Law". Little's Law can now be stated as follows.

*WIP = Output rate * Throughput Time*

Hence, our exercise is an empirical verification of Little's Law . Finally, we reiterate that Little's Law not only holds for serial systems, but for any system or individual Workstation. It is the F=MA equation for manufacturing systems analysis

10.3 An Unbalanced Serial Production Line

The previous example analyzed a serial production line with constant and identical single server Workstations. Consider a second serial production line consisting of five Workstations arranged in a serial manufacturing cell. Workstation data is given in Table 10.2.

This is an unbalanced production line with multiple servers at each Workstation. The Workstation production rate is simply the product of the number of servers times the reciprocal of the server processing time. The bottleneck Workstation is Workstation 3, which is only capable of producing 0.50 parts/hr. Using the notation of Section 10.20, we designate this as the *bottleneck rate*; r_b=0.50 parts/hr.

Table 10.2
An Unbalanced Serial Production Line

Workstation	Number of Servers	Server Process Time (Hrs/Part)	Workstation Production Rate (Parts/Hr)
1	2	3.250	0.6154
2	1	1.250	0.80
3	2	4.0	0.50
4	5	6.50	0.770
5	4	5.0	0.80

The *Raw processing time* of the entire line is still the minimum possible throughput time, or $T_0 = 20$ Hours. The 'Critical WIP" is

$$WIP_C = r_b * T_0 = 0.50(20) = 10 \text{ units.}$$

Recall that the *critical WIP* is the level of work in process allowed in the system (servers plus queues) that will result in minimum cycle time and maximum throughput rate. This level of WIP can be enforced by either a *Kanban pull* system or by a CONWIP control strategy. These concepts are introduced and discussed in Chapter 13. It may be a surprise to some analysts who understand the critical WIP concept that the simple formula used above is an application of Little's Law at the systems level. Put another way, Little's Law applied at either the systems level or at any Workstation says that:

*WIP = System output rate * Total Throughput Time.*

Little's Law theoretically describes an important Lean Manufacturing concept. For fixed product demand (Input rate) and no product loss (Output Rate=Input Rate), when the WIP level increases there is a corresponding increase in System throughput Time. The implication is that WIP should be zero to achieve the minimum cycle time. However, our previous analysis shows that this is not correct. There is a non-zero WIP which will yield minimum throughput time and maximum output rate. This destroys the *myth* of zero WIP as an ultimate goal of Lean Manufacturing. Little's universal law not only applies at the systems level, but also at the cell level or even the Workstation level. For this example, it is instructive and interesting to calculate the WIP distribution at the individual Workstation level. Table 10.3 summarizes these results.

The sum of all individual Workstation WIP values is equal to 10, as it should. Reviewing the results, there are some interesting points that are worth discussing. (1) The bottleneck Workstation is Workstation three. Workstation three is not the slowest group of machines, nor is it the one with the fewest number of machines. (2) Because the system is unbalanced, there is considerable idle time at the non-bottleneck Workstations. Looking at Table 10.3, there is at least 35% idle time at Workstations 2, 4 and 5 and almost 17% at Workstation 1.

Table 10.3
Individual Workstation Behavior

Workstation	Number of Servers	Output Rate	Server Processing Time	Workstation WIP	Utilization
1	2	0.50	3.250	1.625	0.8125
2	1	0.50	1.250	0.625	0.625
3	2	0.50	4.0	2.00	1.00
4	5	0.50	6.50	3.25	0.650
5	4	0.50	5.0	2.50	0.6250

The bottleneck Workstation is running at capacity 100% of the time. (3) System output rate is determined by the bottleneck Workstation, and is ½ units per hour. (4) Workstation Throughput time and Workstation WIP are both inflated at the bottleneck Workstation. (5) It should be noted that Workstation WIP at the non-bottleneck Workstations are non-integer values. Since a fractional part makes no sense, this signals that there is no WIP allocation which will result in "optimal performance". WIP will shift from Workstation to Workstation in steady state with a regular, occurring pattern as blockage occurs. This is a good case for implementing a CONWIP strategy, and simply letting the WIP shift as system performance dictates. It also demonstrates how the Kanban or CONWIP limit can be determined. If a Kanban control strategy is offered, one decision might be to establish Workstation WIP in the following manner.

$$WIP_1 = 2 , WIP_2 = 1 , WIP_3 = 2 , WIP_4 = 4 \text{ and } WIP_5 = 3$$

This strategy will result in system WIP of 12 units, 2 units above optimal WIP. What effect will this have on system performance? By Little's Law, the System Throughput Time $= \frac{WIP_{Sys}}{Output\ Rate} = \frac{12}{0.50} = 24$ Hrs. this is 20% longer than unbalanced theoretical behavior. The question of whether or not Kaizen events should be scheduled for this line is both an operational and economic decision. However, it is correct to categorically state that Kaizen events targeted to non-bottleneck systems will not improve throughput, and may unfavorably impact unit cost. This observation is totally in agreement with the *Theory of Constraints* first proposed by Eli Goldratt.

10.4 A Stochastic Serial Production Line
In the previous two sections we examined idealistic serial production lines in which the service times at each Workstation were a constant. In almost every real world production system, unless highly automated and synchronized, the product service times at each processing /machine center are random variables characterized by the mean, variance and squared coefficient of variation of service times (knowing any two measures of service, one can easily calculate the third) Consider the production line shown in Figure 10.2 composed of three Workstations in series. The number of servers required at each Workstation will be determined from the *offered work load* to each Workstation. Recall that the offered workload is the total input rate of product divided by the average machine processing time.

322

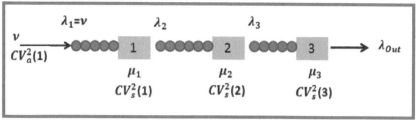

Figure 10.2
Three Workstations in Series

The input rate is $\lambda = v$ (units/hr). The input stream will be defined as a Poisson Process. All service times will be described by an Exponential Density function (hr/part), and waiting space (queue) in front of each machine center will be infinite. The processing centers (Workstations) are each M/M/C with no batching, no reentrant flows, no Quality losses and no limit on queuing space at each processing Workstation. Hence, the arrival rate to each Workstation and the departure rate from each Workstation are identical and equal to the external rate of arrivals; v. This is also the *System Production Rate*. Since the time between arrivals is an Exponential distribution, then the SCV of the inter-arrival time distribution has $CV_a^2 = 1.0$. Since all service times follow an Exponential density function, then $CV_s^2(j) = 1.0$ j=1,2,3. Assume that $v=40$ parts/hr. The processing data for this system is given in Table 10.4.

Table 10.4
Processing Data

Work Workstation	Input Rate (Parts/Hr) λ_i	Server Processing time (Hrs/Part) S_i	Rate of service (Parts/Hr) μ_i
1	40	0.0455	22
2	40	0.02174	46
3	40	0.0667	15

Let the *offered workload* to each service center be:
$$\rho_i = \frac{\lambda_i}{\mu_i} = v * S_j$$

Hence:
$$\rho_1 = v * S_1 = 40(0.0445) = 1.820$$
$$\rho_2 = v * S_2 = 40(0.02174) = 0.870$$
$$\rho_3 = v * S_3 = 40(0.06667) = 2.670$$

The utilization of each Workstation must be less than one to handle the offered workload. Hence, there must be a minimum of two machines at Workstation 1 ($c_1=2$) three at Workstation 3 ($c_3=3$) and one at Workstation 2 ($c_2=1$). The Workstation utilizations using this number of servers and the offered workload are:

$$u_1 = \frac{\rho_1}{c_1} = \frac{1.820}{2} = 0.91 \quad u_2 = \frac{\rho_2}{c_2} = \frac{0.870}{1} = 0.870 \quad u_3 = \frac{\rho_3}{c_3} = \frac{2.670}{3} = 0.890$$

Since each processing center has infinite waiting space, then each can be separately and independently modeled as an M/M/C Workstation. The following (exact) results can be obtained using the formulas in Chapter 9.

Table 10.5
M/M/C Results

Station	Model	Expected No In each Workstation (L_i)	Expected No In each Queue (LQ_i)	Expected Throughput Time At each Workstation (TPT_j)	Expected Waiting Time In Queue At each Workstation (WQ_j)
1	M/M/2	10.528 parts	8.71 parts	0.2632 hr/part	0.2177 hr/part
2	M/M/1	6.667	5.797	0.1674	0.145
3	M/M/3	9.1167	6.45	0.228	0.1615

We can determine the expected number of units in the entire system by summing up the individual expected number in each Workstation.

$$L_{Sys} = L_1 + L_2 + L_3 = 10.528 + 6.667 + 9.1167 = 26.312 \quad \text{Units}$$

The expected time in the system is given by:

$$TPT_{Sys} = TPT_1 + TPT_2 + TPT_3 = 0.2632 + 0.1674 + 0.228 = 0.6586 \text{ Hrs}$$

Of course we know that the System output rate equals the System input rate which is given by λ = 40 Units/Hr. Using Little's Law at the system level: WIP=Throughput Time * Output rate = 0.6586(40) = 26.3 units as before. So, as expected, Little's Law will also hold for the System Performance measures. It is interesting to compare the behavior of this line to optimal performance. Assume that the processing time at each Workstation is a constant. This is an unbalanced line with the following characteristics.

Table 10.6
An Unbalanced Line

Workstation i	Number of Machines C_i	Process Time S_i	Workstation Output Rate
1	2	0.0455	43.96
2	1	0.02174	46.0
3	3	0.0667	44.98

The capacity of the line is determined by the rate of Workstation 1, which is the *Bottleneck* operation. The Bottleneck *rate* is:

$$r_b = 43.96 \text{ Parts/Hr}$$

The Raw Cycle Time is given by:
$$T_0 = 0.13394 \text{ Hrs}$$
Therefore, Critical WIP is given by:
$$W_0 = r_b * T_0 = (43.96)(0.13394) = 5.890 \text{ Units}$$
The results may be somewhat surprising. The shortest throughput time that can be achieved with the same number of machines at each processing Workstation is $TPT_{Shortest} = 0.13394$ hrs, while the modeled (actual) cycle time is $TPT_{Actual} = 0.6586$ hrs., almost a 500% increase. WIP is equally inflated, with an ideal WIP of $WIP_{Ideal} = 5.895$ parts and an actual WIP of $WIP_{Actual} = 26.312$ Parts. The casual observer may ask why there is so much more inefficiency in the actual system, but the informed Lean Engineer will quickly discern the problem…..Variance. The penalties are large, but results to be expected from variance reduction are large

It is important to restate the conditions and assumptions that were imposed upon the problem being addressed. We assumed (1) Infinite WIP at each processing Workstation (2) Poisson arrivals and Exponential service times (3) The first two assumptions allowed us to decompose the serial three Workstation systems into three independent Workstation models, and (4) Then recombine the individual Workstation results to obtain serial system performance measures. In this case, the system was decomposed into one M/M/1 model and two M/M/C models. The following observation is directly related to this result.

If the arrival process to each workstation in an N station serial production system is Poisson, and the service time for each server in an M/M/c workstation with no product losses and infinite queues between each service facility; the system is said to be Markovian and can be decomposed into a series of individual M/M/c Workstation models as long as each server in a multi-server service center has a common processing time distribution.

10.5   Jackson Networks
In the early 1960,s another class of a multiple server queuing system was studied by Jackson (1957, 1963). Jackson studied an open queuing network with the following properties.
   Property 1: The network consisted of N, M/M/1 or M/M/c Workstations with FIFO/infinite queues.
   Property 2: The service times in each service facility followed an Exponential distribution with multiple servers all having the same exponential service time density function.
   Property 3: Arrivals to a Jackson network follow a Poisson distribution.
   Property 4: Upon departure from Workstation i the customer chooses the next Workstation j randomly with probability $p_{i,j}$ or exits the network with the probability $p_{i,x}$ .The model can be extended to cover the case of predetermined routes.

Of course, because a part is randomly routed from one workstation to another, a particular part may or may not visit every workstation in the network. However, once a part enters the system it must have a finite probability of exiting the system. Since there are no product losses, there is conservation of flow.

Jackson proved that under the assumptions stated above, an N station queuing network can be decomposed into N separate workstations and each independently analyzed. System performance measures can then be determined by combining these results. The exact theory of this observation is complex, but the statement holds because in fact it is *not Markovian*, but it *behaves* as if it was *Markovian*. Further, if multiple Poisson arrival streams merge into a single arrival stream, the resulting composite stream is also Poisson. Finally, if a Poisson stream is probabilistically split, the resultant split streams are also individually Poisson. These properties seem to be first put to use by Jackson in the early 1960s, and the resulting network structures which he extensively studied were called *Jackson Networks*. Since Jackson first published his work, a continuing stream of research has been directed to general queuing networks in which Markovian or perceived Markovian properties no longer hold.

If the assumptions of a Jackson network are satisfied, then every Workstation can be represented as M/M/C and analyzed separately. However, there are two additional complexities which must be addressed, both of which are caused by probabilistic branching from a Workstation. The *first* is to determine the steady state rate of product flow into each Workstation. The *second* is that due to probabilistic routing, a particular part routing might result in one or more Workstations being visited more than once, and some may not be visited at all. This will change the way that the system Throughput time has been previously calculated. To recover the Workstation performance measures, each Workstation can be analyzed as an independent M/M/C once the total flow rate into each Workstation is known. However, to recover system performance measures the average Throughput time per part at each Workstation must be multiplied by the expected number of times that each Workstation is visited, and then summed across all Workstations. The computational procedures are as follows.

Theorem 1: For an N Workstation Jackson Network, define the transition probabilities from Workstation i to Workstation j as $p_{i,j}$ and construct a (N x N) matrix **Q** which contains the values of $p_{i,j}$ i =1,...N,   j=1,....N. Define a (N x N) identity matrix **I** with all principle diagonal entries equal to one, and all other entries equal to zero. Finally, define a (N x N) matrix **V** = $(I - Q)^{-1}$. Assume that all external flow (γ) into the Jackson Network occurs at Workstation 1, and let α be a (1 x N) scaling vector such that $\alpha_1 = \gamma_1$ and $\alpha_k = 0$ k=2, 3 .... N. The total product flow into each workstation is given by the components $\lambda_1, \lambda_2 .... \lambda_N$ of the vector λ.
$$\lambda = \alpha \, (I - Q)^{-1}$$

Theorem 2: Let **V** be an (N x N) matrix such that:
$$V = (I - Q)^{-1}$$
The elements in the first row of the matrix **V** represent the expected number of times that each Workstation will be visited by each part provided each part enters the Jackson Network at Workstation 1 in the sequence. Let a row vector $\emptyset$ be designated as $\emptyset = (V_{1,1} , V_{1,2} ..... V_{1,N})$.

Theorem 3: The expected time in the system per part is given by:
$$TPT_{System} = \sum_{k=1}^{N} TPT_k \{V_{1,k}\}$$
Where: TPT_k represents the expected Throughput time per part at Workstation k and $V_{1,k}$ is the expected number of visits per part to Workstation k.

10.5.1 A Jackson Network Example: No Reentrant Flow

Consider the following single product Jackson Flow network with $\gamma_1=20$ parts/hr, $\mu_1=11$ parts/hr, $\mu_2=5.2$ parts/hr, $\mu_3=14.0$ parts/hr and $\mu_4=4.20$ parts/hr. Let $p_{1,2}=0.50$, $p_{1,3}=0.50$, $p_{2,4}=0.75$, $p_{2,3}=0.25$ and $p_{3,4}=1.0$.

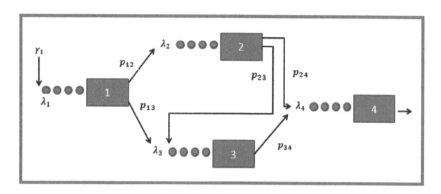

Using Theorem 1: $\quad \boldsymbol{\alpha} = (20, 0, 0, 0)$

$$Q = \begin{bmatrix} 0.00 & 0.50 & 0.50 & 0.00 \\ 0.00 & 0.00 & 0.25 & 0.75 \\ 0.00 & 0.00 & 0.00 & 1.00 \\ 0.00 & 0.00 & 0.00 & 0.00 \end{bmatrix} \quad I = \begin{bmatrix} 1.00 & 0.00 & 0.00 & 0.00 \\ 0.00 & 1.00 & 0.00 & 0.00 \\ 0.00 & 0.00 & 1.00 & 0.00 \\ 0.00 & 0.00 & 0.00 & 1.00 \end{bmatrix}$$

$$(I-Q) = \begin{bmatrix} 1.00 & -0.50 & -0.50 & 0.00 \\ 0.00 & 1.00 & -0.25 & -0.75 \\ 0.00 & 0.00 & 1.00 & -1.00 \\ 0.00 & 0.00 & 0.00 & 1.00 \end{bmatrix} \quad (I-Q)^{-1} = \begin{bmatrix} 1.00 & 0.50 & 0.63 & 1.00 \\ 0.00 & 1.00 & 0.25 & 1.00 \\ 0.00 & 0.00 & 1.00 & 1.00 \\ 0.00 & 0.00 & 0.00 & 1.00 \end{bmatrix}$$

Then: $\quad \boldsymbol{\lambda} = \boldsymbol{\alpha}\,(\mathbf{I} - \mathbf{Q})^{-1} = (20, 10, 12.5, 20)$

The first row of $(\mathbf{I} - \mathbf{Q})^{-1}$ represents the expected number of times that a part entering the system at Workstation 1 will visit Workstation j   j= 1, 2 …. N. Hence:

$$\emptyset = (1.0, 0.50, 0.63, 1.0)$$

The performance measures at each Workstation and the Total System Throughput time can now be calculated. The results are as follows.

Workstation	Arrival Rate	Service Rate	# of Servers	Type	Rho	u	L	LQ	WQ	TPT
1	20	11	2	M/M/2	1.818	0.909	10.528	8.710	0.433	0.524
2	10	5.2	2	M/M/2	1.923	0.962	25.540	23.620	2.362	2.550
3	12.5	14	1	M/M/1	0.893	0.893	8.330	7.440	0.595	0.667
4	20	4.2	5	M/M/5	4.762	0.952	22.500	17.730	0.887	1.125
System							66.898			

One might be tempted to simply sum up the individual Workstation Throughput times to get the average time in the system. However, this is incorrect. Using Theorem 3, the correct average system Throughput time is given by:

$$TPT_{System} = \sum_{k=1}^{N} TPT_K \{ V_{1,K} \}$$
$$TPT_{System} = 0.5264(1.0) + 2.55(0.50) + (0.667(0.625) + (1.125(1.0)$$
$$TPT_{System} = 3.343 \text{ hrs}$$

The average system WIP is still the sum of Workstation WIP values.

$$L_{System} = \sum_{k=1}^{N} L_j = 66.898 \text{ parts}$$

10.5.2 A Jackson Flow Network with Recirculation

In the previous example, note that while individual parts might follow different flow routes through the network, every Workstation will be visited only one time. Suppose we modify the previous example to accommodate re-entrant flows.

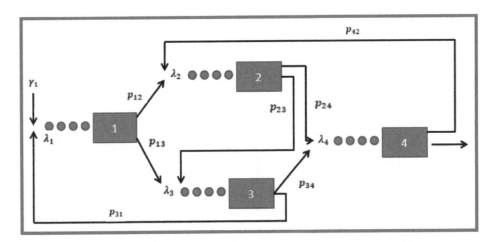

Note that at Workstation 4, the total flow that exits the network is given by $\lambda_4(1- p_{42})$. The basic procedure is to determine $\lambda_1, \lambda_2, \lambda_3$ and λ_4 as before; decompose the network into independent M/M/C Workstations; and calculate the Workstation performance measures. Having done that, the System performance measures can be recovered using Theorem 3.

Let: $v_1=20$ parts/hr, $\mu_1=12.5$ parts/hr, $\mu_2=9.1$ parts/hr, $\mu_3=4.77$ parts/hr and $\mu_4=20$ parts/hr. Let $p_{1,2}=0.50$, $p_{1,3}= 0.50$, $p_{2,4}= 0.75$, $p_{2,3}= 0.25$ and $p_{3,4}= 0.80$, $p_{3,1}= 0.20$ and $p_{4,2}= 0.40$. It follows that: $\boldsymbol{\alpha}= (20, 0, 0, 0)$

$$(I-Q)=\begin{bmatrix} 1.00 & -.50 & -.50 & 0.00 \\ 0.00 & 1.00 & -.25 & -.75 \\ -.20 & 0.00 & 1.00 & -.80 \\ 0.00 & -.40 & 0.00 & -1.0 \end{bmatrix} \qquad V = (I-Q)^{-1} = \begin{bmatrix} 1.181 & 1.257 & 0.905 & 1.667 \\ 0.095 & 1.714 & 0.476 & 1.667 \\ 0.267 & 0.800 & 1.333 & 1.667 \\ 0.038 & 0.686 & 0.190 & 1.667 \end{bmatrix}$$

Hence, by Theorem 1: $\lambda = \alpha\,(I-Q)^{-1} = (23.62, 25.14, 18.10, 33.33)$ Note that the product flow which exits the network is given by $\lambda_{out} = \lambda_4(1-p_{4,2}) = 33.33(1-0.40) = 20$, which verifies the rate-in, rate-out principle at the systems level.

The first row of $(I-Q)^{-1}$ represents the *expected number of times* that a part entering the system at Workstation 1 will visit Workstation j   j= 1, 2 …. N.

Since $\emptyset = (V_{1,1}, V_{1,2} \dots V_{1,N})$
   $\emptyset = (1.181, 1.257, 0.905, 1.667)$

The performance measures at each Workstation and the Total System Throughput time can now be calculated. The results are as follows.

Workstation	Arrival Rate	Service Rate	Rho	C	Mu	Type	L	LQ	WQ	TPT
1	23.620	12.500	1.890	2	0.945	M/M/2	17.650	15.634	0.667	0.747
2	25.140	9.100	2.763	3	0.921	M/M/3	12.773	10.011	0.398	0.508
3	18.100	4.770	0.893	4	0.949	M/M/4	20.274	16.479	0.910	1.120
4	33.330	20.000	1.667	2	0.833	M/M/2	5.503	3.836	0.115	0.165

Finally, again using Theorem 3:

$$TPT_{System} = \sum_{k=1}^{N} TPT_K \{V_{1,K}\}$$
$$TPT_{System} = 0.747(1.181) + 0.508(1.257) + 1.12(0.905) + 0.1667(1.65)$$
$$TPT_{System} = 2.809 \text{ hrs}$$

The average system WIP is given by:

$$L_{System} = \lambda_{Out} * TPT_{System}$$
$$L_{System} = \{0.60*33.33\} * 2.809 = 56.174 \text{ parts}$$

If the Markovian Assumptions just stated are either collectively or individually violated, it is necessary to follow a new line of modeling requirements. In particular, even if infinite WIP is assumed between individual processing Workstations, if there are not Poisson arrivals and Exponential service times, then a more general approach is necessitated based upon approximation techniques. We should add that a common mistake in modeling this class of queuing systems is that if there are multiple machines required at any one service center, and if these machines have different processing time distributions….or if the processing times are step dependent in a routing sheet structure… the system *CANNOT be directly* decomposed and analyzed as independent service centers because the departure process is no longer Poisson. For these systems, the squared coefficients of arrival and service times must be calculated and used to adjust Markovian results. Several Workstation approximation techniques were studied in Chapter 9 to facilitate the analysis of individual, serial M/G/1, G/G/1 and G/G/C service facilities. For integrated networks of G/G/C service facilities with multiple products or single products with reentrant flow, a more general modeling approach is required. In the next section we will show how to analyze single product serial production systems with reentrant flows determined by process routing sheets. We will generalize this approach to multiproduct systems in Chapter 11.

10.6 Single Product Serial Production Lines

For the rest of this Chapter we will be concerned with modeling single product serial production lines with general (Non-Markovian) Workstations. We will expand the notation for single Workstations in Chapter 9 to multiple Workstations in series. The following notation will be used for Single Product, Serial Flow Networks.

v_1	**External flow load for Product 1**
$v_{1,j}$	**The External Demand Rate to Workstation j at Step 1**
M	**The number of Workstations**
λ_i	**Arrival Rate to Workstation i =1,2....N**
μ_i	**Service Rate at Workstation i =1,2...N**
c_i	**Number of Servers at Workstation i =1,2....N**
S_i	**Service Processing Time per server at Workstation i =1,2....N**
WQ_i	**Waiting Time in Queue per unit at Workstation i =1,2....N**
TPT_i	**Workstation Throughput Time at Workstation i =1,2....N**
LQ_i	**Expected Queue Length at Workstation i =1,2...N**
L_i	**Expected Number in System at Workstation i =1,2....N**
CT_i	**Expected Time Between Departures at Workstation i =1,2....N**
ρ_i	**Offered Workload to Workstation i =1,2...N**
u_i	**Utilization of Workstation i =1,2....N**
$CV_a^2(v_1, 1)$	**Squared Coefficient of Variation for External Time Between Arrivals into Workstation 1**
$CV_a^2(j)$	**Squared Coefficient of Variation for Time Between Arrivals into Workstation i =1,2...N**
$CV_s^2(j)$	**Squared Coefficient of Variation for Service Time at Workstation i =1,2...N**
$CV_d^2(j)$	**Squared Coefficient of Variation for Time Between Departures from Workstation i =1,2...N**

The key assumption is that the WIP storage at each processing center is infinite, and multi-server processing Workstations have identical servers. However, Workstation processing times can be step dependent in a multistep routing sequence and individual Workstations can be visited multiple times. These assumptions may appear to be overly restrictive, but in practice yield extremely useful information concerning how to set Workstation WIP levels and what to expect for the system throughput times and Workstation waiting lines. Our approach will be to leverage Queuing theoretic results from a variety of researchers, and *stand on the shoulders of those giants who came before us*. In particular, we will be concerned with a broad class of Production Systems commonly found in typical mass production facilities with *push* order release policies. We will assume (1) Infinite WIP storage between processing centers and in front of each machine. (2) Without loss of generality, we will consider a First-in-first-out processing priority scheme. (3) All process flow sequences will follow *Process Routing Sheets*. We have adopted this method of describing product flow sequences since it is by far the most commonly used routing methodology used by industry. (4) We will restrict our analysis to single product flow in this Chapter, and expand our analysis to multiple product flows in the next Chapter. (5) Re-entrant flows or multiple visits to any one workstation are allowed, with step dependent Workstation processing times. (6) There will be no particular restriction on the number of processing steps in each production sequence (7) Probabilistic branching might occur in the context of rework loops (Poor quality assurance) but we will not allow this in this Chapter. We will show how to model probabilistic branching in Chapter 12. Each processing step in a multistep production sequence can and usually will have a unique processing time with individual mean, variance and SCV. (8) A major restriction, but one that will not severely damage modeling capabilities, is that if a processing Workstation has multiple servers / machines, a composite service time distribution will be developed for each Workstation; although as previously noted the individual Workstation processing times can be product, Workstation and step dependent. (9) Product can enter the network at any Workstation, but for convenience we will usually designate Product 1 entering Step1 at Workstation1. (10) Finally, and most important, as a consequence of the previous assumptions every processing Workstation will (usually) require a G/G/C model representation. Even with infinite WIP between Workstations, these characteristics destroy the *Markovian Assumptions* previously explained. The consequence of this characteristic is that both the arrival stream(s) and the departure stream(s) are no longer Poisson. This will necessitate the calculation of a *Squared Coefficient of Variation (SCV)* for both the arrival and departure streams as previously noted. It is now our objective to present analysis procedures that will allow us to model multi –Workstation production lines for single product flow. The extension to multiproduct flow is straightforward, and will be presented in Chapter 11.

10.7 Characterizing Input and Output Processes from General Service Workstations

One of our main modeling requirements is to characterize the inflow and outflow streams associated with general service facilities (Workstations) by determining the appropriate expected value and squared coefficient of variation. If the arrival process to Workstation j is non-Markovian, then the $CV_a^2(j)$ must be determined from the network structure. Before leaving this observation, we should point out that the characteristics of the external arrival stream (v_1) are usually known. In all of the examples used in this Chapter we will assume without loss of generality that the mean external arrival rate is specified as v. We will usually assume that the mean time between arrivals for external demand follows an Exponential Density function, and the associated SCV is therefore $CV_a^2(v_1, j) = 1.0$, but this is not necessary as long as both parameters are specified.

10.7.1 The Arrival and Departure Process

Figure 10.3 illustrates a single Workstation "j" composed of one or more servers (machines).

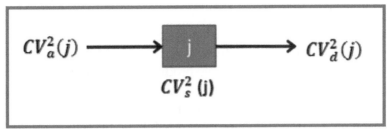

Figure 10.3
The Arrival and Departure Process

Workstation j has three parameters. $CV_a^2(j)$ is the Squared Coefficient of Variation for the time between arrivals, $CV_s^2(j)$ the Squared Coefficient of Variation for the service times and $CV_d^2(j)$ the Squared Coefficient of Variation for time between departures. Recall that the squared coefficient of variation (SCV) is always equal to the variance divided by the mean squared.

Hence, for any Workstation j, $CV_a^2(j) = \frac{\sigma_j^2}{\mu_j^2}$; where σ^2 is the variance of the inter-arrival time distribution function f(t) and μ^2 is the mean of the inter-arrival time distribution squared. $CV_s^2(j)$ and $CV_d^2(j)$ are similarly defined. The distribution function for the inter-arrival times, the service times and the inter-departure times are usually unknown, so it is common to describe these random variables with the associated mean and SCV. We will follow this convention. A relevant question is how the input $CV_a^2(j)$ is transformed into the output $CV_d^2(j)$. For a Markovian process the answer is simple: the output $CV_d^2(j)$ is the same as the input $CV_a^2(j)$ and for serial systems, the next processing Workstation $CV_a^2(j+1)$ is equal to the $CV_d^2(j)$. However, for general G/G/C service systems, things are not so simple. It appears that the first investigation into this problem was by Marshall in 1966. He derived the following general relationship.

$$CV_d^2(j) = CV_a^2(j) + 2\mu^2 CV_s^2(j) - [2\mu(1-\mu)WQ_j]/S$$

For a G/G/1 service facility, using Marshall's result we can easily show that the following equation is correct.

$$CV_d^2(j) = (1-u_j^2)\, CV_a^2(j) + u_j^2\; CV_s^2(j) \qquad C=1 \qquad (10.1)$$

Witt later proposed an approximation for the M/M/C service facility.

$$CV_d^2(j) = 1 + (1-u_j^2)(\,CV_a^2(j) - 1\,) + \frac{u_j^2}{\sqrt{c_j}}(\,CV_s^2(j)-1\,) \qquad C>1 \qquad (10.2)$$

Equation 10.2 reduces to Equation 10.1 when C=1.

10.8 A Three Workstation Non-Markovian Model

Consider once again the three Workstations in series shown in Figure 10.4.

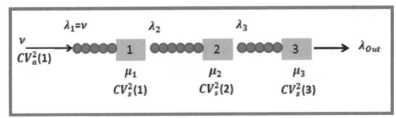

Figure 10.4
Three Workstations in Series

Assume that the arrival rate is $v_1 = 20 \frac{parts}{hr}$. Hence, the inter-arrival time distribution has a mean time between arrivals of $\frac{1}{v_1} = 0.05 \frac{Hrs}{Part}$. The associated $CV_a^2(1) = CV_a^2(v_1, 1) = 1.0$. Recall that the workstation processing times $S_j = \frac{1}{\mu_j}$. Table 10.7 provides the individual machine center's expected processing times S_j and the corresponding $CV_s^2(j)$ j=1, 2, 3.

Table 10.7
A Non-Markovian Model

Workstation (j)	S_j (Hours/Part)	$CV_s^2(j)$
1	.0454	2.10
2	0.087	0.90
3	0.042	1.22

Since v_1 is used to designate the rate of arrival of product to the system, it follows that in this case $\lambda_1 = \lambda_2 = \lambda_3 = \lambda = v_1$. Hence, define the Offered Workload to Workstation j as:
$\rho_j = \lambda * S_j$ j=1, 2, 3. Using this notation and the Workstation operating parameters, $\rho_1 = 0.0454(20) = 0.908$, $\rho_2 = 0.087(20) = 1.74$ and $\rho_3 = 0.042(20) = 0.840$. Hence, Workstations one and three will require only 1 server, and Workstation two will require 2 servers. The utilization of each Workstation is given by $u_1 = 0.908$, $u_2 = 0.87$ and $u_3 = 0.840$. Even though there are infinite queues allowed at each Workstation, the serial system cannot be immediately decomposed and directly analyzed as independent Workstations since the service time Squared Coefficients of Variation are not equal to one. However, since $CV_a^2(1) = CV_a^2(v_1, 1) = 1.0$, the first Workstation is an M/G/1 Workstation, while Workstations two and three are G/G/2 and G/G/1 , respectively. We must determine the $CV_a^2(2)$ and $CV_a^2(3)$ values. Before proceeding, note that in a serial production system, the $CV_a^2(j+1)$ values are the $CV_d^2(j)$ values for j=1, 2...M. In this case, since Workstation 1 requires only one server, $CV_d^2(1)$ will be determined using Equation 10.1

$$CV_d^2(1) = (1-u_1^2) \, CV_a^2(1) + u_j^2 \, CV_s^2(1)$$
$$CV_d^2(1) = (1 - 0.908^2)(1.00) + (0.908^2)(2.10)$$
$$CV_d^2(1) = 1.907$$
$$\text{And} \quad CV_a^2(2) = 1.907$$

For Workstation 2 we must use Equation 10.2 since c_2=2.

$$CV_d^2(2) = 1 + (1-u_2^2)(CV_a^2(2) - 1) + \frac{u_j^2}{\sqrt{c_j}}(CV_s^2(2) - 1)$$

$$CV_d^2(2) = 1 + (1 - 0.870^2)(1.907 - 1) + \frac{0.870^2}{\sqrt{2}}(0.90 - 1.0)$$

$$CV_d^2(2) = 1 + .2205 - 0.054 = 1.167$$
$$\text{And} \quad CV_a^2(3) = 1.167$$

We can now decompose the system into three separate models, one for each Workstation, and analyze each separately. Let the expected time in queue at Workstation j be TQ_j j=1,2,3 ; the expected time in each Workstation be S_j j=1,2,3 ; and the expected queue length at Workstation j be QL_j for j=1,2,3. The expected time in Workstation j is by definition $TPT_j = TQ_j + S_j$ for j=1, 2, 3.

Workstation 1

For the M/G/1 Server at Workstation 1, $u_1 = 0.9080$.
Hence;

$$WQ_1 = [\frac{1 + CV_s^2(1)}{2}][\frac{u_1}{1-u_1}] S_1$$

$$WQ_1 = [\frac{1 + 2.10}{2}][\frac{0.908}{1-.908}] (0.0454) = 0.695 \text{ Hrs/unit}$$

The Expected Time in Workstation 1 is given by:

$$TPT_1 = WQ_1 + S_1 = 0.695 + 0.0454 = 0.740 \text{ Hrs/Unit}$$

Using Little's Law the expected queue length (waiting line) at Workstation 1 and the expected number of parts in Workstation 1 are given by:

$$LQ_1 = \lambda_1 * WQ_1 = 20(0.695) = 13.9 \text{ Units}$$
$$L_1 = LQ_1 + \rho_1 = 14.80$$

Workstation 2:

For the G/G/2 Server, $u_2 = 0.870$

$$WQ_2 = [\frac{CV_a^2(2) + CV_s^2(2)}{2}] WQ \text{ (M/M/2)}$$

Using the Sakasegawa M/M/C approximation presented in Chapter 9:

$$WQ = \frac{s * u^{\sqrt{2c+2}-1}}{c(1-u)} \qquad \text{where:} \qquad u = \frac{\lambda}{c\mu} = \frac{\lambda*S}{c} = \frac{\rho}{c}$$

$$WQ_2 = [\frac{1.907 + 0.90}{2}][0.087(\frac{0.87^{\sqrt{6}-1}}{2.0(1-.87)})] = 0.3838 \text{ Hrs/Unit}$$

The Expected Time in Workstation 2, the expected waiting line and the expected number at Workstation 2 are given by:

$$TPT_2 = WQ_2 + S_2 = 0.384 + 0.087 = 0.471 \text{ Hrs/Unit}$$
$$LQ_2 = \lambda_2 * WQ_2 = 20(0.384) = 7.68 \text{ Units}$$
$$L_2 = LQ_2 + \rho_2 = 9.42$$

Workstation 3:

For the G/G/1 Server, $u_3 = 0.840$

$$WQ_3 = [\frac{CV_a^2(3) + CV_s^2(3)}{2}][\frac{u_3}{1-u_3}] S_3$$

$$WQ_3 = [\frac{1.167 + 1.22}{2}][\frac{0.84}{1-.84}] (0.042) = 0.2632 \text{ Hrs}$$

$$LQ_3 = \lambda_3 * WQ_3 = 20(0.2632) = 5.263 \text{ Unit}$$
$$TPT_3 = WQ_3 + S_3 = 0.2632 + 0.042 = 0.3052$$
$$L_3 = LQ_3 + \rho_3 = 6.103$$

10.8.1 System Performance Measures

The system WIP is recovered by adding up all of the Workstation WIP values.

$$WIP_{Sys} = L = L_1 + L_2 + L_3 = 30.243 \text{ Units}$$

And using Little's Law, the expected time in the system can be calculated as:

$$TPT_{Sys} = WIP_{Sys} / \nu = \frac{30.243}{20} = 1.512 \text{ Hrs}$$

10.9 Serial Systems with Re-entrant Flows

It is very common for a serial production system described by a routing sheet to exhibit re-entrant flows. For example, Semiconductor Manufacturing and Job Shop Models almost always have the same part visiting the same processing center more than once in a single routing sequence, often many times. When re-entrant flow occurs, the common processing Workstation can exhibit one of two processing rules. (1) The Workstation processing time distribution is exactly the same for each visit (2) The Workstation processing time distribution is different at each visit. That is, in a multistep production sequence the processing time parameters can be Workstation and step dependent. We will model this general case. Hence, we will find it necessary to map the required machine processing parameters to each processing Workstation by steps. The specified routing sheet for a particular part will dictate which Workstations are visited, in what sequence they are visited and how many times each Workstation is visited. To capture product flow parameters, special notation is needed. The total flow into a unique manufacturing Workstation j (Lathes, drills, mills, Etc.) will be denoted as usual by λ_j j= 1, 2… N. However λ_j is composed of individual flow rates from Workstation k into Workstation j which will be denoted by $\lambda_{k,j}$. It is also true that Workstation k might be visited at different steps in the transformation process. Hence, the total flow from Workstation K to Workstation j is in turn composed of flow rates from different steps and will be designated as $f_{i,k,j}$, where $f_{i,k,j}$ = expected input to workstation j from Workstation K immediately following Step i. Since we are only modeling single product flow, the SCV parameters and the expected processing times are uniquely defined for each Workstation j at a designated Step i. Let $S_{i,j}$ be the expected processing time at Workstation j on Step i, and $CV_s^2(i, j)$ = Squared Coefficient of Variation of Service Time at Workstation j, Step i for j $\in$ M, i $\in$ N. $CV_a^2(i, j)$ and $CV_d^2(i, j)$ are similarly defined. Table 10.8 summarizes the notation we will use for Single Product, Serial Flow Networks with Workstation re-visitations.

For single product flow networks, we will assume that Workstation j at Step 1 is the point of input for external product flow. The input flow rate (product demand) to Workstation j at Step 1 will be designated as $v_1 j$, the expected processing time at Workstation j, Step 1 as $S_{1,j}$, the squared coefficient of variation for service at Workstation j, Step 1 as $CV_s^2(v_1, j)$, and the squared coefficient of variation for arrivals to Workstation j, Step 1 as $CV_a^2(v_1, j)$ = 1.0. The notation used for input flow into Workstation 1 may seem awkward and not needed, but it will be necessary when multiproduct flows are presented in Chapter 11. Note that once a particular Workstation is visited in a routing sequence, re-entrant flow to that same Workstation can occur at any subsequent step in the routing sequence.

Table 10.8
General Notation for Single Product Flow

Notation	Definition	Index
N	**Number of Steps**	
M	**Number of Workstations**	
v_1	**External flow load for Product 1**	
$S_{i,j}$	**Expected processing time at step i, Workstation j**	i= 1, 2 ... N j=1, 2 ... M
$f_{i,k,j}$	**Following processing at Step i , Flow from Workstation k to Workstation j**	i= 1, 2 ... N j=1, 2 ... M k= 1,2 ... M
$\lambda_{k,,j}$	**Total flow from Workstation k to Workstation j**	J=1, 2 ... M k= 1,2 ... M
λ_j	**Total flow into Workstation j**	j=1, 2 ... M
$CV_S^2(i,j)$	**SCV of Service times at Workstation j, Step i**	i= 1, 2 ... N j=1, 2 ... M
$CV_a^2(i,j)$	**SCV of Time between arrivals to Workstation j at Step i**	i= 1, 2 ... N j=1, 2 ... M
$CV_d^2(i,j)$	**SCV of Time between departures at Workstation j, Step i**	i= 1, 2 ... N j=1, 2 ... M
v_1,j	**Special notation for the demand stream v_1 arriving to Workstation j at Step 1**	j= 1, 2 ... N
$CV_a^2(v_1,j)$	**Special notation for the SCV for the time between arrivals of the demand stream v_1 arriving to Workstation j at Step 1**	j=1, 2 ... M
$CV_S^2(v_1,j)$	**Special notation for the SCV of the Service times for product v_1 arriving to Workstation j at Step 1**	j=1, 2 ... M

This general notation is shown in Table 10.9 for a specific 6 step process involving 4 different Workstations.

Table 10.9
Example Notation for N=6 and M=3 with Reentrant Flows

Process Step	1	2	3	4	5	6
Workstation	1	3	1	3	1	2
$S_{k,j}$	$S_{1,1}$	$S_{2,3}$	$S_{3,1}$	$S_{4,3}$	$S_{5,1}$	$S_{6,2}$
$CV_S^2(i,j)$	$CV_S^2(\gamma_1,1)$	$CV_S^2(2,3)$	$CV_S^2(3,1)$	$CV_S^2(4,3)$	$CV_S^2(5,1)$	$CV_S^2(6,2)$
$CV_a^2(i,j)$	$CV_a^2(\gamma_1,1)$	$CV_a^2(2,3)$	$CV_a^2(3,1)$	$CV_a^2(4,3)$	$CV_a^2(5,1)$	$CV_a^2(6,2)$

In this example, Workstation 1 is used at three different steps in the 6 step process. The total workload on Workstation 1 is composed of (1) external demand into Workstation j at Step 1 $(v_1 j)$, (2) product flow into Workstation 1 following the completion of work at Workstation 3 at Step 2 $(f_{2,3,1})$, and product flow into Workstation 1 following the completion of work at Workstation 3 at Step 4 $(f_{4,3,1})$. For single product production flow networks with an input stream of v_1 with no flow losses, the magnitude of any flow $f_{i,k,j}$ will all be equal to the external

336

flow (demand) rate ν_1. Hence, in this case at Workstation 1 the total input flow load will be the uniquely defined by $\lambda_1 = 3\gamma_1$. In general,

$$\lambda_j = \nu_{1,j} + \sum_{K=1}^{M} \sum_{i=1}^{N-1} f_{i,k,j} \qquad j = 1, 2, 3\ldots M$$

For the process flow sheet shown in Table 10.9, the following flows from Workstation k to Workstation j following step i occur.

$$f_{1,1,3}, f_{2,3,1}, f_{3,1,3}, f_{4,3,1}, f_{5,1,2}$$

Again, for single product flow networks specified by routing sheets in which there is no product loss (to rejects for example), if any flow $f_{i,k,j}$ exists, it is of magnitude ν_1. It follows that the total flow load from any Workstation k to any Workstation j is given by:

$$\lambda_{k,j} = \sum_{i=1}^{N-1} f_{i,j,k} \qquad k \in M, \quad j \in M$$

For this example:

$$\lambda_{1,3} = f_{1,1,3} + f_{3,1,3} = 2\nu_1$$
$$\lambda_{3,1} = f_{2,3,1} + f_{4,3,1} = 2\nu_1$$
$$\lambda_{1,2} = f_{5,1,2} = \nu_1$$

Of course, the total flow load on any Workstation j from all sources is given by:

$$\lambda_j = \sum_{k=1}^{M} \lambda_{k,j} + \nu_{1,j} \qquad j = 1, 2 \ldots M$$

For this example, $\quad \lambda_1 = 3\nu_1 \quad \lambda_2 = \nu_1 \quad \lambda_3 = 2\nu_1$

Since any one part is (possibly) visiting the same Workstation at multiple processing steps….and the processing time parameters can be different at each step….it is necessary to determine a **composite** or average processing time at each machine. The composite processing time at Workstation j will be defined as S_j. This is the weighted average of the step dependent expected processing times at Workstation j. Equation 10.3 gives the desired result.

$$S_j = \left(\frac{\nu_{1,j}}{\lambda_j}\right) S_{1,j} + \sum_{k=1}^{M} \sum_{i=1}^{N-1} \left(\frac{f_{i,k,j}}{\lambda_j}\right) S_{i+1,j} \qquad (10.3)$$

The calculation of S_j is in two parts. The first part is the proportion of external flow ν_j into Workstation j at Step 1, times the expected processing time at Step 1, Workstation j, associated with that input stream. The second part is the weighted average of all input flows from any other Workstation k to Workstation j from any Step i. This notation may seem cumbersome, but will be necessary in the study of multiproduct flow networks in Chapter 11.

It will be also be necessary to calculate the Squared Coefficient of Variation for the composite service time at Workstation j. To calculate the SCV for the composite processing time, we will need the second moment about the origin of the expected processing time. It can be shown that the Equation 10.4 will provide the second moment from known Workstation parameters.

$$S_j^2 = \left(\frac{\nu_{1,j}}{\lambda_j}\right) S_{1,j}^2 \left[CV_S^2(\nu_1, j) + 1.0 \right] + \qquad (10.4)$$

$$\left\{ \sum_{k=1}^{M} \sum_{i=1}^{N-1} \left(\frac{f_{i,k,j}}{\lambda_j}\right) S_{i+1,j}^2 \left[CV_S^2(i+1, j) + 1.0 \right] \right\} \qquad j = 1, 2 \ldots M$$

Then by definition: $\quad CV_S^2(j) = \dfrac{S_j^2}{(S_j)^2} - 1.0 \qquad j = 1, 2 \ldots M \qquad (10.5)$

Finally, since Workstation j might receive workload from several other Workstations, it will also be necessary to calculate the composite Squared Coefficient of Variation for arrivals to Workstation j, $CV_a^2(j)$. This will be a weighted average of the SCV's that are transmitted to Workstation j from any other Workstation k at Step i. Unfortunately, these values are not independent of each other and depend upon the process flow and manufacturing topology. There is a unique $CV_a^2(j)$ value which must be calculated for every Workstation, but they must all be calculated **simultaneously** using a set of linear equations. The following set of M equations will allow us to calculate the $CV_a^2(j)$ values, one for each Workstation j=1, 2…M.

$$CV_a^2(j) = (\frac{v_{1,j}}{\lambda_j})CV_a^2(v_j, j) + \sum_{k=1}^{M}(\frac{\lambda_{k,j}}{\lambda_j}) CV_a^2(k, j) \qquad j=1,2…M \qquad (10.6)$$

Equation 10.6 has two main parts. The first part is a single term which captures the contribution of the external flow SCV to Workstation j at Step 1. The weight used is the proportion of total flow into Workstation 1 due to product demand ($\frac{v_{1,j}}{\lambda_j}$). The second part is the proportion of the SCV's entering any Workstation j from any other Workstation k which is designated as $CV_a^2(k, j)$. The $CV_a^2(k, j)$ term is the proportion of $CV_a^2(k)$ which is split at Workstation k to Workstation j. The determination of $CV_a^2(k, j)$ is given by Equation 10.7.

$$CV_a^2(k, j) = (\frac{\lambda_{k,j}}{\lambda_k}) CV_d^2(k) + (1 - \frac{\lambda_{k,j}}{\lambda_k}) \qquad (10.7)$$

Note that the value of $CV_a^2(k, j)$ transmitted from Workstation k to Workstation j is the full value of $CV_d^2(k)$ multiplied by the proportion of total flow being transmitted to Workstation j from Workstation k, which is given by ($\frac{\lambda_{k,j}}{\lambda_k}$). A second term which is added is $(1 - \frac{\lambda_{k,j}}{\lambda_k})$. This equation has been derived by both Witt and Curry. Finally, the value of $CV_d^2(k)$ has been previously determined to be a function of the operating parameters at Workstation k, and is given by Equation 10.8 if there is only one server at Workstation k

$$CV_d^2(k) = (1-u_k^2) CV_a^2(k) + u_k^2 CV_s^2(k) \qquad k=1,2…M \qquad (10.8)$$

If there are two or more servers at Workstation k, then Equation 10.9 must be used.

$$CV_d^2(k) = 1+ (1-u_j^2)(CV_a^2(k) - 1)$$
$$+ \frac{u_j^2}{\sqrt{c_j}}(CV_s^2(k)-1) \qquad C>1 \qquad k=1, 2…..M \qquad (10.9)$$

The values of $CV_s^2(k)$, u_j^2 and c_j have been previously calculated for each Workstation, so the only set of unknowns is the value of the individual $CV_a^2(k)$'s. Looking at Equations 10.6-10.9, the only other unknown value(s) is that of $\lambda_{k,j}$. But we have previously shown that $\lambda_{k,j}= \sum_{l=1}^{N} f_{i,j,k}$ all k, all j.

To summarize, once the $\lambda_j, u_j, , S_j , CV_a^2(j)$ and $CV_s^2(j)$ values have been determined, the system has been reduced to single station representations with composite, weighted terms. Since there are infinite queues at each Workstation, the system can therefore be decomposed into M separate G/G/c Workstation analysis, and the expected time in queue can be calculated at each Workstation. The other Workstation performance measures and System Performance measures can be recovered using Little's Laws.

10.10   A Five Step, Three Workstation Network Flow Problem

A product is being made in a Job Shop with three work centers, and requires a five step process. The routing structure and processing parameters are given in Table 10.10.

Table 10.10
A Five Step, Three Station Line

Process Step (i)	1	2	3	4	5
Workstation (j)	1	2	1	3	2
$S_{i,j}$	1.60	1.20	2.10	3.20	1.90
$CV_s^2(i,j)$	1.20	0.90	1.40	1.30	1.10

Assume that the external flow load (demand) for the single product is two per 8 hour day. Therefore, $v_1 = 0.25$ parts/hr and enters the system at Workstation 1. We further assume that $CV_a^2(v_1, 1) = 1.0$.

The algorithmic procedure used to model this serial system with re-entrant flows will require several computational procedures previously discussed. (1) The first is to determine the Workstation offered workloads, number of servers and Workstation utilizations. (2) The second is to calculate the expected composite processing time at Workstation j (S_j) averaged across all inputs to Workstation j. (3) The third is to calculate a composite Squared Coefficient of Variation for the composite expected processing time at Workstation j, $CV_s^2(j)$. (4) The fourth is to determine the individual flow rates $f_{i,k,j}$ into each Workstation j from any other Workstation k at any step i. (5) The fifth is to calculate the flow load $\lambda_{k,j}$ from Workstation k to any other Workstation j, and then the total flow load λ_j on Workstation j. These things being accomplished, (6) the system of equations to recover the M linear equations used to calculate the $CV_a^2(j)$ can be constructed and solved. Finally, the system can now be decomposed into individual Workstation analysis, and system performance measures recovered from those results.

10.10.1 Determining Workstation Workload

The offered workloads to each Workstation are a function of the product input rate, and how many times each Workstation is visited in the routing sheet. From Table 5.9, we see that Workstations one and two are each visited twice, and Workstation 3 is visited once. Since there is only one product and external product demand is $v_1 = 0.25$ parts/hr to Workstation 1 at Step 1, then the total product flow into Workstations 1, 2 and 3 are given by:

$$\lambda_j = v_{1,j} + \sum_{K=1}^{M} \sum_{i=1}^{N-1} f_{i,k,j} \qquad j = 1, 2, 3 \ldots M \ .$$

Hence:
$$\lambda_1 = v_{1,j} + f_{2,2,1} = 0.50$$
$$\lambda_2 = f_{1,1,2} + f_{4,3,2} = 0.50$$
$$\lambda_3 = f_{3,1,3} = 0.25$$

The corresponding Workstation processing times are step dependent. Hence, the offered workload to Workstation j must consider the flow load by step.

$$\rho_j = v_1 \{ \sum_{i=1}^{N} \sum_{j=1}^{M} S_{i,j} \}$$

It follows that:
$$\rho_1 = v_1 \{ S_{1,1} + S_{2,1} + S_{3,1} + S_{4,1} + S_{5,1} \}$$
$$\rho_1 = v_1 \{ S_{1,1} + 0 + S_{3,1} + 0 + 0 \}$$
$$\rho_1 = 0.25 \{ 1.60 + 2.10 \} = 0.925,$$

Similarly:
$$\rho_2 = 0.25 \{ 1.20 + 1.90 \} = 0.775$$
$$\rho_3 = 0.25 \{ 3.20 \} = 0.80.$$

Since all loads are less than one, then only one server/machine is needed at each Workstation. Therefore, $C_1=1$, $C_2=1$ and $C_3=1$. It follows that $u_2=0.925$, $u_2=0.775$ and $u_3=0.80$ are the Workstation utilizations.

10.10.2 Analyzing the Product Flow Network

In this example, both Workstation 1 and Workstation 2 are being visited twice, and the service time density functions are not Exponential. It should again be emphasized that this product flow structure negates the previous strategy of directly decomposing the system into independently operating process centers, determining their individual performance measures, and then combining these individual Workstation results to obtain system performance measures. Because this is a system with re-entrant flows, the direct relationship between the Workstation j, $CV_a^2(j)$ value and the arrival CVD to the next Workstation visited, $CV_a^2(j+1)$ is destroyed. That is, $CV_a^2(j+1) = CV_a^2(j)$ no longer holds. In fact, as previously discussed, when Workstation j receives product flow from different process steps, the input $CV_a^2(j)$ is a function of the multiple input flow stream SCV's. When Workstation j sends output (in steady state) to other different processing Workstations, the $CV_a^2(j)$ is transformed into a vector of destination SCV's; $CV_a^2(j,1)$….. $CV_a^2(j,M)$ by a simple branching equation based upon the proportion of product flow which we will present later.

Input to Workstation 1 is from either external flow (v_1), or from Workstation 2 $(f_{2,2,1})$. At steady state, these arrivals occur at random. Of course, when these parts enter the product routing structure they have step-dependent service times. Recall that even for single product flow, product can leave Workstation k and be routed to different destinations at different steps. In steady state, the arrival streams appear to occur at random due to the mixture of different arrival times which are step and Workstation dependent. The system behaves as if it was Markovian, but it is not.

To state our procedure again, our modeling strategy will be as follows. (1) Determine an *average* or *composite* processing time S_j at Workstations 1-3 as a function of the individual part processing times. (2) Calculate an *average* or *composite* $CV_a^2(j)$ for each Workstation j that receives multiple inputs and the corresponding $CV_s^2(j)$ and $CV_a^2(j)$. Having done these things, the system will now behave as if there was only *one* part flowing through Workstations 1, 2 and 3. At this point, the processing sequence can be decomposed into three *pseudo independent* individual processing Workstations, analyzed separately as before, and then the individual Workstation performance measures can be used to recover System performance measures. This approach was first suggested by Witt and proved to be an accurate approximation to exact Workstation and system performance measures. The methodology will now be demonstrated for this example.

10.10.3 Calculating Composite Workstation Processing Times and the Workstation Service SCV Values

The expected processing time at each Workstation j on Step i is defined as $S_{i,j}$ i=1, 2…N steps and j=1,2….M Workstations. The associated SCV values are given by $CV_s^2(i,j)$. Finally, even though there is only one product flowing through the system, the magnitude of product flowing through Workstation j and subsequently being routed to other Workstations is not always just the system input demand rate due to Workstation re-visitation. The magnitude of product flow from

Workstation k to Workstation j will be designated as $\lambda_{k,j}$. Finally, λ_j will designate the *total* product flow into Workstation j from all sources. Of course, without product losses, the rate in-rate out conservation of flow principle will always be true, and the system output will always match system input in steady state. This being the case, Little's Laws will still hold at either the system or Workstation level.

We have already calculated λ_1=0.50, λ_2=0.50 and λ_3=0.25 and determined the $f_{i,k,j}$ values. It is now necessary to determine the individual $\lambda_{k,j}$ values for product flow exiting Workstation k bound for Workstation j. An alternate way to determine the $\lambda_{k,j}$ values for single product flow analysis is to draw a process flow diagram. For this problem, consider the flow representation in Figure 10.5

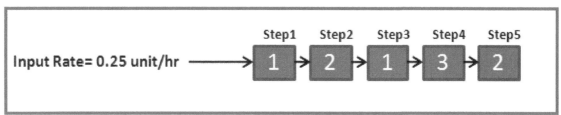

Figure 10.5
A Five Step Flow Representation

The total flow into Workstation 1 is λ_1=0.50, and is composed of $\gamma_{1,1}$=0.25 and $f_{2,2,1} = 0.25$. In a similar manner, $\lambda_2 = f_{1,1,2} + f_{4,3,2} = 0.50$ and $\lambda_3 = f_{3,1,3} = 0.250$ The composite Workstation processing times will be calculated as a weighted average of the individual Workstation, step dependent processing times and the Workstation input flow rates using the following Equation at each Workstation.

$$S_j = (\tfrac{v_{1,j}}{\lambda_j}) S_{1,j} + \sum_{k=1}^{M} \sum_{i=1}^{N-1} \left(\tfrac{f_{i,k,j}}{\lambda_j}\right) S_{i+1,j} \qquad j=1, 2,\ldots M \text{ Workstations} \qquad (10.10)$$

The reentrant product flows from Workstation k to Workstation j are weighted by the magnitude of their flow from Workstation k to Workstation j, relative to the total flow into Workstation j (λ_j) . It is obvious that for Workstation j, $\{ \tfrac{v_{1,j}}{\lambda_j} + \sum_{k=1}^{M} \sum_{i=1}^{N-1}(\tfrac{f_{i,k,j}}{\lambda_j})\} = 1.0$

For Workstation 1: $S_1 = (\tfrac{v_{1,1}}{\lambda_1}) S_{1,1} + (\tfrac{f_{2,2,1}}{\lambda_1}) S_{3,1} = (\tfrac{0.25}{0.50}) 1.60 + (\tfrac{0.25}{0.50}) 2.10 = 1.850$

For Workstation 2: $S_2 = (\tfrac{f_{1,1,2}}{\lambda_2}) S_{2,2} + (\tfrac{f_{4,3,2}}{\lambda_2}) S_{5,2} = (\tfrac{0.25}{0.50}) 1.20 + (\tfrac{0.25}{0.50}) 1.90 = 1.550$

For Workstation 3: $S_3 = (\tfrac{f_{3,1,3}}{\lambda_3}) S_{4,3} = (\tfrac{0.25}{0.25}) 3.20 = 3.20$

It is now necessary to produce a single composite value for CV_S^2 (j) for those Workstations with re-entrant flows. We have previously shown (Feldman, Curry) that the following general equations can be used to define CV_S^2 (j).

Since: $\qquad S_j = (\tfrac{v_{1,j}}{\lambda_j}) S_{1,j} + \sum_{k=1}^{M} \sum_{i=1}^{N-1} \left(\tfrac{f_{i,k,j}}{\lambda_j}\right) S_{i+1,j} \qquad j=1 ,2,\ldots M$

And: $\qquad S_j^2 = (\tfrac{v_{1,j}}{\lambda_j}) S_{1,j}^2 [CV_S^2(\gamma_1, j) + 1.0] +$

$\qquad\qquad \{ \sum_{k=1}^{M} \sum_{i=1}^{N-1} (\tfrac{f_{i,k,j}}{\lambda_j}) S_{i+1,j}^2 [CV_S^2(i+1, j) + 1.0] \} \qquad j=1, 2\ldots M$

341

By definition: $CV_S^2(j) = \frac{S_j^2}{(S_j)^2} - 1.0$ j=1, 2...M

For Workstation 1:

$$S_1^2 = (\frac{v_{1,1}}{\lambda_1})(S_{1,1})^2(1+ CV_S^2(v_1,j)) + (\frac{f_{2,2,1}}{\lambda_1})(S_{31})^2 [1+ CV_S^2(3,1)]$$

$$S_1^2 = (\frac{v_{1,1}}{\lambda_1})(S_{1,1})^2(1+1.2) + (\frac{f_{2,2,1}}{\lambda_1})(S_{31})^2 (1+1.40)$$

$$S_1^2 = (\frac{.25}{.50})(1.60)^2 (2.20) + (\frac{.25}{.50})(2.10)^2(2.40) = 8.108$$

From previous calculations, $S_1 = 1.850$

Therefore: $CV_S^2(1) = \frac{S_1^2}{(S_1)^2} - 1.0 = \frac{8.108}{(1.85)^2} - 1.0 = 1.3690$

Similarly: $CV_S^2(2) = 1.147$
$CV_S^2(3) = 1.30$

10.10.4 Calculating the $CV_a^2(j)$ and the $CV_d^2(j)$ Values

The processing time and the SCV values at Workstations 1, 2 and 3 have now each been reduced to a single number. The remaining task is to determine the values for $CV_a^2(j)$. Since the $CV_a^2(j)$ values are dependent upon the product stream which enters from other, multiple Workstations in the routing sequence, they cannot be independently calculated, but must all be determined simultaneously. The defining equations for the unknown $CV_a^2(j)$ values are as follows (Witt, Curry).

$$CV_a^2(j) = (\frac{v_{1,j}}{\lambda_j}) CV_a^2(v_1,j) + \sum_{k=1}^{M}(\frac{\lambda_{kj}}{\lambda_j}) CV_a^2(k,j) \tag{10.11}$$

Equation 10.11 is identical to Equation 10.6, and uses the same weighted sum approximation methodology. The first term is the impact of the product external flow into Workstation j. Since the external flow load into Workstation j is known, and the SCV for external arrivals $CV_a^2(v_1,j)$ is part of the problem definition, this first term is readily constructed. Since we have defined the demand stream to enter Workstation 1, the first term will only exist for Workstation j at Step 1. *This is not true for multiproduct models.* The second term(s) is the weighted sum of the internal SCV's which arrive from other Workstations k to the target Workstation j. Similar to the weighted Workstation processing time calculation, the weights assigned to each input stream are simply $(\frac{\lambda_{kj}}{\lambda_j})$. This is the proportion of total flow into Workstation j from Workstation k. To determine the values of $CV_a^2(k,j)$, we need to observe that the SCV output from *Workstation k* has a magnitude of $CV_d^2(k)$. This is split into multiple SCV values; one for each destination that can be reached from Workstation k. The portion of $CV_d^2(k)$ sent to each output destination is determined by λ_k and λ_{kj}, and can be calculated by Equation 10.12.

$$CV_a^2(k,j) = (\frac{\lambda_{kj}}{\lambda_k}) CV_d^2(k) + (1 - \frac{\lambda_{kj}}{\lambda_k}) \qquad k \in M \quad j \in M \tag{10.12}$$

Equation 10.6 can now be written as follows.

$$CV_a^2(j) = (\frac{v_{1,j}}{\lambda_j}) CV_a^2(v_1,1) + \sum_{k=1}^{M}(\frac{\lambda_{k,j}}{\lambda_j}) \{(\frac{\lambda_{kj}}{\lambda_k}) CV_d^2(k) + (1 - \frac{\lambda_{kj}}{\lambda_k})\} \tag{10.13}$$

The only remaining term to be determined is $CV_d^2(k)$. As previously discussed, Equations 10.14 and 10.15 can be used to calculate the $CV_d^2(k)$ values for both single server and multiple server processing Workstations.

$$CV_d^2(k) = (1-\rho_k^2) CV_a^2(k) + \rho_k^2\ CV_s^2(k) \qquad\qquad C=1 \tag{10.14}$$

$$CV_d^2(k) = 1 + (1 - u_k^2)(CV_a^2(k) - 1) + \frac{u_k^2}{\sqrt{c_j}} (CV_s^2(k) - 1) \qquad C > 1 \qquad (10.15)$$

The second term(s) in Equation 13 will particularize to one of the forms given by Equation 14 or Equation 15, depending upon whether Workstation k has one or more severs/machines. We are now ready to formulate a set of equations which will yield the values of $CV_a^2(j)$ k=1, 2.....M. There will be exactly M linear equations in M unknowns, one equation for each Workstation j. The set of equations to determine $CV_a^2(1)$, $CV_a^2(2)$ and $CV_a^2(3)$ are best derived as follows. First construct Equations 4-6, then Equation 11. Finally, apply either Equation 12 or Equation 13 to Equation 11 depending upon whether there are one or more servers at Workstation k.

Server 1: $CV_a^2(1) = (\frac{v_{1,1}}{\lambda_1}) CV_a^2(v_1, 1) + (\frac{\lambda_{2,1}}{\lambda_1}) CV_a^2(2,1)$

$\qquad CV_a^2(1) = (\frac{0.25}{0.50})(1.0) + (\frac{0.25}{0.50}) CV_a^2(2,1)$

$\qquad CV_a^2(1) = 0.50 + 0.50\ CV_a^2(2,1)$

$\qquad\qquad CV_a^2(2,1) = (\frac{\lambda_{2,1}}{\lambda_2}) CV_d^2(2) + (1 - \frac{\lambda_{2,1}}{\lambda_2})$

$\qquad\qquad\qquad = (\frac{0.25}{0.50}) CV_d^2(2) + (1 - \frac{0.25}{0.50})$

$\qquad\qquad CV_a^2(2,1) = 0.50\ CV_d^2(2) + 0.50$

$\qquad$ Hence: $\quad CV_a^2(1) = 0.50 + 0.50\ \{0.50 + 0.50\ CV_d^2(2)\}$

$\qquad$ Or: $\qquad CV_a^2(1) = 0.75 + 0.25\ CV_d^2(2)$

Since there is only one server at Workstation 1;

$\qquad CV_d^2(2) = (1 - u_2^2) CV_a^2(2) + u_2^2 CV_s^2(2)$

$\qquad\qquad = (1 - 0.775^2) CV_a^2(2) + (0.775^2)(1.147)$

$\qquad CV_d^2(2) = 0.40\ CV_a^2(2) + 0.689$

This yields: $\quad CV_a^2(2,1) = 0.50 + 0.50\{ 0.40\ CV_a^2(2) + 0.689$

$\qquad$ Or: $\quad CV_a^2(2,1) = 0.845 + 0.20\ CV_a^2(2)$

$\qquad$ Finally: $\qquad CV_a^2(1) = 0.50 + 0.50\ \{ 0.845 + 0.20\ CV_a^2(2)\}$

$\qquad\qquad CV_a^2(1) = 0.923 + 0.10\ CV_a^2(2)$

It will be left as an exercise to the reader to verify that the following equations are correct.

$\qquad\qquad CV_a^2(2) = 0.0361\ CV_a^2(1) + 0.18\ CV_a^2(3) + 0.95836$

$\qquad\qquad CV_a^2(3) = 0.072\ CV_a^2(1) + 1.0855$

Solving these three (linear) equations in three unknowns, we obtain:

$\qquad CV_a^2(1) = 1.043 \quad CV_a^2(2) = 1.205\ CV_a^2(3) = 1.161$

The following results are also available in the solution process.

$\qquad\qquad CV_d^2(1) = 1.322$

$\qquad\qquad CV_d^2(2) = 1.170$

$\qquad\qquad CV_d^2(3) = 1.250$

$\qquad\qquad CV_a^2(1,2) = 1.161$

$\qquad\qquad CV_a^2(1,3) = 1.161$

$\qquad\qquad CV_a^2(2,1) = 1.085$

$\qquad\qquad CV_a^2(3,2) = 1.250$

It is instructive to note that Workstation 1 sends output to both Workstation 2 and Workstation 3. The output $CV_d^2(1) = 1.322$ is split into two pieces, $CV_a^2(1,2) = 1.161$ and $CV_a^2(1,3) = 1.161$. Observe that for this particular example, these two SCV values are equal. However, they are not a 50-50 split of the emerging $CV_d^2(1)$ value. Since the SCV has variance as its numerator, this is a clear proof of how the corrupting influence of variance is propagated throughout the entire production line.

10.10.5 Workstation Analysis

The arrival rates to each Workstation, the server utilizations , the composite average Workstation service times S_j , the Composite CV_s^2 (j) j=1,2,3 values have been calculated and the correct CV_a^2(j) j=1,2,3 values have been determined. The three Workstation serial production lines can now be broken into three independent Workstation analyses. The strategy, as before, is to calculate a single Workstation performance measure for each of the three Workstations, and recover the remaining Workstation performance measures using Little's Law. Finally, the individual Workstation performance measure(s) can be aggregated to calculate system performance measures, primarily system WIP and system Cycle Time. The individual Workstation cycle times and Workstation WIP are determined from the following sets of equations.

Workstation 1:

$$TPT_1 = [\ \frac{CV_a^2(1)+CV_s^2(1)}{2}\ ]\ [\ \frac{u_1}{1-u_1}\ ]\ [S_1] + \ S_1$$
$$CV_d^2(1) = (\ 1 - u_1^2)\ CV_a^2(1) + \ u_1^2\ CV_s^2(1)$$
$$L_1 = TPT_1 * \lambda_1$$

Workstation 2:

$$TPT2_2 = [\ \frac{CV_a^2(2)+CV_s^2(2)}{2}\ ]\ [\ \frac{u_2}{1-u_2}\ ]\ [S_2] + \ S_2$$
$$CV_d^2(2) = (\ 1 - u_2^2)\ CV_a^2(2) + \ u_2^2\ CV_s^2(2)$$
$$L_2 = TPT_2 * \lambda_2$$

Workstation 3:

$$TPT_{Station\ 3} = [\ \frac{CV_a^2(3)+CV_s^2(3)}{2}\ ]\ [\ \frac{u_3}{1-u_3}\ ]\ S_3 + \ S_3$$
$$CV_d^2(3) = (\ 1 - u_3^2)\ CV_a^2(3) + \ u_3^2\ CV_s^2(3)$$
$$L_3 = TPT_3 * \lambda_3$$

All the parameters used in these equations have already been calculated. For example, at Workstation 1:

$$TPT_1 = [\ \frac{1.043+1.369}{2}\ ]\ [\ \frac{0.925}{1-0.925}\ ]\ (1.85) + 1.85 = 29.363\ \text{hrs}$$
$$\text{And:}\quad CV_d^2(1) = (1 - 0.925^2)(1.043) + (0.925^2)(1.369) = 1.322$$
$$L_1 = (29.363)\ (0.50) = 14.681$$

Note also that the Expected number in Queue is readily calculated.

$$LQ_1 = L_1 - \rho_1 = (14.681) - 0.925 = 13.756$$

The reader can verify that:

$$TPT_2 = 7.830 \qquad CV_d^2(2) = 1.170 \qquad L_2 = 3.915 \quad LQ_2 = 3.14$$
$$TPT_3 = 18.950 \qquad CV_d^2(3) = 1.250 \qquad L_3 = 4.738 \quad LQ_3 = 3.938$$

10.10.6 Systems Analysis

The System WIP is the sum of all of the individual Workstation WIP's.

$$WIP_{Sys} = L = L_1 + L_2 + L_3 = 14.681+3.915+4.738 = 23.334\ \text{units}$$

The system Throughput Time can be calculated by applying Little's Law

$$TPT_{Sys} = \frac{WIP_{Sys}}{v_1} = \frac{23.334}{0.25} = 93.336\ \text{hrs}$$

344

10.11 Summary

This Chapter has been concerned with the analysis of single product, multistep production systems. An ideal serial production line with WIP control and constant processing times at each workstation was studied for both balanced and unbalanced serial production lines. The results gave best case performance for an N station, serial production line with balanced workstations. In addition, we provided an intuitive justification for a set of general rules called *Little's Laws*. We have used these laws extensively in both Workstation analysis and systems analysis of key performance indicators. We next presented a special class of queuing networks called a Jackson network which included probabilistic branching.

Single product flow was then described by a *process flow sheet* which mapped multiple processing steps to production centers (machines). An artifact called a *part* was moved through a sequence of processing Workstations to produce a final product. Product could be a finished unit, an assembly or a subassembly depending upon the modeling context. We also assumed that without loss of generality that the product entered the system at Step 1, Workstation 1. Any single processing center might have one or more machines in it, dictated by the capacity required (offered workload) to the processing Workstation. The input data required was the input flow rate (demand), the external demand stream squared coefficient of variation, the expected processing times, and either the variance of the processing time distribution or the squared coefficient of variation for the processing time distribution. Workstation processing times could be step dependent in a multi-station routing sequence. In the next Chapter, we will extend modeling capabilities to multiple product flows and arbitrary points of entry for product demand.

Review Questions

1. What is a "closed queuing system" versus an "open queuing system"?
2. What is a "Bottleneck Workstation"?
3. If several Workstations are in series, why is the "Bottleneck Workstation" not the Workstation with the slowest server?
4. What is "Raw Cycle Time?
5. A Manufacturing Engineer visits a factory with a Lean Engineer, and finds the Workstation with the most WIP in front of it. He turns to the Lean Engineer and says, "This is the Bottleneck Workstation". Discuss this comment.
6. What is a "Balanced Production Line"? Can this ever be achieved? Discuss your answer.
7. What is "Critical WIP" in an Unbalanced Production Line with N Workstations?
8. Under what circumstances will the Critical WIP equal to the number of Workstations? Can this condition ever be achieved? Discuss your answer.
9. What is a classical "CONWIP System"?
10. Using the methodology of Section 10.20 draw a system diagram for a 5-Workstation line with a constant processing time of 5 minutes per part at each Workstation. Manually determine system behavior for a WIP limit of 6 parts and determine the system Throughput time and the output rate. Verify your answer with Little's Law.
11. Write down the Poisson and Exponential density functions. Prove that the Squared Coefficient of variation is equal to one for each density function.
12. Consider the following 4-Workstation Production line (Hopp and Spearman).

Workstation	Machines	Processing Time(Hr/Part)
1	1	2.0
2	2	5.0
3	6	10.0
4	2	3.0

a. What are the Workstation Processing Rates?
b. What is the Bottleneck Rate?
c. What is the Raw Processing Time?
d. What is Critical WIP for this system?
e. What are the individual Workstation WIP values?
f. Verify that Little's Law holds at the Critical WIP value.

13. A manufacturing assembly line consists of 4 Workstations in series, with infinite WIP space between each Workstation. The following data is available.
The average input rate is 24 parts/hr and arrivals follow a Poisson Density function. The rate of service at each Workstation is also Poisson and is given by the following table.

Workstation	Rate of Service(Parts/Hr)
1	8.32
2	12.50
3	6.20
4	25.30

Calculate the (1) WIP at each Workstation (2) The average Waiting time at each Workstation (3) The average Throughput time at each Workstation (4) System WIP and (5) System Throughput time.

14. What is a "Jackson Network"? What assumptions are associated with a Jackson Network? Can a Jackson Network accommodate multiple products? Explain your answer, possibly using the internet for information.

15. Consider the following network of Workstations. Let $p_{1,2}=.45$, $p_{1,3}=.55$, $p_{2,3}=.70$ $p_{2,4}=.30$, $p_{4,2}=.20$, $p_{3,4}=.35$ and $p_{3,1}=.75$. Let the external demand rate be $v_1=20$ parts per hour, and let $\mu_1=5.8$ parts per hour, $\mu_2=10.75$ parts per hour, $\mu_3=12$ parts per hour and $\mu_1=8.25$ parts per hour. Solve for the expected WIP at each Workstation, Throughput time at each Workstation, System WIP and System Throughput time.

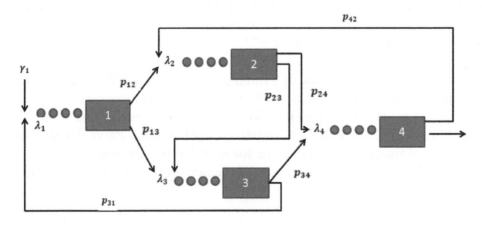

16. Consider the following network of Workstations. Let $p_{1,2}=.45$, $p_{1,3}=.55$, $p_{2,3}=.60$, $p_{2,4}=.40$, $p_{4,2}=.30$, $p_{3,4}=.20$, $p_{3,5}=.30$ and $p_{3,1}=.50$. Let the external demand rate be $v_1=20$ parts per hour, and let $\mu_1=5.8$ parts per hour, $\mu_2=4.75$ parts per hour, $\mu_3=12$ parts per hour, $\mu_4=9.2$ parts per hour and $\mu_5=4.7$ parts per hour. Solve for the expected WIP at each Workstation, Throughput time at each Workstation, System WIP and System Throughput time.

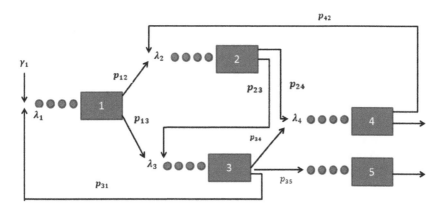

Assume that the product demand is Poisson Distributed with an arrival rate of 0.20 Parts per hour. Determine the following. (1) Offered Workload to each Workstation (2) Workstation Utilization (3) the Product Flow Load on each Workstation (4) the Composite Workstation service time per part (5) the System WIP (6) the System Throughput time (7) the expected queue length at each Workstation and (8) the Expected Throughput time at each Workstation./

17. A Workstation has a Utilization of 0.92 and a single server. Arrivals to the workstation are exponentially distributed and the Service times are exponentially distributed. Calculate the departure stream SCV.

18. A Workstation has a Utilization of 0.95 and three servers. Arrivals to the workstation have an SCV of 1.46, and the service time SCV is 1.28. Calculate the departure stream SCV.

19. Consider a single Workstation which receives input from three different sources. The mean arrival rates for each stream are given by $\lambda_1=12.6$, $\lambda_2=9.75$, and $\lambda_3=5.70$. The SCV for each arrival stream are given by $CV_1^2(1)=2.78$, $CV_2^2(1)=4.12$, AND $CV_1^2(1)=1.67$. (1) Calculate the total flow load on the Workstation. (2) Calculate the average SCV for a composite arrival stream (3) Calculate the Mean and Variance of the Composite Arrival Stream. Hint: Weight the SCV of each individual arrival stream by the ratio of its arrival rate to the total arrival rate to get a composite arrival rate SCV.

20. Using the following product specifications, (1) Draw a Product Routing diagram (2) Determine the Workstation Offered Workload and the Workstation Utilization. (3) Calculate Workstation Performance Measures (Expected time in service, expected time in Workstation, Expected time in Queue, Expected WIP and Expected number in each Workstation (4) Calculate the Expected time to produce a single unit of product and the Expected System WIP.

$$v_1 = 2.0 \text{ Unit/Hr} \qquad CV_a^2(v_1, 2) = 1.10$$

Process Step(i)	1	2	3	4	5
Workstation(j)	2	3	2	1	3
$S_{i,j}$ (Hrs/Part)	1.20	0.75	0.40	1.60	1.10
$CV_S^2(i,j)$	1.20	0.85	1.30	1.45	2.10

21. The following single product serial production system has been presented by Curry and Feldman.

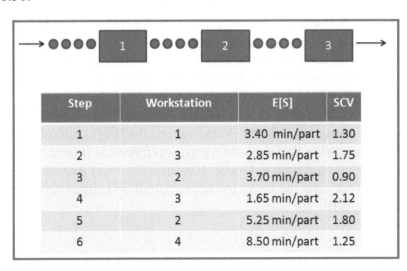

Step	Workstation	E[S]	SCV
1	1	3.0 hr/part	1.00
2	3	2.5 hr/part	0.75
3	2	3.7 hr/part	1.25
4	1	4.0 hr/part	1.75
5	3	3.6 hr/part	1.32

(1) Draw a Product Routing diagram (2) Determine the Workstation Offered Workload and the Workstation Utilization. (3)Calculate Workstation Performance Measures (Expected time in service, expected time in Workstation, Expected time in Queue, Expected WIP and Expected number in each Workstation (4) Calculate the Expected time to produce a single unit of product and the Expected System WIP. $v_1 = 0.20$ Unit/Hr $CV_a^2(1, 2) = 1.0$

22. Compute the System performance measures of Throughput and system WIP for the following Serial System assuming that the external demand rate is 5.0 parts/hr and has an SCV of 1.50.

Step	Workstation	E[S]	SCV
1	1	3.40 min/part	1.30
2	3	2.85 min/part	1.75
3	2	3.70 min/part	0.90
4	3	1.65 min/part	2.12
5	2	5.25 min/part	1.80
6	4	8.50 min/part	1.25

Chapter 11
Multiproduct Factory Flow Models

"The superior man, when he sees what is good, moves toward it; and when he sees his errors, he turns from them"   *The Book of Changes   China, circa 1200 BC*

11.1 Introduction

In Chapter 9 we derived the basic single Workstation service model for an M/M/1 queue, and then presented models for M/M/C AND M/G/1 Workstations. We then extended modeling capabilities to the general G/G/1) and G/G/C) Workstation models. All models made extensive use of Little's Law to produce Workstation and system performance measures. In Chapter 10 we presented factory models for Single Product factory flows, including reentrant flows and step dependent processing times for Non-Markovian Serial Workstations. The special case of Jackson networks with probabilistic branching was introduced and demonstrated. We will now extend modeling capabilities to multiproduct flow systems with product dependent Step/Station processing parameters and reentrant flows.

The decomposition of a multiproduct serial Manufacturing System topology into individual Workstation analysis depends heavily upon the ability to estimate the individual Workstation inflow rates, λ_j j=1, 2….M, composite Workstation processing times S_j j=1,2…..M and calculating the squared coefficients of variation, $CV_a^2(j)$, CV_d^2 (j) and $CV_a^2(j)$ for all Workstations j=1,2…M. These parameters will now need to be estimated for multiproduct flow. Almost all manufacturing facilities are set up to produce multiple products. In addition, it is common practice to create shop floor instructions for each product and route them through the factory using a routing sheet. We will adhere to this practice in this Chapter. Since routing sheets specify with certainty which machine centers will be visited and in what order, it is reasonable to assume for now that there will be no random branching of products to machine center destinations. This will somewhat simplify the modeling approach which we will use, but multiproduct processing through common machining centers with step and product dependent processing times adds an additional degree of complexity. As with single product flow analysis, each type of product may or may not visit every machining center, and each product may visit the same machining center more than once in a routing sequence. We will also assume that different products have different processing time distributions at each Workstation visited, and that they are step dependent. Since multiple products will be flowing through the system simultaneously, we must distinguish between product dependent routing structures and (possibly) different machining times by product by step and by machine in any visitation sequence.

Modeling multiproduct production facilities will require some specialized modeling procedures, but we will see that the estimation of the key parameters $CV_a^2(k)$, CV_d^2 (k) and $CV_a^2(k)$ at each Workstation will follow the same basic procedure as in Chapter 10. First, since the total product flow rate to each Workstation will be the sum of all individual flow rates into each Workstation, it will be necessary to calculate the total flow rate from each of the individual products and also the number of visits to each Workstation by product. The total offered workload to each Workstation is a function of the number of visits to each Workstation by each product, the external demand rate for each product, the processing time required at each processing center by product, and the number of servers at each Workstation. As usual, we will need to determine the $CV_a^2(k)$, $CV_d^2(k)$ and $CV_s^2(k)$ parameters for each Workstation. Second, since each product will usually have a specified mean processing time and CV_s^2 for each product visiting each Workstation by step, it will be necessary to calculate a single weighted processing time for each Machine Center which is a combination of the individual product workloads and their associated processing times. Once the total workload and the composite processing time at each Workstation are known, the required number of servers (machines) and the server utilization u at each machining center can be determined. The final requirement is as in Chapter 10 is to determine values for the $CV_a^2(k)$, $CV_d^2(k)$ and $CV_s^2(k)$ parameters at each Workstation. Having calculated these values, the total manufacturing system can be decomposed into the analysis of individual Workstations and then synthesized to determine system performance measures using individual Workstation results and Little's Law. We will not consider probabilistic branching until Chapter 12.

11.2 Computational Notation

It will now be necessary to identify processing parameters by *product, step* and *Workstation* visited. We will use the same notation as that developed in Chapter 10, but we will add a product identifier. We will again give special consideration and associated notation to outside product flows. Let v_p designate the entry of Product type p=1, 2…P in parts per unit time. Each different part type will be described by its own routing sheet, and parts can enter the manufacturing system at any Workstation. However, we will assume that each part type enters the manufacturing system at the first Workstation in its own routing sheet at Step 1. The associated average time between arrivals and the SCV of inter-arrival times will be given as input data. Table 11.1 summarizes our new and expanded notation.

Table 11.1
Multiproduct Notation

Notation	Definition	Index
P	Number of products	
N	Maximum number of Steps In any one routing sheet	
M	Maximum number of Workstations In any one product sequence	
$f_{i,k,j,p}$	Following Step i, the Flow from Workstation k, to Workstation j, by Product p	p=1,2…P j=1,2…M k=1,2….M
$f_{k,j,p}$	Total Flow from Workstation k, to Workstation j by Product p	j=1,2…M k=1,2….M p=1,2…P
$\lambda_{j,p}$	Total load on Workstation j by Product p	$p \epsilon P$ $j \epsilon M$
λ_j	Total load on Workstation j over all products	$j \epsilon M$
ρ_j	Offered Workload to Workstation j	$j \epsilon M$
u_j	Efficiency of Workstation j	$j \epsilon M$

c_j	Number of Servers at Workstation j	$j \epsilon M$
ν_p	External Load (Demand) for Product p	$p \epsilon P$
$\nu_{1,j,p}$	Special notation : External Load (Demand) of Product p entering Workstation j at Step 1	$p \epsilon P$ $j \epsilon M$
$S_{i,j,p}$	Expected Processing time for Product p on Workstation j at Step i	$p \epsilon P$ $j \epsilon M$ $i \epsilon N \ \ i \neq 1$
$S_{p,j}$	Expected weighted Processing time for Product p at Workstation j	$p \epsilon P$ $j \epsilon M$
S_j	Composite Expected Processing time at Workstation j across all products	$j \epsilon M$
S_j^2	Second moment about origin for the composite Processing time at Workstation j	$j \epsilon M$
$CV_S^2(i,j,p)$	Processing time Squared Coefficient of variation for Product p at Step i on Workstation j	$p \epsilon P$ $j \epsilon M$ $i \epsilon N \ \ i \neq 1$
$CV_a^2(i,j,p)$	Special notation : Time between arrivals Squared Coefficient of Variation of Product p on Workstation j at Step i	$p \epsilon P$ $j \epsilon M$
$CV_S^2(j)$	Composite Processing time Squared Coefficient of Variation at Workstation j across all products	$j \epsilon M$
$CV_a^2(k,j)$	Arrival stream Squared Coefficient of Variation of time between arrivals from Workstation k to Workstation j	k=1,2…M J=1,2…M
$CV_a^2(j)$	Composite Arrival stream Squared Coefficient of Variation of flow into Workstation j	J=1,2…M
$CV_d^2(k,j)$	Time between departures Squared Coefficient of Variation of flow from Workstation k to Workstation j	k=1,2…M J=1,2…M
$CV_d^2(j)$	Composite Time between departures Squared Coefficient of Variation of flow from Workstation j	J=1,2…M

As previously indicated, we must now be able to distinguish between processing parameters by product, step and station. External flow can enter at any Workstation but will be restricted to Step1. The corresponding Squared Coefficient of time between arrivals is equal to $CV_a^2(\nu_p, j, p)$. This is read as *the squared coefficient of variation of the external arrival stream ν_p for part type p at Step 1 into Workstation j*. Hence, it is correct to note that since product flow is defined to enter the manufacturing system at Workstation j on Step1, $\nu_p = \nu_{1,j,p}$ will be more descriptive and consistent with the general notation. Similarly, the expected processing time at Station j on Step 1 in the product routing sheet will be designated as $S_{1,j,p}$, and the squared coefficient of variation as $CV_S^2(1, j, p)$. At any other step in the product routing sheet the expected processing time will be $S_{i,j,p}$ and the squared coefficient of variation of service time as CV_S^2(i, j, p). When analyzing multiple product flow, the flow rate by product p from Workstation k to Workstation j *following* Step i will be designated as $f_{i,k,j,p}$. Using the same logic as that introduced in Chapter 10 it follows that:

$$\lambda_{k,j,p} = \sum_{i=1}^{N} f_{i,k,j,p} \tag{11.1}$$

$$\lambda_{k,j} = \sum_{p=1}^{P} \lambda_{k,j,p} \tag{11.2}$$

$$\lambda_j = \sum_{k=1}^{M} \lambda_{k,j} + \sum_{p=1}^{P} v_{1,j,p} \tag{11.3}$$

The Composite processing time at Workstation j is given by:

$$S_j = \sum_{p=1}^{P} \left\{ \left(\frac{v_{1,j,p}}{\lambda_j}\right) S_{1,j,p} + \sum_{k=1}^{M} \sum_{i=1}^{N-1} \left(\frac{f_{i,k,j,p}}{\lambda_j}\right) S_{i+1,j,p} \right\} \quad j=1,2...M \tag{11.4}$$

And the second moment about the origin for Composite processing time is:

$$S_j^2 = \sum_{p=1}^{P} \left\{ \left(\frac{v_{1,j,p}}{\lambda_j}\right) S_{1,j,p}^2 \left[1 + CV_S^2(1,j,p)\right] \right.$$

$$\left. + \sum_{k=1}^{M} \sum_{i=1}^{N-1} \left(\frac{f_{i,k,j,p}}{\lambda_j}\right) S_{i+1,j,p}^2 \left[1 + CV_S^2(i,j,p)\right] \right\} \quad j=1,2...M \tag{11.5}$$

By definition: $CV_S^2(j) = \dfrac{S_j^2}{[S_j]^2} - 1.0$ (11.6)

From the input data and the values of λ_j $\quad j = 1, 2....K$ the offered workload to Workstation j is given by;

$$\rho_j = \lambda_j * S_j \qquad j=1,2....M \tag{11.7}$$

Workstation Utilization is given by:

$$u_j = \frac{\rho_j}{c_j} \qquad j=1,2....M \tag{11.8}$$

To recover the unknown $CV_a^2(j)$ values we now use the following equations.

$$CV_a^2(j) = \sum_{p=1}^{P} \left\{ \left(\frac{v_{1,j,p}}{\lambda_j}\right) CV_a^2(1,j,p) \right\} + \sum_{k=1}^{M} \left(\frac{\lambda_{k,j}}{\lambda_j}\right) CV_a^2(k,j) \qquad j=1,2....M \tag{11.9}$$

Define: $\quad CV_a^2(k,j) = \left(\frac{\lambda_{k,j}}{\lambda_k}\right) CV_d^2(k) + \left(1 - \frac{\lambda_{k,j}}{\lambda_k}\right)$ (11.10)

And $\quad CV_d^2(k) = (1-u_k^2) CV_a^2(k) + u_k^2 CV_S^2(k) \qquad C=1$ (11.11)

$$CV_d^2(k) = 1 + (1-u_k^2)(CV_a^2(k) - 1) + \frac{u_k^2}{\sqrt{c_k}}(CV_S^2(k) - 1) \quad C>1 \tag{11.12}$$

Having determined the $CV_S^2(j)$ and the $CV_a^2(j)$ composite values, the Throughput time at each Workstation can now be calculated using Little's Law.

$$TPT_j = E\,[Waiting\ Time\ in\ Queue_j] + E\,[Time\ in\ Service_j] \quad j=1,2....M$$

$$TPT_j = WQ_j + S_j \qquad j=1,2....M \tag{11.13}$$

The Workstation and system performance measures can now be recovered using Little's Laws. In principle, Multiproduct flow analysis is no more difficult than single product flow analysis and involves the same basic logic.

11.3 Computational Procedures

We will demonstrate the computational procedures by way of a small but typical example. The Lean Machine Company is a small specialty shop which produces two products using four machining centers. The demand rate for Part 1 is $v_1 = 5$ Units/Hr and for Part 2, $v_2 = 4$ Units/Hr. Product 1 will enter the system at Workstation 1, and Product 2 at Workstation 3. Hence, $v_{1,1,1} = 5$ and $v_{1,3,2} = 4$. The value of CV_a^2 for each product arrival stream will be designated as $CV_a^2(1,1,1) = 1.0$ and $CV_a^2(1,3,2) = 1.0$. The associated $CV_S^2(1,1,1) = 1.4$ and $CV_S^2(1,3,2) = 1.2$. Note that Part 2 enters the system at Workstation 3 and Part 1 at Workstation 1, both at Step 1 in their respective routing sheets. The Manufacturing design team (without a Lean Engineer) has determined that one machine will be allocated to Workstations 1 and 4, three machines to

Workstation 2, and six machines to Workstation 3. The plant manager has asked a Lean Engineer to verify these design specifications and predict System Performance Measures.

The routing sequence for each product is shown in Figure 11.1. Table 11.2 provides the product flow data.

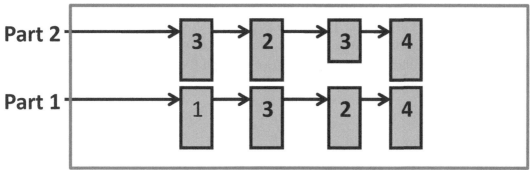

Figure 11.1
Process Flow Diagrams

Table 11.2
Input Data

	Step	1	2	3	4	External Flow
	Station	1	3	2	4	
Product 1	$S_{i,j,1}$	0.1667	0.3846	0.40	0.120	v_1=5 Parts/Hr
	$CV_s^2(i, j, 1)$	1.40	1.30	0.80	1.10	$CV_a^2(1,1,1)$=1.0
	Station	3	2	3	4	
Product 2	$S_{i,j,2}$	0.2632	0.2381	0.40	0.08	v_2=4 Parts/Hr
	$CV_s^2(i, j, 2)$	1.20	1.40	0.90	1.00	$CV_a^2(1,1,3)$=1.0

11.3.1 Calculating the Product Flow loads, Workstation Offered Workloads and the Number of Servers Required at each Workstation

Since the external product demands , the Expected processing times at each Workstation, the Squared Coefficients of Variation and the production sequences are known; we can calculate the product flow load (λ_j) to each work center j from both products. The offered workload to each process step (ρ_j), the minimum number of machines required at each process step/Workstation (c_j) and determine the work center utilization (u_j).

11.3.1.1 Flow Load to each Workstation

In a product routing sheet, every step/processing Workstation j =1,2…M listed in the sequential routing structure for the same product will receive the same amount of flow load from that one product, but each Workstation in the routing sequence can now be visited by different products and can be visited more than once by the same product. The total workload to any Workstation j is now be a mixture of workloads from different products each possibly making multiple visits to the same Workstation. The input rate associated with any one product is v_p . If the same Workstation j is visited more than once in the sequence of steps required to produce product v_p, the total flow load to Workstation j is a sum of the number of times that Workstation j is visited in the routing sheet across all visiting products times the respective input rates. For each visitation in any one product routing sheet, the offered workload is the product flow load v_p multiplied by the expected processing time for product p at each Workstation visited in the routing sequence. The flow load by step by product is a constant across all steps in a particular routing sequence, provided there are no flow losses in the production sequence. We will deal with the effects of flow losses in Chapter 12. The total product flow to each Workstation j is given by λ_j j=1, 2….M. The value of each λ_j can be determined by first listing the flow from any Workstation k to a particular Workstation j *following* Step i for each Product p. Define this flow as $f_{i,k,j,p}$. For this example:

Product 1:	$f_{1,1,3,1}=5$	$f_{2,3,2,1}=5$	$f_{3,2,4,1}=5$
Product 2:	$f_{1,3,2,2}=4$	$f_{2,2,3,2}=4$	$f_{3,3,4,2}=4$

Define: $\lambda_{k,j}=\sum_{i=1}^{N-1}\sum_{p=1}^{P} f_{i,k,j,p}$ $\quad k \in M \quad j \in M$

Then: $\lambda_{1,3}=5 \quad \lambda_{3,2}=9 \quad \lambda_{2,3}=4 \quad \lambda_{3,4}=4 \quad \lambda_{2,4}=5$

Finally: $\lambda_j = \sum_{k=1}^{M}\lambda_{k,j} + \sum_{p=1}^{P} v_{1,j,p}$ $\quad$ j=1, 2….M

Hence: $\lambda_1 = 5$
$\lambda_2 = 9$
$\lambda_3 = 13$
$\lambda_4 = 9$

For example, $\lambda_1= v_{1,1,1}=5$ and $\lambda_3=v_{1,3,2} + \lambda_{1,3}+\lambda_{2,3} =13$. This may seem a bit tedious, but the values of $f_{i,k,j,p}$ will be needed to calculate S_j , and the values of $\lambda_{k,j}$ will be needed to calculate $CV_a^2(j)$ j=1, 2….M.

11.3.1.2 Insight:

The total flow is composed of two parts. External demand ($v_{1,j,p}$) for product p into Workstation j at Step 1 in each routing sheet, and flow from any other Workstation k to Workstation j at Step i by Product p ($f_{i,k,j,p}$). For example, consider *Workstation 3*. Workstation 3 receives flow from the external demand of Product 2 ($v_{1,3,2}$) at Step 1. In the same routing sequence, Product 2 is transmitted from Workstation 2 to Workstation 3 following the completion of Step 2 ($f_{2,2,3,2}$). Finally, Product 1 flows into Workstation 3 from Workstation 1 at the completion of Step 1($f_{2,1,3,1}$). The total product flow into Workstation 3 is given by $\lambda_3=v_{1,3,2} + f_{2,2,3,2} + f_{2,1,3,1} = 5+4+4 = 13$.

The values of λ_j j= 1,2,3,4 can also be determined in a matrix calculation. The matrix is shown in Table 11.3

Table 11.3
Workstation Flows

Workstation	1	2	3	4	
Product 1	1	1	1	1	v_1=5 Parts/Hr
Product 2	0	1	2	1	v_2=4 Parts/Hr
λ_j	5	9	13	9	

The body of the matrix contains the number of times that each product visits each Workstation in its own routing sheet. The last column contains the external demands by product. The values of λ_j are found by multiplying the product demand rates by the number of visitations, and summing across all Workstations by column. For example, $\lambda_3 = v_1(1) + v_2(2) = 5+8 = 13$, which is the same result as before

11.3.2 The Offered Workload to each Workstation, the Required Number of Machines and the Workstation Utilization

To find the offered workload to Workstation j, we must now calculate a Composite Processing time for each Workstation j. The offered workload to each Workstation is determined by the following equations.

$$\rho_j = \sum_{p=1}^{P} \{ v_{1,j,p} \; S_{1,j,p} + \sum_{k=1}^{M} \sum_{i=1}^{N-1} f_{i,k,j,p} \; S_{i+1,j,p} \} \qquad j= 1, 2\dots M \qquad (11.14)$$

Equation 11.14 can be simplified by noting that the value of individual product flow into each Workstation is actually step independent. In other words, if the input flow for Product p is given by $v_{1,j,p}$, then input flow to every Workstation visited at any step in the Product 1 routing sheet will be $v_p = v_{1,j,p}$. Since the Expected processing time $S_{i,j,p}$ at any Step i, Workstation j is specified in the Product p routing sheet, then the partial offered workload to Workstation j is given by $v_p * S_{i,j,p}$ regardless of the position of Workstation j in the routing sequence. Hence, Equation 11.14 can be written as follows.

$$\rho_j = \sum_{p=1}^{P} \{ v_p * S_{1,j,p} \} + \{ v_p \sum_{k=1}^{M} \sum_{i=1}^{N-1} S_{i+1,j,p} \} \qquad j= 1, 2\dots M \qquad (11.15)$$

Machine Center 1:
Machine Center 1 appears only once in both routing sheets, and that is for Product 1 on Step 1 at Workstation 1. Hence:
$$\rho_1 = v_1 * S_{1,1,1}$$
$$\rho_1 = 5.0*0.1667 = 0.8335$$
Machine Center 2:
Machine Center 2 appears twice, once in routing sheet 1 at Step 3, and in once in routing sheet 2 at Step 2. Hence:
$$\rho_2 = v_1 S_{3,2,1} + v_2 S_{2,2,2}$$
$$\rho_2 = 5.0*0.40 + 4.0*(0.2381) = 2.952$$
$$\text{Note that } \rho_2 = f_{2,3,2,1} * S_{3,2,1} + f_{1,3,2,2} * S_{2,2,2} = 5(0.40) + 4(0.2381) = 2.952$$
Machine Center 3:
Machine Center 3 appears three times.
$$\rho_3 = v_1 S_{2,3,1} + v_2 [S_{1,3,2} + S_{3,3,2}]$$
$$\rho_3 = 5.0*0.3846 + 4.0*0.2632 + 4.0*0.40 = 4.576$$
Machine Center 4:
Machine Center 4 appears two times.
$$\rho_4 = v_1 S_{4,4,1} + v_2 S_{4,4,2}$$
$$\rho_4 = 5.0*0.120 + 4*0.08 = 0.92$$

The Offered workload dictates the processing capacity required at each Workstation. The minimum number of processing machines required to satisfy the offered workload at Workstation j is the next highest integer above ρ_j. Hence, Workstations 1 and 4 require only one server ($c_1=c_4=1$), Workstation 2 requires three servers ($c_2=3$) and Workstation 3 requires Five servers ($c_3=5$). This is in contrast to the number of machines allocated to each Workstation by the design team, which has one machine in Workstations 1 and 4, three in Workstation 2 and six in Workstation 3 ($c_1=c_4=1$, $c_2=3$ and $c_3=6$). This clearly points out the need to model production capacity properly before committing to the number of machines required.

As usual; $u_j=\rho_j/c_j$. Therefore:

$u_1=0.8335$
$u_2=0.984$
$u_3=0.9152$
$u_4=0.92$

11.3.3 Calculating the Composite Workstation Processing Times

As previously stated, for multiple product flow networks we can no longer assume that Workstation 1 is the point of input for product flow. Product flow can now enter the production system at any Workstation j in the routing sheet. Of course, this will require a different routing sheet for each product starting at the point of entry which is designated as Step 1. The composite processing time at Workstation j is given by the value of S_j j=1, 2....M. S_j is the weighted average of the step dependent Expected processing times at Workstation j. The following formula gives the desired result for multiproduct flow networks.

$$S_j = \sum_{p=1}^{P} \{ (\frac{v_{1,j,p}}{\lambda_j}) S_{1,j,p} +$$

$$\sum_{k=1}^{M} \sum_{i=1}^{N-1} (\frac{f_{i,k,j,p}}{\lambda_j}) S_{i+1,j,p} \} \qquad j=1, 2...M \qquad (11.16)$$

The calculation of S_j again involves two parts. The first part is the proportion of external flow v_p into any Workstation j at Step 1, times the expected processing time for each part in that input stream. The second part is the weighted average of all input flows from any other Workstation k to Workstation j following any Step i. Note that $S_{k,j,p}$ is the expected processing time for product p at Workstation j at Step k in the routing sequence. Using Equation 11.16 the composite Workstation Service times can be calculated.

$$S_1 = (\frac{v_{1,1,1}}{\lambda_1}) S_{1,1,1}$$

$$S_1 = (\frac{5}{5}) 0.1667 = 0.1667$$

$$S_2 = (\frac{f_{2,3,2,1}}{\lambda_2}) S_{3,2,1} + (\frac{f_{1,3,2,2}}{\lambda_2}) S_{2,2,2}$$

$$S_2 = (\frac{5}{9}) (0.40) + (\frac{4}{9}) (0.2381) = 0.3281$$

$$S_3 = (\frac{v_{1,3,2}}{\lambda_3}) S_{1,3,2} + (\frac{f_{2,2,3,2}}{\lambda_3}) S_{3,3,2} + (\frac{f_{1,1,3,1}}{\lambda_3}) S_{2,3,1}$$

$$S_3 = (\frac{4}{13}) 0.2632 + (\frac{4}{13}) 0.40 + (\frac{5}{13}) 0.3846 = 0.352$$

$$S_4 = (\frac{f_{3,2,4,1}}{\lambda_4}) S_{4,4,1} + (\frac{f_{3,3,4,2}}{\lambda_3}) S_{4,4,2}$$

$$S_4 = (\frac{5}{13}) 0.12 + (\frac{4}{13}) 0.08 = 0.1022$$

11.3.4 Calculating the Composite Workstation Processing Times Squared Coefficient of Variation: $CV_s^2(j)$ j=1, 2…..M

To calculate $CV_s^2(j)$ we will need to calculate the second moment about the origin of the expected processing time at each Workstation j. It can be shown that the following formula is correct.

$$S_j^2 = \sum_{p=1}^{P}\{((\tfrac{v_{1,j,p}}{\lambda_j}) S_{1,j,p}^2 [CV_s^2(v_1,j,p) + 1.0] +$$

$$\sum_{k=1}^{M}\{\sum_{i=1}^{N-1}(\tfrac{f_{i,k,j,p}}{\lambda_j}) S_{i+1,j,p}^2 [CV_s^2(i,j,p) + 1.0]\} \qquad \text{j=1, 2, 3, 4} \qquad (11.17)$$

And by definition:

$$CV_s^2(j) = \frac{S_j^2}{[S_j]^2} - 1.0 \qquad \text{j=1, 2…M} \qquad (11.18)$$

$$S_1^2 = (\tfrac{v_{1,1,1}}{\lambda_1}) S_{1,1,1}^2 [CV_s^2(v_1,1,1) + 1.0]$$

$$S_1^2 = (\tfrac{5}{5})(0.1667)^2(1.40+1) = 0.0667$$

$$S_2^2 = (\tfrac{f_{2,3,2,1}}{\lambda_2})S_{3,2,1}^2 [CV_s^2(3,2,1) + 1] + (\tfrac{f_{1,3,2,2}}{\lambda_2})S_{2,2,2}^2 [CV_s^2(2,2,2) + 1.0]$$

$$S_2^2 = (\tfrac{5}{9})(0.40)^2(0.80+1) + (\tfrac{4}{9})(0.2381)^2(1.40+1) = 0.22047$$

$$S_3^2 = (\tfrac{5}{13})(0.3846)^2(1.30+1) + (\tfrac{4}{13})(0.2632)^2(1.2+1) + (\tfrac{4}{13})(0.40)^2(0.90+1) = 0.27128$$

$$S_4^2 = (\tfrac{5}{9})(0.12)^2(1.10+1) + (\tfrac{4}{9})(0.08)^2(1.0+1) = 0.02249$$

Using Equation 11.18

$$CV_s^2(1) = \frac{S_1^2}{[S_1]^2} - 1.0 = [\frac{.0667}{(.1667)^2}] - 1.0 = 1.40$$

$$CV_s^2(2) = \frac{S_2^2}{[S_2]^2} - 1.0 = [\frac{.224}{(.3281)^2}] - 1.0 = 1.049$$

$$CV_s^2(3) = \frac{S_3^2}{[S_3]^2} - 1.0 = [\frac{.2592}{(.352)^2}] - 1.0 = 1.8964$$

$$CV_s^2(4) = \frac{S_4^2}{[S_4]^2} - 1.0 = [\frac{.02249}{(.1023)^2}] - 1.0 = 1.1532$$

11.3.5 Calculating the Composite Workstation Arrival Time Squared Coefficient of Variations: $CV_a^2(j)$ j=1, 2…..M

The composite values of the S_j and the S_j^2 were used to calculate a single value for the $CV_s^2(j)$ value at each Workstation. This set of $CV_s^2(j)$ values is now product independent. Each Work center behaves as if there was only one product flowing through it. The only remaining task is to construct the equations needed to calculate the $CV_a^2(j)$ j=1, 2….. M for each Workstation. The multiple product factory flow problem will now be reduced to a problem which is very similar to the single product flow problem by calculating a composite $CV_a^2(j)$ for j=1, 2…..M. We will need to initially determine all product flow which occurs between any Workstation k and another Workstation j in the Product routing sheet based upon the individual product input rates and the Workstations visited. For this example, the following values of $f_{i,k,j,p}$ have already been determined.

Product 1: $f_{1,1,3,1}=5$ $f_{2,3,2,1}=5$ $f_{3,2,4,1}=5$
Product 2: $f_{1,3,2,2}=4$ $f_{2,2,3,2}=4$ $f_{3,3,4,2}=4$

It follows that:
$$\lambda_{k,j,p} = \sum_{i=1}^{N} f_{k,j,i,p}$$

Hence: $\quad \lambda_{1,3,1} = 5 \qquad \lambda_{3,2,1} = 5 \qquad \lambda_{2,4,1} = 5$

And: $\quad \lambda_{3,2,2} = 4 \qquad \lambda_{2,3,2} = 4 \qquad \lambda_{3,4,2} = 4$

To obtain the total flow from any Workstation k to any other Workstation j;
$$\lambda_{k,j} = \sum_{p=1}^{P} \lambda_{k,j,p} \qquad k \in M \qquad j \in M$$

It follows that: $\quad \lambda_{1,3} = 5 \quad \lambda_{3,2} = 9 \quad \lambda_{2,4} = 5$
$$\lambda_{2,3} = 4 \quad \lambda_{3,4} = 4$$

Of course: $\quad \lambda_j = \sum_{k=1}^{M} (\lambda_{kj} + v_1, k, j) \qquad j \in M$

So that: $\quad \lambda_1 = 5 \qquad \lambda_2 = 9 \qquad \lambda_3 = 13 \quad$ and $\quad \lambda_4 = 9$

Since there is no probabilistic branching, the general model given by Equations 11.9-11.12 can be used to recover the $CV_a^2(j)$ j=1, 2.....M with only slight modifications. The SCV values $CV_a^2(k,j)$ again represent the squared coefficient of variation for any product leaving Workstation k bound for Workstation j; $j, k \in M$

$$CV_a^2(j) = \{ \sum_{p=1}^{P} \{ (\frac{v_{1,j,p}}{\lambda_j}) CV_a^2(1, j, p) \} + \sum_{k=1}^{M} (\frac{\lambda_{kj}}{\lambda_j}) CV_a^2(k,j) \quad (11.19)$$

Recall that $\quad CV_a^2(k, j) = (\frac{\lambda_{k,j}}{\lambda_k}) CV_d^2(k) + (1 - \frac{\lambda_{k,j}}{\lambda_k}) \quad (11.20)$

Substituting Equation 11.20 into Equation 11.19

$$CV_a^2(j) = \{ \sum_{p=1}^{P} (\frac{v_{1,j,p}}{\lambda_j}) CV_a^2(1, j, p) \}$$

$$+ \sum_{k=1}^{M} (\frac{\lambda_{k,j}}{\lambda_j})[(\frac{\lambda_{k,j}}{\lambda_k}) CV_d^2(k) + (1 - \frac{\lambda_{k,j}}{\lambda_k})]$$
$$(11.21)$$

Finally, the value of $CV_d^2(k)$ depends upon whether Workstation j has one server or multiple servers.

$$CV_d^2(k) = (1 - u_k^2) CV_a^2(k) + u_k^2 CV_s^2(k) \qquad\qquad c_k = 1 \qquad (11.22)$$

$$CV_d^2(k) = 1 + (1 - u_k^2)[CV_a^2(k) - 1] + \frac{u_k^2}{\sqrt{c}}[CV_s^2(k) - 1] \qquad c_k > 1 \qquad (11.23)$$

Since the composite expected service time S_k at Workstation k has already been calculated as a weighted average of all the different products processed at Workstation j, the system now *behaves* as if it were processing a single product. This *behavior* is called a *renewal process* and is of course an approximation. However, it turns out to be quite good. When Equation 11.21 is combined with either Equation 11.22 or Equation 11.23, we have an expression in terms of only the (unknown) CV_a^2 (j) values. There are exactly M of these equations in M unknowns, and they are all linear. We note that Equation 11.23 with $c_k = 1$ reduces to Equation 11.22. In the event that Workstation j has Poisson Input and Exponential Service times, then $CV_a^2(k)$ and $CV_s^2(k)$ both equal to 1.0, so that $CV_d^2(k)$ also equals to one. This is due to the *Markovian Properties* of the M/M/1 and M/M/C queuing models. When there are infinite queues in front of each Service Workstation and the preceding properties hold, both the single product model and the multiproduct model can be immediately decomposed into separate M/M/1 and M/M/C independent Workstation analysis. One might be tempted to call this class of models a Jackson Network, but Jackson Networks cannot accommodate Multiproduct or Step dependent processing times.

As before, the approximation estimates $CV_a^2(j)$ as a weighted sum of the individual squared coefficient of arrivals for composite product flow that enters Workstation j for service. This approximation is comprised of two parts. The first is a weighted sum of the external flow into Workstation j; $CV_a^2(j)$. For the external flow into Workstation j, both the $v_{1,j,p}$ and the $CV_a^2(v_1, j, p)$ values are specified as input data. The second summation term represents the weighted contribution to $CV_a^2(j)$ by each external flow stream into Workstation j from any other Workstation k. The internal flow streams from any Workstation k to Workstation j bring along $CV_a^2(k, j)$, which is the weighted proportion of total flow leaving Workstation k to Workstation j that enters Workstation j; ($\frac{\lambda_{k,j}}{\lambda_k}$). It follows that: $\sum_{p=1}^{P}(\frac{v_{1,j,p}}{\lambda_j}) + \sum_{k=1}^{M}(\frac{\lambda_{k,j}}{\lambda_j}) = 1$. The $CV_a^2(k, j)$ is a function of the composite departure time squared coefficient of variation at Workstation k. Curry has shown that the $CV_a^2(k, j)$ values for product leaving Workstation k for Workstation j is given by a remarkably simple relationship, which is Equation 11.24.

$$CV_a^2(k, j) = (\frac{\lambda_{k,j}}{\lambda_k})\, CV_d^2(k) + (1 - \frac{\lambda_{k,j}}{\lambda_k}) \qquad (11.24)$$

Finally, as previously shown in Chapter 10, the value of $CV_d^2(k)$ is calculated as a weighted sum of the (unknown) value of $CV_a^2(k)$ and the calculated values of the composite squared coefficient of variation for service times at Workstation k, $CV_s^2(k)$. For single server Workstations, Equation 11.22 can be used; and for multiple server Workstations Equation 11.23 must be used. The final set of equations are linear in the M unknown values of $CV_a^2(j)$ j=1, 2…M. In order to understand the procedure of how to calculate the $CV_a^2(j)$ values, Figure 11.2 and Table 11.4 will be useful. Recall that $v_1 = 5$ and $v_2 = 4$.

Figure 11.2
Two Product Flow Diagrams

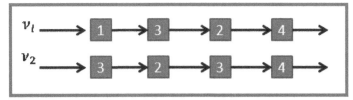

Table 11.4
Two Product Flow Data

	Step	1	2	3	4	External Flow
	Station	1	3	2	4	
Product 1	$S_{i,j,1}$	0.1667	0.3846	0.40	0.120	v_1=5 Parts/Hr
	$CV_s^2(i, j, 1)$	1.40	1.30	0.80	1.10	$CV_a^2(1,1,1)$=1.0
	Station	3	2	3	4	
Product 2	$S_{i,j,2}$	0.2632	0.2381	0.40	0.08	v_2=4 Parts/Hr
	$CV_s^2(i, j, 2)$	1.20	1.40	0.90	1.00	$CV_a^2(1,1,3)$=1.0

$\lambda_{1,3} = 5 \qquad \lambda_{3,2} = 9 \qquad \lambda_{2,4} = 5 \qquad \lambda_{2,3} = 4 \qquad \lambda_{3,4} = 4$

Workstation 1:

The only flow entering Workstation 1 is external (Demand). Hence, we know that $CV_a^2(1) = 1.0$. Note that Equation 11.19 will always hold:

$$CV_a^2(1) = \left(\frac{v_{1,j,p}}{\lambda_1}\right) CV_a^2(1,1,1) = \left(\frac{5}{5}\right)(1) \tag{11.25}$$

$$CV_a^2(1) = 1 \tag{11.26}$$

Since there is only one machine at Workstation 1,

$$CV_d^2(1) = (1-u_1^2)\, CV_a^2(1) + u_1^2\, CV_s^2(1)$$
$$CV_d^2(1) = (1-.8335^2)CV_a^2(1) + (.8335^2)(1.4)$$
$$CV_d^2(1) = 0.30528\, CV_a^2(1) + 0.97261$$

But, since Workstation1 has only Product1 flowing through it at Step1: $CV_a^2(1)$ is only 1.0.

$$\text{Hence,} \quad CV_d^2(1) = 1.2789 \tag{11.27}$$

Workstation 2:

Workstation 2 has no external flow, but receives input from Workstation 3 from both Product 1 and Product 2. There is no need to distinguish Product 1 flow from Product 2 flow, only the total product flow. Hence, combining product flow rates and recognizing that Workstation 2 has no other input flows, we obtain:

$$CV_a^2(2) = \left(\frac{\lambda_{32}}{\lambda_2}\right)\left[\left(\frac{\lambda_{32}}{\lambda_3}\right) CV_d^2(3) + \left\{ 1 - \left(\frac{\lambda_{32}}{\lambda_3}\right)\right\}\right]$$

$$CV_a^2(2) = \left(\frac{9}{9}\right)\left[\left(\frac{9}{13}\right) CV_d^2(3) + \left(1 - \left(\frac{9}{13}\right)\right)\right]$$

$$CV_a^2(2) = 0.6923\, CV_d^2(3) + 0.3077 \tag{11.28}$$

Since there are 4 servers at Workstation 2,

$$CV_d^2(2) = 1 + [1 - u_2^2] [CV_a^2(2) - 1] + \frac{u_2^2}{\sqrt{c_2}} [CV_s^2(2) - 1.0]$$

$$CV_d^2(2) = 1 - [1 - (0.9841)^2] [CV_a^2(2) - 1] + \frac{(0.9841)^2}{\sqrt{3}} [1.049 - 1]$$

$$CV_d^2(2) = 0.99585 + 0.03155\, CV_a^2(3) \tag{11.29}$$

Workstation 3

Workstation 3 receives input from Workstations 1 and 2, and from external flow, v_2.

$$CV_a^2(3) = \left(\frac{v_{1,3,2}}{\lambda_3}\right) CV_a^2(1,3,2) + \left(\frac{\lambda_{13}}{\lambda_3}\right)\left[\left(\frac{\lambda_{13}}{\lambda_1}\right) CV_d^2(1) + \left\{ 1 - \left(\frac{\lambda_{1,3}}{\lambda_1}\right)\right\}\right]$$
$$\qquad + \left(\frac{\lambda_{23}}{\lambda_2}\right)\left[\left(\frac{\lambda_{23}}{2}\right) CV_d^2(2) + \left\{ 1 - \left(\frac{\lambda_{23}}{\lambda_2}\right)\right\}\right]$$

$$CV_a^2(3) = \left(\frac{4}{13}\right)(1) + \left(\frac{5}{13}\right)\left[\left(\frac{5}{5}\right) CV_d^2(1) + [1 - (1)]\right] + \left(\frac{4}{13}\right)\left[\left(\frac{4}{9}\right) CV_d^2(2) + \left\{(1 - \left(\frac{4}{9}\right)\right\}\right]$$

$$CV_a^2(3) = 0.4787 + 0.3846\, CV_d^2(1) + 0.1367\, CV_d^2(2) \tag{11.30}$$

From 11.27: $\qquad CV_d^2(1) = 1.278$

And from 11.29: $\qquad CV_d^2(2) = 0.99585 + 0.03155\, CV_a^2(3) \tag{11.31}$

It follows that we can solve directly for $CV_a^2(3)$.

It is easy to determine that: $CV_a^2(3) = 1.111$ and $CV_d^2(2) = 1.0292$

Workstation 4:

$$CV_a^2(4) = \left(\frac{\lambda_{24}}{\lambda_4}\right)\left[\left(\frac{\lambda_{24}}{\lambda_2}\right) CV_d^2(2) + \left(1 - \left(\frac{\lambda_{24}}{\lambda_2}\right)\right)\right] + \left(\frac{\lambda_{34}}{\lambda_4}\right)\left[\left(\frac{\lambda_{34}}{\lambda_3}\right) CV_d^2(3) + \left\{ 1 - \left(\frac{\lambda_{34}}{\lambda_3}\right)\right\}\right]$$

$$CV_a^2(4) = \left(\frac{5}{9}\right)\left[\left(\frac{5}{9}\right) CV_d^2(2) + \left\{1 - \left(\frac{5}{9}\right)\right\}\right] + \left(\frac{4}{9}\right)\left[\left(\frac{4}{13}\right) CV_d^2(3) + \left\{1 - \left(\frac{4}{13}\right)\right\}\right]$$

$$CV_a^2(4) = 0.308\, CV_d^2(2) + 0.1372\, CV_d^2(3) + 0.556 \tag{11.32}$$

From 11.30 and 11.31; $CV_a^2(3) = 1.111$ and $CV_d^2(2) = 1.0292$

It follows that: $CV_a^2(4) = 1.022$

We have now recovered all the $CV_a^2(j)$ j= 1,2,3,4
$$CV_a^2(1) = 1.0$$
$$CV_a^2(2) = 1.062$$
$$CV_a^2(3) = 1.111$$
$$CV_a^2(4) = 1.022$$
And from previous analysis we have $CV_s^2(j)$ j= 1,2,3,4
$$CV_s^2(1) = 1.4$$
$$CV_s^2(2) = 1.0487$$
$$CV_s^2(3) = 1.1896$$
$$CV_s^2(4) = 1.1522$$

11.3.6 Workstation Analysis

Since all of the $CV_a^2(j)$ and $CV_s^2(j)$ values are known for Workstations j=1, 2 ,3 ,4 the system can now be decomposed and analyzed as four separate Workstation models using the following approximation from Chapter 9.

$$WQ_j = [\frac{CV_a^2(j) + CV_s^2(j)}{2}] WQ_j \text{ (M/M/C)} \quad J = 1, 2, 3, 4 \qquad (11.33)$$

The following specialized versions of Equation 11.33 can also be used.

Recall that: $\rho_j = \lambda_j * S_j$ and $u_j = \frac{\lambda_j * S_j}{C_j} = \frac{\rho_j}{C_j}$ j=1, 2, 3, 4

For the M/M/1 Server:
$$WQ_j = [\frac{\rho_j}{1-\rho_j}] * S_j \qquad j=1, 2, 3, 4 \qquad (11.34)$$

For the M/G/1 Server:
$$WQ_j = [\frac{1 + CV_s^2(j)}{2}] [\frac{\rho_j}{1-\rho_j}] * S_j \qquad j=1, 2, 3, 4 \qquad (11.35)$$

For the G/G/1 Server:
$$WQ_j = [\frac{CV_a^2(j) + CV_s^2(j)}{2}] [\frac{\rho_j}{1-\rho_j}] * S_j \qquad j=1, 2, 3, 4 \qquad (11.36)$$

For the G/G/C Server:
$$WQ_j = [\frac{CV_a^2(j) + CV_s^2(j)}{2}] [\frac{u^{\sqrt{2c+2}-2}}{c}] [\frac{u}{1-u}] * S_j \qquad j=1, 2, 3, 4 \qquad (11.37)$$

Workstation 1:
Workstation 1 is an M/G/1 System since the external arrival stream has $CV_a^2(1,1,1) = 1.0$, which is the SCV for Exponential time between arrivals. Hence, the Pollaczek-Khintchine formula (Equation 11.35) can be used.

The Expected time in Queue
$$WQ_1 = [\frac{1 + CV_s^2(1)}{2}] [\frac{\rho}{1-\rho_1}] * S_1$$
$$WQ_1 = [\frac{(1+1.4)}{2}] [0.8335(.1667)/.1665] = 1.001$$

The Expected time in Workstation 1
$$\text{TPT}_1 = WQ_1 + S_1 = 1.168$$
   The Expected Queue Length
$$LQ_1 = \lambda_1 * WQ_1 \quad \text{(Little's Law)}$$
$$LQ_1 = 5*1.001 = 5.007$$
The Expected number in Workstation 1
$$L_1 = WQ_1 + \text{E [No in service]} = WQ_1 + \rho_1$$
$$L_1 = 5.005 + 0.8335 = 5.8405$$

Workstation 2:

Workstation 2 is a G/G/C System with $c_2 = 3$ servers.

   The Expected time in Queue

$$WQ_2 = [\frac{CV_a^2(j) + CV_S^2(j)}{2}] [\frac{u^{\sqrt{2c_2+2}-2}}{c_2}] [\frac{u}{1-u}] * S_j \qquad (11.38)$$

   Since: $c_2=3$, $CV_a^2(2)=1.064$, $CV_S^2(2) = 1.049$, and $u_2=0.984$

   Then: $WQ_2 = 7.062$

The reader can verify that:

   The Expected time in Workstation 2 = TPT_2 = 7.390
   The Expected Queue Length = QL_2 = 63.561
   The Expected number in Workstation 2 = L_2 = 66.513

Workstation 3:

Workstation 3 is a G/G/C queue with $c_3 = 5$ servers.

   The Expected time in Queue = WQ_3 = 0.76715
   The Expected time in Workstation 3 = TPT_3 = 1.1191
   The Expected Queue Length = QL_3 = 9.9723
   The Expected number in Workstation 3 = L_3 = 14.549

Workstation 4:

Workstation 4 is a G/G/1 queue with $c_4 = 1$ server.

   The Expected time in Queue = WQ_4 = 1.2774
   The Expected time in Workstation 4 = TPT_4 = 1.380
   The Expected Queue Length = QL_4 = 11.4971
   The Expected number in Workstation 4 = L_4 = 12.417

11.3.7 System Analysis

The system level measures of performance require a synthesis of Workstation model results, and the application of Little's Law at the systems level. The system WIP is determined by adding together all of the individual Workstation WIP'S. This total is $WIP_{sys}=$ 5.841+66.513+14.549+12.417=93.32 units. The total external flow load and therefore the system throughput is $v_1+v_2=9$. Hence, applying Little's Law at the system level we obtain the average throughput time across all products. $TPT_{avg}= 93.32/9 = 11.0355$ hours.

11.3.8 Product Analysis

The System level analysis does not provide any information about the individual product flows. To recover those system performance measures we need to compute the throughput time for each product. The procedure is to calculate the expected throughput time at each Workstation by individual product type. Once this is calculated for every Workstation in the individual's routing sheet, the total throughput time for that product can be determined. Once this is known, we can apply Little's Law to that particular product. To do this, we need to know the throughput time for each individual product at each Workstation. This product throughput time is the expected waiting time in queue plus the expected service time for that particular product at that particular Step and Workstation. The input rate for product Type p is simply γ_p. The system product throughput time is relative to the total number of products that enter the factory (product demand) per unit time. Hence, the individual product throughput times must be divided by the total product flow rate per unit time. Note that since there is only one entry point for Type 1 and Type 2 product flows, the individual product flow load to every Workstation in the routing sequence is the same at every Workstation, and is also the external demand rate, ν_i i=1,2. Let TPT^j_{sys} be the system sojourn (throughput) time for product j = 1, 2. Hence, the TPT^1_{sys} for Product 1 is simply the sum of all the queue wait times and product service times at each Workstation in the product routing structure. The Throughput Time in the system by product Type p can be computed from Equation 11.39.

$$TPT^p_{sys} = \sum_{i=1}^{N} \{ \sum_{j=1}^{M} WQ_j + S_{i,j,p} \} \qquad p=1, 2, P \qquad (11.39)$$

Product 1
$$TPT^1_{sys} = (1.001 + 0.1667) + (7.0623 + 0.3846) + (0.7671 + .40)$$
$$+ (1.2774 + 0.120)$$
$$TPT^1_{sys} = 11.1791 \text{ Hours}$$

Product 2
$$TPT^2_{sys} = (0.7671 + 0.02632) + (7.062 + 0.2381) + (0.7671 + 0.40)$$
$$+ (1.2774 + 0.08)$$
$$TPT^2_{sys} = 10.618 \text{ Hours}$$

The composite average System Throughput Time for any one part is given by:

$$TPT_{Sys} = (\tfrac{5}{9}) 11.1791 + (\tfrac{4}{9}) 10.618 = 11.0355 \text{ Hours}$$

This verifies the previously calculated result for the average system throughput time. It is also possible to recover the Throughput Time and WIP in queue by product at the Workstation level. For example, consider Workstation 4. The individual product loads on Workstation 4 are $\nu_1 = 5$ and $\nu_2 = 4$. The individual Cycle times at Workstation 4 are the average queue wait times plus the product dependent service times.

Let TPT^p_j be the Throughput Time at Workstation j for Product p. It follows that:
$$TPT^p_j = WQ_j + \sum_{i=1}^{N} S_{i,j,p} \qquad j = 1, 2.....M, \quad p=1, 2...P$$
For example, consider Product 1 at Workstation 4 and Product 2 at Workstation 4.
$$TPT^1_4 = 1.277 + 0.120 = 1.397, \text{ and } TPT^2_4 = 1.277 + 0.08 = 1.357$$

Note that the expected Throughput time across both products is the weighted average of the Throughput for each individual product.

$$TPT_j = \sum_{p=1}^{P} \left(\frac{\lambda_{j,p}}{\lambda_j}\right) TPT_j^p \qquad j=1, 2 \dots M$$

For Workstation 4;
$$TPT_4 = (\tfrac{5}{9}) \, 1.397 + (\tfrac{4}{9}) \, 1.357 = 1.3792$$

This verifies the result obtained previously.

The total WIP in the Workstation by product can also be determined By Little's Law.
$$WIP_j^p = \lambda_j * TPT_j^p$$

For Product 1 at Workstation 4 and Product 2 at Workstation 4 :
$$WIP_4^1 = 5(1.397) = 6.985, \text{ and } WIP_4^2 = 4(1.357) = 5.428$$

Of course, the average WIP at Workstation 4 is given by:
$$WIP_4 = WIP_4^1 + WIP_4^2 = 6.985 + 5.428 = 12.418$$

Which confirms the previous result of $L_4 = 12.418$

11.4 A Flow Shop Model

Consider a flow shop consisting of four shops which manufacture three different products (Feldman and Valdez-Flores). There are three machining centers: Lathes (L), Milling machines (M), Drilling machines (D) and a final Packaging (P) area. The facility produces three types of different products (P1, P2, and P3). The routing sheets for each product are shown in Table 10.6 with their hourly demand rates.

Table 11.6
A Flow Shop Model

Product	Routing Structures	Product Demand
P1	L⇨M⇨D⇨P	2.0 Parts/Hr
P2	M⇨L⇨M⇨D⇨L⇨P	3.0 Parts/Hr
P3	L⇨M⇨P	2.5 Parts/Hr

Figure 11.3 is a process flow diagram for all three products. The processing times and CV_s^2 values by product by shop are shown as Table 11.7. Note the identification of machines by numbers.

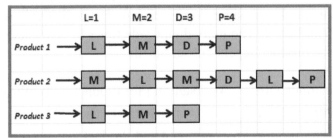

Figure 11.3
Flow Shop Process Flow Diagrams

Table 11.7
Table 11.7
Processing Times and CV_s^2 Values

	Step	1	2	3	4	5	6
Product	Workstation	1	2	3	4		
1	$s_{i,j,1}$	.250	.50	.333	.40		
	$CV_s^2(i,j,1)$	1.10	.50	0.60	.20		
Product	Workstation	2	1	2	3	1	4
2	$s_{i,j,2}$	.333	.40	.350	.350	.310	.60
	$CV_s^2(i,j,2)$	.80	1.20	.80	1.05	1.40	1.30
Product	Workstation	1	2	4			
3	$s_{i,j,3}$	.750	.667	.250			
	$CV_s^2(i,j,3)$	1.50	.30	1.30			

$$CV_s^2(1,1,1) = 1.0 \quad CV_s^2(1,2,2) = 1.0 \quad CV_s^2(1,1,3) = 1.0$$
$$v_{1,1,1} = 2.0 \text{ parts/hr} \quad v_{1,2,2} = 3.0 \text{ parts/hr} \quad v_{1,1,3} = 2.5 \text{ parts/hr}$$

11.4.1 The Product Flow Rates

Product 1: $f_{1,1,2,1} = 2 \quad f_{2,2,3,1} = 2 \quad f_{3,3,4,1} = 2$

Product 2: $f_{1,2,1,2} = 3 \quad f_{2,1,2,2} = 3 \quad f_{3,2,3,2} = 3 \quad f_{4,3,1,2} = 3 \quad f_{5,1,4,2} = 3$

Product 3: $f_{1,1,2,3} = 2.5 \quad f_{2,2,4,3} = 2.5$

Since: $\quad \lambda_{k,j} = \sum_{i=1}^{N} \sum_{p=1}^{P} f_{i,k,j,p}$

Then: $\quad \lambda_{1,2} = 7.5 \quad \lambda_{2,3} = 5 \quad \lambda_{3,4} = 2 \quad \lambda_{2,1} = 3 \quad \lambda_{3,1} = 3 \quad \lambda_{1,4} = 2 \quad \lambda_{2,4} = 2.5$

And: $\quad \lambda_j = \sum_{k=1}^{M} \lambda_{k,j} + \sum_{p=1}^{P} v_{1,j,p}$

Therefore: $\quad \lambda_1 = \lambda_{2,1} + \lambda_{3,1} + v_{1,1,1} + v_{1,1,3} = 3+3+2+2.5 = 10.5$

Similarly: $\quad \lambda_2 = 10.5 \quad \lambda_3 = 5 \quad \lambda_4 = 7.5$

11.4.2 Calculating the Offered Workloads, Server Utilization and the Number of Servers Required at each Workstation

Since the Expected processing time $S_{i,j,p}$ at any Step i, Workstation j is specified in the Product p routing sheet, then the partial offered workload to Workstation j is given by $v_p * S_{i,j,p}$ regardless of the position of Workstation j in the routing sequence. Hence:

$$\rho_j = \sum_{p=1}^{P} \{ v_p * S_{1,j,p} \} + \{ v_p \sum_{k=1}^{M} \sum_{i=1}^{N-1} S_{i+1,j,p} \} \qquad \text{j= 1, 2... M} \qquad (11.40)$$

Lathes: $\quad \rho_1 = v_1 * S_{1,1,1} + v_2 \{ S_{2,1,2} + S_{5,1,2} \} + v_3 * S_{1,1,3}$
$\quad\quad\quad\quad \rho_1 = 2.0(.250) + 3.0(.40) + 3.0(.310) + 2.5(.750) = 4.505$

Mills: $\quad \rho_2 = v_1 * s_{2,2,1} + v_2 \{ s_{1,2,2} + s_{3,2,2} \} + v_3 s_{2,2,3}$
$\quad\quad\quad\quad \rho_2 = 2.0(.50) + 3.0(.333) + 3.0(.35) + 2.5(.667) = 4.7175$

Drills: $\quad \rho_3 = 2.0(.333) + 3.0(.35) = 1.717$

Packaging: $\quad \rho_4 = 2.0(.40) + 3.0(.60) + 2.5(.250) = 3.225$

The Lathe and Mill cells require 5 machines ($C_1=C_2=5$), the Drill cell 3 machines ($C_3=3$) and the Packaging cell 4 machines ($C_4=4$). The Workstation utilizations are given by:

$$u_j=\rho_j/c_j \quad j=1, 2, 3, 4$$
$$u_1 = 4.505/5 = .9010 \quad \text{and} \quad u_2=0.943 \quad u_3=0.858 \quad u_4=0.806$$

We will show once again how to use the matrix approach to calculate the total Workstation input flow rates by product, Table 11.8 shows the number of visits to each Workstation (shop) by product, the input (v_i) loads, and the total flow load on each Workstation (λ_j)

Table 11.8
Total Flow Loads to each Workstation

Workstation	L	M	D	P	
Product 1	1	1	1	1	$v_1= 2.0$
Product 2	2	2	1	1	$v_2= 3.0$
Product 3	1	1	0	1	$v_3= 2.50$
λ_j	10.50	10.50	5.0	7.50	

$$\lambda_1= 10.5 \quad \lambda_2= 10.50, \quad \lambda_3= 5.0, \quad \lambda_4= 7.50$$

This verifies the values of $\lambda_1, \lambda_2, \lambda_3$ and λ_4 previously determined.

11.4.3 Calculating the Composite Workstation Processing Times (S_j) and the

$CV_S^2(j)$ for j=1, 2, 3, 4

The calculation of S_j involves two parts as usual. The first part is the proportion of external flow v_{1_p} into Workstation j at Step 1 times the expected processing time of each part in that input stream. The second part is the weighted average of all input flows from any other Workstation k to Workstation j at any Step i. Note that $S_{i,j,k}$ is the expected processing time for Product p at Workstation j at Step k in the routing sequence.

$$S_j=\sum_{p=1}^{P}\{ (\tfrac{v_{1,j,p}}{\lambda_j}) S_{1,j,p} + \sum_{k=1}^{M} \sum_{i=1}^{N-1}(\tfrac{f_{i,k,j,p}}{\lambda_j}) S_{i+1,j,p} \} \qquad j=1, 2...M$$

$$S_j^2 =\sum_{p=1}^{P}\{ (\tfrac{v_{1,j,p}}{\lambda_j}) S_{1,j,p}^2[CV_S^2(1,j,p) + 1.0] \qquad\qquad (11.41)$$

$$+ \sum_{k=1}^{M} \sum_{i=1}^{N-1} (\tfrac{f_{i,k,j,p}}{\lambda_j}) S_{i,k,j,p}^2 [CV_S^2(i, j, p) + 1.0] \} \qquad j=1,2...M$$

And by definition: $CV_S^2(j) = \dfrac{S_j^2}{[S_j]^2} - 1.0 \qquad j=1, 2 ...M$

Hence: (See Tables 11.6 and 11.7)

Lathe: Workstation 1

$$S_1 =(\tfrac{v_{1,1,1}}{\lambda_1}) * S_{1,1,1} +(\tfrac{v_{1,1,3}}{\lambda_1}) * S_{1,1,3} + (\tfrac{f_{1,2,1,2}}{\lambda_1}) * S_{2,1,2} + \left(\tfrac{f_{4,3,1,2}}{\lambda_1}\right) * S_{5,1,2}$$

$$S_1 = (\tfrac{2}{10.5}) 0.25 + (\tfrac{3}{10.5}) 0.40 +(\tfrac{3}{10.5}) 0.310 + (\tfrac{2.5}{10.5}) 0.75 = 0.429$$

$$S_1^2 = (\tfrac{v_{1,1,1}}{\lambda_1})* (S_{1,1,1}^2) * [CV_S^2(1,1,1) + 1.0]$$

$$+ \left(\frac{v_{1,1,3}}{\lambda_1}\right) * s_{1,1,3}^2 \quad * \quad [CV_s^2(1,1,3) \;+\; 1.0\,]$$

$$+ \left(\frac{f_{1,2,1,2}}{\lambda_1}\right)S_{2,1,2}^2 \;[CV_s^2(2,1,2) \;+\; 1] + \left(\frac{f_{4,3,1,2}}{\lambda_1}\right)S_{5,1,2}^2 \;[CV_s^2(5,1,2) \;+\; 1.0\,]$$

$$S_1^2 = \left(\frac{2}{10.5}\right)(0.25)^2\,(1.10+1) + \left(\frac{3}{10.5}\right)(0.40)^2\,(1.20+1)$$

$$+ \left(\frac{3}{10.5}\right)(0.310)^2 + (1.40+1) + \left(\frac{2.5}{10.5}\right)(0.750)^2\,(1.50+1) = 0.526$$

Hence: $CV_s^2(1) = \dfrac{S_1^2}{[S_1]^2} - 1.0 = \left\{\dfrac{.526}{(.429)^2}\right\} - 1.0 = 1.859$

Mill: Workstation 2

$$S_2 = \left(\frac{v_{1,2,2}}{\lambda_2}\right)*s_{1,2,2}] + \left(\frac{v_{2,2,1}}{\lambda_2}\right)*s_{2,2,1} + \left(\frac{f_{1,1,2,3}}{\lambda_2}\right)*s_{2,2,3} + \left(\frac{f_{2,1,2,2}}{\lambda_2}\right)*s_{3,2,2}$$

$$S_2 = \left(\frac{2}{10.5}\right)0.50 + \left(\frac{3}{10.5}\right)0.333 + \left(\frac{3}{10.5}\right)0.350 + \left(\frac{2.5}{10.5}\right)0.667 = 0.449$$

$$S_2^2 = \left(\frac{v_{1,2,2}}{\lambda_2}\right)* s_{1,2,2}^2 \quad [CV_s^2(1,2,2) \;+\; 1.0]$$

$$+ \left(\frac{f_{1,1,2,1}}{\lambda_2}\right)S_{1,2,1}^2 \;[CV_s^2(1,2,1) \;+\; 1.0\,]$$

$$+ \left(\frac{f_{2,1,2,2}}{\lambda_2}\right)S_{2,2,3}^2 \;[CV_s^2(2,2,3) \;+\; 1.0]$$

$$+ \left(\frac{f_{1,1,2,3}}{\lambda_2}\right)S_{2,2,3}^2 \;[CV_s^2(2,2,3) \;+\; 1.0\,]$$

$$S_2^2 = \left(\frac{2}{10.5}\right)(0.50)^2\,(0.50+1) + \left(\frac{3}{10.5}\right)(0.333)^2\,(0.80+1)$$

$$+ \left(\frac{3}{10.5}\right)(0.350)^2\,(1.05+1) + \left(\frac{2.5}{10.5}\right)(0.667)^2\,(0.30+1) = 0.329$$

$$CV_s^2(2) = \frac{S_2^2}{[S_2]^2} - 1.0 = \frac{.329}{(.449)^2} - 1.0 = 0.631$$

Drills: Workstation 3

$$S_3 = \left(\frac{2}{5}\right)0.333 + \left(\frac{3}{5}\right)0.350 = 0.343$$

$$S_3^2 = \left(\frac{2}{5}\right)(0.333)^2\,(0.60+1) + \left(\frac{3}{5}\right)(0.350)^2\,(1.05+1) = 0.222$$

$$CV_s^2(3) = \frac{S_3^2}{[S_3]^2} - 1.0 = \frac{.222}{(.343)^2} - 1.0 = 0.881$$

Packing: Workstation 4

$$S_4 = \left(\frac{2}{7.5}\right)0.40(0.20+1) + \left(\frac{3}{7.5}\right)0.60(0.30+1) + \left(\frac{2.5}{7.5}\right)0.250(1.30+1) = 0.430$$

$$S_4^2 = \left(\frac{2}{7.5}\right)(0.40)^2\,(1.20) + \left(\frac{3}{7.5}\right)(0.60)^2\,(2.30) + \left(\frac{2.5}{7.5}\right)(0.25)^2\,(2.30) = 1.315$$

$$CV_s^2(4) = \frac{S_4^2}{[S_4]^2} - 1.0 = \left\{\frac{.430}{(.430)^2}\right\} - 1.0 = 1.327$$

11.4.4 Calculating the $CV_a^2(j)$ values for j=1, 2 , 3 ,4

We have now reduced the individual product service times by Workstation and by step to a single composite value at each Workstation, and calculated the $CV_s^2(j)$ values for each Workstation. The only tasks that remain are to determine the $CV_a^2(j)$ values and calculate the System Performance Measures by Workstation and for the system. Since every processing center has multiple servers, we must use Equations 11.21 and 11.23. For convenience, we summarize results to this point and add the pair wise flow values for easy reference.

Table 11.9
Calculated Values

Workstation Number	Server u_j	Servers c_j	Server $CV_s^2(j)$	Total Flow λ_j
1	.901	5	1.859	$\lambda_1 = 10.5$
2	.943	5	.631	$\lambda_2 = 10.5$
3	.858	2	.881	$\lambda_3 = 5.0$
4	.806	4	1.327	$\lambda_1 = 7.5$

$\lambda_{21} = 3 \quad \lambda_{31} = 3 \quad \lambda_{12} = 7.5 \quad \lambda_{32} = 2.5 \quad \lambda_{34} = 2 \quad \lambda_{14} = 3 \quad \lambda_{24} = 2.5$

Figure 11.4
Multiproduct Flow Diagrams

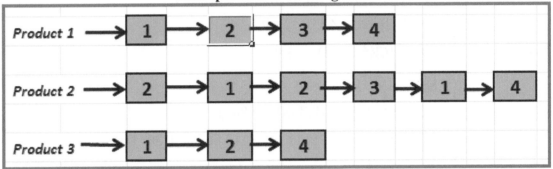

Workstation 1:

$$CV_a^2(1) = (\tfrac{v_{1,1,1}}{\lambda_1})\, CV_a^2(1,1,1) + (\tfrac{v_{1,1,3}}{\lambda_1})\, CV_a^2(1,1,3) + (\tfrac{\lambda_{21}}{\lambda_1}) \{ (\tfrac{\lambda_{21}}{\lambda_2})\, CV_d^2(2) + (1 - \tfrac{\lambda_{21}}{\lambda_2}) \}$$
$$+ (\tfrac{\lambda_{31}}{\lambda_1}) \{ (\tfrac{\lambda_{31}}{\lambda_3})\, CV_d^2(3) + (1 - \tfrac{\lambda_{31}}{\lambda_3}) \}$$

$$CV_a^2(1) = (\tfrac{2}{10.5})(1) + (\tfrac{2.5}{10.5})(1) + (\tfrac{3}{10.5}) \{ (\tfrac{3}{10.5})\, CV_d^2(2) + (1 - \tfrac{3}{10.5}) \}$$
$$+ (\tfrac{3}{10.5}) \{ (\tfrac{3}{5})\, CV_d^2(3) + (1 - \tfrac{3}{5}) \}$$

$$CV_a^2(1) = 0.747 + 0.0816\, CV_d^2(2) + 0.1714\, CV_d^2(3)$$

$$CV_d^2(2) = 1 + (1 - u_2{}^2)(\, CV_a^2(2) - 1\,) + \tfrac{u_2{}^2}{\sqrt{c_2}} [\, CV_s^2(2) - 1\,]$$

$$CV_d^2(2) = 1 + [1 - (.943)^2\,][\, CV_a^2(2) - 1\,] + \tfrac{(.943)^2}{\sqrt{5}} (\,0.631 - 1\,)$$

$$CV_d^2(2) = 0.7425 + 0.1108\, CV_a^2(2)$$

$$CV_d^2(3) = 1 + (1 - u_3{}^2)(\, CV_a^2(3) - 1\,) + \tfrac{u_3{}^2}{\sqrt{c_3}} [\, CV_s^2(3) - 1\,]$$

$$CV_d^2(3) = 1 + [\,1 - (.858)^2\,][\, CV_a^2(3) - 1\,] + \tfrac{(.858)^2}{\sqrt{2}} (\,0.881 - 1\,)$$

$$CV_d^2(3) = 0.7012 + 0.2368\, CV_a^2(3)$$

The remaining Equations for the $CV_a^2(2)$, $CV_a^2(3)$ and $CV_a^2(4)$ values are constructed in the same way. These expressions will be left as a homework exercise at the end of this Chapter. Once these Equations are constructed, there will be 4 linear equations in 4 unknowns. Once solved, the following results are obtained.

$$CV_a^2(1) = 0.976 \quad CV_a^2(2) = 1.157 \quad CV_a^2(3) = 0.938 \quad CV_a^2(4) = 1.017$$
$$CV_d^2(1) = 1.307 \quad CV_d^2(2) = 0.870 \quad CV_d^2(3) = 0.922 \quad CV_d^2(4) = 1.00$$

The reader can verify that these values are indeed correct by substituting them into the above equations and showing that they hold.

11.4.5 Workstation Performance Measures

Since all the processing Workstations contain multiple servers, the following approximation must be used to recover the Time in Queue at each Workstation j.

$$WQ_j = [\frac{CV_a^2(j) + CV_s^2(j)}{2}] \{WQ_J \text{ (M/M/C)}\} \tag{11.42}$$

For the G/G/C processing centers,

Let $u_j = (\frac{\lambda_j}{c_j}) * S_j$

Then: $$WQ_j = [\frac{CV_a^2(j) + CV_s^2(j)}{2}][\frac{u_j^{\sqrt{2c_j+2-2}}}{c_j}][\frac{u_j}{1-u_j}] * S \tag{11.43}$$

The time in system is easily recovered by adding the expected processing time at each Workstation J times to the expected queue time. As usual, Little's Law can be used to recover other Workstation performance measures and the System Performance Measures.

Workstation 1:

The Waiting time per part in Queue 1 is given by;

$$WQ_1 = [\frac{CV_a^2(1) + CV_s^2(1)}{2}][\frac{u_1^{\sqrt{2c_1+2-2}}}{c_1}][\frac{u_1}{1-u_1}] * S$$

$$WQ_1 = [\frac{0.976 + 1.859}{2}][\frac{0.901^{\sqrt{2*5+2-2}}}{5}][\frac{0.901}{1-0.901}] *(0.429)$$

$$WQ_1 = 0.95 \text{ hrs}$$

The WIP in Queue 1 can be calculated using Little's Law

$$LQ_1 = \lambda_1 * WQ_1 = 10.5(0.950) = 9.978$$

It follows that the expected time in Workstation 1 is the sum of the expected Queue waiting time and the expected service time.

$$TPT_1 = WQ_1 + S_1 = 0.950 + 0.429$$
$$TPT_1 = 1.379$$

The WIP at Workstation 1 (Queue + service) can also be calculated using Little's Law.

$$L_1 = \lambda_1 * TPT_1 = (10.50)(1.379)$$
$$L_1 = 14.483 \text{ units}$$

Note that $L_1 = LQ_1 + \rho_1 = 9.978 + 4.505 = 14.483$ as above.

The procedure for Workstations 2-4 is exactly the same, since every Workstation has more than one server. For Workstations 2-4 we obtain:

369

$$WQ_2 = 1.228 \qquad TPT_2 = 1.677 \qquad LQ_2 = 12.889 \qquad L_2 = 17.606$$
$$WQ_3 = 0.884 \qquad TPT_3 = 1.227 \qquad LQ_3 = 4.418 \qquad L_3 = 6.135$$
$$WQ_4 = 0.408 \qquad TPT_4 = 0.838 \qquad LQ_4 = 3.061 \qquad L_4 = 6.286$$

11.4.6 System Performance Measures

The system WIP can be recovered by summing all the Workstation WIP.

$$WIP_{Sys} = \sum_{j=1}^{4} L_j = 14.483 + 17.606 + 6.135 + 6.286$$
$$WIP_{Sys} = 44.510$$

Since the total system outflow ($\lambda_{Sys\,out}$) equals the total system inflow
($v_1 + v_2 + v_3$) = 7.50, then the system average flow time for each part is given by:

$$TPT_{Sys} = \frac{WIP_{Sys}}{\lambda_{Sys\,out}} = \frac{44.510}{7.50} = 5.935 \text{ Hours}$$

11.4.7 Individual Product Performance Measures

The performance measures by individual part type can be obtained with a little more work by applying Little's Law to each Workstation by product. Let the load at Workstation j by product i be denoted by $\lambda_{i,j}$. The throughput time for product i at Workstation j will be denoted by $TPT_{i,j}$. The $TPT_{i,j}$ will be composed of two parts, the expected time in queue $WQ_{i,j}$, and the expected service time $S_{i,j}$ for each visit to Workstation j by product i. The queue waiting time delay will be Workstation, product and Step dependent. All delays at a single Workstation must be determined by product, by step in the routing sheet. The necessary calculations are summarized in the following three tables, one for each product.

Table 11.10
Workstation Data: Product 1

Workstation	Queue Time per visit	Number of Visits	Total Queue Time per Workstation	Total Service Times on Routing Sheet	Service Time Per Part	Total Time in Workstation
1	0.95	1	0.95	0.25	0.25	1.20
2	1.228	1	1.228	0.50	0.50	1.7228
3	0.8840	1	0.8840	0.333	0.333	1.217
4	0.448	1	0.448	0.40	0.40	0.808

The Total Throughput Time for Product 1 ={Total queue time} + {Total time in Workstations}=
$TPT_1 = 4.953$ Hrs
Since the arrival rate to Workstation 1 is λ_1=2.0, then by conservation of flow, this is also the output rate. By Little's Law:
$WIP_{System}^{Product\ 1} = TPT_1 *$ Rate of Input = 4.953(2) = 9.906

370

Table 11.11
Workstation Data: Product 2

Workstation	Queue Time Per visit	Number Of Visits	Total Queue Time per Workstation	Service Time by Step Visit	Total Service Time	Time in System
1	0.95	2	1.90	(1)*0.40 + .310	0.710	2.610
2	1.228	2	2.456	0.333 + .30	0.633	3.1390
3	0.8840	1	0.8840	0.350	0.350	1.234
4	0.408	1	0.408	0.60	0.60	1.008

Total Throughput Time for Product 2 = 7.991 Hours and $WIP_{System}^{Product\ 1} = 23.97$ Units

Table 11.12
Workstation Data: Product 3

Workstation	Queue Time Per visit	Number Of Visits	Total Queue Time per Part	Service Time per Part	Total Service Time	Time in System
1	0.95	1	0.95	0.75	0.75	1.70
2	1.228	1	1.228	0.667	0.667	1.895
3	0.8840	0	0.00	0.00	0.00	0.00
4	0.408	1	0.408	0.250	0.250	0.658

Total Throughput Time for Product 3= 4.253 Hrs and $WIP_{System}^{Product\ 1}= 10.633$ Units

Note that: $WIP_{System}^{Product\ 1}+ WIP_{System}^{Product\ 2}+ WIP_{System}^{Product\ 3}= L_{System}= 44.510$ Units as expected.

11.5   Group Technology and Focused Factories

In the mid-1970s a movement to form manufacturing cells which were designed around part families was popularized and advanced by Burbridge and several other researchers under the general umbrella of Production Flow Analysis (PFA). The general idea of PFA is to divide the factory into manufacturing cells configured for a single product or for a family of products with similar processing requirements…hence the term *Focused Factory* or a *Factory within a Factory*. Each cell possesses all of the equipment and resources necessary to complete the part for which it was configured to produce. Since first introduced, the term PFA has come to be associated with largely independent manufacturing cells, dedicated to one or a few products which share the same production flow routing sequence, or jointly use most if not all of the cell's machining capabilities. Forming a Focused Factory involves three main engineering activities. (1) Determining a part or group of parts that can be isolated into a single integrated manufacturing cell (2) Determining how that cell should be aligned, arranged and operated in terms of shape, structure, automation and operation. (3) Designing the cell to take advantage of similarities between operations and part routing structures. One of the outgrowths of a Focused Factory has been the emergence of U-Shaped manufacturing cells designed in such a way that output is determined by labor assignment and not strictly by machine capacity. This type of manufacturing cell is discussed in Chapters 1-6. There are three basic approaches to forming a family of products; (1) Visual analysis (2) Coding and classification methodologies and (3) Cluster analysis. Each approach looks for commonalities in product routing structures, product features, raw materials required and common manufacturing processes required to produce a final product. Visual analysis can be used for a small number of manufactured products (3-5). Coding and Classification methods tend to be very expensive and depend heavily upon preexisting descriptive data. Burbridge developed a technique called Cluster Analysis which did not require extensive data descriptions and used shop floor knowledge readily available from shop floor supervisors and operators. We will not explore any of these techniques in this section, but refer the interested reader to almost any Production Planning and Control textbook for detailed treatment of this subject. The issue of concern in this section is how to analyze and compare a Focused Factory to a traditional manufacturing layout. The tools which we have developed in this Chapter are more than adequate for this task. The main reason for considering the formation of a Focused Factory is to improve production performance measures, such as Throughput time and Production Rates. These goals will be extended in Chapters 12 and 13 to minimize work in progress, reduce queue time and support Pull philosophies. We have discussed elsewhere using Little's Law how WIP reduction is positively correlated to machine utilization, which correspondingly affects both throughput time and output rates. Since a key factor in minimizing WIP and reducing Throughput time is maintaining Workstation utilization below 90%, these performance measures at key machines can be reduced by how a cell partitions itself in terms of both machine configuration and usage. It should be recognized that like all *Lean Manufacturing Facilities*, the efficient and effective operation of a Focused Factory requires a goal of zero defects (TQM), just in time material delivery (JIT), setup time reduction (SMED) and high machine availability (RAM). These issues are addressed elsewhere in this book. We will concern our self here with process modeling.

11.5.1 Modeling a Focused Factory

Consider a Focused Factory proposed by Curry. The manufacturing facility has been designed to produce 4 different products using five different types of machines. Table 11.13 has been constructed using Production Flow Analysis. The "1" in each column indicates that Part type "p" requires Machine "k" in its process sequence. The actual step on which Machine k is being used is not important to group identification. The product routing sheets are shown in Figure 11.13.

Table 11.13
Machines Used by Part Type

Machines	1	2	3	4	5
Part Type					
1	1	1	1	0	0
2	1	1	1	0	0
3	0	0	1	1	1
4	0	0	1	1	1

It is obvious that Parts 1 and 2 can each be produced in one Manufacturing cell, and Products 3 and 4 in a second dedicated cell. The issue concerns the dispensation of Machine three, which would be required in both cells. There are two design alternatives which might be considered: (1) Share machine 3 with both cells. This will require a Kaizen event to reduce the set up and changeover time to a minimum, so that valuable production time in each cell would not be lost. Of course, high availability and zero defects at machine three will need to also be a primary goal. (2) The second alternative is to purchase or procure duplicates of machine Type 3, and locate one or more in both cells. Due to the dynamics of cell and product interactions, the correct decision should rest upon how each alternative can be characterized in terms of system performance measures. This will require a multiproduct flow analysis. Of course, the economics of balancing operating efficiency against capital investment costs will also be a determining factor, as well as automation and Material Handling requirements, but this will not be a subject of this analysis. Figure 11.5 illustrates the Process Flow diagram for each product. Table 11.14 contains product processing requirements. In order to focus on system configuration options, we will use $CV_s^2(i, j, p) = 1.20$ for each processing step time and $CV_a^2(i, j, p) = 1.0$ for all external flow streams.

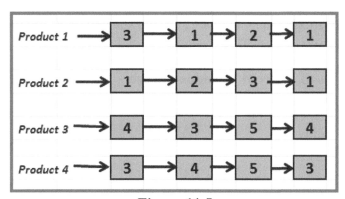

Figure 11.5
Process Flow Diagrams

Table 11.14
Processing Times and CV_s^2 Values

	Step	1	2	3	4
Product 1	Workstation	3	1	2	1
	$s_{i,j,1}$	8.0	6.0	4.50	6.0
	$CV_s^2(i,j,1)$	1.20	1.20	1.20	1.20
Product 2	Workstation	1	2	3	1
	$s_{i,j,2}$	5.0	6.0	8.0	4.0
	$CV_s^2(i,j,2)$	1.20	1.20	1.20	1.20
Product 3	Workstation	4	3	5	4
	$s_{i,j,3}$	2.0	4.0	8.0	4.0
	$CV_s^2(i,j,3)$	1.20	1.20	1.20	1.20
Product 4	Workstation	3	4	5	3
	$s_{i,j,4}$	7.0	3.0	2.0	4.0
	$CV_s^2(i,j,4)$	1.20	1.20	1.20	1.20

$CV_a^2(1, 3, 1) = 1.0 \quad CV_a^2(1, 1, 2) = 1.0 \quad CV_a^2(1, 4, 3) = 1.0 \quad CV_a^2(1, 3, 4) = 1.0$

$$v_{1,3,1} = 0.064 \text{ parts/hr} \quad v_{1,1,2} = 0.096 \text{ parts/hr}$$
$$v_{1,4,3} = 0.080 \text{ parts/hr} \quad v_{1,3,4} = 0.10 \text{ parts/hr}$$

We will construct and analyze the two different process flow configurations that could be used to manufacture these four products. The first is an integrated Factory Model in which products 1 and 2 are produced in one manufacturing cell, and products 3 and 4 in a second cell. The cells are not completely independent, since they will share machine number 3. The second alternative is to completely split products 1 and 2 from products 3 and 4, forming two independent cells. Each cell will need to have enough Machine 3 capacity to operate independently, and as we will see this will require a capital investment decision.

Model 1: Two Manufacturing Cells Sharing Machine 3

The models can be easily constructed and analyzed using the modeling methodologies previously introduced in this Chapter. The 4 process flow diagrams shown in Figure 11.5 will be used to describe the independent product routings for each part. The difference in modeling strategies rests in the combined or independent analysis of two dedicated cells and the use of Machine 3. Model 1 will represent both cells sharing machine 3.

Product Flow Load, Offered Workload Servers Required and Machine Utilizations

The Flow rates are as follows.

$$\begin{array}{lll}
f_{1,3,1,1} = 0.064 & f_{2,1,2,1} = 0.064 & f_{3,2,1,1} = 0.064 \\
f_{1,1,2,2} = 0.096 & f_{2,2,3,2} = 0.096 & f_{3,3,1,2} = 0.096 \\
f_{1,4,3,3} = 0.08 & f_{2,3,5,3} = 0.08 & f_{3,5,4,3} = 0.08 \\
f_{1,3,4,4} = 0.10 & f_{1,4,5,3} = 0.10 & f_{3,5,3,3} = 0.10
\end{array}$$

$$\lambda_{k,j} = \sum_{i=1}^{N} \sum_{p=1}^{P} f_{i,k,j,p}$$

$$\lambda_{3,1} = f_{1,3,1,1} + f_{3,3,1,2} = 0.064 + 0.096 = 0.160$$
$$\lambda_{1,2} = f_{2,1,2,1} + f_{1,1,2,2} = 0.064 + 0.096 = 0.160$$

Similarly: $\quad \lambda_{2,1} = 0.064 \quad \lambda_{2,3} = 0.096 \quad \lambda_{4,3} = 0.08 \quad \lambda_{3,5} = 0.08$
$$\lambda_{5,4} = 0.08 \quad \lambda_{3,4} = 0.10 \quad \lambda_{4,5} = 0.10 \quad \lambda_{5,3} = 0.10$$

$$\lambda_j = \sum_{k=1}^{M} \lambda_{k,j} + \sum_{p=1}^{P} v_{1,j,p}$$

$$\lambda_1 = \lambda_{2,1} + \lambda_{3,1} + v_{1,1,2} = 0.064 + 0.160 + 0.096 = 0.320$$
$$\lambda_1 = 0.320$$

Similarly: $\quad \lambda_2 = 0.160 \quad\quad \lambda_3 = 0.440 \quad\quad \lambda_4 = 0.260 \quad\quad \lambda_5 = 0.180$

Using a Product flow matrix.

Table 11.15
Total Flow Loads to each Workstation

Workstation	1	2	3	4	5	
Product 1	2	1	1	0	0	$v_1 = .064$
Product 2	2	1	1	0	0	$v_2 = .096$
Product 3	0	0	1	2	1	$v_3 = 0.08$
Product 4	0	0	2	1	1	$v_4 = 0.10$
λ_j	.320	0.160	0.440	0.260	0.180	

$$\lambda_1 = 0.320 \quad \lambda_2 = 0.160, \quad \lambda_3 = 0.440, \quad \lambda_4 = 0.260, \quad \lambda_5 = 0.180$$

The offered workload to each Workstation is the product of the input demand streams at each Workstation times the expected processing times by step. These results confirm the flow rates previously calculated.

The Offered Workload for this example is as follows.
$$\rho_j = \sum_{p=1}^{P} \{v_p * S_{1,j,p}\} + \{ v_p \sum_{k=1}^{M} \sum_{i=1}^{N-1} S_{i+1,j,p}\} \qquad j= 1, 2 \dots M$$
$$\rho_1 = \{v_1 * S_{1,2,1}\} + v_2 \{ S_{2,1,2} + S_{5,1,2}\}$$
$$\rho_1 = 0.064 (6.0+6.0) + 0.096 (5.0+4.0) = 1.632$$
Similarly: $\rho_2 = 0.864 \quad \rho_3 = 2.70 \quad \rho_4 = 0.780 \quad \rho_5 = 0.840$

Hence; Workstation one requires 2 servers, Workstations two, four and five 1 server, and Workstation three 3 servers.

The Workstation Utilizations are given by: $u_j = \rho_j / c_j$ j=1, 2, 3, 4
$$u_1 = 0.816 \quad u_2 = 0.864 \quad u_3 = 0.90 \quad u_4 = 0.78 \quad u_5 = 0.847$$

The Composite Processing Workstation Service Times and the $CV_s^2(j)$ j=1,2...5
The composite service Workstation Processing times are given by:

$$S_j = \sum_{p=1}^{P} \{ (\frac{v_{1,j,p}}{\lambda_j}) S_{1,j,p} + \sum_{k=1}^{M} \sum_{i=1}^{N-1} \left(\frac{f_{i,k,j,p}}{\lambda_j}\right) S_{i+1,j,p}\} \qquad j=1, 2 \dots M$$

$$S_j^2 = \sum_{p=1}^{P} \{ (\frac{v_{1,j,p}}{\lambda_j}) S_{1,j,p}^2 [CV_S^2(v_1, j, p) + 1.0]$$

$$+ \sum_{k=1}^{M} \sum_{i=1}^{N-1} (\frac{f_{i,k,j,p}}{\lambda_j}) S_{i,k,j,p}^2 [CV_S^2(i, j, p) + 1.0] \} \qquad j=1, 2 \dots M$$

And by definition: $CV_S^2(j) = \frac{S_j^2}{[S_j]^2} - 1.0 \qquad j=1, 2 \dots M$

The calculated S_j values are as follows.
$$S_1 = 5.10 \quad S_2 = 5.40 \quad S_3 = 6.130 \quad S_4 = 3.00 \quad S_5 = 4.667$$
The S_1^2 and the $CV_s^2(1)$ values are:
$$S_1^2 = 40.05 \quad \text{and} \quad CV_s^2(1) = 0.540$$
The reader will be asked to verify as homework exercise that:
$$CV_s^2(2) = 0.528 \quad CV_s^2(3) = 0.631 \quad CV_s^2(4) = 0.603 \quad CV_s^2(5) = 1.112$$

The CV_a^2 (j) Values for j=1, 2...5

The $CV_a^2(j)$ values for the arrival stream to each service Workstation are calculated as before by formulating and solving a system of linear equations.

$$CV_a^2(j) = \sum_{p=1}^{P} \left(\frac{v_{1,j,p}}{\lambda_j} \right) CV_a^2(1,j,p)$$

$$+ \sum_{k=1}^{M} \left(\frac{\lambda_{kj}}{\lambda_j} \right) \left[\left(\frac{\lambda_{kj}}{\lambda_k} \right) CV_d^2(k) + \left(1 - \frac{\lambda_{kj}}{\lambda_k} \right) \right]$$

Finally, the value of $CV_d^2(k)$ depends upon whether Workstation j has one server or multiple servers.

$$CV_d^2(k) = (1-u_k^2) \, CV_a^2(k) + u_k^2 \, CV_s^2(k) \qquad\qquad c_k = 1$$

$$CV_d^2(k) = 1+(1-u_k^2) \left[CV_a^2(k)-1 \right] + \frac{u_k^2}{\sqrt{c_k}} \left[CV_s^2(k) - 1 \right] \qquad c_k > 1$$

Workstation 1

$$CV_a^2(1) = \left(\frac{v_{1,1,2}}{\lambda_1} \right) CV_a^2(\gamma, 1, 2) + \left(\frac{\lambda_{31}}{\lambda_1} \right) CV_a^2(3,1) + \left(\frac{\lambda_{21}}{\lambda_1} \right) CV_a^2(2,1)$$

$$CV_a^2(1) = 0.30 + 0.50 \, CV_a^2(3,1) + 0.20 \, CV_a^2(2,1) \qquad\qquad (11.44)$$

$$CV_a^2(3,1) = \left(\frac{\lambda_{31}}{\lambda_3} \right) CV_d^2(3) + \left(1 - \frac{\lambda_{31}}{\lambda_3} \right)$$

$$CV_a^2(3,1) = \left(\frac{.16}{.14} \right) CV_d^2(3) + \left(1 - \frac{.16}{.14} \right) = 0.3636 \, CV_d^2(3) + 0.6364 \qquad (11.45)$$

$$CV_d^2(3) = 1 + (1-u_3^2)(\, CV_a^2(3) - 1) + \frac{u_3^2}{\sqrt{c_3}} \left[CV_s^2(3) - 1 \right]$$

$$CV_d^2(3) = 1 + (1-.90^2)(\, CV_a^2(2) - 1) + \frac{.90^2}{\sqrt{3}} \left[CV_s^2(2) - 1 \right]$$

$$CV_d^2(3) = 0.6374 - 0.19 \, CV_a^2(3) \qquad\qquad (11.46)$$

$$CV_a^2(2,1) = 0.40 \, CV_d^2(3) + 0.60 \qquad\qquad (11.47)$$

Combining equations by back substituting, a single equation in the unknown variables $CV_a^2(1)$ and $CV_a^2(2)$ is obtained. Following the same procedure, the other four equations in the unknown $CV_a^2(j)$ j=1, 2….5 variables can be generated. These five equations are linear in the five unknown $CV_a^2(j)$ values. Solving these equations will yield the following values.

$$CV_a^2(1) = 0.9360 \qquad CV_a^2(2) = 0.8810 \qquad CV_a^2(3) = 0.9440$$
$$CV_a^2(4) = 0.992 \qquad CV_a^2(5) = 0.933$$

System Performance Measures

The system can now be decomposed into individual Workstation analysis.

$$THP_j = \left[\frac{CV_a^2(j) + CV_s^2(j)}{2} \right] \left[\frac{u_j}{1-u_j} \right] \left[\frac{u_j \sqrt{2c_j+2-2}}{c_j} \right] + S_j \qquad j=1,\dots,5$$

To demonstrate the calculations, the Cycle Time in Workstation 1 is as follows.

$$THP_1 = \left[\frac{CV_a^2(1) + CV_s^2(1)}{2} \right] \left[\frac{u_1}{1-u_1} \right] \left[\frac{u_1 \sqrt{2c_1+2-2}}{c_1} \right] + S_1$$

$$THP_1 = [0.936 + 0.540] \left[\frac{0.816}{1-0.816} \right] \left[\frac{0.816 \sqrt{2(2)+2-2}}{2} \right] + 5.1 \qquad THP_1 = 12.716 \text{ hrs}$$

By Little's Law;
$$L_1 = \lambda_1 * THP_1$$
$$L_1 = 0.32(12.716) = 4.069$$

The values of L_2, L_3 and L_4 can be calculated in the same way. It will be left as an exercise to show that:
$$WIP_{Sys} = \sum_{j=1}^{5} L_j = 25.608 \text{ units}$$

The total external flow into the system is:
$$\lambda_{Sys} = \sum_{i=1}^{4} \nu_i = 1.36$$

Hence, the average time that a job spends in the system is:
$$\text{TPT}_{Sys} = \frac{WIP_{Sys}}{\lambda_{Sys}} \qquad TPT_{Sys} = \frac{25.680}{1.36} = 18.88 \text{ hrs}$$

Model 2: Two Manufacturing Cells Operating Independently

The alternative that we now consider is to separate the four products into two independent cells. The performance characteristics we will compare are System WIP and the System Throughput time. Products 1 and 2 will be placed in Cell 1, and Products 3 and 4 into Cell 2. It is obvious that both Cells will need stand-alone Machine 3 capacity. From Table 11.15, we can easily determine that Cell 1 will have an offered flow load to Machine 3 of 1.28 units/hr, and Cell 2 will have a load of 1.421 units/hr to Machine 3. It is obvious that both Cell 1 and Cell 2 will require two machines of Type 3. Since it is only necessary to have three machines of type 3 in a shared arrangement, there are two issues to be addressed. The first is to determine if there is an operational advantage in operating two independent cells. The most reasonable way to answer this question is to compare performance measures between the two options. This is precisely what we plan to do. The second issue is an economic one. Should we purchase or procure another machine of Type 3? The economic justification is clearly one of capital expenditure, but should also be based upon the operational impact of improved performance. Modeling both Cells as separate facilities, we obtain the following results.

Table 11.16
Cell Comparison

	Cell 1 Machines 1 and 2	Cell 2 Machines 3 and 4	Combined Operation
System WIP (Jobs)	10.543	10.943	25.608
System Cycle Time (Hrs/Job)	65.896	60.792	75.530
System Throughput (Parts/Hr)	0.160	0.180	18.88

The analysis is revealing. The independent operation of two cells yields a lower Throughput Time in each cell than the Focused Factory using two dependent cells which share Workstation 3. There is a significant improvement in all system performance measures using independent cells. The system output rate is exactly the same since there are no product losses. The improvement in system Throughput times is largely due to the reduced WIP in the independent cell operations. Since in-process inventory costs time, money and floor space, this could be a significant advantage. Reexamining the system configuration requirements, one might be tempted to look at the combined cell operation using four machines rather than the three required. Since machine three utilization is $u_3 = 0.90$, we might suspect that more processing capacity would reduce both Throughput time and system WIP. This being the case, we proceed with this new analysis. The new machine three utilization is now $u_3 = 0.675$. The modeling results are as follows.

Table 11.17
Single Factory, Four Machines of Type 3

	Combined Operation
System WIP	**20.582 parts**
System Cycle Time	**60.535 hours**
System Throughput	**0.34 Parts/Hour**

The results are enlightening, but not unexpected to the Lean Engineer. As conjectured, both the System WIP and the System Throughput Time per part are lower than the two original system configurations. One of the things that should be noted is that by lowering the utilization of Machine 3, all system performance measures improve. This configuration also has more capacity to expand from the standpoint of the *key machine*, which is Machine 3. This provides a cushion against market growth and periodic fluctuations in product demand. Another important point is that simply partitioning the factory into separate cells will always result in better system performance measures, but this is not the only reason to embrace such an organizational change.

Elsewhere in this book we point out that this now paves the way to implement a *Pull System*, and to minimize cell WIP and Throughput time by letting each cell be *labor paced* and not *machine paced*. Ultimately, WIP and Throughput time can be further reduced with Kaizen events, setup time reduction, workplace ergonomic studies and operator training. Of course, these potential gains do not negate the necessity to execute modeling and analysis of both the *to- be* and *as-is* operations to make intelligent decisions. An alternate, simpler way, to quickly assess system performance is through the use of *Value Stream Mapping*, a topic we have covered in Chapter 8.

11.6 Summary
This Chapter has developed a modeling methodology for general Manufacturing Systems based upon Queuing Network Analysis. The modeling paradigm is targeted to multiple product flow systems for which the routing and processing characteristics are based upon Process Routing Sheets, which are commonly used in most manufacturing industries. A wide variety of problems were solved, but each was directed to determining the WIP levels, throughput times and other performance measures for steady state performance. A common objective was to demonstrate that the design of Greenfield facilities and the analysis of an existing Manufacturing facility can be guided by modeling analysis to both predict system performance and to determine the level of critical resources required to satisfy product demand(s). Personal intuition and experience can

play a major role in the Lean transformation process, but intuition can often fail to capture the wide variety of interaction effects and product dependencies that exit in a multi-Workstation, multi-product manufacturing environment. The basic modeling methodologies in this Chapter can be extended to include the impact of poor quality, maintenance, batching and set-up times. This is the subject of Chapter 12.

Review Questions

1. Data for a three product flow system is shown below with $v_1=6.2$ parts/hr, $v_2=4.0$ parts/hr and $v_3=2.0$ parts/hr. Write down the values for the $f_{i,k,j,p}$ values for i=1,5 j=1,4 k=1-4 and p=1-3. Using these values, calculate the $\lambda_{k,j,p}$, $\lambda_{k,j}$, and the λ_j value

	Step	1	2	3	4	5	External Flow
	Station	3	2	3	2	4	
Product 1	$S_{i,j,1}$	0.18	0.33	0.22	0.42	0.28	$v_1=6.2$
	$CV_S^2(i,j,1)$	1.15	0.98	1.62	2.34	1.11	$CV_a^2(1,3,1) = 1.5$
	Station	1	3	1	3	4	
Product 2	$S_{i,j,1}$	0.22	0.25	0.15	0.19	1.12	$v_2=4.0$
	$CV_S^2(i,j,1)$	1.0	1.24	1.87	1.25	0.98	$CV_a^2(1,1,2) = 1.5$
	Station	3	1	1	2	3	
Product 3	$S_{i,j,1}$	0.31	0.44	0.25	0.38	0.23	$v_3=2.0$
	$CV_S^2(i,j,1)$	1.35	1.56	0.98	2.10	3.2	$CV_a^2(1,3,3) = 1.5$

2. Calculate the Offered Workload and the Workstation Utilization for each Workstation in Problem 1.0.

3. Calculate the Composite Workstation Processing times for each Workstation in Problem 1.

4. Calculate the Composite Workstation Squared Coefficient of Variation for Service Times at each Workstation in Problem 1.0.

5. Explain Equations 11.22 and 11.33. How are they related?

6. Equation 11.21 is created by substituting Equation 11.20 into Equation 11.19. Write out the general form of Equation 11.21 assuming that there is only one server at Workstation j. Write the general form of Equation 11.21 for multiple servers at Workstation j. Verify that the general multiple server equation yields the general single server equation when c=1.

7. Write out and solve the set of equations that are required to solve for the $CV_a^2(j)$ j= 1,2, 3,4 in Problem 1.0.

8. Using the results of Problems 1-7, calculate the (1) Average System Throughput time and (2) the Average System WIP.

9. Calculate the Average System Throughput time and the average system WIP by individual product for the Problem 1 data set. Verify the results of Problem 8.0 by using these results.

10. Calculate the Average Workstation Throughput time and the average system WIP at each individual Workstation by product for the Problem 1 dataset. Calculate the results of Problem 9.0 by using these results.

11   Consider the following data for the "Three Amigos" , three product , multiproduct flow manufacturing facility. (1) How many machines are needed at each Workstation and in the entire manufacturing facility? (2) How much square foot of WIP storage is needed in front of all the Workstations if each unit of Product 1 requires 5 ft^2 of floor space, Product 2 requires 8 ft^2 of floor space and Product 1 requires 6.4 ft^2 of floor space? If each machine requires 20 ft^2, how much floor space is required at each Workstation.

	Step					External Flow
	Station	2	1	3	2	
Product 1	$S_{i,j,1}$	0.26	0.47	0.22	0.52	$v_{1,1,1}$=4.2 Parts/Hr
v_1=4.2 Parts/Hr	$CV_S^2(i,j,1)$	1.25	1.98	1.30	1.34	$CV_a^2(1,1,1) = 1.0$
	Station	3	2	1	3	
Product 2	$S_{i,j,1}$	0.32	0.45	0.35	0.20	v_2=2.0 Parts/Hr
v_2=2.0 Parts/Hr	$CV_S^2(i,j,1)$	1.0	1.54	1.27	1.25	$CV_a^2(1,3,2) = 1.5$
	Station	2	1	1	2	
Product 3	$S_{i,j,1}$	0.31	0.44	0.25	0.38	
v_3=1.0 Parts/Hr	$CV_S^2(i,j,1)$	1.25	1.56	1.98	1.0	$CV_a^2(1,2,3) = 2.5$

12   Define a *Focused Factory*. What is the difference in a Focused Factory and a *Manufacturing Cell*? What would you define as the *Critical Machine* in this context? Discuss the reasons to consider a Focused Factory.

13   For the Focused Factory example in Section 5.7.2, (1) verify that S_1=5.10 and that S_3=6.13 (2) Verify that S_1^2=40.50 and $CV_S^2(1)$=0.540 (3) Verify that $CV_S^2(3)$=0.631 .

14   In Section 10.2.7.4, verify the result given for System WIP.

15   There are two tables that provide the results of a product coding and classification exercise involving 5 products (Nicolas). Determine the key operating performance measures for a combined Manufacturing facility. Suggest Manufacturing Cell configurations, analyze and compare performance measures.

	Step	1	2	3	External Flow
	Station	1	2		
Product 1	$S_{i,j,1}$	0.48	0.80	.42	$v_{1,1,1}$=2 Parts/Hr
	$CV_S^2(i,j,1)$	1.40	1.30	1.00	$CV_a^2(1,1,1) = 1.0$
	Station	1	2		
Product 2	$S_{i,j,2}$	0.30	0.28		$v_{1,3,2}$=3 Parts/Hr
	$CV_S^2(i,j,2)$	1.0	1.54		$CV_a^2(1,1,2) = 1.0$
	Station	1	2	3	
Product 3	$S_{i,j,3}$	.25	.20	.62	$v_{1,1,3}$=3.5 Parts/Hr
	$CV_S^2(i,j,3)$	1.25	.90	1.10	$CV_a^2(1,1,3) = 1.0$
	Station	3	4	5	
Product 4	$S_{i,j,4}$	.78	.69	.35	$v_{1,1,4}$=2.5 Parts/Hr
	$CV_S^2(i,j,4)$	2.2	1.7	1.30	$CV_a^2(1,3,4) = 1.0$
	Station	4	5		
Product 5	$S_{i,j,5}$	.35	.50		$v_{1,1,5}$=5.4Parts/Hr
	$CV_S^2(i,j,5)$	0.85	1.10		$CV_a^2(1,1,5) = 1.0$

Products	1	2	3	4	5
Station					
1	1	1	1		
2	1	1	1		
3			1	1	
4				1	1
5				1	1

380

Chapter 12
The Disruptive Effects of Variance

"There is more to optimizing a system than finding a set of numerical recipes"
Foundations of Optimization, II Edition   Beightler, Phillips and Wilde

12.1 Introduction

We have previously suggested that most problems experienced in manufacturing systems manifest themselves in long throughput times, excessive WIP or high utilization of scarce resources. When these poor performance measures arise, it must be stressed that these are not problems in themselves, but simply symptoms of deeper problems. In most cases, these problems can be traced to operating practices which inflate variance in the system. Until recently, the full impact of variance components inherent in key manufacturing operating policies have either been ignored or poorly understood. Classical statistical analysis provides a way to computationally measure variance.

$$\sigma^2 = \sum_{all\ x} (x - \mu)^2\ p_x \qquad \text{Discrete Random Variables}$$
$$\sigma^2 = \int_{all\ x} (x - \mu)^2\ f(x)dx \qquad \text{Continuous Random Variables}$$

These definitions of variance will measure the magnitude, but provide no help in identifying the *source* of variance or the ***variance components***. To find this out, the Lean Engineer must execute ***Root Cause Analysis***. In Chapters 9-11 of this textbook, we have been concerned with the Mean (μ), Variance (σ^2) and the Squared Coefficient of Variation (SCV) to describe system operating characteristics, including time between arrivals and service times arrival streams and departure streams from individual Workstations. Let us assume that an engineered standard time or a time study analysis has been conducted to determine Workstation Throughput Time in a Manufacturing Workstation. If the Throughput time is a constant, then both σ^2 and the associated SCV are both zero. However, this is rarely if ever the case, even in highly automated systems. Any deviation from either standard times or standard procedures cause *Variance*, and variance can be manifested in many ways. A few operating practices which cause variance to occur are:

- Poor Training of Employees
- Fatigue
- Non-homogeneous material
- Tool wear
- Batch arrivals and batch processing
- Set-up times
- Load and unload times
- Quality problems
- Production delays
- Prioritization and expediting
- Machine Failures
  ....*and many more*

In general, variance components can arise from three main sources. (1) External sources (2) Internal sources and (3) Random influences. **External Variability** is usually due to *poor management control practices*. Raw material deliveries, wildly fluctuating order release quantities, excessive large production lot sizes, poor scheduling practices, and reactive supply chain management are good examples. We have often been told that *I cannot dictate demand patterns, I can only respond to fluctuating customer demand(s) as quickly as possible*. But in most cases, level and balanced production can be achieved by developing consistent order release policies for the shop floor, and design buffer stock levels to offset fluctuating demand patterns. The use of *Supermarkets* to buffer inconsistent input and output flow often help, but the real solution is to work hard with both suppliers and customers (external and internal) to smooth flow. Admittedly, this is a hard row to hoe, but the payoff justifies the effort. **Internal Variability** is usually the result of *poor production control practices*. This is *induced variation*, and usually occurs when the principles of Lean Engineering are either ignored or violated. Examples of poor production control practices abound, and their elimination is in fact the reason why we wrote this textbook. Poor order release practices, large batch moves, inconsistent material handling practices, lack of Lean work cell design principles, expediting/priority routing schemes, poor Quality Control and failure to execute real preventative maintenance practices are a few common examples. The third category we will simply call *random variation*. No matter how much effort is expended to reduce variance, there will always be interruptions and influences which simply cannot be eliminated. In this Chapter we will concentrate on demonstrating the corrupting influence that these three categories of variance exert on system performance measures, and suggest that these variance components can be reduced and controlled.... sometimes eliminated. We will attempt to characterize the impact of variance on our well known system performance measures of WIP, Throughput Time and Output rate. In order to demonstrate the corrupting influence of variance, we will focus on only a few specific case studies due to time and space. We call these detractors our *Big Six* since they are almost always present in a manufacturing environment, even one that engages in Lean Engineering practices. The *Big Six* in our opinion are as follows.

1.0   *Failure to reduce variance in customer demand rates or time between arrivals*
2.0   *Inducing Variance by Poor Order Release Policies*
3.0   *The disruptive effects of poor maintenance practices*
4.0   *The disastrous effects of bad quality and rework*
5.0   *The Consequences of Batch Processing*
6.0   *The effects on capacity of long Set-Up and changeover Times*

We will investigate each of these topics and demonstrate the effect that these practices have on variance components. Remember....Lean Manufacturing *Enemy number one* is not variety, but the effects of **variance.**

12.2 Failure to Reduce Variance in Time Between Arrivals

Since our definition of a *system* can accommodate a single machine, a subassembly cell or even an entire factory; the concept of an *arrival* and a *service* is equally robust. In general, arrivals can be from *External or Internal* sources. Some authors have ventured to state that external arrivals are beyond management control, and so we simply accept and try to manage *random variations*. While this is sometimes true, it does not mean that we cannot try to understand and address the problems which uncertainty creates, and the associated impact on key performance metrics. The impact that variance has on Workstation and System performance measures will vary from system to system, but the impact is always the same. As variance components are increased, the system will experience increased Throughput times, inflated WIP and reduced output capacity. The important fields of *Supply Chain Management* and *Just-in-Time* are largely concerned with controlling and managing external flows. In our previous models, we have captured variability in the demand/supply stream to individual Workstations with the CV_a^2, which is defined by σ_a^2 and μ_a^2. The point is that while μ_a^2 might vary considerably in response to changing customer demands, σ_a^2 can always be reduced with sound production practices and targeted Kaizen events. Since the value of σ^2 always appears in the numerator of any SCV, if standard practices and standard processing times are established by a industrial engineering, any reduction in the Variance will always have a beneficial effect on WIP, Throughput time and output. If the arrival stream is internal (machine to machine), then, σ_a^2 can *always* be reduced using the Lean Engineering practices discussed in this book. In fact, whether implicitly or explicitly stated, variance reduction is at the heart of all principles presented in this book. In our analysis of either individual Workstations or manufacturing networks of Workstations, we have relied heavily upon first calculating the Expected Time in Queue, and then recovering other performance measures by using Little's Laws. Consider again the expected time in queue for a G/G/1 service system.

$$WQ \ (G/G/1) = \{ \frac{CV_a^2 + CV_s^2}{2} \} \ (\frac{u}{1-u}) * S \qquad (12.1)$$

Equation 12.1 has several interesting properties. It yields an exact solution for the M/M/1 queue since $CV_a^2 = CV_s^2 = 1.0$. Observe that this equation can be nicely divided into three terms which we will call the T^3 influence.

A *Variance* Component: $\qquad T_1 = \{ \frac{CV_a^2 + CV_s^2}{2} \}$

A *Workstation* Component: $\quad T_2 = (\frac{u}{1-u})$

A *Service* Component: $\qquad\quad T_3 = \ S$

Consider the influence of T_1. Since $CV^2 = \frac{\sigma^2}{\mu^2}$, it is clear that the expected waiting time in queue can be reduced by reducing either σ_a^2 or σ_s^2. Any variance reduction in either the time between arrivals or the service times will reduce the queue waiting time. Since Little's Law is $L_q = \lambda * W_q$, the expected number of units in the queue will also be reduced for a fixed demand rate. However, note that in the CV_s^2 term, the expected service time is in the denominator, and any reduction in the mean service time without reducing variance will inflate CV_s^2. This justifies the necessity to clearly develop best work practices and standard times for all workstation activities. Now consider T_2. From this term, it can be deduced that as u approaches one, the expected time in queue is drastically increased, particularly for $0.80 < u < 1.0$. We have already noted this phenomenon in Chapter 9. But note that the expected service time is also part of the server utilization. Finally, consider the term T_3. Any direct reduction in the expected time in service seems to indicate that the expected time in queue is directly reduced. However, as previously noted, this is not a linear benefit because both CV_s^2 and u both depend upon the expected service time, S. Consider a numerical example. Assume that $\lambda = 17$ arv/hr, $\mu = 18$ serv/hr, $CV_a^2 = 1.20$ and $CV_s^2 = 1.40$. Since the Server utilization is $\rho = u = \frac{\lambda}{\mu}$, then $u = \frac{17}{18} = 0.944$. Suppose a Kaizen study increases μ to 24 serv/hr, a 25% increase in rate of service. In this case, $u = \frac{17}{24} = 0.7083$. The corresponding $CV_s^2 = 2.4832$ Comparing the old and new values of the Expected Time in Queue, we obtain:

Old: $\quad WQ_{Old} = \{ \frac{1.20+1.40}{2} \} (\frac{0.9444}{1-0.9444}) (\frac{1}{18})(60) = 73.05$ min

New: $\quad WQ_{New} = \{ \frac{1.20+2.4832}{2} \} (\frac{0.7083}{1-0.7083}) (\frac{1}{24})(60) = 16.33$ min

The results are unpredictable and dramatic to say the least. Even though the CV_s^2 has increased by 1.0832, the expected waiting time in queue has significantly decreased from 73.05 min to 16.33 min. This points out the benefits in understanding how variance components influence system behavior and the need to model workstation behavior.

12.3   Inducing Variance by Poor Order Release Policies

In Chapters 9-11 we often made the assumption that time between arrivals followed an Exponential distribution. We have seen that since the Exponential assumption results in $CV_a^2 = 1.0$, the arrival stream exhibits moderate variation. Aside from the nice properties which the Poisson density function exhibits in Markovian systems, the Poisson assumption is often justified in actual practice. Job shop operations and remanufacturing operations make a variety of products from multiple sources, so that the Poisson/Exponential assumption is often appropriate. Service systems such as hospitals drive in banks and fast food services can also typically justify Poisson arrivals. Poisson arrival rates are the direct result of either single or multiple merging demand streams to a Manufacturing System or Service System. Arrival rates that fluctuate through time inevitably result in increased Throughput times and long waiting lines due the variance components induced. These effects are recognized and often mitigated by what the Lean Engineer would refer to as *leveling* or *smoothing production*. Smoothing is recognizing the variation in daily/weekly/monthly demand rates and rather than *chasing demand* produces at an *average or smoothed rate*. It is interesting to note that many Production Control departments fail to understand the need to smooth order release, and simply release work to the shop floor as soon as an order is received, in the quantity ordered. Indeed, this is a basic premise of an MRP driven order release system which uses little or no capacity planning. Scheduling production in reaction to changing demand may seem to be a reasonable thing to do, but it usually results in either *slugging* or *starving* the shop floor with varying order sizes. Chasing demand and releasing work to the shop floor by order quantity often results in *batching*. The Lean Engineer knows that following fluctuating demand rates and slugging or starving are not the primary problem. The problem is in creating periods of high equipment utilization which causes long queues in the system. It is also quite common to make *Move lot sizes* either the same or directly proportional to *production lot sizes* (batches). This is sometimes casually called *maintaining order integrity*. Such policies cause WIP to uncontrollably rise and fall, which in turn tremendously increases variance components which adversely affects both Throughput Time and server utilizations. To compensate, the Manufacturing Engineer will usually try to project or forecast anticipated demand, and release orders to the shop floor at regular intervals, in quantities that are based upon the projected capacity to produce. As good as this sounds, the traditional Manufacturing Engineer often fails to fully complement this basic notion with SMED (short set up times), enforcing push vs. pull strategies when possible, creating the flexibility to absorb short time imbalances between planned production and actual production requirements with *Supermarkets*, and fully engaging in Lean practices. Of course, smoothing will create the need to buffer demand fluctuations with some in-process inventories, but that is the price to be paid for consistent and constant work schedules. As it turns out, the disadvantages of increased inventories are usually significantly offset by reducing variance in the demand stream. The Lean Engineer fundamentally understands that the secret is in the ability to manage and eliminate variance. Variance in the order release process is particularly damaging, since it propagates through the entire manufacturing system. But how do we convince management? In order to demonstrate the impact of poor order release, consider the following scenario. Our Lean Engineer has taken the monthly demand over the last 3 months, adjusted for seasonality and decided to recommend a level production release of 100 units/month. This is further refined to a release rate of 25 units per week or 5 units per day. Of course, there will need to be Kaizen events to investigate the use of *Supermarkets*, make-to- stock policies and pull mechanisms to absorb short term imbalances, but we will focus only on the impact that level order release will have on system performance and variance components. All business functions such as sales, purchasing, delivery, procurement and production control should work in harmony to try and smooth order release quantities. A smooth order release policy implies an order release strategy which certainly might change from week to week, but within an established closely bound maximum and minimum. The use of a Continuous Uniform Density Function to control order release reflects a smoothing of erratic demand and is a first step in the prevention of induced variation. Recall that the Continuous Uniform density function is as follows.

$$f(x) = \frac{1}{B-A} \qquad B \leq X \leq A$$

$$\mu = \frac{B+A}{2}$$

$$\sigma^2 = \frac{(B-A)^2}{12}$$

$$and \qquad CV_x^2 = \frac{1}{3}\left(\frac{B-A}{B+A}\right)^2$$

The SCV can also be written as $\quad CV_x^2 = \frac{1}{3}\left\{\frac{1-\frac{A}{B}}{1+\frac{A}{B}}\right\}^2$.

This distribution function bounds order release by a specified maximum and minimum. Looking at the CV_x^2, it should be noted that if A=0, an upper bound is obtained of 1/3. This is significantly lower than the SCV of the Exponential Density function which is 1.0. Suppose that the upper and lower bounds on expected demand are given by B = 200 $\frac{Units}{Wk}$ and A = 100 $\frac{Units}{Wk}$. This results in a CV_x^2=0.037. The average order release rate would be 150 units. Note again that if the actual demand/order release policy followed a Poisson Density function (Releasing orders as received and not to an average), then the SCV would equal 1.0. By following a Uniform release policy, production control would reduce the SCV by a factor of 27. It should be clearly

385

understood that even if demand is smoothed, there could still be a variation or difference in the actual order release from week to week if necessary or desired, but not as widely varying as a *release order quantity as received* policy. In either case, some *make to stock* inventory practices and some nonzero WIP will be required to keep production running at the steady, planned rate. Workstation utilization should also be kept between 80%-85% to accommodate periodic changes in the planned production requirements. It is also true that the average order release should consider Taktr time. Taktr time is simply the available working time per planning period divided by the demand over that same time period. For example, assume that there are 2400 minutes of production time available each week and product demand is 1200 units per week. Taktr time would be $\left(\frac{2400\ min/wk}{1200\ unit/wk}\right) = 2$ min/unit. In other words, a unit of finished product must come from the manufacturing line at an average rate of one unit every 2 minutes which would require a planned production rate of no more than 240 units per day working a full 8 hour shift. But how rigid is this requirement in the face of changing demand rates? Hendrix reports that based upon his experience at AT&T, most manufacturing facilities can absorb up to a 10% swing in order release from day to day without severely impacting production. If the manufacturing system can be *leaned* to the point that it can consistently handle a 10% to 15% spread, an extremely low impact on variance components would be observed. The advantages in creating uniform order release practices are significant, and in all cases a positive impact on WIP and throughput is observed due to the variance inherent in *make-to order* policies.

12.4 The Disruptive Effects of Poor Maintenance Practices

In most manufacturing facilities which we have visited, the Maintenance organization is viewed as a *Fire station.* When someone rings the bell, run to put out the fire. It is also true that when employees are laid off to reduce direct costs, the maintenance organization is often the first and hardest hit. Both practices represent a tremendous error for more reasons than might be obvious to the casual observer. Our focus here will be to show how poor Preventative Maintenance policies adversely affect production. It has often been stated that *an hour lost to production maintenance during working hours can never be fully recovered.* Our motivation here is not to prove or disprove this statement, but to show how variance behaves as a function of how often PM occurs, and how long it takes. Machines are subject to periodic failures since they are mechanical devices. To mitigate or mostly eliminate machine failures during normal production cycles, it is important that a sound Total Productive Maintenance program (TPM) be put into place. Indeed, the twin practices of TPM and Total Quality Management (TQM) are two cornerstones of the now familiar Toyota Production System. This is well recognized by most Lean engineers, but there is a question that is not normally considered. Should maintenance be scheduled frequently and of short duration, or less frequently and of long duration? Here we assume that the production equipment is required at all times, and that any unplanned interruption without proper buffering will have a significant impact on normal production. Assume that a machine will make a part on the average every t_0 hrs. Further assume that this machine will produce N parts between machine maintenance, and that it takes an average of t_r time units (hrs) to bring the machine back on line, with a SCV of CV_r^2. The total *Cycle Time* between machine startups is $T = N*t_0 + t_r$. The average machine availability is given by A $= \frac{N*t_0}{N*t_0+t_r}$. We are interested in determining the impact that both the duration and frequency of a planned maintenance policy has on normal production mean, variance and SCV parameters. The *effective processing time per part* for a Cycle of Operations required to produce N parts is given by:

$$t_e = t_0 + \frac{t_r}{N}$$

The associated *effective variance* and the *effective SCV* are given by:

$$\sigma_e^2 = \sigma_0^2 + \frac{\sigma_r^2}{N} + \left(\frac{N-1}{N^2}\right) t_r^2$$

$$CV_e^2 = \frac{\sigma_e^2}{t_e^2}$$

Where: t_0=Processing time per unit
σ_0^2=Processing time variance per unit
t_r = Expected repair/maintenance time
σ_r^2= Variance of repair/maintenance time
N=Number of parts produced until repair/maintenance

These results can be found in Hopp and Spearman, or on their website at *WWW.FactoryPhysics .com*.

Suppose we provide a steady flow of parts to a manufacturing cell, and each part takes an average of t_0 =30 min to produce with σ_0 =0.10* t_0 min or 0.05 hrs. Hence, CV_o = 0.10. Maintenance is scheduled after every 500 parts and takes an average of t_r = 7.5 hours, with a variance of 0.10*t_r hrs. In this case, we would take an average of 250 hrs of machine time to make 500 parts. Assume that a Preventative Maintenance (PM) occurs immediately after a production run of 500 parts. We will call the time it takes to produce 500 parts plus PM downtime the *Actual Production Cycle (APC)*. In this case, APC=257.5 hrs. If there was no downtime, the APC would be only 250 hrs. The extra 7.5 hours required for PM is lost capacity to produce. Perhaps more damaging; by looking at the equation for σ_e^2 the raw processing time variance is inflated in two ways. The first is in the second term and is reflected by σ_r^2 , which is the variance of the maintenance time. The variance impact is inversely proportional to the frequency N, or ($\frac{\sigma_r^2}{N}$) . The second is in term 3 which represents the repair time squared, weighted by a complex combination of the frequency N or ($\frac{N-1}{N^2}$) t_r^2. The equation for σ_e^2 seems to imply that longer, infrequent maintenance would be preferred to shorter, frequent maintenance. For this scenario, the following results are obtained. t_e=0.515 hrs, σ_e^2=0.1159 and CV_e^2=0.437 Note that this is a significant increase from t_o = 0.50, σ_o^2= 0.0025 and CV_O^2=0.05 Consider the following alternative scenario. Suppose that we wanted to let machine availability remain the same, hence the same APC, but schedule maintenance after every 100 parts are produced. Amortizing downtime in equal increments, downtime would last about t_r=1.5 hrs, with σ_r = 0.10*1.5 = 0.15   Since N=100: t_e=0.515, σ_e^2=0.025 and CV_e^2= 0.0943. Table 12.3 shows the behavior of t_e , σ_e^2 and CV_e^2 as a function of the number of parts produced before a downtime cycle.

Table 12.3
Problem Data

N	t_e	σ_e^2	CV_e^2
500	0.515	0.1159	0.437
400	0.515	0.0932	0.3513
300	0.515	0.0705	0.2656
200	0.515	0.0477	0.1799
100	0.515	0.025	0.0943

Note that by construction, t_e has a constant value of 0.515, which is a 3.1% increase in the standard time, t_o. This is because we maintained the same APT for comparative purposes. These results show that *all things being equal*, shorter, more frequent maintenance times are preferred over longer, infrequent maintenance times based on their impact on variance. However, we offer a word of caution. If the decision is to compare one production machine against another, this may not be the case. In general, both machines would need to be evaluated based upon their maintenance requirements and the performance measures being compared. For example, suppose that two machines designated as A and B are both being considered for use on the shop floor. The operating characteristics of both machines are shown in Table 12.4.

Table 12.4
Operating Characteristics

	N	t_o	CV_o	t_r	CV_r
Machine A	600	1.00 hrs	1.1	15 hrs	1.0
Machine B	200	1.00 hrs	1.0	8 hrs	1.5

The comparative results are shown in Table 12.5.

Table 12.5
Comparative results

	Machine A	Machine B
t_e	1.025	1.075
σ_e^2	1.9594	2.8394
CV_e^2	1.865	2.457

For these operating parameters, the machine with both the largest N and the longest repair time would be preferred. In general, the Lean Engineer should always evaluate alternatives with an appropriate decision model and not rely upon intuitive reasoning.

12.5 Random Failures and Procurement

The seasoned veteran may diligently inquire; Is *there any relationship between the concepts of random failures, maintenance policies and procurement*. To this question we venture to say *yes indeed*, and proceed to show the linkage. An aging Lathe needs to be replaced, and procurement has released bid specifications to potential suppliers. There are two important specifications in the bid package. Procurement specifies a mean processing time of 30 minutes per part for the item being manufactured, with a variance of no more than 2 minutes per part. In addition, the machine has to be available for processing at least 95% of the time. Two bids have been received. Both machines admit to random failures, but the operating profiles are distinctly different. *Old Blue* has a Mean Time to Failure (MTTF) of 160 hours, with a mean time to repair (MTTR) of 5 hours. *Old Red* has a mean time to failure of 640 hours, and a mean time to repair of 20 hours. We will assume that the repair time for both machines follow an Exponential distribution function, so that the $CV_R^2 = 1.0$ for both Old Blue and Old Red. The advertised processing capability of Old Blue is 27 min/part with a standard deviation of 1.5 minutes, and for Old Red it is 26.5 min/part with a standard deviation of 1.5 minutes. It should be noted that Old Blue has a short MTTF, but a short MTTR. Old Red has a long MTTF but a corresponding long MTTR. Production control uses a classical definition of how to compute the availability of both machines.

$$\text{Machine Availability: } A = \frac{MTTF}{MTTF+MTTR} \qquad (12.2)$$

$$\text{Old Blue:} \quad A_b = \frac{MTTF}{MTTF+MTTR} = \frac{160}{160+5} = 0.9697$$

$$\text{Old Red:} \quad A_r = \frac{MTTF}{MTTF+MTTR} = \frac{640}{640+20} = 0.9697$$

Both machines have an availability of A = 96.97 %, which exceeds the design specification. A Manufacturing Engineer is asked to determine the capacity to produce for both machines. The actual average processing times (PT) adjusted for availability is:

$$\text{Old Blue:} \quad PT_A = \frac{27}{.9697} = 27.884 \text{ min/part or } 2.155 \text{ parts/hr}$$

$$\text{Old Red:} \quad PT_B = \frac{26.5}{.9697} = 27.328 \text{ min/part or } 2.196 \text{ parts/hr}$$

Both machines exceed the design specification of 30 min/part or 2 parts/hr.
Since both machines can meet design specifications related to both Availability and Capacity to produce, and the actual processing rates are for all practical purposes are the same, this data cannot help procurement determine which machine to buy. Enter the Lean Engineer. The Lean Engineer is always concerned about the impact of each machine on Variance components, and how WIP might be affected. The Lean Engineer also knows that if the CV_S^2 could be calculated, this would provide an indication of which machine would be more/less disruptive. After looking in the technical literature, the Lean Engineer discovers that Hopp and Spearman have derived the necessary equations to calculate the adjusted CV_S^2 to account for required maintenance. The *adjusted* CV_S^2 for both machines can be calculated from the following equation.

$$CV_S^2(Adjusted) = CV_S^2 + (1+CV_R^2)A(1-A)\left\{\frac{MTTR}{s}\right\} \qquad (12.3)$$

Where: A= machine availability
MTTR=Mean Time To Repair
s= Expected Processing time per unit
$CV_S^2 = \sigma_S^2 * s^2$
CV_R^2 = Squared Coefficient of variation of the repair time

389

The construction of $CV_S^2(Adjusted)$ is informative. The first term is due to natural variation as measured by the CV_S^2. The second term is a complex function of Availability, Mean Time to Repair, the expected processing time per unit and CV_R^2. Note that even if the MTTR is a constant and CV_R^2 is zero, the second term still has a (possibly) significant contribution to $CV_S^2(Adjusted)$. In this case, since the repair times follow an Exponential density function, the $CV_R^2=1.0$ for both machines. For each machine we obtain:

Old Red: $CV_S^2(Adjusted) = CV_S^2 + (1+CV_R^2) \, A \, (1-A) \, \{\frac{MTTR}{s}\}$

$\quad\quad CV_S^2(Adjusted) = (\frac{1.5}{2.7})^2 + (1 + 1.0) \, (0.9697)(0.0303) \, (\frac{300}{27})$

$\quad\quad CV_S^2(Adjusted) = 0.6561$

Old Blue: $CV_S^2(Adjusted) = CV_S^2 + (1+CV_R^2) \, A \, (1-A) \, \{\frac{MTTR}{s}\}$

$\quad\quad CV_S^2(Adjusted) = (\frac{1.5}{26.5})^2 + (1 + 1.0)(\, 0.9697)(0.0303) \, (\frac{1200}{26.5})$

$\quad\quad CV_S^2(Adjusted) = 2.664$

This analysis yields a result which favors *Old Red* over *Old Blue*. The SCV of *Old Blue* is over four times that of *Old Red*. We know that this will result in proportionately higher WIP and Increased Throughput time by Little's Law. Examining the equation for $CV_S^2(Adjusted)$, it can be seen that due to the multiplicative effect of the MTTF in the second term, *all other things being equal,* a machine with more frequent failures and short repair time will *always* be preferred to a machine with long time between failures and long repair time. In general, if two machines are significantly different in their key operating parameters, this may or may not be true. However, two or more competing machines will usually be fairly equal in their processing capabilities. What will always be true is that the machine with the lowest SCV will always be preferred due to the adverse effect that an increased SCV will have on system performance measures. What is worse, if the machine(s) being evaluated are near the front of a line, an increase in variance will impact every Workstation down the line.

12.6 The Disastrous Effects of Bad Quality and Rework

One of the cornerstones to Lean Production is the notion of *Just in Time.* Although sometimes overused and misunderstood, JIT means having everything needed, in the right quantities, at the right place at the correct point in time. Inherent in this definition is that when work is scheduled, it should proceed uninterrupted at the planned pace, in the planned quantity. There is nothing more disruptive to Lean Engineering and just in time production than poor quality and rework. There is simply nothing good about having to rework items or compensate for bad quality by increasing lot sizes. Scrap and rework introduce multiple disruptive effects into the system when they occur. Among these factors are:

- Overproduction to compensate for yield losses
- Loss of capacity to execute planned orders
- Reprioritization and expediting to "catch up"
- Increased material handling
- Slippage of delivery dates
- Lot splitting to compensate for losses

- Increased raw material requirements
- Shorter planned tool life
- Starvation of downstream processes
    ………………..*and many more*

The inevitable conclusion and the effect that we are concerned with here is that poor quality creates an increase in variance in almost every operation in the production cycle. In a dedicated production line or *Focused Factory* dedicated to a single product, small amounts of rework and scrap can often be tolerated, but whether it is recognized or not variance will still increase both Throughput time and WIP. The effects of poor quality are even more serious in *Mixed Model* production systems, but in *all* serial production systems or assembly line manufacturing the effects can be disastrous. Again we state. . .*there are no redeeming outcomes in poor quality.* Deming and his Japanese followers seriously recognized this fact in the Japanese Industrial Revolution. There are two disastrous effects on a manufacturing line. (1) Intermittent gaps or shortages to upstream processes and (2) increased capacity requirements on upstream processes. To offset these types of disturbances, Production Control often fails to address the root cause problem and simply starts additional units to meet anticipated or scheduled demand…throwing away bad product as it occurs. Let us assume that demand calls for producing N products with the probability that a good unit will be produced equal to p. (This will oversimplify the total impact because we have not considered the effect of rework on the capacity to produce). The probability that n good units will be produced follows a Poisson distribution function. The density function describes the expected number of good units to be produced out of n units started. The mean, variance, and the SCV are given by:

$$p_x = \frac{\lambda^x e^{-\lambda}}{x!} \qquad x = 0, 1, 2 \ldots \ldots \infty$$
$$\mu = E(x) = np$$
$$\sigma^2 = np(1-p)$$
$$CV_x^2 = \frac{1-p}{np} = \frac{1}{n}\left[\frac{1-p}{p}\right]$$

If the number of parts demanded is n_d and yield is given by **(1-p)**, if production control releases n parts adjusted by a factor of $\frac{1}{P}$, the Squared Coefficient of Variation is increased by the same factor, which demands an increase in capacity or higher server utilization . This also leads to more WIP and increased Throughput Time. If a dedicated Workstation is set up to produce m good units every week and the probability that any one unit will be scrapped is P, Production Control will need to start $n=m/(1-p)$ units . Suppose that we want to deliver m =200 units to the customer from a process with a 10% defect percentage. In this case, we will need to start n=223 units of work. The 23 units lost to poor quality are pure waste. Also be warned that this practice will require constant monitoring of both the in-process inventory and output finished goods inventory. Raw material consumption will increase, machine utilization will increase and production capacity will be decreased. The only real solution is that a program of Total Quality Control and Lean Thinking must be continually operational to improve quality and eliminate scrap and reworked items. Since our purpose here is to model the effects of poor quality and to predict the impact on variance, we will proceed to do just that.

In Chapter 10 we presented and demonstrated a modeling procedure for analyzing single product flow through serial systems, where the sequence of machining operations was designated by a *routing sheet*. In order to model the effects of poor quality (rework and scrap), we need to be able to model *probabilistic branching.* We will present the general procedures for modeling probabilistic flows, but we will use this theory within a narrow context specifically directed to modeling inspection Workstations. Consider the following (simplified) process flow diagram which shows the first Workstation installing printed circuit boards into a metal case. The second Workstation solders all leads and connects all the boards to form a unified set of operations. Workstation 3 is an inspection Workstation which tests all the circuits for continuity and the solder joints for shorts. After inspection, there is a probability of α that the board will need to be returned to Workstation 2 for rework. There is also a probability of β that the board cannot be reworked and will need to be scrapped. Upon scrapping a board, Workstation 3 immediately contacts Production Control to schedule a replacement board and case. In the long run, these two actions result in an increased workload on Workstations 1, 2 and 3. The question to be answered is: *How does this effect system WIP and Throughput time?*

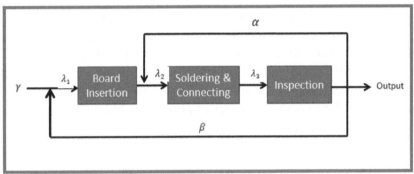

Figure 12.1
An Inspection Workstation

12.6.1 The Effects of Poor Quality

There are several new procedures to consider in this example. It is recommended that the student review Chapter 10 before proceeding. The first thing to notice is that λ_1, λ_2 and λ_3 are not independent flow rates. Because of probabilistic branching, product flow into each Workstation is composed of an initial required flow rate (v) and additional flow caused by the feedback loops into Workstations 1 and 2. Of course, v represents independent outside demand. For modeling purposes, we will need to determine the *offered flow load* to each Workstation, and then determine the number of servers needed to accommodate that load. Since there are probabilistic branching points, the values of λ_1, λ_2 and λ_3 will have to be determined simultaneously. To calculate λ_1, λ_2 and λ_3, observe that even when there is recirculation, in steady state what goes into a Workstation is what comes out of a Workstation. Using this principle we can write a set of *rate-in and rate-out equations*

Workstation 1: $v + \beta \lambda_3 = \lambda_1$
Workstation 2: $\lambda_1 + \alpha \lambda_3 = \lambda_2$
Workstation 3: $\lambda_2 = \lambda_3$

Since v, α and β are known, this is a set of 3 linear equations in 3 unknowns. The solution to these equations is as follows.

$$\lambda_1 = \frac{v(1-\alpha)}{1-\alpha-\beta} \qquad (12.4)$$

$$\lambda_2 = \frac{v}{1-\alpha-\beta} \qquad (12.5)$$

$$\lambda_3 = \frac{v}{1-\alpha-\beta} \qquad (12.6)$$

The following data is known for this three server system.

Table 12.6
A Three Server System with Poor Quality

Workstation k	Service time (S_k) for Workstation k	Service Rate (μ_k) for Workstation k	$CV_s^2(k)$
1	0.1667	6.00	2.0
2	0.1818	5.50	1.4
3	0.1429	7.00	0.90

Assume that $v=5.0$, $CV_a^2=1.50$, $\alpha=0.10$ and $\beta=0.05$. Using this data, we obtain the following values for λ_1, λ_2 and λ_3.

$$\lambda_1 = 5.294118$$
$$\lambda_2 = 5.882353$$
$$\lambda_3 = 5.882353$$

These values can also be obtained by using the matrix inversion methods presented in Chapter 10 which were used in analyzing Jackson Networks. As in previous Chapters, we now need to calculate the *offered flow load* to each Workstation and the required number of servers.

12.6.2 Offered Flow Loads

Workstation 1: $\quad \rho_1 = \dfrac{\lambda_1}{\mu_1} = \dfrac{5.294118}{6.0} = 0.882353$

Workstation 2: $\quad \rho_2 = \dfrac{\lambda_2}{\mu_2} = \dfrac{5.882353}{5.50} = 1.069519$

Workstation 3: $\quad \rho_3 = \dfrac{\lambda_3}{\mu_3} = \dfrac{5.882353}{7.0} = 0.84336$

Since ρ_1 and ρ_3 are both less than one, only one machine is required at Workstations 1 and 3. Hence, $u_1 = 0.882353$ and $u_3 = 0.84336$. Workstation 2 requires two machines so $u_2 = 0.53476$.

12.6.3 Solving for the Arrival SCV's

We are now ready to construct the equations required to solve for CV_1^2, CV_2^2 and CV_3^2. Note that the magnitude of flow transmitted from Workstation k to Workstation j is now determined by probabilistic branching from Workstation 3. In Chapters 10 and 11, we introduced General Network Equations needed to solve for the SCV values associated with manufacturing networks whose product flow is specified by routing sheets with no probabilistic branching. The new defining equations for the unknown $CV_a^2(k)$ values are as follows, appropriately modified for probabilistic branching (Witt, Curry).

$$CV_a^2(j) = \left(\frac{\gamma_j}{\lambda_j}\right) CV_a^2(v, j) + \sum_{k=1}^{M}\left(\frac{p_{kj}\lambda_k}{\lambda_j}\right) CV_a^2(k, j) \qquad (12.7)$$

$$CV_a^2(k, j) = p_{kj}\, CV_d^2(k) + (1\text{-}p_{kj}) \qquad (12.8)$$

Where: $\quad CV_d^2(k) = (1\text{-}u_k^2)\, CV_a^2(k) + u_k^2\, CV_s^2(k) \qquad$ C=1 $\qquad (12.9)$

$or\quad CV_d^2(k) = 1 + (1-u_k^2)(\, CV_a^2(k) - 1) + \frac{u_k^2}{\sqrt{c_k}}(\, CV_s^2(k)\text{-}1) \qquad$ C>1 $\qquad (12.10)$

We can combine Equation 12.8 with Equation 12.7 to form a more compact representation.

$$CV_a^2(j) = \left(\frac{v}{\lambda_j}\right) CV_a^2(v) + \sum_{k=1}^{M}\left(\frac{p_{kj}\lambda_k}{\lambda_j}\right)\{\, p_{kj}\, CV_d^2(k) + (1\text{-}p_{kj}) \} \qquad (12.11)$$

As usual, the value of $CV_d^2(k)$ depends upon whether Workstation j has one server or multiple servers (Equations 12.9 and 12.10, respectively)

These Equations are very similar to those used for deterministic flows in Chapter 10. The difference is that product flow into Workstation j from any other Workstation k is now weighted by the probability of that flow occurring from Workstation k by the probability p_{kj}. The expression for $CV_a^2(k, j)$ also includes the probability p_{kj}. It is instructive to explain the logical structure of Equation 12.11. The first term is the impact of the on the $CV_a^2(j)$ from external flow(v) carrying $CV_a^2(v)$. The second term is the proportion of the $CV_d^2(k)$ that is delivered to Workstation j from Workstation k; $CV_a^2(k, j)$. Note that input flows into Workstation j from any other Workstation k are appropriately weighted by the following term, $\{\frac{p_{kj}\lambda_k}{\lambda_j}\}$. The value of λ_j is the total inflow into Workstation j. The value of λ_k is the total flow into Workstation k, and is therefore the total flow leaving Workstation k. This flow carries with it the value of $CV_d^2(k)$. The following relationship will always hold. $\sum_{k=1}^{M}(\frac{p_{kj}\lambda_k}{\lambda_j}) + (\frac{v}{\lambda_j}) = 1$. Finally, as previously noted the flow from Workstation k to Workstation j carries with it a weighted portion of $CV_d^2(k)$. This has been determined to be (Curry):

$$CV_a^2(k, j) = p_{kj}\, CV_d^2(k) + (1 - p_{kj}) \qquad (12.12)$$

Substituting $CV_a^2(k, j)$ into Equation 12.7, we obtain Equation 12.11. Finally, it should be observed once again that if there are N Workstations in the manufacturing system, there will be exactly N linear equations in the form of Equation 12.11, in exactly N+1 unknowns: $CV_a^2(j)$ j=1,2…N, plus $CV_a^2(v)$. Both v and $CV_a^2(v)$ are both known from the problem input data, so the N values of $CV_a^2(j)$ j=1,2…N are easily recovered by any linear equation solving methodology. We will now analyze the system shown in Figure 12.1, and recover system performance measures of WIP at each Workstation, System WIP and Throughput time. The determined $CV_a^2(j)$ will be compared to the case where there is no rework or scrap to identify the impact that scrap and rework will have on these variables. We will need to construct three defining equations, one each for Workstations 1, 2 and 3. We will write these equations in general form, and then solve using the input data.

Workstation 1:
$$CV_a^2(1) = \left(\frac{v}{\lambda_1}\right) CV_a^2(v) + \sum_{k=1}^{3}\left(\frac{p_{k1}\lambda_3}{\lambda_1}\right)\{\, p_{k1}\, CV_d^2(3) + (1\text{-}p_{k1}) \}$$

$$CV_d^2(3) = (1-u_3^2)\, CV_a^2(3) + u_3^2\, CV_s^2(3)$$

$$CV_a^2(1) = (\frac{5}{5.294})(1.50) + [\frac{0.05(5.882)}{5.294}]\{\, 0.05\, CV_d^2(3) + (1-0.05)\, \}$$

$$CV_d^2(3) = (1-0.882^2)\, CV_a^2(3) + 0.882^2\ (0.95)$$

$$CV_d^2(3) = 0.2944\, CV_a^2(3) + 0.635$$

Workstation 2:

$$CV_a^2(3) = CV_d^2(1)$$

Workstation 3:

$$CV_a^2(3) = \sum_{k=1}^{3}(\frac{p_{k1}\lambda_k}{\lambda_1})\{\, p_{k1}\, CV_d^2(3) + (1-p_{k1})\, \}$$

$$CV_d^2(3) = (1-u_3^2)\, CV_a^2(3) + u_3^2\, CV_s^2(3)$$

$$CV_a^2(3) = (\frac{0.05(5.882)}{5.882})\{\, 0.05\, CV_d^2(3) + (1-0.05)\, \}$$

$$CV_d^2(3) = (1-0.84336^2)\, CV_a^2(3) + 0.84336^2\, CV_s^2(3)$$

$$CV_d^2(3) = 0.289\, CV_a^2(3) + 0.71\, CV_s^2(3)$$

These equations can be reduced to:

$$CV_a^2(1) = 1.4695*0.00278\, CV_d^2(3) \tag{12.13}$$
$$CV_a^2(2) = 0.90\, CV_a^2(1) + 0.01\, CV_d^2(3) + 0.90 \tag{12.14}$$
$$CV_a^2(3) = CV_d^2(2) \tag{12.15}$$
$$CV_d^2(1) = 0.2221\, CV_a^2(1) + 1.556 \tag{12.16}$$
$$CV_d^2(2) = 0.714\, CV_a^2(2) + 0.3670 \tag{12.17}$$
$$CV_d^2(3) = 0.2944\, CV_a^2(3) + 0.635 \tag{12.18}$$

At this point, Equations 12.16-12.18 can be substituted into Equations 12.13-12.15, resulting in three equations in the three unknowns: $CV_a^2(1)$, $CV_a^2(2)$, and $CV_a^2(3)$. If desired, the values of $CV_d^2(1)$, $CV_d^2(2)$ and $CV_d^2(3)$ can be recovered by back-substitution. The required SCV values are as follows.

$$CV_a^2(1) = 1.4726, \quad CV_a^2(2) = 1.7961 \text{ and } \quad CV_a^2(3) = 1.6493$$

12.6.4 Workstation Analysis

The individual Workstations can now be analyzed separately. For reference purposes, we reproduce the following results from Chapter 9.

$$WQ\ (G/G/1) = (\frac{CV_a^2+CV_s^2}{2})(\frac{u}{1-u}) * S \tag{12.20}$$

$$WQ\ (G/G/C) = (\frac{CV_a^2+CV_s^2}{2})[\frac{u^{\sqrt{2\,C+2}\,-\,2}}{c(1-u)}] * S \tag{12.21}$$

Workstation 1: (G/G/1)

$\lambda_1 = 5.294118 \quad CV_a^2(1) = 1.4726 \quad CV_s^2(1) = 1.4726$

$S_1 = 1/6 = 0.667 \quad \rho_1 = 0.8824 \quad C_1 = 1 \quad u_1 = 0.8824$

$WQ_1 = (\frac{1.473+1.40}{2})(\frac{0.882}{1-0.882})\,(0.1667) = 2.17 \text{ hrs}$

$TPT_1 = \text{Time in Queue}_1 + S = 2.17+0.1667 = 2.337$

$L_1 = TPT_1 * \lambda_1 = 12.2734 \text{ units}$

$LQ_1 = L_1 - \rho = 12.2734-0.8824 = 11.490$

Workstation 2: (G/G/2)

$\lambda_2 = 5.8824$ $CV_a^2(2) = 1.7961$ $CV_s^2(2) = 1.40$

$S_2 = 1/5.5 = 0.1818$ $\rho_2 = 1.07$ $C_2 = 2$ $u_2 = 0.5348$

$WQ_2 = (\frac{1.7961+1.40}{2})[\frac{u^{\sqrt{2*2+2}}-2}{2(1-.5348)}] (0.1818) = 0.1260$ hrs

$TPT_2 = WQ_2 + S = 0.1260 + 0.1818 = 0.3078$ hrs

$L_2 = TPT_2 * \lambda_2 = 1.8109$ hrs

$LQ_2 = L_2 - \rho_2 = 0.7414$ unit

Workstation 3: (G/G/1)

$\lambda_3 = 5.8824$ $CV_a^2(3) = 1.6493$ $CV_s^2(3) = 0.90$

$S = 1/7 = 0.1429$ $\rho_3 = 0.8824$ $C_3 = 1$ $u_3 = 0.8413$

$WQ_3 = 0.9584$ hrs

$TPT_3 = 1.1012$ hrs

$L_3 = 6.4779$ units

$LQ_3 = 5.6376$

12.6.5 System Performance Measures

The total System WIP is given by:

$WIP_{System} = \sum_{j=1}^{3} WQ_J = 20.6612$ units

The System Throughput Time is given by:

$TPT_{System} = \frac{WIP_{System}}{\gamma} = \frac{20.6612}{5} = 4.1322$ hrs

12.6.6 No Rework or Scrap

In order to quantify the impact of poor Quality on the system, we will recalculate the System Performance Measures assuming no rework or scrap.

Figure 12.13
No Rework or Scrap

In this case, $\lambda_1 = \lambda_2 = \lambda_3 = 5.0$ This is a simple one product serial system typical of those analyzed in Chapter 10, but since all the servers are G/G/C, , the values of the $CV_a^2(j)$ and $CV_d^2(j)$ j=1,2,3 must all be determined using either Equation 12.9 or Equation 12.10 at each Workstation. Note there is no probabilistic branching since there is no rework and no scrap. The values are as follows.

$CV_a^2(1) = 1.50$ $CV_a^2(2) = 1.8472$ $CV_a^2(3) = 1.731$

$CV_d^2(1) = 1.8472$ $CV_d^2(2) = 1.731$

$CV_s^2(1) = 2.0$ $CV_s^2(2) = 1.40$ $CV_s^2(1) = 0.90$

396

Workstation 1:

$S_1 = 1/6 = 0.667$ $\rho_1 = 0.833$ $C_1=1$ $u_1 = 0.833$

$WQ_1 = 1.4583$ hrs

$TPT_1 = 1.625$ hrs

$L_1 = 8.125$ units

$LQ_1 = 7.2917$

Workstation 2:

$S_2 = 1/5.5 = 0.18182$ $\rho_2 = 0.91$ $C_2=2$ $u_2 = 0.4545$

$WQ_2 = 0.0863$ hrs

$TPT_2 = 0.2681$ hrs

$L_2 = 1.3406$ units

$LQ_2 = 0.4315$

Workstation 3:

$S_3 = 1/7 = 0.1429$ $\rho_3 = 0.7143$ $C_3=1$ $u_3 = 0.7143$

$WQ_3 = 0.4698$ hrs

$TPT_3 = 1.1012$ hrs

$L_3 = 3.063$ units

$LQ_3 = 2.3488$

12.6.7 System Performance Measures

$TPT_{System} = 2.5057$ hrs

$WIP_{System} = 12.5286$ units

12.6.8 Comparing the Two Systems

By comparing the two sets of system performance measures, the devastating influence of poor quality control cannot be more revealing. With scrap and rework, the inflows into Workstations 1, 2 and 3 are given by $\lambda_1 = 5.294$, $\lambda_2 = 5.882$ and $\lambda_3 = 5.882$. Without any rework or scrap, the inflow rates are given by $\lambda_1 = \lambda_2 = \lambda_3 = 5.0$. The increase in flow rates represents lost capacity, increased material handling, expediting, and use of more raw materials. The system parameters are even more revealing.

For the scrap and rework system:

The total System WIP is given by:

$WIP_{System} = 20.6612$ units

The System Throughput time is given by:

$THP_{System} = 4.1322$ hrs

For the no scrap and no rework system:

The total System WIP is given by:

$WIP_{System} = 12.5286$ units

The System Throughput time is given by:

$THP_{System} = 2.5057$ hours

Another important observation that should be restated is that a decrease in rework and elimination of scrapped parts *always* creates more processing capacity, and the results are often significant. In this example, note that once scrap and rework is eliminated, the arrival rate to

Workstation 2 is reduced to $\lambda_2 = 5.0$. With this reduction, it is now possible to satisfy Workstation 2 demand with only one server, since $\rho_2 = u_2 = \frac{5.0}{5.5} = 0.9091$. The new System performance measures using only one server at Workstation 2 are as follows.

The total System WIP is given by:
$$WIP_{System} = 14.7202 \text{ units}$$
The System Cycle Time is given by:
$$TPT_{System} = 2.944 \text{ hrs}$$

The system performance measures using only one server at Workstation 2 are close to the previous results. The primary reason why WIP_{System} and TPT_{System} are inflated is due to the fact that Workstation 2 utilization rose from $u_2 = 0.4545$ to $u_2 = 0.91$ when one server was removed. This is just on the edge of when WIP begins to increase dramatically . If the increase in System Throughput time from 2.5067 hrs to 2.944 hrs and System WIP from 12.5286 units to 14.7202 units can be tolerated, the capital expense of staffing and maintaining a second machine at Workstation 2 can be avoided. Finally, it is intuitively clear that if our Lean Engineer conducts a Kaizen event at Workstation 2, it should not be a major accomplishment to reduce server utilization. Minor changes in server utilization will bring large dividends in this case.

12.7   The Effects of Batching

In Section 12.6 we saw that poor quality control can have dramatic effects upon system performance. The impact of rework and rescheduling due to scrap can significantly increase the variance components in a production system, which we know will in turn increase both WIP and Throughput time. Another operating policy that can significantly impact Throughput Time and WIP is the use of *batching*. There are two primary mechanisms by which batches are formed in a typical manufacturing operation. The first is what we will call *process batching*. There are two basic types of process batching procedures. Parts arrive as single units and are combined into lots for batch processing, or parts arrive in batches and are processed as a batch. The difference in these two types of process batching procedures is where the batch is formed. In the first case, the batch is formed at the destination Workstation while in the second case it is usually formed at a prior originating Workstation. The first case will require significantly more material handling than the second case. Both cases will often cause starvation and delays waiting for the batch to be formed. The second case illustrates what we will call *Move batching*. Move batching is widely used to try and minimize the impact of material flow time and part movement on the material handling functions. In manufacturing systems where two sequential process steps cannot be directly linked with a conveyor or a gravity chute, it is common practice to move parts in groups of size **k** on a pallet or in containers. Forklifts, pusher dogs and AGV's are often used to transfer materials, and the material handling functions are often labor intensive. It is also common practice to have multiple lots of size k waiting in line, and when a lot reaches the server the items are then processed one at a time. Whether or not the items are then re-batched or resume unit flow is determined by process planning. A more general and comprehensive discussion of batching and the impact of different types of batching policies on system performance measures are extensively discussed by Hopp and Spearman in their excellent textbook, Factory Physics (3rd edition). Curry and Feldman have also derived several interesting queuing results for batch processing. Both sources are recommended reading for understanding the general application of queuing theory to this set of problems.

In general, we offer the following set of observations based upon Hopp and Spearman.
- Batching will *always* degrade system performance measures.
- Batching may be unavoidable in some environmentally sensitive operations such as cleaning and plating.
- The JIT community is fond of insisting upon unit lot sizes to support Lean Manufacturing, but a unit lot size may not optimize system performance, and lot sizes greater than one may be preferred.
- Batching *usually* decreases material handling requirements by manual labor, but can be ergonomically unsound due to heavy loads in manual transfers.
- Batching may require more capital investment in Materials Handling equipment.
- Batching at process Workstations causes cycle time at those Workstations to grow proportionately to batch size.
- Workstation throughput time can usually be minimized by some batch size other than one.
- If setup and changeover times cannot be drastically reduced by SMED, then batch sizes greater than one might be necessary or even preferred.
- Batch size decisions should always be based upon a behavioral model of the system.

12.8 Analysis of a Serial Batching System

As an example of how batching can seriously degrade manufacturing system performance, consider the following two Workstation production line, which may be imbedded in a larger production system.

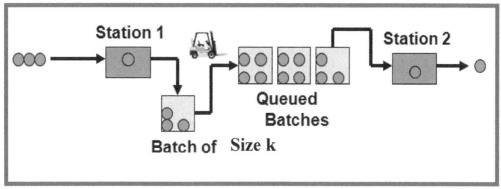

Figure 12.4
Batching Example

Workstation 1 (WS1) is a single server processing Workstation which produces parts one at a time for use at Workstation 2. Parts which exit Workstation1 are placed on a pallet at a location immediately following WS1 until a batch size of k has been accumulated. A fork lift truck then moves the pallet and the batch of k parts to Workstation 2, where it is placed in a queue directly in front of Workstation 2 (WS2). Workstation 2 processes units one at a time. The queued batches await service on pallets, until a pallet is moved to the front of the line. Parts are then removed one at a time for processing. After processing, parts move immediately to the next Workstation. We will model the effects that batch processing has on *System Throughput Time*. System Throughput time for each **part** is composed of six separate time components.

399

$$THP_{System} = \text{Part Queue Wait Time}_{Station\,1} + \text{Part Processing Time}_{Station\,1} +$$
$$\text{Batch Forming Time}_{Station1} + \text{Batch Queue Time}_{Station\,2} +$$
$$\text{Unbatching Time}_{Station2} + \text{Part Processing Time}_{Station\,2}$$

We will assume for simplicity that the time required to move between Workstations is zero to focus on batching effects on Workstation performance. We will now show how to calculate the six components of Throughput time per part.

Let : k=Batch size

λ_1=Part Arrival rate to Workstation 1   (Parts/Min)

t_1 = Time between part arrivals to Workstation 1   (Min/Part)

S_1=Expected time to process one part at Workstation 1 (Min/Part)

$\rho_1 = \lambda_1 * S_1$

WQ_1= Part Waiting time in Queue 1   (Min/Part)

$CV_a^2(1)$=SCV of time between arrivals to Workstation 1

$CV_s^2(1)$=SCV of service times at Workstation 1

$CV_d^2(1)$=SCV of departures from Workstation 1

BF_1= Time to form one batch of size k (Min/Batch)

λ_2=Part Arrival rate to Workstation 2 (Parts/Min)

T_B = Time between batch arrivals to Workstation 2 (Min/Batch)

$WQ_2(B)$= Batch Waiting time in Queue 2   (Min/Batch)

UB_2 = Average time to unbatch parts at Workstation2   (Min/Part)

S_2 =Expected Time to process one part at Workstation 2   (Min/Part)

$\rho_2 = \lambda_2 * S_2$

$CV_a^2(2)$=SCV of individual part arrivals to Workstation 2

$CV_s^2(2)$=SCV of individual part service times at Workstation 2

$CV_{aB}^2(2)$=SCV of batch (pallet) arrivals to Workstation 2

$CV_{sB}^2(2)$=SCV of batch (pallet) service times at Workstation 2

The total throughput time for the entire system per part is given by:

$$TPT_{System} = TPT_{WS1} + TPT_{WS2}$$
$$TPT_{WS1} = \text{Queue Time}_{WS1} + \text{Processing Time}_{WS1} + \text{Batch Form Time}_{WS1}$$
$$TPT_{WS2} = \text{Batch Queue Time}_{WS2} + \text{Unbatch Waiting Time}_{WS1} + \text{Processing Time}_{WS2}$$

12.8.1   Time in Queue at Workstation 1

Since we assumed a single server system at Workstation 1, the time in queue is given by Equation 12.22 for a G/G/1 Workstation.

$$\text{TQ}_1 = \left[\frac{CV_a^2(1) + CV_s^2(1)}{2} \right] \left(\frac{\rho_1}{1-\rho_1} \right) * S_1 \qquad (12.22)$$

12.8.2   Time in Queue plus Processing Time at Workstation 1

The Queue waiting time plus processing time at Workstation 1 is given by Equation 12.23.

$$TPT_{WS1} = \left(\frac{CV_a^2 + CV_s^2}{2} \right) \left(\frac{\rho_1}{1-\rho_1} \right) * S_1 + S_1 \qquad (12.23)$$

12.8.3 Time to form a Batch at Workstation1

The time to form a batch of size k at Workstation 1 can be calculated by observing that by the average time it takes to add one part to a batch of size k at Workstation1 is t_1 time units. Note that t_1 is *not* the service time, but the time between arrivals to WS1. The first part added to the batch must wait a total of $(k-1)* t_1$ time units for the batch to form. The second part must wait $(k-2)* t_1$ time units, and so on until the last part finishes service. The last part will experience no delay time. Hence, the average time T to form a batch can be determined as follows.

$$T=\sum_{j=1}^{k}(k-j)\, t_1 = t_1 \sum_{j=1}^{k}(k-j)= t_1 \{ (k-1) +(k-2)+\ldots+(1)\}$$
$$T=t_1 \{ 1+2+3+\ldots(k-1)\}$$
$$T + t_1 * k = t_1 \{1+2+3+\ldots k\} = t_1 \{\frac{k(k+1)}{2}\}$$

Hence,

$$T= t_1 \{\frac{k(k-1)}{2} \} \tag{12.24}$$

The average time to form a batch *per part* is therefore given by:

$$BF_1= \frac{(k-1)\, t_1}{2} \quad \text{time units} \tag{12.25}$$

12.8.4 Throughput Time at Workstation 1 per Part

Since there is only one server at Workstation 1, The server utilization is: $\rho_1=\lambda_1 * S_1$

The *total time spent per part* at *Workstation 1* can now be expressed as follows.

$$TPT_{WS1} = Queue\ Waiting\ Time_{Station1} +$$
$$Service\ Time_{Station1} \qquad\qquad + Batch\ Form\ Time_{Station1}$$
$$TPT_{WS2} = [\frac{CV_a^2(1)+CV_s^2(1)}{2}](\frac{\rho_1}{1-\rho_1}) * S_1 + S_1 + \frac{(k-1)t_1}{2} \tag{12.26}$$

12.8.5 Cycle time at Workstation2

To calculate the Cycle Time at Workstation 2, we will need to determine three things. (1) The average time that *each part* spends in a batch waiting at Queue 2 (Batch waiting time) , (2) the average time that a part spends waiting to be un-batched at server 2 and (3) the time spent in the service facility *per part*. The expected time that *each part* spends in the server is simply S_2. The waiting time in queue for each *batch* can be determined using either the G/G/1 or the G/G/C equation in Chapter 9 in a batch processing mode, provided that the parameters are properly adjusted to reflect batch moves. Since all items wait in a batch, the batch waiting time is also the average waiting time in queue per part.

12.8.6 Determining the $CV_{aB}^2(2)$ into Workstation 2

The total throughput time in Workstation 2 per part can be determined by observing that individual parts arrive to Workstation 2 in batches (lots) of size k. Each part waits in a FIFO queue until that batch reaches the front of the batch queue. At that time, the parts are *un-batched* and processed one at a time. In order to calculate the throughput time in Workstation 2, the individual CV_a^2 and CV_s^2 and μ_2 parameters must be adjusted to reflect batch behavior. Let $CV_{aB}^2(2)$, $CV_{sB}^2(2)$ and ρ_B be the values corresponding to the **batch** parameters. The appropriate formula to calculate the expected time in queue at WS2 per batch is given by:

$$WQ_B(2) = [\frac{CV_{aB}^2(2)+CV_{SB}^2(2)}{2}](\frac{\rho_{B2}}{1-\rho_{B2}}) * S_B$$

The calculation of $CV_{aB}^2(2)$ rests upon the mean and variance of a batch departure from the batching station. A batch can be formed in $TB_k = k*t_1$ time units. Note that individual units depart from the workstation that immediately precedes the batch operation every t_1 time units and not s_2 (why?). The expected value of $TB_k = E(k*t_1) = k*E(t_1)$. The variance of $TB_k = \text{Var}(k*t_1) = k*\text{Var}(t_1)$. Hence, the squared coefficient of variation is:

$$CV_{aB}^2(2) = k*\text{Var}(t_1) / [k * E(t_1)]^2 = \text{Var}(t_1)/k * E(t_1).$$
Hence: $CV_{aB}^2(2) = CV_a^2(1)/k$

12.8.7 Time to Un-batch at Workstation 2

We need to calculate the average waiting time for those units which are removed from a full lot of size k to be individually processed. The waiting time for un-batching parts depends upon where the individual part might be among the batched items. The first item experiences no delay in moving into the service activity, the second experiences a delay of one service time, the third two, etc. This is a mirror image of the time to batch at WS1, and the average time to un-batch is determined in the same manner as the average time to batch.

Therefore , $UB_2 = \{\frac{k-1}{2}\}*S_2$.

12.8.8 Determining the $CV_{SB}^2(2)$ at Workstation 2

The calculation of $CV_{SB}^2(2)$ is straightforward. The processing time per batch is given by: $t_B = k * t_2$. Hence, $CV_{SB}^2(2) = \text{Var}(t_B) / [E(t_B)]^2 = \text{Var}(k*t_2)/[E(k*t_2)]^2 = k*\text{Var}(t_2) / \{k^2 E(t_2)\}^2$

Hence, $CV_{SB}^2(2) = CV_S^2(2)/k$.

12.8.9 Determining the Waiting Time per Part at Workstation 2

We can now use our standard procedure to calculate the *Batch* waiting Time in Queue 2 which we will call $WQ_B(2)$. This is also the time spent waiting for each part in the batch.

$$WQ_B(2) = \{ \frac{CV_{aB}^2(2)+CV_{SB}^2(2)}{2} \} * \{ \frac{\rho_B}{1-\rho_B} \}*S_B \qquad (12.27)$$

$$WQ_B(2) = \{ \frac{\frac{CV_a^2(2)}{k} + \frac{CV_S^2(2)}{k}}{2} \} * \{ \frac{\rho_2}{1-\rho_2} \}*(kS_2) \qquad (12.28)$$

$$WQ_B(2) = \{ \frac{CV_a^2(2)+CV_S^2(2)}{2} \} \{ \frac{\rho_2}{1-\rho_2} \}*S_2 \qquad (12.29)$$

This is a surprising result which is counter intuitive. It says that the time an individual item waits in line as part of a batch of size k is exactly the same as the time an individual item released from Workstation 1 would wait in queue at WS2, since the k values cancel out.

12.8.10 Server Utilization at Workstation 2

To finish our analysis, we need the server utilization at WS2. This utilization is the average batch arrival rate to Workstation 2 multiplied by the average batch service time. We have shown the average time to process a batch at WS2 is $S_B = \{k*S_2\}$. Recall that $\lambda_B = \left(\frac{1}{TB}\right)$. It follows that

$\rho_2 = \{\frac{\frac{\lambda_2}{k}}{\mu_B}\} = \{\frac{\frac{\lambda_2}{k}}{\frac{1}{k*S_2}}\} = \lambda_2 * S_2$. But conservation of flow demands that $\lambda_2 = \lambda_1$

This is another surprising result, which says that the average server utilization for batch arrivals to Workstation 2 is the same as that incurred for individual part arrivals.

12.8.11 Throughput Time at Workstation 2

The total throughput time *per part* at Workstation 2 is given by the following formula.

TPT_{WS2} = Batch Waiting Time$_{Station2}$ + *Average* Unbatching Time per part $_{Station2}$ +
Service Time per part $_{Station2}$

It follows that the Total Throughput Time at Workstation 2 for a single server system is given by Equation 12.30.

$$TPT_{WS2} = \{\frac{CV_a^2(2)+CV_s^2(2)}{2}\}[\frac{\rho_2}{1-\rho_2}]*S_2$$
$$+ \{\frac{k-1}{2}\}*S_2 + S_2 \qquad (12.30)$$

Note that the Throughput time at Workstation 2 is composed of three individual terms. The first and third terms are exactly the same as if individual parts were arriving to Workstation 2, but the second term is strictly due to batching. Note also that if the batch size is one, we get the usual G/G/C model results.

The *Total Throughput Time in the System per Part* can now be determined as follows.

$$TPT_{System} = TPT_{WS1} + TPT_{WS2}$$
$$TPT_{System} = \{\frac{CV_a^2(1)+CV_s^2(1)}{2}\} (\frac{\rho_1}{1-\rho_1}) * S_1 + \frac{(k-1)*S_1}{2} + S_1 \qquad (12.31)$$
$$+ \{\frac{CV_a^2(2)+CV_s^2(2)}{2}\} \{\frac{\rho_2}{1-\rho_2}\}*S_2 + \frac{(k-1)*S_2}{2} + S_2$$

Finally, note that the increase in System Throughput time due to batching is $[(k-1)/2]*[S_1 + S_2]$, which is a linear combination of the batch size and the service times at Workstations 1 and 2. This is proof positive that in this case, batching seriously inflates system throughput time. This scenario is very common in many manufacturing systems today.

12.8.12 Batching Numerical Example

Figure 12.5 illustrates the same batch move scenario with associated parameters.

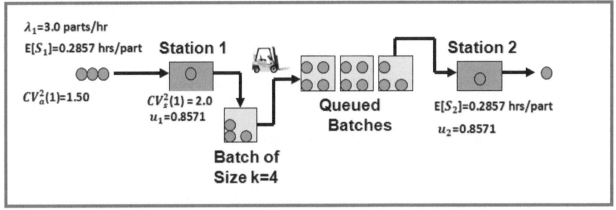

Figure 12.5
Batching Numerical Example

We will assume that the batch move time from Workstation 1 to Workstation 2 is negligible.
   Server Utilizations
$\rho_1 = \lambda_1 * S_1 = 3.0(0.2587) = 0.8571$. We have assumed *Balanced Servers*, so that $S_1 = S_2 = 0.2587$. Hence, $\rho_1 = \rho_2 = 0.8571$. The first Workstation is M/G/1, and the $CV_d^2(1)$ value will need to be calculated as usual from the following equation.

$$CV_d^2(1) = (1-u_1^2)\, CV_a^2(1) + u_1^2\; CV_s^2(1) \qquad (12.32)$$

For this example,
$$CV_d^2(1) = (1-0.8571^2)\,(1.5) + 0.8571^2\,(2) = 1.8673$$

The SCV values for the arrival and service of batches at Workstation 2 are therefore:
$$CV_{aB}^2(2) = \frac{CV_d^2(1)}{k} = \frac{CV_a^2(2)}{k} = \frac{1.8673}{4} = 0.4668$$
$$CV_{SB}^2(2) = \frac{CV_s^2(2)}{k} = \frac{1.50}{4} = 0.375$$

The System Throughput time *per part* at Workstation 1 is as follows.

$$TPT_{WS1} = \text{Queue Wait Time} + \text{Service} + \text{Batch Form Time}$$
$$\text{TPT}_{WS1} = \left(\frac{CV_a^2 + CV_s^2}{2}\right)\left(\frac{\rho_1}{1-\rho_1}\right) * S_1 + S_1 + \frac{(k-1)*S_1}{2}$$
$$\text{TPT}_{WS1} = 3.0 + 0.2857 + 1.286 = 4.5712$$

The System Throughput time *per part* at Workstation 2 is as follows.
$$TPT_{WS2} = \text{Batch Wait Time} + \text{Unbatch Time} + \text{Service Time}$$
$$\text{TPT}_{WS2} = \left\{\frac{CV_a^2(2) + CV_s^2(2)}{2}\right\}\left\{\frac{\rho_2}{1-\rho_2}\right\} * S_2 + \left\{\frac{k-1}{2}\right\} * S_2 + S_2 \qquad (12.33)$$
$$\text{TPT}_{WS2} = 2.885 + 1.286 + 0.2857 = 4.457$$

We now have all the results we needed to calculate the Total *System Throughput Time per part.*
$$TPT_{System} = TPT_{WS1} + TPT_{WS2}$$
$$TPT_{System} = 9.028$$

It is interesting and instructive to break the TPT_{System} time into two groups. That which would be present if there was no batching (k=1), and that which add to the cycle time because of batching.

$$\text{TPT}_{System} = TPT \ Components_{No \ Batching} + \ TPT \ Components_{Batching}$$

$$\text{TPT}_{System} = TPT \ Components_{No \ Batching} + \frac{(k-1)*S_1}{2} + \frac{(k-1)*S_2}{2} \quad (12.34)$$

It is now obvious that as k gets larger, the Throughput Time per part grows as a function of both the batch size k and the service time components. For this example, if no batching is used (k=1) then the Throughput Time per part would be:

$$TPT_{part} \ (k=1) = 6.4562$$

Batching causes an increase in Throughput time per part of 2.572 min/part, or about 40%. It should also be noted that batching also effects the WIP and Throughput times at downstream processes. Curry and Phillips reported that the departure stream from a Workstation receiving batch arrivals which are then processed as individual items and released is given by:

$$CV_d^2(2)=(1-\rho_2^2) \ CV_d^2(1) + \rho_2^2 CV_s^2(2) + (k-1) \ (1-\rho_2)^2 \quad (12.35)$$

In this example, the $CV_d^2(2) = 3.910$ Note that the departure stream $CV_d^2(2)$ is always inflated by (k-1) $(1-\rho_2^2)$ due to batch arrivals. The $CV_d^2(2)$ value for a single server G/G/1 system without batching is 1.51930. The increase due to batching at Workstation 2 is 2.389, which is a 258 % increase! This increase will be propagated to following workstations.

For this example, batching causes a 40% increase in Throughput time and a 258% increase in the departure stream SCV. Even using a balanced line, there is *ALWAYS* an increase in the average cycle time per unit if batch moves are used. This seems to be a direct verification of the *LO-CO-MO-MOO* philosophy of **L**oad **O**ne, **M**ake **O**ne, **C**heck **O**ne, and **M**ove **O**ne **O**n. Before we go gung ho and move to unit lot sizes, beware of other problems not uncovered in this example. First, *beauty is in the eye of the beholder*. If unit lot size is used, cycle time per unit will go down *ALL OTHER THINGS BEING EQUAL*. But, try to put yourself in the eyes of the material handling function. If we move lots in size of k=10, and it takes 15 minutes to make a part, a material handling function operating "just-in-time" must have a conveyance device at the supply Workstation at every 2.5 hours. If we assume that to move a batch of k=10 parts to the next Workstation takes 30 minutes, the material handling function will be idle for 1.5 hours between moves assuming that it takes 30 minutes to return. Great you say, it will give the MH function time to do other things. This is true, but the system will pay for it in the time it takes to fill a batch for the move. As cycle time goes up, so will system WIP by Little's Law. Oh-Oh! ; maybe we should reconsider and run batch sizes of k=5, which we can do if no other demands are placed on the MH function. However, our Lean Engineer understands that if move batch sizes are greater than one, then there will be additional time components introduced at the batching Workstation. Hey, let's use unit lot sizes. Ok, but now MH needs to be in place at the supply Workstation and ready to move a single product as soon as it leaves the machine: In this case, every 15 minutes. Since the roundtrip time for pickup and delivery is 30 minutes, then we have a problem. As a minimum we will now need one more MH device and possibly one more employee. All of this point out the need to execute a total evaluation of the system and make feasible choices that will minimize Throughput times as much as possible, consistent with the "GOAL" of maximizing profits and customer satisfaction. Our motivation in this section was just to show that batching *always* increases Throughput times all other things being equal. We also want to point out that we studied a balanced system since we were studying the impact on Throughput time all other things being equal. However, a balanced production system should be a goal of our Lean Engineer anyway. If the system is unbalanced and the SCV's involved have a wide range of variation, it is even more necessary to use this same modeling approach to determine what lot size will increase system performance. Hate to tell you, but in these cases the optimal lot size is rarely one! Finally, the Lean Engineer should be acutely aware of a common practice that can either help or degrade system performance depending upon the situation. *Never blindly* assume that the Manufacturing Lot Size and the Delivery Lot Size need to be the same. This is rarely the best policy, and large dividends can *sometimes* result from *Lot Splitting* provided that part integrity is maintained. In this case there will be an optimum splitting policy which may require a simulation model to determine the lot size. In the end, each situation must be modeled and studied in its own right. If *Lean Engineering* was easy and intuitive, then anyone could do it overnight.

12.8.13 Final Comments on Batching

Batching can exert enormous influence on throughput times and inventory. In addition, it is the single largest cause of starvation to downstream processes waiting for a batch to be delivered. Large move batch sizes complete destroy the concepts of Balancing and Leveling. As we have seen, larger than needed batch sizes are often used to buffer poor quality, accommodate the material handling function or to support excessive make-to-stock policies. The usual rationale for batching is to try and minimize the effect that long set up times have on available Workstation capacity. It seems natural to assume that by executing production in large lot sizes, that the set up time can be spread over the batch size, but this is usually a poor and unjustified policy, and was recognized as such early on by the Toyota Production System pioneers. In fact, they applied enormous resources to the problem of minimizing set up times, and called this event SMED. At Toyota, SMED was successful in reducing set up times by factors of 50% to 90%, and paved the way for Mixed Model Production and minimum lot sizes. In the Western hemisphere across the pond, the amount of effort expended on set up time and the associated lost production was often overlooked and largely ignored. The strategy was to automate and add capacity, not engage in Lean thinking. The associated problems of large lot sizes were often addressed by adopting a form of JIT. There are several other relevant operating philosophies which should accompany any move to reduce lot or batch sizes. Direct as much effort as possible to minimizing set up times.

- Invest in setup jig & fixture design. Use rapid mount and dismount mechanisms. Implement Flexible Manufacturing Policies
- Standardize…standardize…standardize
- Reduce Manufacturing Lot sizes (Batching) concurrently with reducing setup time
- Avoid excessive material handling times, long material transfer distances and reliance on manual labor
- Form Lean Manufacturing Cells which are U-shaped or close together

- Abolish the traditional policy that move batch sizes and manufacturing batch sizes must be the same
- Understand the effects of batching…learn the rules….make more money

For a more complete treatment of batching and its disruptive effects, the reader is encouraged to see Curry and Feldman or Hopp and Spearman.

12.9 Summary and Conclusions

This Chapter has been concerned with the detrimental effects of variance upon the behavior and operation of general manufacturing operations. The influence and degrading effects of variance have been studied for six common practices which we call the *Big 6*.We have tried to quantify the effects of increased variance components on Throughput times, production rates, WIP and lost capacity. We hopefully have discussed and analyzed a number of current common manufacturing practices which are not clearly understood with regards of how they can inflate Workstation utilization, WIP, Cycle Time and Throughput. We must admit that our treatment of this subject may have been at best superficial. The methods used to model general Lean Manufacturing Systems are just now beginning to emerge, and present a wide variety of research topics for the new Lean Engineer. The best that we can hope for is to create an awareness that the corrupting influence of variance is widespread and devastating to system performance measures and will always result in lower profit margins and customer dissatisfaction.

Review Questions

1. Show that the following formula for the variance of a random variable
$\sigma^2 = \sum_{all\ x}(x - \mu)^2\ p_x$ can be represented as $\sigma^2 = E[X^2] - E[X]^2$.

2. A statement was made in Section 12.0 that to determine the source of variance in a production process, the Lean Engineer must engage in " root cause analysis" Explain what this means? What tools could be used?

3. Discuss the context of variance on Workstation behavior for the following scenarios. (1) How does variance affect the average Workstation Throughput time? From the Workstation modeling perspective, explain how increasing variance in either the arrival process or the service times will affect WIP ?

4. Explain the difference in External Variability and Internal Variability. Which one has the largest impact and why in general?

5. What is the difference in Variance and Variation? Do they have the same effect on the manufacturing system performance measures of throughput, production rate and cycle time?

6. The equation for a G/G/1 queue expected time in queue has three components. A *variance* component, a *Workstation* component and a *Service* component. Discuss each of these and the influence that each term has on the expected time in queue. Compare these three components for a M/M/1 queue waiting time and a M/G/1 waiting time.

7. Assume that a single server Workstation has parts arrive at a rate of 10 per hour, and the service rate is 11 per hour. The $CV_a^2=1.62$ and the $CV_s^2=1.25$. What is the effect on the Average time in the System of changing the average service rate to 13 per hour? Starting with the base case, plot the Expected Time in the System as the expected service rate is

increased by 5% from 11 per hour to 15 per hour. What conclusions can you draw from this exercise?

8. A production control system is currently releasing work to the shop floor every Monday morning based upon the orders received over the previous week. There can be a 20% to a 30% swing in order quantities from week to week. Comment on this practice in terms of its potential Consider the positive and negative effects on the shop floor.

9. The order quantities for Acme Widgets over the past 6 weeks are as follows. Q_1=120 units, Q_2=100 units, Q_3=95 units, Q_4=120 units, Q_5=105 units, Q_6=120 units. Predict the Expected time in Queue and the expected WIP for a G/G/C Workstation with a CV_s^2=1.65 that can produce 60 units per week.

10. For the model in Question 9.0, assume that an Engineer wants to smooth production by releasing the expected number of units determined by a Continuous Uniform Density function. Compute the expected time in queue and the expected WIP for this policy.

11. Compare the results of Problem 9.0 to those of Problem 10.0 and discuss. What changes might need to be made at the Workstation to implement a Uniform Order Release policy?

12. A preventative maintenance organization works 24 hours per day in a large 3-shift factory. They have steadfastly maintained that they cannot plan and execute preventative maintenance because there is no time. They claim that "respond as quickly as possible" is the only reasonable policy, and have asked for a larger maintenance crew. Analyze this scenario.

13. The traditional way to measure machine availability (A) has been with the following equation; $A=\frac{MTBF}{MTBF+MTTR}$, where MTBF is the "mean time between failure" and MTTR is the "mean time to repair". As this formula suggests, the Availability is increased by either *increasing* the MTTF or *decreasing* the MTTR. If MTBF is 120 hours and the MTTR is 12 hours, what is the machine availability? For this data, plot Availability against. MTBF holding MTTR constant, while increasing MTBF by 5% up to 50% Now hold the MTBF constant and plot Availability while decreasing the MTTR by 5% up to 50%. Compare these two cases and comment.

14. Consider the following two machines. Machine one has a MTBF of 500 hours and a MTTR of 50 hours. Machine two has a MTBF of 50 hours and a MTTR of 5 hours. What is the traditional value of Availability for both machines? Without going through a complex analysis, comment on which machine would be preferred in terms of Availability, WIP and Throughput times.

15. The traditional formula for Availability (See Problem 13) has been criticized by claims that it does not distinguish between being inactive due to *repair* or *non-repair* down times. An alternate measure is $A=\frac{Actual\ time\ running}{Total\ time\ it\ could\ be\ running}$. A third way to calculate Availability might be to define $A=\frac{Planned\ run\ time}{Actual\ run\ time}$, where Planned run time =Total time machine should be running minus planned downtime (PM, lunch breaks, bathroom rest breaks, etc.). Actual runtime is Planned runtime minus set up times, test runs, QC checks, etc. Comment on these two new measures of Availability versus the traditional formula.

16. Calculate the *Effective Variance* and the *Effective SCV* for a machine which is required to make a part every 1.5 hours with an SCV of 1.50. A Preventative Maintenance is scheduled after every 25 parts and it lasts an average of 4 hours with a variance of 0.50 hours. . Calculate the Effective Variance and the Effective SCV for this machine. What if maintenance is scheduled every 50 parts? After 100 parts?

17. Discuss how the random failure rate of two machines (A and B) might influence the purchase of either A or B.

18. Assume that purchasing has identified three machines (M1, M2 and M3) which are capable of replacing an older, inefficient machine. Machine M1 costs $50,000 and can produce at a rate of up to 100 parts/hr. M1 has a MTTF of 150 hours with a MTTR of 5 hours. Machine M2 costs $52,000 and can produce up to 105 parts/hr with a MTTF of 500 hours and a MTTR of 20 hours. M3 costs $51,000 and can produce up to 120 parts/hr, with a MTTF of 300 hours and a MTTR of 30 hours. All machine repair times follow an Exponential repair time distribution. Which machine would you recommend to purchase based upon this data?

19. If you have a versatile machine with high variance in both its processing and repair times, where in a serial production line would you recommend placing the machine....in the front, at the end or in the middle ?

20. Define the concept of "Just-In Time" as it relates to Lean thinking strategies.

21. List 15 disruptive effects of "poor quality" on a production facility.

22. The Quality Control manager estimates that his production process has about a 5% reject rate. If the density function that describes the number of bad units produced is Poisson, (1) What is the variance of defective units produced? What is the Squared Coefficient of variation? If 15 good units per day are to be produced, what is the probability that if 15 per day are started they will all be good? What is the probability of 5 units or less? 10 units or more? How many units will need to be started to produce 15 good units per day?

23. What is the difference between Process Batching and Move Batching? Why are these two types of batching often the same size? What are some of the production line problems that batching often creates? Under what situations would batching actually be the only option?

24. List 10 bad effects of using batching in serial production lines.

Thoughts and Things.......

Chapter 13
Push and Pull Systems

Push, Push, Push to the limit, never quit for even a minute, just do it, do it!
Push, Push, Push !!   Unknown Poet

13.1   Push vs. Pull: What are the issues?

The purpose of this textbook is to create a framework for the role of a Lean Engineer in the Factory with a Future. A cornerstone of Lean Thinking is the deliberate and aggressive migration from traditional Push Manufacturing philosophies to Pull Manufacturing philosophies. But what exactly are the basic concepts of push and pull? In this Chapter we will try to de-mystify these two terms and show why pull systems are superior to push systems by almost any measureable yardstick. This Chapter will not be concerned with the execution and management of pull manufacturing; this will be discussed elsewhere.

.

If you ask any student who is studying manufacturing systems or any engineer working in a manufacturing industry, they will acknowledge that they have heard of *push vs. pull*. However, if the issue is pressed to explain why one is superior to the other and fully discuss how, you might get 10 different answers from 10 different people. In recent years, the authors of this textbook have observed a growing body of confusion and diversity as to how a pull system can be defined and what actually makes it work. In a recent technical presentation, a practicing *Lean expert* related a story of how he posed this question to a line foreman in a company which had only recently began a lean movement and started to implement Pull systems. The foreman's response was short and to the point.

"I don't do nothin at any time unless someone has authorized me to do it with a card"

Amusing as this may seem, it is indicative of how some professionals view pull systems. This prompts us to address the following set of questions.

- What is a formal definition of a Push and Pull system?
- How does Pull actually improve system performance measures?
- Why is Pull superior to pure Push methods?
- How does Lean Engineering relate to pull strategies?

To fully understand and answer these questions it is appropriate to quickly address the historical development of Push vs. Pull, and try to unify concepts.

13.2   Historical Perspectives

In their landmark textbook, *The Machine that Changed the World*, by Womack, Jones & Roo, the authors put forth the concept that the modern automobile was the catalyst that changed the entire landscape of manufacturing. We all agree that the father of mass production and assembly line manufacturing was Henry Ford. His dream was to build an automobile which any American citizen could afford to own. Currently, the average American household has 1.7 cars, so someone must have been successful.

In order to accomplish this goal, he developed the modern mass production facility built around an assembly line. Ford constantly sought ways to produce a high quality product in a minimum amount of time at the lowest cost possible. In this respect, he was the first *Lean Engineer*. If Henry Ford was alive today, he would undoubtedly be at the forefront of Lean Thinking, and a champion of Lean Engineering principles. Womack, Jones and Roo were correct, the automobile did change the manufacturing world, but Ford initially failed to recognize that the customer will eventually expect a wide range of choices and options to suit individual wants and needs. Ford's philosophy was: *I will sell you any color of automobile that you might want as long as it is black.* This short sided philosophy was what caused William Durant in 1920 to purchase several automobile manufacturing firms and incorporate General Motors. Durant added diversity and choice to the automobile, and the race was on. Today, there are literally thousands of shapes, sizes and colors for the consumer to choose from. The system of *Mass Customization* that exists today was built upon the shoulders of Ford & Durant, but as complexity and variety increased the role of Production Planning and Inventory Control increased proportionately. It is our belief that the *Machine that Transformed the Manufacturing World* was undoubtedly the advent of the mainframe computer and desktop computing. Without the local and distributed powers of desktop computing, coupled with globalization, managing a modern Lean Enterprise would be a daunting if not impossible task.

To fully understand the pivotal role of Pull Systems, it is necessary to understand how distributed computing was an enabling factor. Prior to computerized Production and Inventory Control, manufacturing systems were largely controlled by individuals in remote offices (sound familiar?) using inventory control techniques such as Reorder Point (ROP) and Reorder Quantity (ROQ) methodologies (sound familiar?). As computational power increased and things got increasingly complex, enter a new corporate strategy called *Material Requirements Planning* or MRP. MRP Fueled by IBM and the APICS Society, the MRP movement swept through manufacturing like a hurricane. By the mid-1980s, yearly sales of MRP software and technical support exceeded $1 Billion annually. MRP was a good idea. It attempted to integrate three Production Planning Functions into one package: (1) The quantity, type and mix of raw materials needed to support production, (2) the schedule of when to release production requirements and (3) the timing of release to operational components. MRP established time-sensitive production completion dates for each manufacturing component based upon both forecasted and actual demand. A bill-of-materials usually accompanied these releases specifying what resources were required to support production, when they were needed, the quantities required and a production schedule. As good as these sounds, initial MRP systems either failed to recognize or ignored the shop floor status of machines, facilities and personnel; and virtually ignored real-time production capacity. Worse, they depended heavily upon achieving an accurate forecast of product demand. In a nutshell, initial MRP systems initially failed to address several critical Manufacturing and Inventory Control functions.....and it was terribly labor intensive and expensive.

To plug these gaps, Oliver Wight in the late 1980s proposed an expansion of MRP to accommodate all of the functions of the manufacturing enterprise; including product costing, capacity planning, inventory Control, order release and scheduling. While this new system might sound a lot like MRP, it went far beyond initial vision and implementation. The new system was envisioned as a corporate planning tool to execute ALL of the production planning and control functions of the manufacturing enterprise, a goal which was fueled by distributed processing and LAN lines. Wight proposed that the new system be called MRP-II or *Manufacturing Resource Planning* …and the new name stuck. MRP-II was soon stream rolling American manufacturing, but it became evident that to support and use such a monster required serious capital investment and integrated, real-time data. Today, MRP-II has evolved into what is called Enterprise Resource Planning (ERP).

Both MRP II and ERP require millions of lines of computer code, large corporate staffs and a virtually unlimited array of *plug-ins* to support applications. Like its forerunner MRP, MRP-II and ERP systems attempted to unify three important components of manufacturing management and control (Sipper and Bulfin). These are broadly speaking (1) Management planning (2) Operations planning and (3) Operations execution strategies. The management planning functions are strategic in nature but include product development, economic analysis, workforce planning and capital acquisitions. Operations planning involve sales, forecasting, production planning, production control and order release. Operations execution involves shop floor control, sequencing, scheduling and WIP management. Business objectives are translated into corporate goals, corporate goals into production planning and production planning into execution. The lines between strategic and tactical planning were ideally meant to be transparent. All Manufacturing Planning systems attempt to establish a forecast for product consumption and set due dates for product delivery; using a break-back, hierarchical planning module to determine when each step of the manufacturing process needs to be completed, what raw materials are required and when to release work to the shop floor with appropriate shop floor instructions. The product start date is determined by estimating the product due date, and then subtracting the manufacturing, waiting and move times plus any *lead time* factors.

In addition to requiring accurate, up to date, shop floor data; this planning cycle time is subject to our old enemy **variance** and Murphy's Law. The product flow time or throughput time is how long product actually takes from order release to final delivery, and this can be significantly different than the planned time. The result is slipped schedules, reprioritization, expediting, variable lot sizes, lot splitting and scheduling overtime to make up for missed due dates. Once released, work usually flows through the system as fast as it can from one operation to the next, *pushed* from Workstation j-1 to Workstation j , sometimes regardless of WIP status. Hence, MRP/MRP II systems have become known prototypically as *Push Production Systems*.

As MRP-II and ERP began to define American manufacturing, there were other things developing across the *Big Pond*; particularly in Japan. Supported by a universal drive to higher quality products by Juran, Deming, Feigenbaum and others (America initially rejected this group), the Japanese automobile industry set their site on gaining a share of the American Automobile Industry. As late as 1955, the *big three* American automobile companies accounted for 95% of approximately $7 billion in annual sales. The Japanese had limited resources, and were not as yet on the cutting edge of computer technologies worldwide. The Japanese were not interested in controlling production in *corporate offices* but wanted to develop a shop floor system that would rely heavily upon skilled labor, shop floor status and customer demand…….not forecasted schedules or make-to stock policies. A deliberate decision was made to empower the average worker to manage and execute shop floor control.

Enter two historical figures that more than any other force were responsible for a new industrial revolution called the *Toyota Production System* (TPS). These two individuals were Mr. Taiichi Ohno and Mr. Shigeo Shingo of the Toyota Automobile Corporation. To be sure, they were supported by a talented set of dedicated manufacturing engineers, but it is generally conceded that Ohno should be credited with conceptualizing the TPS, and Shingo was the engineering genius who made it work. It is recorded that Ohno struggled for some time to develop a system which would/could be executed by people on the shop floor, and would quickly respond to system status and resource availability. The story goes that he visited America to get fresh ideas, but found American Manufacturing mired in a *push world* with excessive work-in-progress. He also observed that enormous resources were being consumed to develop and execute *super* manufacturing machining centers which were very expensive and highly automated. One day he visited a modern American supermarket and was impressed by how goods and services were supplied to the customer in real-time, and groceries replenished according to customer induced stock levels and not ROP/ROQ rules. Customers would take only what they needed in the quantities desired from shelf stock; variations in customer demands were buffered by *Supermarket storage areas* at the store level; and the store would order only what was needed when it was needed from a host of outside suppliers to replenish low Supermarket inventories. The customers essentially **pulled** items through the system according to whatever the demand pattern might dictate, including seasonal variations and holidays. Hence, Ohno galvanized his thinking along one line of thought…*pull not push*. With a push system, work-in-process is released into and moved through the system according to a predetermined schedule …*damn the torpedoes (system status), full speed ahead*. Conversely, in a pull system, manufacturing activities are in *theory* initiated at the final stage of the process, hopefully in direct response to a customer request. There is no forecasting, scheduling or *slack time* used. Ideally, parts or groups of parts (small lots) are moved only when needed in response to shop floor conditions. Hence, the term *Just –in-Time* became almost synonymous with pull manufacturing. As Ohno later admitted, the term Just-in-time was fabricated to confuse the real meaning of what we now call *Lean Thinking*, which embraced all of the principles described in this book. JIT can be implemented in many ways without a pull system, but a pull system is strategically dependent upon JIT philosophies. The entire set of production, management and control strategies became known as the Toyota Production System (TPS). Many of the elements of TPS are discussed in this book: TQM, TPM, Poka-yoke, SMED, 5-S, PDCA, Autonomation, Kaizen events and Jidoka to name a few.

13.3 The Characteristics of Pull Systems

Pull systems are a way to control and coordinate shop floor activities that reduce both WIP and Throughput time. The production control vehicle to achieve these goals is called a **Kanban**, which is usually associated with a card system, but can in fact use lights, colored balls, or any other communication device to trigger shop floor production, material handling, raw material supply and a host of other functions. The Japanese word *Kanban* literally means a *visible record* or a *card*. Quoting Spearman, there is a fundamental difference in push vs. pull systems.

"A push system controls order release to the shop and material movement based upon schedules, while pull systems control order release and material movement by system status".

To illustrate this point and explore how Kanban based pull systems work, we will consider a typical two-card Kanban pull system shown in Figure 13.1. Workstation k is producing parts for use at Workstation k+1. These parts are being stored in a container which holds N items on the output side of Workstation k (Point A). As long as there is space available in the container, Workstation k will produce parts using the raw materials stored in its inbound stock point (Point C) as long as it is authorized to do so. When the container at Point A is full, it is either moved or production stops. The full container cannot be moved without authorization from a *Move card*, which authorizes the material handling function to move the lot to a location specified on the move card. Once the full container is ready to move, another card called a *Production Card* is issued to Workstation k authorizing additional production. Production cannot begin at Workstation K without this authorization.

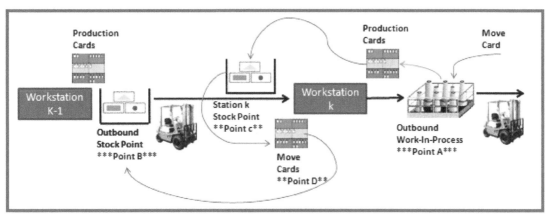

Figure 13.1
A Two Card Kanban System

When Stock Point C reaches a certain level called a *reorder point*, a card is posted to Point D which authorizes material handling to go and get more production resources, possibly those setting in the outbound Stock Location B at Workstation k-1. The entire production line operates in the same way. Ideally, the only schedule to authorize production is issued to the last Workstation in the line...everything else moves smoothly in relation to line status, requirements and authorizations. The principles of JIT are obvious. Less obvious is the absolute requirement to move only good product and have each machine available when needed. Hence, the production practices of TQM and TPM are fully imbedded in the system. For multiple product lines, set up and change over times must be minimized, and production smoothing techniques such as those presented in Chapters 2 and 7 must be implemented with Mixed Model Production (MMP) sequencing rules. Finally, if the Workstations are close together there may not be a need to employ both a move card and a production card. System status and a single move card may be sufficient to control product flow and authorize production. This is called a *one card Kanban system*.

Through time, many cards were developed to trigger manufacturing activities, including conveyance Kanbans, signal Kanbans, SP-Kanbans and a host of others. The interested reader is directed to the excellent textbook of Nicholas for a comprehensive discussion. The issue of lot sizing is an operational issue, and will not be covered here. Drawing heavily from the work of

Suzaki, Hall and especially Nicholas we offer the following characteristics of a pull production system.

> - The quantity of items in a Lot is pre- determined. Ideally, they are unit lot sizes but in practice they hold a small number of parts and can be easily moved.
> - Downstream workstations only request and receive the amount of items specified on the move card
> - Upstream suppliers only produce when authorized, and in the quantity specified
> - Sequencing is specified on each move card.
> - Any container in the system will have at least one card; possible more.
> - Defective items cannot be moved, and high quality is required
> - Machines must always be available when required
> - Material Handling must be responsive and JIT
> - Raw materials needed for production must be at the right place, in the right quantities at the right time
> - WIP is controlled by the number of items either being worked on or being stored in the system. This is in turn dictated and controlled by the number of Kanban cards in the system
> - WIP is continually reduced by reducing the number of cards. Recall that neither zero WIP nor excessive WIP is desirable. There is an *optimal* level.
> - Product management and control basically resides on the shop floor, not in an *ivory tower*
> - Continuous improvement is everyone's responsibility.
> - Worker idle time is spent on housekeeping, process improvement and brainstorming
> - Set up and changeover times must be minimized
> - Level production and work smoothing are each a goal and an operational requirement
> - Workers must be cross trained and understand their role in the *big picture*
> - The *external supply chain* must be fully integrated into the manufacturing system with JIT requirements
> - Expediting and reprioritization are to be avoided if not totally

13.4 Why Pull??

Now that we have characterized, discussed and defined a typical pull system, what is it that is so special about this system? By construction, a typical pull production line is nothing more than a variation of what academics refer to as a *closed queuing system* with finite WIP, as opposed to an *open queuing system* with infinite WIP. Why would the Toyota system strictly limit the amount of WIP? To fully understand why and how this is done, we again state that ideally nothing is allowed to be produced or moved without authorization, and the amount of material in the system cannot be more than the sum of all the material specified on all Kanban cards. We build on Hopp and Spearman to list several benefits of a finite WIP system.

(1) Lower WIP and shorter cycle times are achieved by limiting the total amount of items in the system. By Little's Law, for a fixed (demand) rate this translates into shorter cycle times.

(2) By strictly controlling the amount of WIP by product and material type, a degree of production smoothing is achieved and excess inventory is minimized.

(3) A system with fewer parts in process, generally promotes visual control and higher quality levels. Large levels of WIP often bury parts ready to be moved or processed.

(4) High WIP always translates directly into longer throughput times and usually promotes expediting and reprioritization.

(5) Less WIP and shorter cycle times directly translate into higher profits. By any Economic Lot Size formula, the cost of raw materials, storage space and carrying cost all go down. Of course, this will be directly reflected in lower lot sizes. Note that to achieve this goal, lower set up times are necessary.

(6) Less WIP and smaller lot sizes are usually reflected in more predictable Material Handling times

(7) Excessive WIP usually results in the shop floor personnel working *faster* not *smarter* because of a perception that *hurry up and catch up* is needed.

(8) Traditional monolithic monsters called "ASRS" systems are more easily controlled and managed at lower inventory levels

(9) Less WIP means less congestion and more working space

One might argue that these benefits can also be achieved with push systems if time and effort is applied. This is fundamentally true, but the very nature of push systems makes it difficult to achieve the previous benefits. The real secret is in the one fundamental difference in operating strategies. ***A Kanban pull system which strictly limits the amount of product in a system automatically provides a "WIP Cap", while the traditional push system does not***. In short, a WIP Cap *prevents a WIP explosion.* In its simplest terms, a Pull system will *not* authorize production or material movement until it is asked for by upstream or downstream processes. Hence, WIP is precisely maintained at predetermined levels throughout the entire manufacturing system. Further, the concept of *pacing the production process* at a rate called Takt time regulates production in response to customer demands. Lean Production means doing things at the right pace, at the right time in the right way. Having said this, one might infer that production in a Lean system is strictly *make to customer order*, but in reality this is almost impossible except in dedicated production lines with stable demand requirements. In actual practice, the modern Lean Engineer has learned that even the best Lean systems sometimes need to *build to stock* or *make to forecast* to some degree.

13.5 Is it Push or Pull???

Consider the following example (HTTP://ELSMAR.COM) of an ordinary vending machine. The vending machine contains product (WIP) in varying quantities, but because there is only a limited amount of storage for each product, the amount of WIP is either constant at storage capacity or somewhat less due to periodic withdrawal or demand. Once the door is opened, the maximum allowable stock level for each product is a visible tray or hangers which limit the amount of product in that slot. If the tray is at or below a *replenish stock level*, the provider (Supplier) is authorized to bring the stock level up to the *lot size* dictated by the vending machine storage rack. The customer interface is strictly *pull*, and without adequate consumption (demand) the supplier is not authorized to add stock.

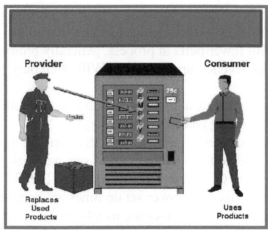

Figure 13.2
A Typical Vending Machine

This is a pure pull system with limited buffer size. However, the supplier in this case is just a material handler and a *manual Kanban system*….The source of raw materials or product is a supply source (the provider) which buffers demand with finished goods. The Supplier does not carry an unlimited supply of each stocked product; rather he/she is allocated work-in-process according to daily, seasonal or special event requirements. The manufacturing source undoubtedly has to forecast demand for such events and *pushes* product to the supplier, which in turn is *pulled to the vending machine* by the end user. In this case, it is clear that a *pull-push interface* exists and it is at the Vending machine itself. The entire system is a combination of push vs. pull. This operating scenario is actually the norm rather than the exception. Although the *Lean Wizards* will rarely admit it, pure pull systems are rare and *zero inventories* is a myth. Of course, almost all subassembly cells and Mass Production lines *often behave as a pure pull system, but between cells there is almost always a make*-to stock storage area called a *Supermarke*t to buffer temporary imbalances and whatever variance exists. To buffer material shortages, machine breakdowns, temporary spikes in demand and raw material shortages; a *Supermarket* is almost always used to protect lines or processes….Particularly if a *Bottleneck Process* is involved.

Consider another typical example, Fitzwilly's Hamburger Emporium. Fitzwilly's is a local hamburger joint which specializes in custom made burgers. The system is viewed as a strict pull system because they make burgers upon request, with condiments mixed in different ways depending upon taste and preference. The system operates as follows.

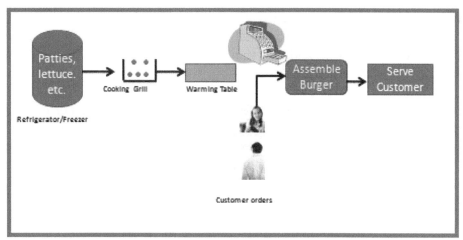

Figure 13.3
Fitz Hamburger Emporium

The burger emporium appears to be a classic pull system to the casual observer, but in reality it is not. The customer orders one or more burgers as a pull agent, and *manufacturing* responds accordingly. As orders are passed to the cook a burger is assembled, packaged and sent to the customer. The ingredients are from a set of *Mini-supermarkets* which are pickles, onions, buns lettuce, tomatoes etc. The *raw materials* are not pulled from upstream, but are in fact stored in small trays. The stocking of these trays is done early in the workday, and the *stock levels* are determined by forecasted demand and are critical. If not enough lettuce is cut up, operations must be interrupted to get some more. If too much is cut up early, there will be stock left over and will need to be thrown away, cutting into marginal profits. The critical ingredient is of course the hamburger patty. Patties, lettuce, tomatoes etc. are *pushed* into the Fitzwilly system on a daily basis and, patties are pulled into the cooking area in small lots (boxes). Total stock quantities are based upon either daily or short term forecast. More is needed on a football day, or on a Friday. This push is also critical, and can result in either an over or under stocked situation. But there is more. In reality, in heavy traffic situations, hamburger patties are not immediately cooked as requested. Instead, they are retrieved in a box of N pre-prepared meat patties from a storage device (cooler), broken into small groups (lot sizes), and cooked as a group on a grill (parallel processing). They are then stored on a warming table (intermediate WIP Supermarket). This is a typical push system influenced by demand fluctuation.

The point in both of these examples is that no real system actually operates as a pure push or pull system, but has *push-pull interfaces* which must be managed. Some are more important than others. In the case of the vending machine, the critical push-pull interface is at the point of supply to the roving vendor-person. In the Fitzwilly's burger barn, it is at the warming table. A common misconception has been that MRP/MRP-II systems should be scrapped and abandoned, but this is not true. MRP-II systems can and should play a large role in supporting Lean manufacturing…they just should not run the manufacturing shop floor. In this context, our understanding of a pull system is expanded to embrace the fundamental issues of Lot Sizing, Production Smoothing, Make-to Stock and Make-to Order…..and thus it should.

419

13.6 CONWIP: A Hybrid Push-Pull System

It should be clear from previous discussions that there is a fundamental difference between how Push and Pull systems are designed and how they operate. To emphasize these differences we offer the following definition of a Push system.

Push systems try to schedule production according to demand (Forecasted or actual), and execute order release to the manufacturing system based upon a Master Production Schedule which attempts to schedule production based upon a break-back structure with appropriate manufacturing, move and slack time components. The slack times are used to buffer uncertainty. Push systems specifically try to dictate throughput and observe the resulting WIP levels. Pull systems schedule to intermediate or final assemblies based upon customer demands (Takt time) and strategic Make-to-Stock policies. All pull system manufacturing or move activities are authorized by coordinated signals such as Kanban cards. Pull systems strictly enforce a WIP Cap, and control throughput by adjusting the WIP cap and manufacturing capacity to satisfy demand (Takt time) requirements.

Push systems typically release work to the shop floor in large batches. Order, production and move batches are often the same. Material movement occurs by pushing work in progress through the system as an opportunity arises, as quickly as possible. Most of the problems associated with push systems are twofold. First, all operations are scheduled in advance and will be executed as soon as possible. If an order gets behind schedule, the order may be split, reprioritized, *red-tagged* or expedited. The second major problem is that any of these remedies create variance components which corrupt every operation; decrease throughput; and increase WIP. Conversely, in a pull production system there are no pre-determined move schedules. Product is pulled from Workstation to Workstation based upon resource availability and system status.

The key issues are WIP control and WIP levels. By Little's Law, for a designated input rate, an increase in WIP will increase throughput time. The *Magic* of a pull system is the *WIP cap* which is strictly enforced by Kanban. The amount of WIP in any pull system can never exceed the amount of WIP allowed by the amount of product authorized by Kanban. A typical push system like MRP/MRP II will attempt to release work to the shop floor based upon a set schedule and assumed shop floor capacity. If too much work is released and/or capacity changes, a WIP explosion will occur with associated increase in cycle time. If too little work is released, then starvation can occur at bottleneck Workstations. A properly configured and managed pull system cannot exceed a WIP cap, and work will not be scheduled to either be moved or to start processing unless specifically authorized by shop floor personnel. There is simply no scheduling required except at final assembly. Sequencing should follow predetermined MMP strategies if there are multiple products; workload leveling, Workstation balancing, predetermined product mix sequencing and a FIFO discipline should always be strictly observed.

A relatively new and interesting shop floor control system is called *CONWIP*, a term coined by Spearman. CONWIP attempts to leverage the best properties of a push system with those of a pull system. CONWIP is based upon what we would call a *Closed Product Flow Network* (CPFN). A CPFN is a manufacturing cell, system or subsystem which is only authorized to hold a specific amount of WIP, and as with all pull systems that level can be changed but cannot be violated. A CONWIP CPFN is shown in Figure 13.4.

The role of the Kanban cards and the way product arrives to the system serve to define whether the system is operating as a classic Push system or as a classic CONWIP system. A push system would assume that there are an infinite number of Kanban cards, and that parts arrive for processing according to some order release strategy. If the arrival stream is random and Poisson, and each Workstation service time density function is Exponential, then we have a classical three Workstation serial push Markovian network. Note that in any push system with infinite storage at each workstation, and assuming that no product is lost to poor quality, the system output rate is always the same as the system input rate.

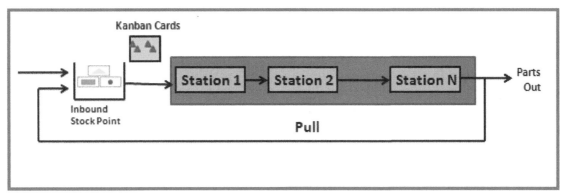

Figure 13.4
A Typical CONWIP System

A CONWIP system behaves in an entirely different way than a Push system. A typical CONWIP system might operate as follows. Assume that there are only **w** unit release Kanban cards available. A part cannot enter the serial system unless authorized by a Kanban card. As long as cards are available, a part can seize one and immediately enter the system. When there is no Kanban card available, no part can enter the system. When a finished part leaves the system, a Kanban card is released back to the inbound stock point where it is available to authorize another part to enter the system. Once a part enters the system at Workstation 1, it is pushed through Workstations 1-3. If there are **w** Kanban cards authorizing unit entry to the system, then system WIP is bounded by **w**. A relevant question is; *where will WIP be in this system*? After a short period of operation the work-in-process will tend to gravitate or spread to each of the Workstations according to system service parameters, bottleneck processes and other interactive parameters.

Hopp and Spearman in their book *Factory Physics* show that a CONWIP system is more *efficient* than a corresponding pure push system. Being efficient simply means the ability to consistently produce product with shorter throughput time and at a higher release rate than a push system with the same number and type of workstations. System WIP is determined by the capacity to produce and the differences in maximum output rate at each Workstation. Note that in a pure push system, parts arrive to each Workstation at a rate which is difficult to control once an order is released. In a CONWIP strategy, parts are *pulled* through the system at a rate which is set by demand and available working time (Takt time). If one is to compare these two systems, the concept of operational efficiency can be formalized as follows.

> *For a given push system configuration with an observed output rate, the same output rate can be achieved with a lower WIP level using a properly configured CONWIP system.*

Hopp and Spearman also state that, in general, CONWIP systems exhibit less variability in Throughput times, which is a persistent goal of Lean Engineering. This benefit is a direct result of negative correlation between WIP levels at each Workstation.

To test the claim of efficiency, we will study the three Workstation production lines shown in Figure 13.4. The *Push* model will have Exponential time between arrivals and Exponential service times at each server. We will assume that there is an infinite queue in front of each Workstation, and that each workstation is M/M/1. Further assume that the arrival rate to this system is Poisson with parameter $\lambda=5.0$ parts/hr, and that the Workstation service times are Exponential with 0.17 hrs/part at Workstation 1; 0.18 hrs/part at Workstation 2; and 0.19 hrs/part at Workstation 3. For the Push analysis, we will set the number of Kanban cards to infinity. Because each of the Workstations is M/M/1, the expected WIP in each individual Workstation can be easily calculated using the formulas given in Chapter 9.

For the CONWIP strategy, we will set the number of Kanban cards to a value of **w**, and assume that parts are always available to enter the system when a Kanban card is available. The question we are trying to answer is as follows. *"What constant (finite) WIP Cap can be placed upon the CONWIP system that will achieve an equivalent output rate as the Push system?* "If the same output rate can be achieved in both (Push and Pull) systems, the question of *efficiency* hinges on which system will have a lower WIP level. We know that WIP is directly related to Throughput (sojourn) time by Little's Law. Hence, we can also calculate the corresponding Throughput times. Note that this latter calculation is somewhat academic but highly relevant, since for a fixed output rate the system with the lowest WIP level will also have a lower Throughput time.

13.7 A Pure Push Model
For the push model, system performance measures are easily calculated by decomposing the three Workstation serial systems into three individual M/M/1 Workstations; calculating Workstation performance measures; and then synthesizing individual results to recover system performance measures. The results are shown in Table 13.1

Table 13.1
The Push System

Standard Queuing 'Push' Analysis... Unit of time is Hours					
	k=1	k=2	k=3	For Lamda=	For C=
Lamda(k)=	5.0000	5.0000	5.0000	5	1
Service Time=	0.1700	0.1800	0.1900		
Throughput Time(k)=	1.1333	1.8000	3.8000	Throughput-System)=	6.7333
L(k)=WIP(K)=	5.6667	9.0000	19.0000	WIPsystem=	33.6667
LQ(K)=	4.8167	8.1000	18.0500		
Rho(K)=	0.8500	0.9000	0.9500		
U(K)=	0.8500	0.9000	0.9500		
System Output Rate=	5.0000	Parts/Hr			
System Throughput Time =	6.7333	Hrs			
WIP(System)=	33.6667	Parts			

The calculated System Throughput time is 6.7333 hrs and the System WIP is 33.667 parts, which includes the individual Workstation waiting lines and the expected number in service at each workstation.

13.8 The CONWIP Model

We now need to examine how a CONWIP system would behave with a WIP cap of **w**, forcing the same required output rate (λ_{Out}=5.0), and the same Workstation service rates (S_1=0.17 hrs/part, S_2=0.18 hrs/part, S_3=0.19 hrs/part). The conjecture is that if a CONWIP system can be found that enforces a WIP limit of less than w=33.667 parts with the same output rate; that system will be more efficient by our definition. By Little's Law it will also produce a lower system Throughput time. To compare the two systems, we are faced with a complex mathematical problem. A CONWIP system is identical to a closed queuing network with limited customers allowed in the system. Closed queuing networks are much harder to analyze than open queuing networks. Fortunately, for a closed queuing network with single server workstations and Exponential service times, a relatively simple algorithmic procedure is available (Curry and Feldman).

13.9 Closed Queuing Networks

Closed queuing networks are special queuing systems that cannot have more than **w** parts in it at any one time. The concept of an input flow rate does not exist, and the output flow rate is strictly dependent upon the number of Kanban (control) cards (**w** units) allocated to the system, one card for each part in the system. Parts are allowed entry into the system until the system part limit **w** is reached. At that point, no other part is allowed in the system until one exits the system. Even if infinite queues are allowed between Workstations, the system will always operate with w units in the system at any one point in time. Once the system is fully loaded, it will operate at this level of WIP, maintaining a constant *WIP cap* of w… hence the term CONWIP. Analysis of this (closed loop) queuing system can be done by using a branch of queuing theory known as *Mean Value Analysis*, or MVA. When all Workstations have Exponential, single server configurations the MVA procedure will yield *exact* results. The MVA procedure is recursive in nature, and depends upon the observation that if system performance measures for a closed queuing network can be determined for a WIP level of (k-1), then the system performance measures can be recursively calculated for a system with a WIP level of k…. hence the recursive nature of a computational procedure is suggested. System performance measures are clearly a function of the WIP limit, w. We will need to distinguish between these measures using the notation in Table 13.2.

Table 13.2
MVA Notation

Notation	Definition or Formula	Index
N	Number of Workstations	
k	Index of Workstation	k=1,2….N
w	WIP cap	
$\lambda_k(w)$	Arrival rate to workstation k	k=1,2….N
$L_k(w)$	Total WIP in Workstation k	k=1, 2… N
$LQ_k(w)$	WIP in Queue at Workstation k	k=1, 2… N
$TPT_k(w)$	Throughput time at Workstation k	k=1,2…N
$WQ_k(w)$	Queue Waiting time at Workstation k	k=1,2…N
S_k	Service Time at Workstation k	k=1, 2… N

Theorem 1: For a closed network with N workstations each having Exponential service times, single servers and infinite queues with no probabilistic branching or product loss, the Throughput time at each Workstation is a function of w , and is given by:

$$TPT_k(w) = S_k + S_k \left\{ \frac{(w-1)\, TPT_k(w-1)}{\sum_{j=1}^{K} TPT_j(w-1)} \right\} \qquad k=1,2… N$$

Theorem 2: For a closed network with N unique workstations each having Exponential service times, single servers and infinite queues with no probabilistic branching or product loss; the arrival rate to each Workstation is identical and is given by:

$$\lambda_k(w) = \frac{w}{\sum_{j=1}^{N} CT_j(w)} \qquad k=1,2… N$$

Theorem 3: For a closed network with N unique workstations, each having Exponential service times, single servers and infinite queues with no probabilistic branching or product loss, the output rate from the last Workstation N is given by:

$$\lambda_{Out} = \lambda_N(w)$$

Theorem 4: By Little's Law, the following performance measures can be calculated. Recall the CONWIP system limit will be set at w units.

$$LQ_k(w) = \lambda_k(w) * TPT_k(w) \qquad k=1,2… N$$
$$WIP_{System}(w) = \sum_{k=1}^{K} LQ_k(w)$$
$$TPT_{System}(w) = WIP_{System}(w) / \lambda_{Out}$$

The recursive procedure is easily programmed in EXCEL, and proceeds as follows.

Step1:   Set $TPT_j(1) = j$ j=1,2... K   and the WIP Cap at "w"
   Set wp=2

Step 2:   Determine $TPT_k(wp) = \dfrac{(wp-1) * TPT_k(wp-1)}{\sum_{j=1}^{K} TPT_j(wp-1)}$ k=1,2... N

   And

Step 3:   If wp = w, then stop;
   otherwise set wp = wp+1 and return to Step 2.

The computational results for this procedure are shown in Table 13.3. It is instructive to note just how the results of Table 13.3 were determined. Using the computational procedure for MVA, the sequence of recursive results were calculated for WIP=1, 2 3... W and are shown in Table 13.4 on the next page. For computational details, see Curry and Feldman.

The $S_1, S_2,$ and S_3 values are the service times in hours per part. The R1, R2 and R3 values are scaling factors which should always be one. The "w" values represent the WIP cap on the system, and the TPT (1), TPT (2) and TPT (3) values are the corresponding Throughput times from Theorem 2. The System Throughput times and the System output rate are also displayed. Table 12.6 actually provides the behavior of this particular CONWIP system for values of w=1, 2.... 25. The results for row "n" are based upon the results for row "n-1". Note that for w=1, the

Table 13.3
CONWIP Mean Value Analysis

CONWIP Mean Value Analysis

	k=1	k=2	k=3	For w=	For C=	Output Probability=
Lamda(k)=	5.0028	5.0028	5.0028	20	1	1.00
Throughput Time(k)=	0.9465	1.2757	1.7756	Sum of TPT Times=	3.9978	
WIP(K)=	4.7353	6.3820	8.8828	Sum of WIP=	20.0000	
Rho(K)=	0.8505	0.9005	0.9505			
U(K)=	0.8505	0.9005	0.9505			
System Output Rate(w)=	5.0028					
System Throughput Time(w)=	3.9978					
System WIP=	20.0000					

Workstation throughput times are simply the individual Workstation service times, and the system Throughput time is the sum of these three times. We are particularly interested in the level of "w" which will yield λ_{Out}=5. From Table 13.4, we see that the output rate for λ_{Out}=5 is found in the row corresponding to w=20 to three decimal places. The results of both models are compared in Table 13.5

425

Table 13.4
MVA Recursive Results

	S1	S2	S3		
	0.17	0.18	0.19		
	r1	r2	r3		
	1	1	1		
w	TPT(1)	TPT(2)	TPT(3)	TPT(Sys)	Output Rate
1	0.1700	0.1800	0.1900	0.5400	0.2501
2	0.2235	0.2400	0.2569	0.7204	0.5003
3	0.2755	0.2999	0.3255	0.9009	0.7504
4	0.3260	0.3598	0.3959	1.0817	1.0006
5	0.3749	0.4195	0.4682	1.2626	1.2507
6	0.4224	0.4790	0.5423	1.4437	1.5008
7	0.4684	0.5383	0.6182	1.6250	1.7510
8	0.5130	0.5974	0.6960	1.8065	2.0011
9	0.5562	0.6562	0.7756	1.9881	2.2513
10	0.5981	0.7147	0.8571	2.1699	2.5014
11	0.6386	0.7729	0.9405	2.3519	2.7515
12	0.6777	0.8307	1.0258	2.5341	3.0017
13	0.7156	0.8880	1.1129	2.7165	3.2518
14	0.7521	0.9450	1.2019	2.8990	3.5019
15	0.7875	1.0014	1.2928	3.0817	3.7521
16	0.8216	1.0574	1.3856	3.2646	4.0022
17	0.8546	1.1128	1.4803	3.4476	4.2524
18	0.8863	1.1677	1.5768	3.6309	4.5025
19	0.9170	1.2220	1.6753	3.8142	4.7526
20	0.9465	1.2757	1.7756	3.9978	5.0028
21	0.9750	1.3288	1.8777	4.1815	5.2529
22	1.0024	1.3812	1.9817	4.3653	5.5031
23	1.0288	1.4329	2.0876	4.5494	5.7532
24	1.0542	1.4840	2.1953	4.7335	6.0033
25	1.0787	1.5343	2.3048	4.9179	6.2535

Table 13.5
Comparison: Pull vs. Push

Comparison		
	Pull	Push
System Output Rate(w)=	5.0028	5.0000
System Throughput Time=	3.9978	6.7333
System WIP=	20.0000	33.6667

The CONWIP model has achieved an identical output rate with a WIP cap of 20 units, compared to a WIP requirement of 34 units (rounded up) for the corresponding push model. This is a 41% reduction in system WIP. It might be interesting to see what output rate could be accomplished using a WIP cap of w=34 units. It will be left as a homework exercise to show that λ_{Out}= 8.5047 units/hr for this WIP level. Based upon this empirical study, it appears that for each infinite WIP system there is a corresponding CONWIP system which is more efficient provided there is no system starvation. This can be proved true using more advanced analysis. It is also true that there is a finite WIP level which is *optimum* with respect to achieving maximum throughput, and it is certainly not zero.

13.9.1 Optimum System Performance

To achieve *optimum* WIP and *optimum system performance*, it is necessary to do two things; (1) Balance server workload and (2) eliminate all variance. For this case, the Workstation service rates will all be a constant and system optimum throughput time will occur at the optimum WIP level. Recall in Chapter 10 we introduced Little's Law by analyzing a four Workstation serial production line with these characteristics. We are now in a position to extend our understanding of how any system might behave with a balanced workload and zero variance. From a theoretical perspective, such a system would exhibit *best case operating characteristics* and serve as a *benchmark* for corresponding real systems. Using the results in Chapter 10, we can now state the following proposition.

Proposition I

In a serial production line with N Workstations, where each processing station is balanced in terms of constant and identical processing times $S=S_1, S_2, S_3, \ldots S_N$, the following results hold for a specified WIP cap of w :

The minimum possible throughput time is given by:
$$TPT_{Best} = N*S \quad for\ w \le N$$
$$w*S \quad\quad Otherwise$$

The maximum possible output rate is given by:
$$\lambda_{Out} = \frac{w}{N*S} \quad for\ w \le\ N$$
$$\left(\frac{1}{S}\right) \quad\quad Otherwise$$

Table 13.6 summarizes the general behavior of an optimum WIP capped system consisting of N=4 serial workstations with S= S_1=S_1= S_3=S_4=2.0 In terms of WIP vs. System Throughput time, and WIP vs. Output Rate.

Table 13.6
Optimum System Performance

	W=1	W=2	W=3	W=4	W=5	W=6	W=7	W=8	
TPT_{Max}	8	8	8	8	10	12	14	16	Hrs/part
λ_{Out}	1/8	2/8	3/8	4/8	4/8	4/8	4/8	4/8	Parts/hr

From Table 13.6, it is clear that as WIP increases there is a corresponding increase in output rate up to a point which we have identified as **Critical WIP (CW)**. Critical WIP for an N Workstation serial system with *balanced* servers is given by CW=N. After this level of WIP is reached, the rate of output will not increase no matter how much WIP is allowed in the system. From the first row, as WIP is increased the System Throughput time will remain a constant until Critical WIP is reached, and then it will increase in a linear fashion as more WIP is added to the system. Optimum performance can also be calculated for unbalanced systems with multiple servers.

From Chapter 10 define the **raw throughput time as T_0** and the **bottleneck processing rate as b_r**. The number of servers at each Workstation is given by C_j j=1, 2... N.

Let: $T_0 = \sum_{k=1}^{N} S_k$

$b_r = \min \left[\dfrac{C_1}{S_1}, \dfrac{C_2}{S_2}, \dfrac{C_3}{S_3} .... \dfrac{C_N}{S_N} \right]$

The level of specified WIP which leads to optimum performance will be called the *Critical WIP*, and is given by:

$$w_0 = b_r * T_0$$

We can now state the following proposition.

Proposition II

In an unbalanced serial production line with N Workstations, where the Workstation processing time are given by $S_1, S_2, S_3, ... S_N$, define $T_0 = \sum_{j=1}^{N} S_j$ and $b_r = \min \{ S_1, S_2, S_3, ... S_N \}$. For a specified WIP cap of w:

The minimum possible throughput time is given by:

$TPT_{Best} = T_0 \quad$ *for* $w \leq N$

$\quad\quad\quad\quad \dfrac{w}{b_r} \quad$ *for* $w > N$

The maximum possible output rate is given by:

$\lambda_{Out} = \dfrac{w}{T_0} \quad\quad$ *for* $w \leq N$

$\quad\quad\quad b_r \quad\quad$ *for* $w > N$

Proposition II reduces to Proposition I if all of the service rates are equal, and all Workstations have single servers. Consider the four Workstation unbalanced line summarized in Table 13.7.

Table 13.7
A Four Workstation Unbalanced Line

Workstation	Number of Servers	Service Time Per Server	Output Rate
1	4	7 Hr/Part	4/7 Part/Hr
2	3	9 Hr/Part	1/3 Part/HR
3	5	6 Hr/Part	5/6 Part/Hr
4	6	4 Hr/Part	3/2 Part/Hr

For this example the raw Throughput time is $T_0 = 29$ Hours. The bottleneck Workstation is Workstation 3, and the bottleneck rate is $b_r = \frac{1}{3}$ Part/Hr. Hence, the critical WIP is:

$$w_0 = b_r * T_0 = (\tfrac{1}{3}) * 29 = 9.670.$$

There are several observations that should be made.

(1)   The bottleneck workstation is not the Workstation with the slowest processing time per server. That Workstation is number one.

(2)   The bottleneck workstation is not the Workstation with the minimum number of servers. That is also Workstation number one.

(3)   Critical WIP is given by 9.67 Parts.

(4)   The minimum possible system throughput time is 29 hours per part.
(5)   Critical WIP is not the number of Workstations, which would be four.
(6)   The system output rate is given by one part every three hours.

The critical WIP is given by 9.67 parts, which is a fraction. This indicates that there is not a unique WIP level for the system which will yield minimum throughput time. In practice, either 9 or 10 should be used. The *optimum* critical WIP would be four parts in the system, and this would be possible if each workstation is producing at a constant rate of 1/3 part/hr. The optimum system throughput time is a function of the optimum Workstation processing capability, and is given by $TPT_{Best} = N*S$. One might be tempted to say; *The problem is one of Variance…every Lean Engineer would know that*. However, note that the service times at each Workstation are all a constant, although they are different. The correct answer is of course in how one defines the concept of variance. In the balanced system, the Workstation processing times possess no statistical variance either within or across Workstations. In the unbalanced Workstations, there is no statistical variance within a Workstation but there are different (constant) processing times across Workstations. We would prefer to say that the unbalanced Workstations have *Variabilty* and this variability is what is causing the problem. The message is as follows. Isolated Kaizen events directed to the processing times at each individual Workstation may possibly result in a series of processing steps without any individual Workstation variance, but without line balance. If Kaizen events or continuous improvement efforts are not directed to bottleneck or critical operations, there may be no *system* improvement at all.

This example clearly indicates that a CONWIP system with a specified WIP cap will always dominate a corresponding Push system with infinite WIP provided that the system is never starved. This is easily accomplished by providing some work- in- process at the front of the CONWIP line. In addition, increasing WIP by either fiat or by push philosophies beyond Critical WIP will result in both unnecessary work in progress and a corresponding increase in system throughput time. This is a killing blow to any form of MRP scheduling system that release orders to the factory in a manner that exceeds Critical WIP. This is another example of the *Magic of Pull*…it is not so much the pull system itself as it is the *WIP Cap*. Finally, as previously observed, a WIP cap will obviously result in fewer requirements for expensive floor space, and less capital investment in inventory and raw materials. As already stated, another advantage is that a closed loop Pull system will by design exhibit less variance than a corresponding push system. This will be reflected in the squared coefficient of variation for time between part departures for the CONWIP system.

13.10   Observations on Variability and its Influence
The key issues of variance reduction, continuous improvement and JIT should continue to dominate Lean Thinking. Pull is only a part of the larger scheme of things, but a big part. Finally, we directly quote Hopp and Spearman concerning variability.

*"Regardless of (its nature, magnitude) or source, all variability will be buffered. A fundamental principle (of the manufacturing rules) of Factory Physics is that there are three types of **variability buffers**, (1) **Inventory** (2) **Capacity** and (3) **Time**. For example, safety stock represents inventory buffering against variability in either demand requirements or production capability. While the exact mix and content of buffers is basically a management decision, the decision of whether or not buffering is required, while dealing with variability is not. If*

*variability exists, it will be buffered. If you will not or refuse to invest significant time and resources into variability reduction, you **will** pay for it in one or more of the following ways.*

- *Long cycle times*
- *High WIP levels*
- *Wasted and underutilized capacity*
- *Lower throughput capabilities*
- *Loss of customer satisfaction*
- *Lost revenue*

13.11 Summary and Conclusions

In Chapters 10 and 11 we presented a methodology which could be used to model any open Queuing Network with single or multiple products with routing sheets, including Workstation dependent processing and reentrant flows. In Chapter 12 we extended these models to include quality reject and rework. These systems were characterized by *push philosophies*, and formed a unified theory by which traditional and many modern production facilities can be modeled and studied. In this Chapter, we built a case for transforming push systems into pull systems, and explored the difference in the two operational philosophies by using a CONWIP strategy. CONWIP is just one of many possible pull strategies, but the results which were given were typical of those expected as one transitions from a push to pull operating strategy. CONWIP was chosen for this study because it is easy to implement and is essentially self-regulating in a manufacturing cell. A major contribution to our discussion was to substantiate that a pure pull system is always superior to a pure push system in terms of required throughput time and WIP. This fact was fully exploited and explored in Chapters 1-8, which were concerned solely with the design and operation of U-Shaped manufacturing and assembly cells. In this Chapter we have also pointed out that most operational systems have both push and pull embedded in their operation. Finally, we stated that the *Magic of Pull* is not a set of operating rules, system configuration or operating principles; but the simple fact is that a Kanban pull system of any type puts a *WIP Cap* on the amount of inventory in the system.

Review Questions

1. Write out your own definition of a Lean Engineer. Provide a definition of *Lean Thinking*. Contrast Lean Engineering to Lean Thinking.
2. If a company wanted to hire a Lean Engineer, provide a description of what he/she would do, what expectations and what sort of projects.
3. Explain the difference in *push* production control systems and *pull* systems.
4. What was the most important thing(s) that Henry Ford accomplished? What important operating principles and management philosophies eventually caused problems and why?
5. Contrast the manufacturing and corporate objectives of Henry Ford to William Durant. What were the commonalities?
6. List and discuss three major problems and three good things about the original concepts and implementation of MRP.
7. Discuss the major differences in MRP and MRP II.
8. Why are MRP/MRP II Production Planning & Control Systems often referred to as "push" systems?

9. Discuss the fundamental differences in the Toyota Production System and MRP/MRP II.
10. What is the fundamental difference in a "push" versus a "pull" system?
11. Using textual material and the World Wide Web, discus and contrast the difference between a "two card pull" system and a "one card pull" system.
12. List 5 major benefits in using a pull system versus a push system.
13. Consider a major United States retail store such as KMART or Sears. . Discuss how such a store represents a typical "push-pull system" Where are the pull components, the push components and the push-pull interface? How could you improve that system?
14. Discuss a general CONWIP system which places a WIP Cap on the total amount of work-in-progress for a serial, three Workstation production line. What prevents all of the product from building up at the bottleneck workstation?
15. Consider the CONWIP example given in Section 7.60. (1) Verify the results shown in Table 13.10 using the formulas given in Chapter 9. (2) Using Theorems 1-4, construct an EXCEL spreadsheet and verify the results shown in Figure 7.5 and Table 13.6. What would happen if the input demand rate increased by 25%? 50%? 100%
16. For the following four Workstation production line: calculate (1) System WIP, (2) Workstation WIP, (3) System Throughput time and (4) Individual Workstation WIP and Throughput times. (5) Verify that Little's law holds at the system level. Each Workstation has an infinite WIP holding area and the system arrival rate is $\lambda=10$ parts/hr.

Workstation ID	Service Rate(Parts/hr)	SCV
1	10.8	1.25
2	5.5	0.85
3	3.4	1.50
4	11.0	1.0

17. Using the data from Problem 13.16, and the EXCEL spreadsheet developed in Problem 13.15; find an equivalent CONWIP system that has the same output rate as problem 13.16. Compare the System Throughput times and the total WIP in each system.
18. What two things must be true for an "N" Workstation serial system to achieve "optimum" system performance?
19. Calculate the Throughput time and output rate for a five server , serial Workstation production line in which each server can process at a constant rate of 5 parts per hour by capping the system WIP level from W=1 to W=10 units. Plot these results. What can you infer or deduce from this exercise?
20. Discuss the concept of *Critical WIP*.
21. Consider the following four Workstation production line with no variance in the processing rates.

Workstation ID	Number of Servers	Service Time
1	2	5
2	4	7
3	1	1.5
4	3	10.0

(1) Calculate the raw throughput time for this system.
(2) What is the Bottleneck Rate?
(3) What is the "Critical WIP"?
(4) What is the expected WIP level at each Workstation?

22. Explain the following statement: *Variability in a Manufacturing System will always be buffered.* How will it be buffered? Give three examples illustrating this principle.

Chapter 14
Integrated Quality Control

"The least cost method of manufacturing
is the one that produces NO defects."   *Dennis Butt*

14.1   Introduction

In lean manufacturing systems, quality is everybody's business. The cost of quality is the expense of doing things wrong (making defects) plus the cost of finding the defects, fixing the problem and the cost of reworking the defects (the hidden factory). The real secret is to *insure quality at the source* to prevent a defect from occurring in the first place. The cost of controlling quality is the expense of finding and reworking defective products. One of the most respected Quality Control pioneers in this country was Crosby, who said that *quality is free*. Believe me; to enforce zero defects in a Lean factory is neither easy nor free. Crosby implied that everyone can be responsible for high quality products as part of their daily job. To economically achieve a high quality level, a product must be designed to be manufactured without defects. Then manufacturing systems must be configured and built to achieve superior quality at the least cost in a flexible way. The latter usually means that the company has designed and built its own manufacturing equipment and understands that linked-cell manufacturing systems, then institutionalized the MO-CO-MOO methodology to directly involve shop floor workers

Many companies in the United States are working hard to improve product quality. After World War II, the United States, one of the few countries with manufacturing facilities undamaged by the war prospered. Fueled by postwar economy, the United States became a world leader in productivity and quality. Across the big pond, before World War II the quality of Japanese products was poor and they were difficult to sell, even at extremely low prices. As part of the Marshall plan, a National Productivity Center was developed in Japan. Experts in statistical quality control, like Deming, Juran and Feigenbaum went to Japan and aided Japanese industry by teaching statistical quality control (SQC) methods.

The Japanese took the advice of these men very seriously. Believing that everyone in the United States practiced SQC, they not only acted on their advice, but elaborated and improved upon it. They taught SQC methods to their engineers and quality control departments. Later they expanded the quality training program to include managers and supervisors at all organization levels and in all company departments. The seeds for total quality management programs were being sown. Then they did something that American companies never thought of doing. They educated the shop floor workers in process quality control fundamentals and techniques. The internal customers learned the now famous 7 tools of quality (Montgomery), and thus the people who ran the processes learned how to control the quality of the processes. The primary message of Total Quality Control (TQC) was that the responsibility for quality rests with the makers (the internal customers) of the part, which was almost a direct Quote from A. V. Feigenbaum's *Total Quality Control* book. Japanese managers, production/workers, and engineers became the best-trained people in the world in quality control. The TQC training encompassed all departments, not just manufacturing but even the lunch room. Improvements were made in all functions, including product design and field service. Because the training was carried on at all levels,

Japanese managers were able to utilize the experience of the entire company, including the work force.

While traditional SQC methods were being used by U.S. companies, the Japanese developed many new methods of quality improvement and control that depended less on sampling, statistics, and probabilistic approaches and more on self-checking and defect prevention. The integration of the quality control into the manufacturing system begins with giving workers the responsibility and authority to make good products. This is the key to attacking the source of the defects in component manufacturing. The fundamental idea is worth repeating: *Prevent the defect from occurring in the first place rather than to find the defect after it* has *been made.* Putting this simple idea into practice, however, is not so simple. It requires that the manufacturing system be changed to accommodate the techniques and methodologies of Total Quality Control. In the 1980s, Motorola developed the six sigma program, a top-down approach to reduce the process variance. Many companies now use lean and 6-σ methods to achieve quality goals. Table 14.1 provides a brief history of how Quality Control evolved in Japan after WWII.

14.2 Statistical Quality Control

Statistical quality control began at the Bell Telephone Laboratories in the 1920s. Since that time, it has become very popular throughout the world, being used by a multitude of industries. There is a variety of statistical quality control tools. The two most popular techniques are *acceptance sampling* and *control charts*. Both these methods use inferential statistics, which means that a small amount of data (a sample) is used to draw conclusions about a much larger amount of data. This large amount of data is often called the *parent population*. The decisions based on the sample cannot be stated with absolute certainty. Therefore, uncertainties are encountered, calling for the mathematics of probability and statistics.

14.2.1 Acceptance Sampling

The purpose of sampling is to draw a conclusion about a process by examining only a fraction of its output. Sampling inspection is needed when it is difficult, costly, or impossible to measure an entire population. For example, when you are making razor blades, the expense involved in observing every blade may be prohibitive. Alternatively, the required inspection process may destroy the product. Traditionally, acceptance sampling required that someone decides: *What is the maximum percentage defective that can be considered satisfactory?* That is, by definition some level of defectives must be accepted. This percentage defective is called the *acceptable quality level (AQL).* After determining how many samples need to be taken to achieve this level of acceptance (or rejection), sampling inspection is carried out. Samples should offer a true, unbiased representation of the parent population, but this depends on many factors, such as the size of the sample and the way it was collected. Usually it is difficult to obtain a truly unbiased sample of a population. For example, if the inspectors always draw parts from the top of each box of parts, the operators quickly learn to put the best parts on top so that their best work is inspected and the entire lot is accepted. But it is the AQL concept of satisfactory defects that creates the most problems for the lean engineer. By the definition of the AQL, a certain level of defectives are tolerated or accepted. In our world of the Lean Engineer defectives are not acceptable, in any number, period!

Table 14.1
Levels of Quality with Toyota/Japan

A History of Quality in Japan	
Late 1940s	U.S. occupation forces showed the Japanese how to use statistical sampling tables (MIL-STD-105D, etc.)
1945-1950	Post-World WAR II - Japanese industry had to rebuild with only one resource -people. Quality of most products was very poor.
1950	Deming invited to Japan - introduces SQC to design engineers and manufacturing engineers
1953	End of Korean War -Industry was rocked by the termination of U.S. military contracts. Japanese industry was reacquainted with the reality that Japan had few resources other than its people, and that they must develop manufactured goods for Export to raise capital and the standard of living.
1953	Juran invited to Japan - introduces management QC to top and middle management
1950s - 60s	Massive QC training of Japanese executives and managers
1961	Total Quality Control, a book in English, by AV. Feigenbaum
1962	Gemba for QC (QC for foremen) began publication Shop Floor Techniques – Foremen learn about QC
1965	Total Quality Control (TQC) * started at Toyota * Changed inspection focus from acceptance to the manufacturing process * Just-in-Time marries up with TQC at Toyota. Goes to all departments. Companywide QC implemented everywhere. Objective - zero defects.
1966	Quality Control To Expand a Company - Successful QC, a book in Japanese by H.Karatsu (sometimes considered the bible of TQC by the Japanese)
1960s - 1970s	TQC concepts and techniques developed by trial and error on the floor shop
1970	Vendors Quality Control - subcontractors (small to medium companies)
1973	OPEC Oil Crisis - Before the crisis, TQC stagnated. This is most apparent in the reduction of Deminq Awards prior to the crisis. But the crisis reignited the TQC movement as a way for industry to stay competitive
1975	Construction Quality Control - nonmanufacturing industry
1980	White Collar Quality Control - service industry
1990	Taguchi methods widely used in America and Japan
2000	Six Sigma projects where defects are measured in parts per million
2007	Reduction in variability prevent defects Perceived quality, meaning visual appearance and tactile senses Performance quality meaning power, smoothness, stability, safety

14.2.2 Control Charts

Control charts are used to track the accuracy (via the mean) and the variance (via the range or standard deviation) of a process by plotting selected sample statistics. When a process produces a product, no two parts are exactly alike because of variations in the manufacturing processes, in the materials, and in operator performance. Variability, whether large or small, is always present, and many sources can contribute to it. In 1924, W. A. Shewhart of Bell Telephone Laboratories developed statistical charts for process control. This is called *Statistical Quality Control* (SQC).

By traditional SQC thinking, the factors that contribute to product variation can be classified as either inherent (random or chance) causes or assignable causes. *Inherent* or *Chance causes* are considered to be a natural part of the process, difficult to isolate or eliminate or too small to worry about. Some examples are small variations in material chemistry or properties, age of the machine being used, machine vibrations, and variations in human performance. *Assignable causes* are events that produce detectable changes in the behavior of the process that can be identified and possibly eliminated. These changes are usually large in magnitude and controllable. Examples of assignable causes are changes in tool geometry due to tool wear, use of the wrong cutting tool, temperature fluctuations, and pressure variations. When only chance causes are present, the process is considered to be under statistical control. However, when assignable causes occur, the process must be analyzed to determine the source of the assignable error, the problem eliminated, and the process controlled.

Shewhart realized that it should be possible to determine when variations in product quality are the result of random chance cases or due to an assignable cause. He developed control charts for this purpose. There are several different types of control charts, but only the charts for variable data will be discussed in this Chapter. The $\bar{X}$ chart monitors the process mean, and the R chart Range chart) monitors the process variance. Control limits for both charts are usually set at three standard deviations above and below the mean. An example of an $\bar{X}$ and R chart is shown in Figure 14.1. In these charts, sample statistics ($\bar{X}$ and R values) are plotted based upon a random sample of size n from the process. The horizontal axis is time. In this example, 25 samples of size n= 5 were taken. From these 25 samples, an average of each sample was taken , and then each one of these x-bar values was averaged to get an "x-double bar"; these values are then plotted. This is called a *preliminary control chart*. At this point, the 25, x-bar and r values, are plotted on the chart. If each x-bar and r value falls within the upper and lower control limits, the chart is ready to use. If one or more points fall outside the control limits; search for assignable causes. Throw away the points that fell outside the control limits, and retain the points for which no assignable cause was found. Repeat the process until all points (x-bar values) fall within the control limits.

Figure 14.1 shows a plot of 25 x-bar values on a preliminary x-bar chart. Point 15 is very close to being outside the upper Control Limit, but it is not. The chart is now accepted and can be put into use.

Shewhart based his charts on the knowledge that sample x-bar values will be normally distributed about the population mean regardless of the shape of the population from which the samples were drawn. This is based upon the well-known Central Limit Theorem (CLT). Using a typical chart construction procedure with 3-sigma limits, false alarms will occur approximately only 2.5% of the time.

Another indication of an assignable cause is unusual run conditions or patterns of non-random behavior. For example, a run of seven points consecutively either up or down is certainly non-random behavior, and indicates a trend. This should be investigated when it occurs. A run of eight points all above or all below the central line is also an unlikely event, indicating that the process average has changed. Did this occur in the $\bar{X}$ chart shown in Figure 14.1? Yes it did! Samples 18 to 25 all fell below the center line ($\bar{X}$) so it is highly likely the mean of the process has shifted down. In 1956 Western electric published a set of 4 rules to detect non-random behavior. Later, Nelson and Kodak added 6 more rules. This collective set is known as sensitizing rules. We will not discuss them here in any great detail: See Montgomery.

The historical function of x-bar and r charts has been to control the mean of the process (x-bar chart) and the variance of the process (R-chart) after all variance except that produced by chance causes has been removed. However, the charts have not traditionally been operated by the people who run the processes (the users of the manufacturing system). Rather, the charts have been kept by people in the quality control departments. Inspectors were sent to the factory floor to gather and analyze sample data. The inspector's job became that of the quality enforcer or process controller. The production worker viewed his/her job as meeting production rate standards, regardless of quality. If the quality was bad, so what? It could be reworked or scrapped. Over the years, an adversarial relationship developed between manufacturing and quality control. Let's examine the decision-making situation a bit further. As shown in Table 14.2, deciding about how the entire parent population is behaving based upon a small random sample can result in two kinds of errors (as well as two correct decisions):

>   *Type I Error*   Deciding the process is out of control, when it is not making defects. The
>   probability of this occurring is given by α
>   *Type II Error*   Deciding the process is in control, when it is actually making defects. The
>   probability of this occurring is given by β

The standard American control chart sets its control limits at μ plus or minus 3σ. Sigma is usually estimated from the sample range or the sample standard deviation. By using 3σ to establish the upper and lower control limits, the probability of making a Type I error is very small (about 0.00135) for an x=bar chart while the probability of making a Type II error is usually quite large for moderate changes in the process. Why did the chart designers do this?

Psychologically speaking, when the person who makes decisions about the quality of a process is not the same as the person who runs the process, the decision maker will set the probability of a Type I error so as to not lose face. This practice is quite common, and is in fact the basis for the widespread use of 3-sigma limits in the United States. The European community often chooses the level of a Type I error which can be tolerated, and then determines the upper and lower control limits using statistical techniques. Note that the when a Type I error occurs, the immediate action is to stop the process and try to find an assignable cause for an out-of-control point. This can severely disrupt the flow of production. If a statistical sample indicates that the process is producing defects, the decision maker (DM) has to take action, maybe even recommend stopping the process. If no problems are found, the DM looks bad in the eyes of those who run the processes in the manufacturing system. On the other hand, not properly diagnosing

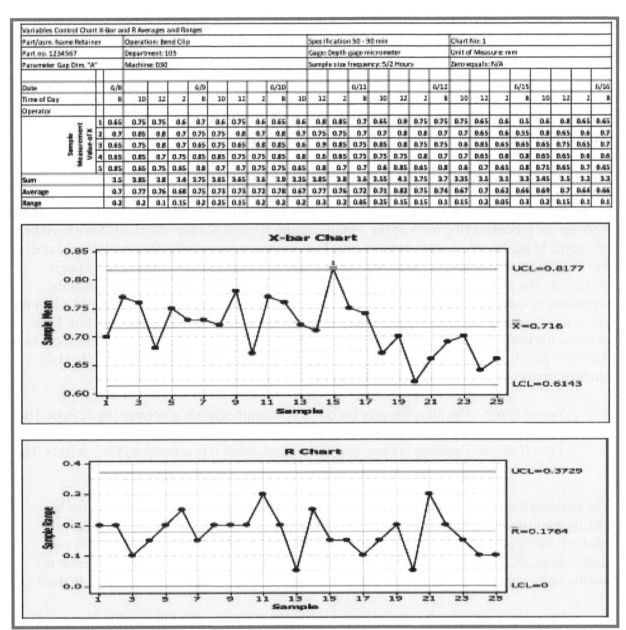

Figure 14.1

Example of $\bar{X}$ and R control charts, $\bar{\bar{X}}$=0.716 and $\bar{R}$=0.1764 (Source Ford Motor Company Statistical Methods Office)

Type II error requires no action by the DM and therefore no blame is usually assigned. The *do nothing decision* shifts the blame for making the defects to the production worker. Of course, both the external and internal customer suffers because defects are permitted to leave the process.

In Lean manufacturing cells, the workers are given the tools and responsibility to control the quality of the processes. The workers use the control charts to monitor the process and they are used in real-time. While some methods of inductive statistics do integrate the worker into the quality control system, they do not guarantee zero defects or even extremely low defect rates. A

438

common misconception about control charts is that they are used to indicate *what* went wrong in a process. Control charts should only be used as detection devices, to indicate *when* something goes wrong, but not what went wrong. If control charts are not used correctly, they are nothing but a waste of time. For example, some companies actually take all the data and then wait until the end of the shift to plot the points! By this time, it is too late to react to trends and out-of-control points. All this changes when the internal customer runs the control charts in real time.

Table 14.2
Statistical Sampling can result in two kinds of errors in the decision making process

	A sample was taken and it was Decided that:	
	The process *had not* changed	The process *had* changed
The truth was that the process had *not* changed	DM takes no actions as nothing is wrong. The correct decision	DM takes action but nothing can be found to be wrong with the process; Type I error. Production time lost.
The truth was that the process *had* changed	DM takes no action and the process continues making defects; A Type II error.	DM takes action, finds what caused the out-of-control point. The correct decision.

14.2.3 Integrated Quality Control (IQC)

Poor quality has been the demise of more than one Lean implementation plan. The integrated approach to quality is sometimes called *Total Quality Control* (TQC) or *Company-Wide Quality Control (CWQC)* because all departments in the company participate in QC, as do all types of employees. Schonberger used the term *Total Quality Control* in his first book on JIT manufacturing techniques. The idea is that if a company takes care of quality, profit will increase. A total quality commitment in all production departments and at all levels of management is the requirement for TQC. Every person must have a complete understanding of quality control, the methods used, the benefits and the expectations. Large, central quality control departments are not the answer. More inspectors working on checking quality is not the answer. Actual control must be integrated into the manufacturing-system. We call this *Lean Integrated Quality Control* (LIQC). LIQC is not a series of specialized techniques, but a procedure that defines a manufacturing-based corporate strategy that incorporates quality control at all levels of an organization. Line personnel must be given the necessary training and initial responsibility to carry out quality control functions. Eventually, IQC is extended to include all the vendors, suppliers, and subcontractors in order to improve the quality of supplies and materials.

14.3 Process Analysis: Tools and Techniques

Table 14.3 Lists some common techniques which can be used for continuous improvement. Remember, in a pull system problems (quality and other types) can be exposed through the deliberate removal of inventory from the links, so inventory is used as a quality control device.

Table 14.3
Quality Control: Concepts and Categories (Schonberger, 1982)

Quality Control: Concepts and Categories (Schonberger, 1982)	
LIQC Category	*LIQC concept*
1. Organization	Production is responsibility for quality - Individual workers, Quality Circles
2. Goals	Habit of improvement by everyone Perfection- & Zero defects – not a program, a goal
3. Basic Principles	Process control – defect *prevention* not *detection* Easy to see quality – quality on display so buyers can see and inspect – easy to understand Insist on compliance Line stop when something goes wrong 100 percent check: MO-CO-MOO Project-by-Project improvements
4. Facilitating concepts	QC department as facilitator not the driving force Audit suppliers Help in quality improvement projects Train workers, supervisors, suppliers Use Small lot sizes Housekeeping Less than full capacity scheduling Check machines daily, use check lists like an airplane pilot Total Preventive maintenance (see Chapter 16) 8-4-8-4 two shift scheduling
5. Process Analysis Tools and Techniques	Expose problems, solve problems Defect prevention, poke-yokes for checking 100 percent of parts Analysis tools such as…… Cause and effect diagrams (Fishbone or Ishikawa) Histograms and run charts and check sheets Control Charts (X bar and R charts) Scatter diagrams (X-Y correlation chart) Pareto charts Process flow charts Taguchi and DOE methods

Specification Limit (LSL) and the Upper Specification Limit (USL). The dashed vertical line shows the nominal or desired mean of the process. Several disadvantages of the histogram include its inability to show trends, and its inability to take into consideration any correlation. If data is plotted against time, this is called *a run chart*.

What is not commonly recognized is that *inventory is an independent control variable* rather than a dependent variable. Problems are discovered and analyzed using:

- *Control charts/run charts* are to monitor process behavior, not control quality
- *Process flow charts* are used to define all the steps in the process, identify waste and uncover problem areas
- *Check sheets and histograms* are used to analyze data from the process.
- *Pareto charts* are used to identify the most Important problems
- *Scatter plots* are used to find correlations between factors.
- *Cause and Effect / lshikawa / Fishbone diagrams* are used to determine the causes of the problem that resulted in a lack of quality.

In recent years, Taguchi methods, design for manufacturing, 6-Sigma and Experimental Design techniques have also been used under the general identification of 6-sigma techniques to improve process quality by reducing variance. Integrated techniques are used to sustain continuous improvements in the processes. Quality circles are used to find (and sometimes to solve) the problems.

14.4   The 7 Tools of Lean QC

14.4.1   The Histogram

The histogram is a plot in which data is grouped into cells (or intervals) and the frequency of observations falling into each interval is visually displayed. When grouped into a cell, all the observations within a cell are considered to have the same value, which is the midpoint of the cell. A histogram is simply a *visual picture* that describes the variation in a process. It is good to have this visual impression of the distribution of the data along with the mean and standard deviation. Histograms are used in many ways in IQC.

- Visually accessing central tendency, shape and dispersion
- Comparing the process with the specifications
- Suggesting the probability density function that might describe the data
- Indicating discrepancies in data, such as gaps.

Frequency of numbers within each cell is usually shown as vertical rectangles. The lengths of the rectangles are proportional to their numerical values while the widths are proportional to their class intervals. There are several different types of histograms. A histogram shows either the absolute frequency (actual occurrence) or relative frequency (percentage), Histograms can also be cumulative. Each type has its own use for different situations. Figure 14.2 shows a frequency distribution for 45 measurements. One observes that the goal of the process is on the low side of all the observations (13.21), but all the data is within tolerance as indicated by the Lower Specification Limit (USL) of 13.06

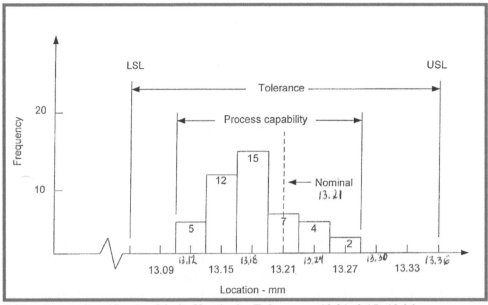

LSL = Lower Specification Limit=Nominal – Tolerance =13.21- 0.15=13.06
USL = Upper Specification Limit=Nominal + Tolerance =13.21+0.15=13.36

Figure 14.2
This Histogram Shows Frequency versus Location for 45 Measurements. One Observes the Aim of the Process is Low but All the Data is within Tolerance

14.4.2 The Run Chart

A *run chart* is a plot of a quality characteristic as a function of time. It provides some idea of general trends, and degrees of correlation. Run charts reveal information that a histogram cannot, such as certain trends over time or cyclic behavior at certain times of the day. In Figure 14.3 the lengths of manufactured rods are measured, and then plotted through time as a run chart. Individual measurements are taken at regular time intervals, and the points are plotted on a connected line graph as a function of time. The graph is then used to notice any obvious trends in the data. Run diagrams are helpful to identify the basic time-varying behavior of a process. Without this knowledge, one might interpret histogram data incorrectly. For example, the histogram might hide tool wear if frequent tool changes and adjustments are made between groups of observations. As a result, a run diagram (with 100% inspection where feasible) should precede the use of both control charts and histograms which use the average from a set of data.

14.4.3 The Process Flow Chart

Perhaps the first task of the lean engineer should be flow charting. A flow chart is a pictorial representation of a process flow, and the inter-connectivity between the steps in the process. Flow charts provide excellent documentation of how a multistep process is connected at any point in time, and can be used to show the relationships between each process step. The Lean tool of Value Stream Mapping (VSM) is a specialized process flow modeling technique which is intended to show WIP buildup, process throughput time and non-value added activities. See Chapter 8 for more details concerning VSM. Flow charts are constructed using easily recognized symbols to represent the type of activities performed. There are as many symbols as there are types of manufacturing systems, and there is little standardization.

442

First shift :	35	40	27	30	30	34	26	31
Second shift :	24	23	20	15	23	17	16	21
Third shift :	15	13	28	8	20	9	5	11
Fourth shift :	16	5	9	13	16	10		

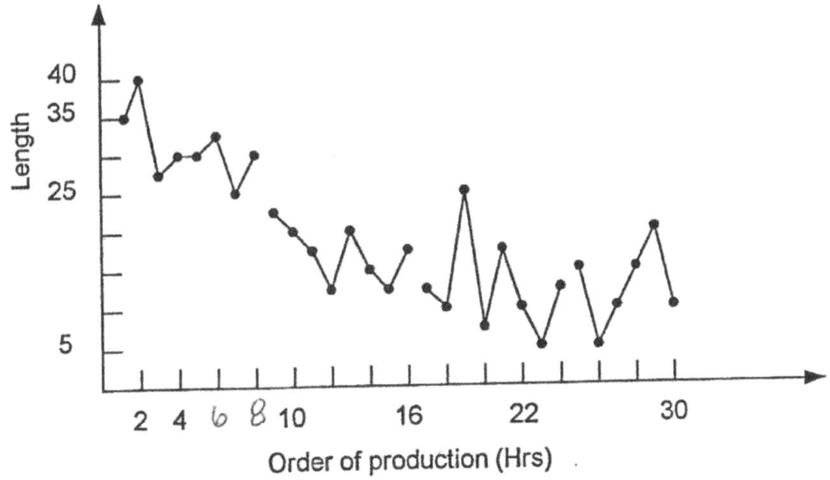

Figure 14.3
Length of Bolts vs Time are Plotted in this Run Chart
There is an Observable Downward Trend

Constructing a flowchart will often lead to the discovery of steps that not everyone thought were taking place, or a sequence of operations that does not make sense when analyzed. An example of a flow chart is shown in Figure 14.4. This example is part of a more complex process flow diagram in the manufacture of refrigerators.

This is but one example. Flowcharts can be used at any level within an organization, from the highest level management process, to the process used to ship an order to a customer. There are a few simple rules to follow when constructing a flow chart:

1) List all known steps in the process,
2) Use the simplest symbols possible,
3) Make sure that every manufacturing loop has an escape,
4) Make sure that every process is logically connected to both the preceding and succeeding processes in the process flow.

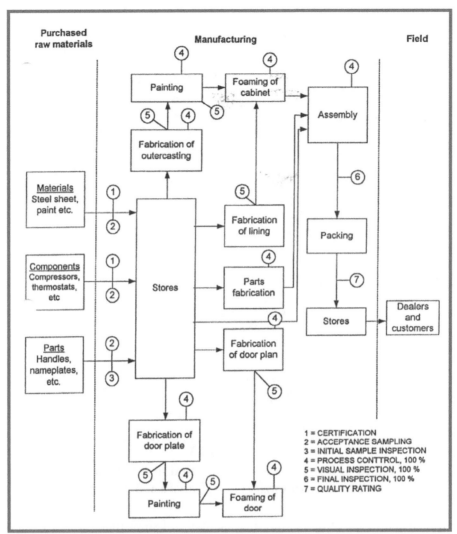

Figure 14.4
Process Flow Diagram for Refrigerator Manufacturing

14.4.4 The Pareto Diagram

Pareto diagrams are a type of bar chart or histogram which displays the frequency with which a particular phenomenon is occurring relative to the occurrence of others. Such diagrams help focus everyone's attention on the most frequently occurring problems and prioritizing which is the most important problem to solve. This type of diagram is styled after the *Pareto Principle*. The Pareto principle is that only a few things will cause the majority of the disturbances. These are areas where quality improvement must focus. A plot of this nature shows the *biggest bang for your buck* when trying to determine the most important Kaizen events to launch. A Pareto chart helps to establish top priorities and is visually very easy to understand. A Pareto analysis is also fairly easy to execute. The only expertise which may be needed is in the area of data gathering, Figure 14.5 shows a Pareto analysis of the usage rate of various TQC methods reported by attendees at a QC conference

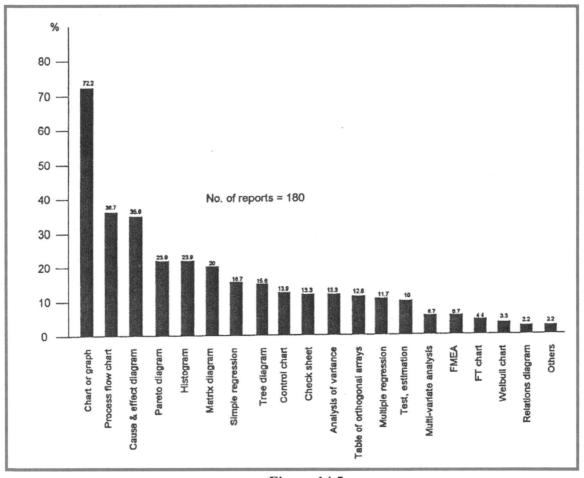

Figure 14.5
Pareto Diagram for the usage Rate of QC methods and techniques reported at a Recent National Conference

14.4.5 The Cause and Effect Diagram

One of the most effective methods for improving quality is the cause -and-effect diagram, also known as a *fishbone diagram* because of its structure. Initially developed by Kaoru Ishikowa in 1943, this diagram helps to identify the probable cause of a problem. The center line of the chart represents a quality problem that needs to be investigated. Slanted *Fishbone lines* are drawn from the main line. These lines represent the main factors that could have caused the problem (Figure 14.6). Branching away from each of the Fishbone lines are even more detailed factors. This decomposition can be continued for several cause and effect levels, but no more than 3-4 decompositions is recommended. Everyone taking part in making a cause and effect diagram contributes and gains new knowledge of the process. When a diagram serves as a focus of the discussion, everyone knows the topic, and the conversation does not stray. A *manufacturing* Fishbone diagram is often structured around four main branches; (1) the machine tools (or processes), (2) the operators (workers), (3) the method, and (4) the material being processed. Another version of the diagram is called the CEDAC; a cause-and-effect diagram on which details are recorded on cards. The effect is often tracked and monitored with a control chart

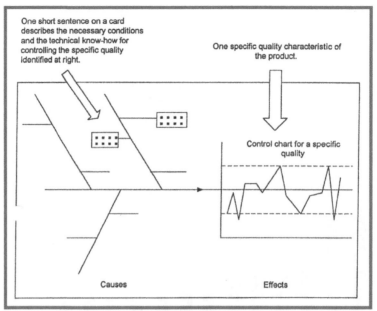

One short sentence on a card describes the necessary conditions and the technical know-how for controlling the specific quality identified at right.

One specific quality characteristic of the product.

Control chart for a specific quality

Causes

Effects

Figure 14.6
A Cause and Effect Diagram Where the Effect is Shown with a Run Chart

14.4.6 The Scatter Diagram

The scatter diagram is a graphical representation which is helpful in identifying any correlation that might exist between a quality or productive performance measure and a factor that might be driving the measure. Figure 14.7 shows the general structure that is commonly used for a scatter diagram and some typical patterns.

14.4.7 Check Sheets

Check sheets are used to record data on a given process to determine the problem areas. It follows the Pareto principle in that it tries to help locate the defects, the symptoms of the defects, and the causes of the defects. To determine the defects in each area, a tally of problems can be kept for each step of process. A check sheet is shown in Figure 14.8, which represents the time for an operator to perform a manual task in a subassembly cell. The check sheet is also an excellent way to view data while it is being collected. It can be constructed to detect the magnitude of suspected problems and the frequency they occur. This behavior is observed as the data is being collected. The check sheet can provide basically the same information as the histogram.

14.5 Monitoring Process Quality

The field of Statistical Quality Control (SQC) is full of techniques and methodologies to monitor and control process quality. There are literally hundreds of books and articles addressing this subject. The interested reader is referred to Montgomery for a detailed coverage. In the context of Lean engineering, we will present only two techniques which have widespread applicability: (1) Control charts and (2) Process capability analysis. We will conclude by discussing the effects of accuracy and precision.

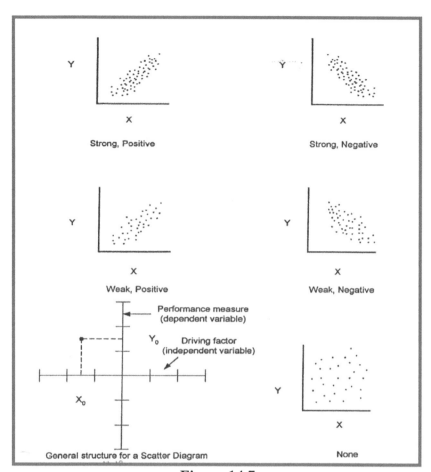

Figure 14.7
Typical Patterns of Correlation Found in Scatter Diagrams

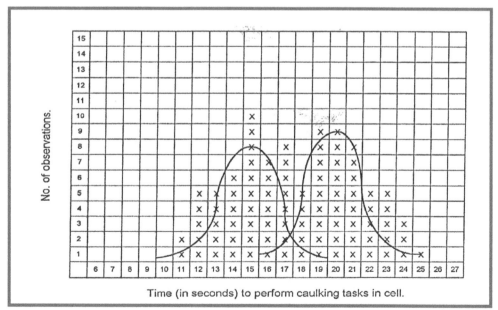

Figure 14.8
A Check Sheet to Graphically Display Machine Adjustments

14.5.1 Control charts

In Lean manufacturing, it is extremely important to produce parts and components which are defect free and conform to design specifications. The relationship between design specifications and actual production capability is often poorly understood. The process design engineer will provide three parameters for every manufactured part or component: (1) The Upper Specification Limit (USL), (2) The Lower Specification Limit (LSL), and the Target Value. For example, a 4 inch stainless steel bolt is being manufactured. A bolt longer than 4.1 inches (USL) or shorter than 3.9 inches (LSL) must be scrapped. The process which is used to manufacture the bolt has inherent production capabilities which can be characterized statistically. The two important statistical properties of the process are the *mean* and the *variance*. The mean is the average bolt length produced over a representative period of time and the variance is a measure of the spread. The square root of the variance is called the *standard deviation*. For a stable and mature process in which the process parameter (bolt length) follows a normal distribution function, about 99.75% of all manufactured parts will fall between plus and minus 3 standard deviations of the mean. It is important to monitor both the mean and variance to insure that only product within the design specifications will be produced. This is done by monitoring the process with *control charts*. The two most commonly used control charts are the *x-bar chart* for monitoring the *mean* of the process and a *range chart* for monitoring the *standard deviation*.

Statistical procedure

(1)   Collect m samples of size n and: (a) Calculate an x-bar value for each sample
(b) Calculate the range of values in each sample.

$$\overline{X}_j = (\textstyle\sum_{i=1}^n x_i)/n \quad \text{and} \quad R_j = (\text{Max } x_i - \text{Min } x_i) \qquad j = 1, 2 \ldots\ldots \text{ n}$$

(2)   Average all the x-bar and range values to get:

$$\overline{\overline{x}} = \textstyle\sum_{j=1}^m \overline{X}_j \quad \text{and} \quad \overline{R} = \textstyle\sum_{j=1}^m R_j$$

The range is clearly related to the process standard deviation. For an extremely large sample, the range is about equal to $6\sigma_x$. For small sample sizes, the range can be adjusted to closely approximate the process standard deviation using an adjustment factor of d_2.

(3)   Estimate the process Standard Deviation by $\sigma_x = \overline{R}/d_2$
The value of d_2 is a function of the sample size n and can be found in any quality control textbook such as Montgomery, Appendix 6. A partial list of d_2 values is as follows.

n	3	4	5	6	8	10
d_2	1.693	2.059	2.326	2.534	2.847	3.078

(4)   The Control Charts for x-bar and the range are given by:

$$CL = \overline{\overline{x}}$$
$$UCL = \overline{\overline{x}} + 3/\sqrt{n}\,(\overline{R}/d_2)$$

$$CL = \overline{R}$$
$$UCL = \overline{R} + 3(\overline{R}/d_2)$$

$$LCL = \overline{\overline{x}} - 3/\sqrt{n}\,(\overline{R}/d_2) \qquad\qquad LCL = \overline{R} - 3(\overline{R}/d_2)$$

Where: CL=Center Line, UCL = Upper Control Limit and LCL = Lower Control Limit

A Preliminary control chart is now constructed using these values, and each $\overline{X}_j$ and R_j values, are plotted on the $\overline{x}$ and R charts. Two different things might occur. (1) If all of the plotted points in both charts are within the control limits, then accept both charts and begin to plot periodic sampled values of $\overline{x}$ and R. (2) If one or more of the values used to construct the preliminary charts are outside any upper or lower control chart limit, look for an assignable cause. If an assignable cause is found, discard the point(s) and recomputed the preliminary control chart limits. If no assignable cause is found, retain the point(s) and assume the cause was a random occurrence. Once all points are within the control chart limits process monitoring can begin.

Suppose that a Lean engineer wants to monitor the Takt time for a new subassembly cell. To establish both the $\overline{x}$ and R charts, 15 samples of size n=6 were taken during each hour. The sampled data is shown in the following table.

$\overline{x}$ - Chart	R-Chart
CL=20.078	CL=2.834
UCL=21.448	UCL=5.68
LCL=18.708	LCL=0

Sample Taken	Sample Number 1	2	3	4	5	6	Sample Mean $\overline{X}$	Sample Range R
1	19.91	20.70	19.50	18.03	20.29	18.44	19.48	1.47
2	19.54	21.65	21.32	20.58	21.46	19.68	20.71	0.14
3	20.06	20.94	19.93	20.18	21.23	20.02	20.40	0.04
4	20.98	19.31	21.52	19.27	19.78	21.05	20.32	0.07
5	18.91	19.01	20.45	19.92	21.46	19.28	19.84	0.37
6	19.24	18.16	19.38	20.71	21.58	20.49	19.92	1.25
7	19.16	19.01	20.38	20.88	20.47	20.41	20.05	1.24
8	17.70	20.16	21.43	20.74	18.31	19.06	19.57	1.37
9	18.87	20.21	20.86	18.32	22.08	20.02	20.06	1.15
10	19.89	20.75	20.79	20.02	18.97	21.16	20.26	1.28
11	19.31	20.09	21.84	19.39	20.50	19.30	20.07	0.01
12	19.67	21.32	20.02	17.96	20.31	21.40	20.11	1.74
13	20.29	20.54	20.27	19.38	18.42	21.96	20.14	1.67
14	18.11	20.48	22.37	20.02	19.82	19.53	20.06	1.41
15	20.95	20.56	21.06	17.47	20.38	20.66	20.18	0.29

$\overline{\overline{X}}$ 20.08 $\overline{R}$ 1.21

For a sample size of n = 6, the value of d_2 = 2.534. The sampled data was used to compute the Upper Control Limit, the Lower Control Limit and the Center Line for both an $\overline{X}$ and R chart. The 15 sampled values of $\overline{X}$ and R were then plotted. Since all the x-bar and range values used

449

to construct the two charts are within the upper and Lower Control limits, the Lean engineer now implements the two control charts to monitor subassembly cell Takt time (Figure 14.9)

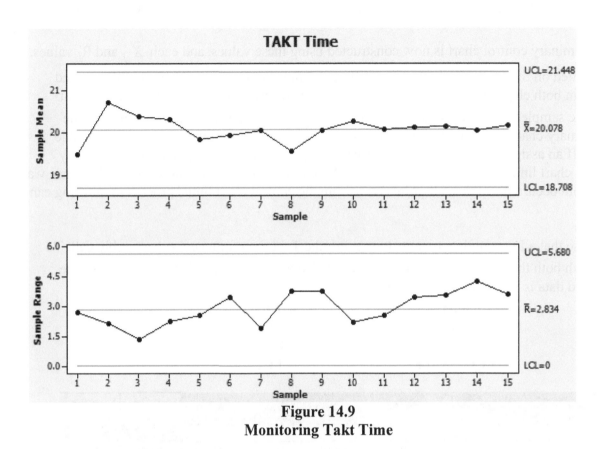

Figure 14.9
Monitoring Takt Time

14.6   Process Capability of a Centered Process

It is common practice to not only monitor process mean and standard deviation, but also to periodically report on the capability of the process to meet the three process standards: The Upper Specification Limit, the Lower Specification Limit and the Target value. There are three different relationships which can exist between actual production capabilities and the design specifications. Consider the three cases shown in Figure 14.10.

Invoking the Central Limit Theorem, the distribution of x-bar values plotted on the x-bar chart is approximately normally distributed, and is exactly normally distributed if the sampled values are taken from a population of normally distributed random variables. Hence the area under the normal probability density function within s-sigma limits is close to 0.9973. Hence, only about 0.0027 of sampled x-bar values will randomly fall outside the upper and lower control limits. Hence, when the actual production process is centered on the Design Target Value and the Upper/Lower Control limits correspond to the Upper/Lower Specification limits; the process is operating within 99.75% of the process design specifications. This case is shown in Figure 14.10-C. However, to the Lean Engineer this is unacceptable. Put another way, out of 1 million parts produced from such a process, about 2,700 will be outside the design specification limits. A more favorable situation is shown in Figure 14.10-A. If the design specification limits are more than plus or minus 4 standard deviations from the actual process average, then virtually no

450

process output would be defective. In fact, when there are 12 standard deviations between the Upper and Lower Specification units, the process will produce 0.0018 out -of -specification units per 1 million observations. This is called a *six-sigma process*. Note that in Figure 14.10, the actual process mean ($\bar{\bar{x}}$) is shown centered at the design specification Target Value for all three cases, and there is a value of c_p shown in each diagram. c_p is a *quantitative measure of process capability*. The value of c_p shown in each diagram is calculated as follows.

$$c_p = \frac{(USL-LSL)}{6\sigma}$$

Note that the true process standard deviation (σ) is in the denominator. The value of σ is rarely known, so it is estimated by $\sigma' = \bar{R}/d_2$ as before. Hence:

$$c_p = \frac{UCL-LCL}{6\,\bar{R}/d_2}$$

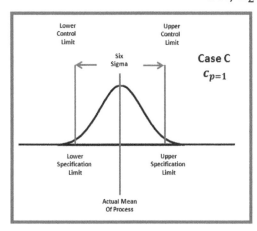

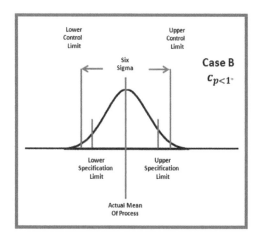

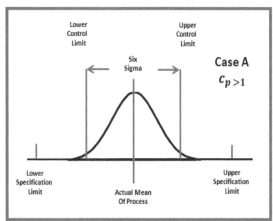

Figure 14.10
Relationship Between Design Specifications and Actual Processing Capability

Note that if the parameter being monitored follows a normal distribution function with the Upper and Lower Control limits calculated as the mean plus or minus 3 sigma, the total spread between the natural Upper and Lower Control limits (UCL-LCL) is 6 sigma. In Figure 14.10 case C

$c_p = 1.0$. It follows that in Figure 14.10 Case A, $c_p > 1.0$ and $c_p < 1.0$ in Figure 14.10 Case B. It is clear that the case shown in Figure 14.10 Case A is preferred; the case shown in Figure 14.10 Case B is totally unacceptable and the case shown in Figure 14.10 Case C is not desirable. It is generally conceded that a $c_p > 1.3$ is good, and any $c_p > 1.9$ is world class. A $c_p = 2.0$ is commonly called a six-sigma process.

14.7 Process Capability for an Off-Centered Process

In order to use $c_p = \frac{(USL-LSL)}{6\sigma}$ to measure process capability, *the process must be centered*. It has been observed by the authors that c_p is regularly reported and posted in many manufacturing companies without realizing that if the process is not centered on the target value, this value is meaningless at best and totally misleading at the worst. Consider the three situations shown in Figure 14.11.

In Case A, the actual process is producing exactly on Target. In Case B, the actual process mean has drifted to the right, producing more defects than Process A. In Case C, the true process mean has moved outside the Upper Specification Limit and is producing well over 50% defects. Note that in all three cases, the value of c_p is 1.0. The problem lies in the formula used to calculate c_p, which fails to recognize that in Cases B and C, the process mean is off-center. To accurately reflect the true process capability, the following procedure must be used.

$$c_{p,k} = \text{minimum} \{ c_{p,u}, c_{p,l} \}$$

$$\text{Where: } c_{p,u} = \frac{USL - \text{Target Value}}{3\sigma}$$

$$\text{And: } c_{p,l} = \frac{\text{Target Value} - LSL}{3\sigma}$$

For example, consider a process that has design specifications of USL=40.0, LSL=30.0 and Target Value = 35.0. From 30 samples of size n=5, the process was found to produce at $\bar{\bar{x}} = 37.0$ with σ_x estimated with $\bar{R}/d_2$, as 1.0. The process is clearly producing product off centered from the target value of 35. If the process capability index is reported as $c_p = \frac{(USL-LSL)}{6\sigma}$, then $c_p = \frac{(40-30)}{6(1)}$ or $c_p = 1.6$; which is world class production. The actual process capability is:

$$c_{p,k} = \text{Min} \{ c_{p,u}, c_{p,l} \} = \text{Min}\{ \frac{(40-37)}{3(1)}, \frac{(37-30)}{3(1)} \} = \text{Min}\{ 1.0, 2.33\} = 1.0$$

In this case, the process is only marginally acceptable. Of course, this is an opportunity for the Lean engineer. If the process can be centered without increasing the process standard deviation, then world class production can be achieved.

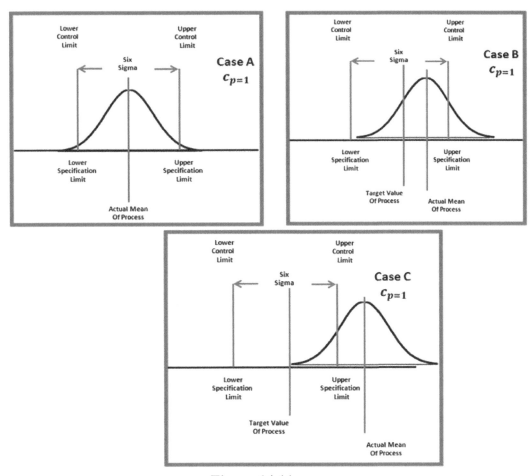

Figure 14.11
Off Centered Processes

14.8 Precision versus Accuracy

The upper and lower control limits on the x-bar and R control charts represent *the actual production capability* for a mature, stable process. The engineering design function will independently provide three design parameters for any part produced by the process: (1) the target value of the process, (2) the upper specification limit (USL) and (3) a lower specification limit (LSL). Any part which is not produced between (USL-LSL) is considered unusable or defective. A good process will make parts well within the specification limits prescribed by the design engineer. Although sometimes misunderstood, there is no relation whatsoever between the USL/LSL and the UCL/LCL on a control chart. The former are product design specifications, and the latter is what the actual process or machine is capable of producing.

All manufacturing processes display some level of random behavior, referred to as *inherent capability* or *inherent uniformity*. For example, suppose that we perform the following experiment: *shooting at the bull's eye on a metal target*. I hand you the gun and tell you to take nine shots at the bulls-eye. Thus you are the operator and the process. To measure the *process capability* (PC) that is your ability to consistently hit what you are aiming at (the bull's-eye) the target is inspected after you have finished shooting. So the capability of a manufacturing process is determined by measuring the output of the process. In *quality control* (QC), the product is examined to determine whether or not the process is accomplishing what was specified in the design....the *nominal value,* and the *Upper and Lower Tolerance Limits.* Inherent capability can be measured by the dual concepts of *accuracy* and *precision.* Accuracy is how close all outcomes come to the desired target. Precision is how closely bunched are all possible outcomes. A precise result may or may not be accurate, and an accurate result may or may not be precise. Figure 14.12 graphically depicts four possible outcomes.

The *accuracy* of *the process* refers to both the *uniformity* and the *aim* of the process. Thus in the target-shooting example, a perfect process would be capable of placing all nine shots right in the middle of the bull's-eye, one right on top of the other. The process would display no random behavior with perfect *accuracy.* Such performance would be very unusual in a real industrial process. Almost all production processes exhibit some degree of variability. This variability may have *assignable causes* which can be corrected if the cause can be found and eliminated. Variability for which no cause can be assigned and which cannot be eliminated is *inherent* in the process and is called *random.* The goal of an x-bar control chart is to make sure that the process is centered or running as close as possible to the target value in the design specifications, and that the variance of the produced part or process is low enough to keep the parts produced within the upper and lower tolerance limits. In addition, the actual process capability must also be well within the upper and lower design specifications. The best process performance will always occur when the process mean corresponds to the design target value. Such a process is said to *be centered.* If the distance between the process mean and the Upper / Lower Specification limits is 6 -sigma, this is called a *6-sigma process.* Such a process will produce only 34 defective parts in a production run of one million parts; even if the process mean *drifts* plus or minus 1.5 sigma from the target value.

Quality Control analysis frequently uses other statistical and analytical tools other than Control charts. Going back to our example, a QC study might be directed toward quantifying the inherent accuracy and precision in the process. This particular QC program might be launched to root out problems than can cause defective products during production. Traditionally, the objective of QC studies has been to monitor and control processes. The more progressive point of view is to use statistical analysis and tools to both reduce variance and prevent defects from occurring. Variance reduction is at the heart of 6-sigma programs.

Some examples of *assignable variability* that can be eliminated include not using multiple machines for the same components; different operators with different levels of training; operator self-induced mistakes; defective or non-homogeneous raw materials; and progressive wear in tools used during machining variance induced by these factors can be found and eliminated. Sources of *inherent variability* in the process include variation in material properties; operator variability; vibrations, chatter, and the wear in the machine tool components. Such variance components cannot be eliminated. These variance components constitute what we have called *random* or inherent variance components. Almost every process has multiple causes of variability occurring simultaneously, some may even be correlated. The new Lean engineer will be a champion of both Statistical quality Control charts and 6-sigma initiatives. Both are necessary to eliminate all defective products and rework.

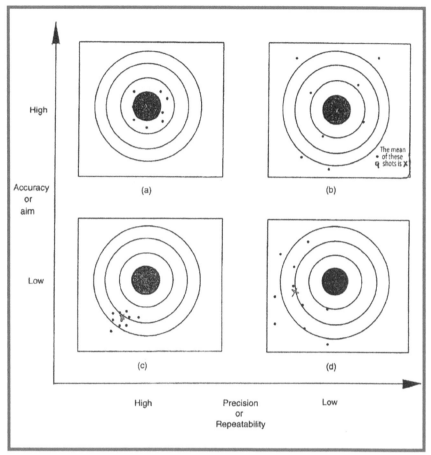

Figure 14.12
The Concepts of Accuracy and Precision are Shown in the Four Target Outcomes: (a) The process is accurate and precise; (b) Accurate but not precise; (c) Not accurate but precise; (d) Neither accurate nor precise.

14.9   Additional Benefits of Quality Control (QC) Studies

The application of Control Charts is extremely useful in making sure that the actual mean and standard deviation of the process under study is acceptable and exhibiting only random behavior. Out-of-Control points caused by non-random behavior immediately signal a problem in the process. Monitoring Process Capability with the correct Process Capability Ratio is a good way to report process behavior relative to design specifications. Periodic checks on process precision and accuracy can help insure a consistent, repeatable process. QC studies can also tell the lean engineer how pilot processes compare with existing production processes, and vice versa. The time when the sample was taken is carefully recorded, so information about process time-varying behavior can also be generated. The consistency and repeatability of the process with respect to time (time-to-time variability) is often ignored or overlooked in QC studies. Of course, patterns in control chart points (runs, cyclic behavior, etc.) will also help to identify changes in time-varying behavior. Most important, when processes are duplicated, QC studies generate information about machine-to-machine variability. Going back to our target-shooting example, suppose that two different guns were used, all of the same make and type. The results would have been different, just as having two marksmen use the same gun would have resulted in yet another outcome. Thus QC studies generate information about the homogeneity and the differences in multiple machines and operators.

It is quite often the case in QC studies that one variable dominates the process. Repeatable and accurate processing is often directly related to the skills and capabilities of the operator. Processes that are not well engineered or not automated, or in which the worker is viewed as highly skilled, are usually operator dominated. Processes that change or shift uniformly with time, but has good repeatability in the short run are often machine dominated. For example, the mean of the process will usually shift after a tool change, but the variability may decrease or remain unchanged. Machines tend to become more precise (to have less variability within a sample) after they have been broken in (i.e., the rough contact surfaces have smoothed out because of wear), but will later become less precise (will have less repeatability) due to poor fits between moving elements (called *backlash)* of the machine under varying loads. Other variables that can dominate processes are setup, input components, and even information interpretation by the operator.

In many CNC or automatic machining processes, the task of tool installation and alignment has been replaced by an automatic tool-positioning capability, which means that one source of variability in the process has been eliminated and the process becomes more repeatable. In the same light, it will be very important in the future for manufacturing engineers to know the process capability of robots they want to use in the workplace.

The discussion to this point has not assumed that the *parent population* is normally distributed, because the distribution of the sampled mean for reasonable sized samples (n>3) are approximately normally distributed regardless of how the underlying population random variables are distributed. However, if the underlying f(x) is not Normal, then the inherent capability of the process to consistently follow predicted behavior may be influenced. In particular, if the underlying f(x) is highly skewed one way or another, there may be an underlying tendency for the process to produce defects in that direction. The shape of a histogram from individual measurements may reveal the nature of the process to be skewed to the left or the right (unsymmetrical), often indicating some natural limitation in the process or machine. Drilled holes exhibit such a trend as the drill tends to make the hole oversize but not undersize. Another possibility is a bimodal distribution (two distinct peaks), often caused by two processes being mixed together. The possibilities are endless and require a careful recording of all the sources of the data to track down the factors that result in loss of precision and accuracy in the process. Remember that an X-bar chart is based upon a sampled average value and is intended to only monitor the mean of the process. An R-Chart must be run concurrently with an X-bar chart using the same set of sampled data to monitor process variance. Always check the range chart first for out of control conditions, since the sampled standard deviation is used in calculating the UCL and LCL in an x-bar chart. Avoiding a time delay between when the sample is taken and when it is plotted and analyzed is perhaps the most important factor.

14.10 Quality Redefined

It is always cheaper to do the job right the first time. As Phil Crosby (1979) would say, *quality is free, but it is not a gift.* The cost of quality is the expense of doing things wrong or doing the right thing with a defective part. Simply allowing defects to occur and adjusting output to compensate for poor quality is a recipe for disaster. All defects must be found and eliminated, and this costs money and time. G. Taguchi (1986) has provided yet another definition of quality: Deviation from a target or goal. This definition uses the cost of not hitting the nominal (the target) value.

Quality can be defined as conformance to specifications or requirements. This means that the standards of conformance must be precisely stated. Failure to meet all standards of conformance *always* cost the company money. The fastest and surest path to low-cost operations is to make the product right the first time, thereby eliminating rework and scrap. Studies done on the amount of rework done in a typical factory are scary, and often range as high as 30-40 percent, meaning that 30-40 percent of what the company makes requires some rework. Some authors refer to this as the *hidden factory*; meaning the *rework factory* within the factory.

14.11   Internal Quality Checks (IQC) in Lean Manufacturing Cells

IQC goes hand-in-hand with the cellular manufacturing systems concept. Cell operators control the quality in the cells. The rule is *Make One-Check One-Move One On.* It's that simple. Perform a step in the process; Check the product to ensure that step was done correctly; Move on to the next step. Between the processes in the cell, devices can be added to assist the worker in checking the part, thus ensuring that defective products do not get passed on. This is called *Poka-yoke*. Checking can be manual or automatic. Automatic checking forms the basis for *autonomation.* This is a very important concept, though the word is often confused with *automation.* Autonomation refers to the *autonomous control of both defects and quantities.* For manned (and unmanned) cellular systems, this means that the storage areas or the *decouplers* between the processes are equipped with sensors to detect one or all of three things

(1)   When a defective item has been produced
(2)   When something has gone wrong with the process
(3)   When something is changing that will eventually lead to failure in meeting product specifications (defect prevention)

Figure 14.13 shows the relationship between autonomation and lean manufacturing. Physically, sensors and devices are incorporated into the machines and the decouplers to automatically check the critical aspects of the parts at each stage of the process. Causes of defects are investigated immediately, and corrective action is implemented.

14.12   IQC Concepts

Within the organization of a company, quality must be the responsibility of everyone involved in manufacturing. The involvement of production line workers is absolutely necessary. This means that the primary responsibility for quality is assigned to the people who make the product. These workers must develop the responsibility of continuous improvement and the desire for perfection. (They must strive for zero defects.) This requires that the desire to strive continually for perfection in quality must be instilled in the workers. Remember, quality depends on the efforts of everyone, from sales through design and purchasing, manufacturing, shipping, and so forth. When changing the design of the manufacturing system it is critical to change the attitude of the workers. The effects of defective products on the L-CMS cannot be ignored. Quality is just as important as output rates.

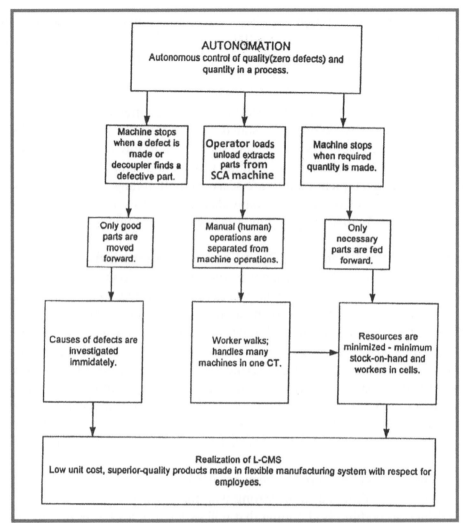

Figure 14.13
How Lean Manufacturing is Realized by Autonomation (Toyota)

In fact, it is usually not recognized that by eliminating defects and rework at a workstation will actually increase *capacity*. Machine capacity is usually increased by: (1) Working overtime (2) subcontracting out work or (3) Purchasing additional equipment. Notice that all three options for creating additional capacity will cost money, thereby decreasing profits. Of course, one could argue that to eliminate all defects and rework will also cost the company money, but this is probably an illusion. If a company has a Quality Control function, it should be their job to improve quality. After all, this is their reason for existence. Second, compensating for poor quality by subcontracting, working overtime or purchasing more equipment is a continual expense; resolving quality issues is likely a permanent solution.

14.13 Basic Principles of IQC

1.   Design and operate the process to prevent defects, rather than inspect after the fact to find defects. The Japanese term for defect prevention is *Poka-yoke*. At every stage, the product must be checked; thus every worker must be an inspector. Quality is controlled at the source.

2.   Production workers correct their own errors, and there are no separate rework

lines. This requires one-piece flow in the cells and immediate feedback to the place where any problem might surface on the final assembly lines. This does not necessarily mean that workers inspect their own work. The next worker can check the work pulled from the previous worker or an automatic inspection device can be placed between the processes or the workers to check the quality characteristics. This is one form of autonomation using *decouplers*.

3.   Make quality easy to see. Display boards and highly visible charts on the plant floor. Boards and charts detail the quality factors being measured, the state of recent performance, current quality improvement projects, recent award winners for quality improvement, and so forth. Quality and its characteristics must be clearly and simply defined.

4.   Insist on compliance to the quality standard. Conformance to the quality standards must come first, ahead of output.

5.   Give workers the authority to stop the process when something goes wrong. Mechanized processes can have devices to do this automatically (in-process inspection). More refined systems may have the ability to adjust or modify the process to correct the problem. The machine must be programmable. Equip the machine or process to prevent the defect from occurring using Poka-yoke devices or methods.

6.   For the inspection of finished goods, make it a rule to check 100 percent of the *critical* attributes. This is part by part inspection and not lot inspection.

7.   Make your *modus operandi* a constant succession of projects for quality improvement in every work area. Continuous improvement should be a routine way of life. This is one of the main functions of *quality circle* groups.

8.   Eliminate incoming inspection. The objective is to move toward no inspection of incoming goods. This requires the buyer to work closely with one vendor, to the exclusion of all other vendors. Ultimately all vendors should be lean manufacturers and be an extension of the lean system.

9.   Eliminate setup time to enable unit flow. The drive toward small lot size requires the elimination of setup time, which in turn makes it possible to reduce lot size. The concepts of EOQ and EPQ are limiting. The theoretical assumptions are not correct. In addition, it is easier to spot problems. The control of inventory levels is discussed in more detail in Chapter 15.

10.   The optimum lot size is one. This is readily achievable within cells. Obviously, between cells it may not be possible to transfer material in unit lot sizes, and in most cases there is no real advantage in doing so.

11.   Keep the workplace clean. Good housekeeping is fundamental and absolutely necessary in order for a plant to improve quality and foster better work habits. Housekeeping is the responsibility of everyone, from plant manager to foremen and workers. Housekeeping is needed to improve and maintain safety in a dangerous environment and to maintain pride and company morale. Nobody really wants to work in a dirty place. A dirty workplace is obviously a deterrent to superior-quality performance in electronics fabrication, painting, and finishing areas, but other areas of the plant must also be well maintained.

12.   Organize the workplace. This is really important in areas where there are many loose tools and multiple components. The motto is *A Place for Everything, Everything in its Place.*

14.14 Six Sigma Manufacturing

In order to meet the challenge of the Japanese, an American electronics company, Motorola, developed the six sigma concept. What is six sigma? *Six Sigma is* an aggressive and systematic program aimed at reducing the variability in key product characteristics, thereby reducing to reduce or completely eliminating any product outside of design specifications.

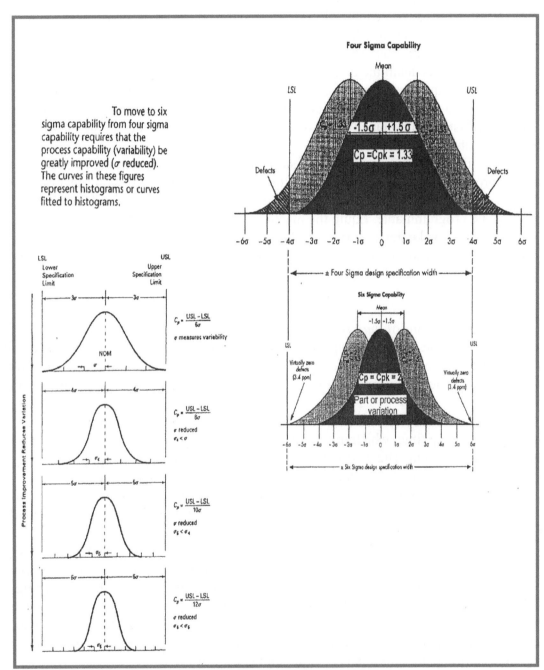

Figure 14.14
A ±3 sigma Process Compared to a ±6 Sigma Process.

More simply put, the goal of Six Sigma is to reduce process variance so that even a significant shift in the process mean results in very few defects. The six sigma concept is shown two ways in Figure 14.14. The process is *centered* on the target value (center line) in the top curve. In the picture just below that one, the process is allowed to drift either 1.5 standard deviations to the

460

right or 1.5 standard deviations to the left. Even so, the Upper and Lower Specification Limits limits are still 4.5 standard deviations from 99.75% of all values. A process operating in this manner will only have 0.2 defective parts per million, compared to 2,700 parts per million for a standard 6-sigma process. The advantage in reducing process variance to achieve 6-sigma status cannot be overstated. It is crucial to Lean systems design.

The true (unknown) process variance can be estimated by $\frac{\bar{R}}{d_2}$. The distance between the USL and LSL does not change (decrease) when improving from three sigma limits to four, five or six sigma levels. Variance is decreased through process improvements. Notice again that the actual distance from the LSL to USL is 12 sigma, not 6. Table 14.5 provides some practical meaning or examples of sigma levels of defects compared to a 99.75% normal 3 sigma process.

Table 14.5
The Practical Meaning of Six Sigma

99% GOOD (3 Sigma)	99.99966% GOOD (6 sigma)
20,000 lost articles of mail per hour	Seven articles of mail lost per hour
Unsafe drinking water for almost 15 minutes per day	One unsafe minute every 7 months
5,000 incorrect surgical operations per week	1.7 incorrect operations per week
Two short or long landings at most major airports each day	One short or long landing every five years
200,000 wrong drug prescriptions each year	68 wrong prescriptions each year
No electricity for almost 7 hours each month	One hour without electricity every 34 years

14.15 Six Sigma Methodologies

Six sigma uses a five-phase approach to affect change: ***Define, Measure, Analyze, Improve and Control*** (DMAIC) as shown in Table 14.6. DMAIC is a structured problem solving life cycle with standard procedures which were adopted from the military.

Employees and suppliers of the company are often trained in Six Sigma methodologies. There are three different levels of training: The lowest level is called *Green Belt*, the next highest level is called Black *belt* and the highest level is called *a Master Black Belt*. (Table 14.7). Once trained, employees seek improvement opportunities in their functional areas of work. With experience, an employee can achieve progressively higher levels of certification and company-wide recognition. Lean/Six Sigma engineers gain experience as keen listeners and problem-solvers, not strictly as *tool masters*. They use creativity and leadership to discover and execute challenging projects to successfully reduce variance in a wide variety of production settings.

14.16 Lean Six Sigma Tools

Lean/Six Sigma methodologies include a variety of tools such as: Brainstorming, Affinity Analysis, Process Mapping, Pareto Charts, Fishbone Diagrams, Kaizen, Mistake Proofing, Point of Use Storage, SQC X bar and R Control Charts for process monitoring, Taguchi Methods, Hypothesis Testing , ANOVA, Regression, Design of Experiments, Monte Carlo Analysis, Correlation and Variance reduction techniques. Unfortunately, the full potential of 6-sigma has not been realized because 6-sigma training often fails to explore and appreciate the role of variance reduction and process improvement within the context of Lean Engineering. We advocate a full marriage of Lean Engineering with 6-sigma methodologies to achieve excellence.

Table 14.6
Six Sigma DMAIC to Reduce Variability

Six Sigma Steps to Reduce Variability (DMAIC)	
Define	**Define** the project goals and customer deliverables
	• What is the problem I am trying to fix?
	• What is the CTQ?
	• What is the customer's expectation?
Measure	**Measure** the process to determine current performance; quantify the problem
	• Can I measure it?
	• How is my process performing today?
Analyze	**Analyze** and determine the root cause(s) of the defect
	• What is my current process?
	• What are the root causes of defects?
Improve	**Improve** by eliminating the root causes
	• How can I "fix" those root causes?
Control	**Control** future process information
	• How can I keep those root causes in control, so I don't have to fix them again next year?

14.17 Taguchi Methods

The drive towards superior quality has led to the introduction of Taguchi methods for improvement in products, product design, and processes. Basically, SQC is focused to process monitoring and defect detection, not defect prevention or variance reduction. Taguchi methods, however, span a much wider scope of functions and include the design aspect of product development and the use of cost components in guiding process analysis. The pioneering work of Taguchi has not been properly recognized by the theorists. Taguchi was a practicing engineer in a real company. He was interested more in cost effective solutions than theoretical correctness. His work has created another threshold in quality control, witnessed by the continuing application of his contributions to the expanding role of quality in the production of goods and services. The consumer is the central focus of attention on quality, and the methods of quality design and control have been incorporated into all phases of production.

Table 14.7
The Hierarchy of Six Sigma Practitioners

Training Levels for Six Sigma	
Champion • Top Leadership • Identifies key business issues • Allocates resources	**Master Black Belt** • Full-time deployed resource • Master of tools/creative solutions • Experienced in leading change • Trainer and coach
Black Belt • Dedicated project team leader • Skilled in tools/problem-solving techniques • Trainer	**Green Belt** • Executes projects related to role • Knows most tools • Project facilitator
Six Sigma Employees • Participates on process improvement teams in their work Area	

Taguchi methods incorporate the following general features:
- Quality is defined in relation to total loss to the consumer or society from less than-perfect product quality. Methods include placing a monetary value on quality loss. Anything less than perfect is waste.
- In a competitive society, continuous quality improvement and cost reduction are necessary to stay in business.
- Continuous quality improvement requires continuous reduction in the variability of the product-performance characteristics. This observation by Taguchi is perfectly in line with the basic philosophies of Lean Engineering presented in this book.

Taguchi methods cannot be used to executing process capability studies, and should not replace statistically correct process capability analysis (PCA). PCA is a daily, ongoing activity and Taguchi methods are more directed to changes in design for manufacturing or basic process changes. One Taguchi approach uses what is called a *truncated experimental design (TCD),* which is a special application of factorial designs. The Taguchi TCD is called an *orthogonal array,* and determines which process inputs have the greatest effect on process variability. Inputs with the greatest influence are set at various levels to determine their effect on process variability. As shown in Figure 14.15, factors A, B and C have an effect on the process variable, V. By selecting a high level of A and low levels of B and C, the inherent variability of the process can be reduced. Those factors that have little effect on the process variable V (D in this case) can be used to adjust or re-center the process without increasing variability. In other words, Taguchi methods seek to discover and eliminate causes of variability, and, thus reduce total process variability.

The following important observations can be made.

- The Engineering Design function and the associated manufacturing systems largely determine the quality and cost of a manufactured product.
- Exploiting and mitigating the nonlinear, interactive effects of a process or product parameters on performance characteristics reduces the variability in product/process performance characteristics.
- Statistically planned designs of experiments or Taguchi experiments determine the settings for processes and other parameters that reduce variance.
- Design and improvement of products and processes can make them robust, and thus insensitive to uncontrollable or difficult-to-control variations (called *noise* by Taguchi).
- Taguchi methods are more than just statistical procedures. They infuse an overriding new philosophy into manufacturing management, basically making quality the primary issue in manufacturing with the focus on profit and the cost of poor quality. The manufacturing world needs to clearly understand that the consumer is the ultimate judge of quality. Continuous quality improvement toward perfect quality is a necessary and achievable goal. Finally, it is recognized that final quality and cost of a manufactured produce are determined to a large extent by the engineered designs of:
  - ➤ The product
  - ➤ The manufacturing processes
  - ➤ The tooling used
  - ➤ The methods used

Lean statistical analysis must be used to:
  - ➤ Statistically monitor and analyze every manufacturing process
  - ➤ Uncover the nonlinearities and interactions of every component in the manufacturing system (integration of the product and the process)
  - ➤ Determine the critical role that *Variance* plays in quality, system performance measures, customer satisfaction and product sales price

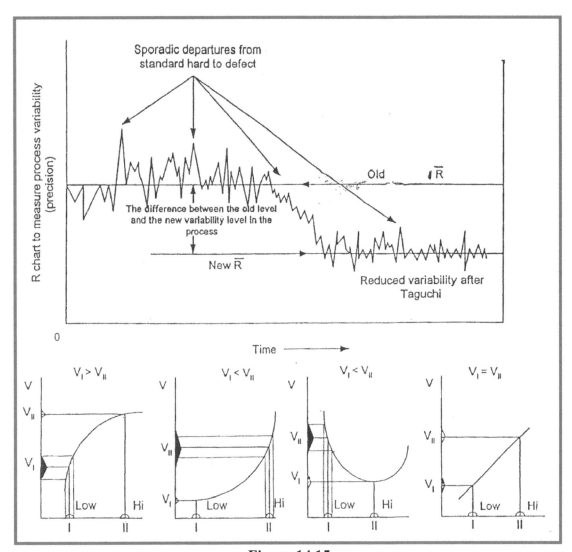

Figure 14.15
The Use of Taguchi Methods can Reduce Process Variability as Shown in the Figure Above; Factors A, B, C and D versus the Process Variable V are Shown in the Lower Figures

A new understanding of quality has emerged. Process variability is not an uncontrollable factor that must be lived with instead of corrected. The variance of a process can be reduced by exploring the nonlinear and interactive effects of the products (or process) parameters on performance characteristics and product quality.

It is predicted that the future manufacturing management will include the following changes:
- Continual training and massive implementation of statistical process control (focusing on company- wide Quality Assurance programs
- Use of *statistical process Control* as necessary part of producing a final product. Training and implementation of Six Sigma and Taguchi methods for process design and improvement of products and processes
- Concurrent (or simultaneous) engineering of products and processes to reduce the

time needed to bring new high quality, low cost designs to the customer.

- Attitude adjustment and worker empowerment to make quality the primary consideration
- Institutionalize continuous improvement of all processes, procedures, work elements and work practices
- Make Variance reduction a primary goal of the Manufacturing Engineering led by a fully trained Lean Engineer
- Aggressively and relentlessly attack and eliminate all forms of waste and non-value added time

The unifying concept of all these methods is to continuously improve the Manufacturing System Design, particularly the individual processes involved. (Figure 14.16). However, be cautioned that process improvement and Quality Control Kaizen events need to be carefully chosen. There a multitude of continuous improvement projects which will look good *on paper* but *may or may not improve systems performance.* Millions of dollars can be wasted on improvement projects which later will be found to have no significant impact on profit or systems performance measures. It is recommended that the system be modeled to assess impact before actually engaging in a Kaizen event. This requires time and hard work on the part of everyone.

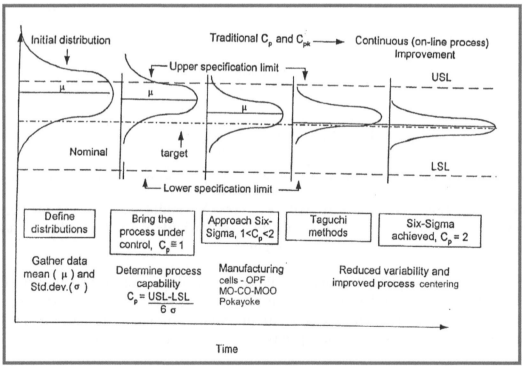

Figure 14.16
Continuously Improving the Processes is Part of the Continuous Improvement Effort for the Lean System.

14.18 Teaming and Quality Circles

A number of Quality Control programs are built upon the concept of participative management; such as quality circles, improvement teams, and task groups. These programs have been very successful in many companies but have failed miserably in others. The difference is often due to the way management implemented the program and the crucial role that the internal customer plays in Quality Improvement. All quality improvement programs must be integrated and managed within the context of a Lean manufacturing system design strategy. For example, asking an employee for a suggestion that management does not use or will not use defeats a suggestion system. Management must learn to trust the employees' ideas and decisions and move the decision making to the factory floor. Even if quality improvement suggestions offered by process operators fail, this serves a useful purpose. We all learn by our mistakes.

The *quality circle* movement started in Japan in 1962 and grew rapidly. Quality circles are a popular form of participative management. However, other countries have often experienced problems duplicating the QC circle movement. Perhaps it is because quality circles are the last brick in the IQC wall, and other countries have not taken the necessary preparatory steps. A quality circle is usually a group of employees within the same department with an organization structure composed of members, a team leader, a facilitator, often a lean or manufactured engineer, and a steering committee.

Quality circles usually have the following main objectives. They provide all employees with a chance to demonstrate their ideas; raise employee morale; and encourage them to use their knowledge to solve quality problems. Table 14.8) lists some typical production problems which might be addressed by quality Circles. Quality circles also unify company-wide QC activities, clarify managerial policies, and develop leadership and supervisory capabilities Quality circles have been implemented in U.S. companies with limited success when they are not part of a Lean strategy. It is possible for quality circles to work in the United States, but they must be encouraged and supported by management. Everyone must be taught the importance and benefits of integrated quality control and given the opportunity to make a difference.

Table 14.8
Typical QC Opportunities for Quality Circles

Typical QC Improvement Opportunities	
Paperwork	Materials
Hardware	Software
Communication	Tooling
Service	Material Handling
Processes	Delays
Scrap Reduction	Cost Reduction
Productivity	

14.19 Poka-yokes

Many companies have developed an extensive QC program based on having many inspections. However, all inspections can do is find defects, not prevent them. Adding more inspectors and inspections merely uncovers more defects, but does not prevent them. Clearly, the least costly system is one that produces no defects. But is this possible? Yes, it can be accomplished through two methods: Defect prevention by using Poka-yoke, and root-cause analysis or fixing Quality problems at the source.

Table 14.9
Quality Circle Objectives

	Quality Circle Objectives
1	Develop workers' skills and knowledge
2	Introduce a team effort among workers, supervisors and mangers
3	Unlock the creativity inherent in workers
4	Improve quality consciousness
5	Create a more harmonious work force leading to high morale
6	Encourage commitment and contribution to corporate goals of better quality and higher productivity
7	Encourage leadership qualities in circle leaders
8	Improve communications and extend recognition

Many people do not believe that the goal of zero defects is possible to achieve, but many companies have operated with near- zero defects for a considerable length of time, or have reduced their defect level to virtually zero using Poka-yokes and Root-cause analysis. *Poke-yoke* is a Japanese word for defect prevention. Poka-yoke devices and procedures are often devised mainly to improve the safety of operations. However, this is extremely short sighted since Poka-yokes can also be used to significantly improve process quality. The idea is to develop a method, mechanism, or device that will prevent the defect from occurring, rather than to find the defect after it has occurred. Poka-yokes can be designed and attached to any process to automatically check the products or parts in a process. Poka-yokes differ from source inspections in that they are *proactive* and instantaneous, not *reactive* and time delayed. The production of a bad part is prevented by the device. Some Poka-yoke devices automatically shut down a machine if a defect is produced, preventing the production of an additional defective part. Poka-yokes use 100 percent inspection/detection to guard against unavoidable human and machine errors. Figure 14.17 illustrates some Poka-yokes. One device ensures that the worker will apply a label to all finished products, thus insuring that each part is used in the right place at the right time.

Manual inspections work very well when *physical* detection is needed, but many items can be checked only by *sensory* detection methods, such as monitoring for tool wear or ensuring that the surface finish is correct. For such problems, a system of automation and sensory checks need to be used. *Source inspection*s using manual and visual inspections are often effective in looking for obvious problems before they cause defects. Internal machining characteristics and adherence to standards are difficult manual tasks. Detection of these types of processing problems are best performed by automated Poka-yoke devices which either stops the system or automatically makes corrections before a defective item can be made. The common term for automated inspection in manufacturing processes is adaptive control (A/C). There are two ways to characterize inspections relative to an individual machining center or process step: *vertically* and *horizontally*. *Vertical inspections* try to control *upstream processes* that can be the source or the cause of downstream defects. *Horizontal inspections* try to control the way current processes effect downstream manufacturing. Here is an example. Steel bars were being cylindrically ground. After grinding, about 10% of the bars warped (bent longitudinally) and were rejected. The grinding process was studied extensively and no cause was found. The problem was with the

Before improvement

The worker had to remove a label from a tape and place it on the product.

The operation depended on the worker's vigilance.

After Improvement

The tape fed out by the labeler turns sharply so that the labels detach and project out from the tape. This is detected by a photoelectric tube and, if the label is not removed and applied to the product within the cycle time of 20 seconds, a buzzer sounds and the labeler stops.

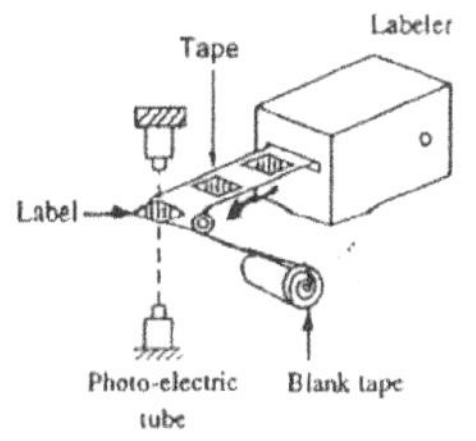

Before Improvement

Since there are as many as 100 kinds of parts involved on a single line, workers take parts indicated by lights. Assembly defects would sometimes occur because workers would take the wrong parts.

When assembly kahban were inserted in a reader, lights above the parts would go on as signals to workers. (Workers sometimes took the wrong parts even with the signal lights.)

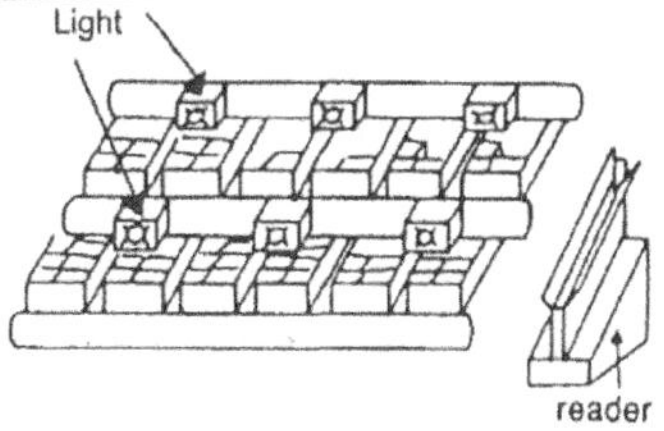

After seemd Improvement

When an assembly kanban is inserted in the reader, air cylinders push out only the parts boxes needed. (Parts cannot be taken out of boxes that are not protruding.)

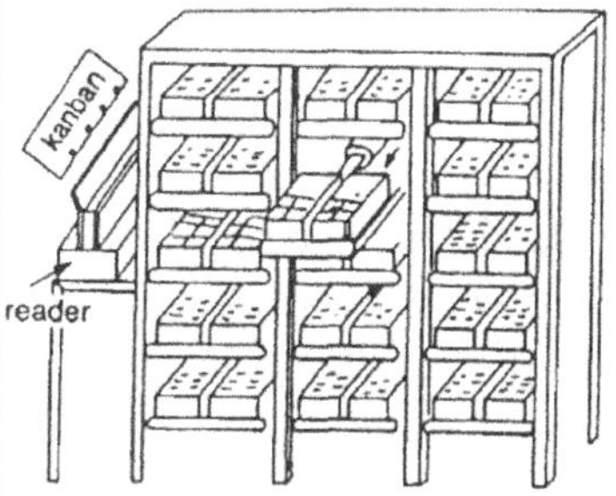

Figure 14.17
Examples of Poka-yoke's for Preventing Operator Errors (Sekine)

heat-treat process that preceded cylindrical grinding. About 10% of the bars were not getting a complete, uniform heating prior to a quenching operation. These bars lay close to the door of the oven, which was not properly sealed, resulting in a temperature gradient inside.

Quenching of the bars induced a residual stress that was released by the grinding and caused warping. Thus, horizontal source inspections should detect poor manufacturing practices or physical problems that could or would result in downstream problems. Corrective actions should then be taken to eliminate the problem.

Manual operations can present some additional psychological problems which are not always recognized. If a worker inspects each part immediately after it finishes processing, this is called *self-checking*. There is an immediate feedback to the worker on quality. However, it would be difficult for many workers not to allow a certain degree of bias to creep into their inspections, whether they were aware of it or not, since they are inspecting their own work. Within cells operated by multiple workers, the operator of the *downstream* station or process can inspect the parts produced by the *upstream* operator. If there is a problem with the parts, the defective item is then immediately passed back to the worker at the previous station. There, the defect is verified and the problem corrected. Action is immediately taken to prevent any more defective parts. While this is going on, the line is shut down. This is called *successive checking*.

In order for *successive checks* to be successful, several rules should be followed. All the possible variables and attributes should not be measured. This would eventually lead to errors and confusion in the inspection process. The part should be analyzed so that only one or two critical attributes are inspected. Only the most important elements are inspected or perhaps only the features more prone to error. Another important rule is that immediate feedback of a defect must lead to immediate action. Since the parts are produced in an integrated manufacturing production system, this will be very effective in preventing the production of more defective parts. Suppose that the cell has only one or two workers and they are not in a position to directly check each other's work after each step. What can be done in this situation? Here is where the *decouplers* scan play a major role by providing automatic checking of the parts' critical features before proceeding to the next step. Poka-yokes often play a major role in decoupler inspections. Only perfect parts are pulled from one process to the next through the decoupler.

14.20   Line Stops

A pair of yellow and red lights hanging above the workers on the assembly line can be used to alert everyone in the area to the status of the processes. Many companies use *Andon boards* which hang above the aisles. The number on the board reflects stations on the line. A worker can turn on a yellow light when assistance is needed, and nearby workers will move to assist. The line will keeps moving until the product reaches the end of the station. Only then is a red light turned. If the problem cannot be solved quickly and the line needs to be stopped for a long period of time, a Lean engineer should be immediately called to the workstation. When the problem is resolved and everyone is ready to go again, the red light goes off and everyone starts back to work, all in synch.

Every worker should be given the authority to stop the production line to correct quality problems. In systems using Poka-yoke or autonomation, devices may stop the line automatically. The assembly line and in synch subassembly lines and cells should be stopped immediately and started again only when the necessary corrections have been made. Although stopping the line takes time and money, it is advantageous in the long run. Problems can be found immediately, and the workers have more incentive to be attentive because they do not want to be responsible for stopping the line. It is also important to hold an *after-action meeting* of all engineers and line workers involved to discuss how/why the problem occurred, and how it can be prevented in the future.

14.21   Properly Implementing QC

The basic idea of QC integration is to shift functions that were formerly done by a staff organization, often called the Quality Control Division, onto the manufacturing floor if QC inspections and responsibilities shift to the factory floor, what happens to the quality control department? The department serves as the facilitator and therefore acts to promote quality concepts throughout the plant. In addition, its staff educates and trains the workers in statistical and process control techniques and provides engineering assistance on visual and automatic inspections. The quality control department also performs complex or technical inspections, total performance checks (often called end item inspection), chemical analysis, X-ray analysis and destructive testing. One of its most important functions may be to educate and train the entire company in Quality Control and Lean Engineering practices.

Another important function of the QC department will be to work with the external *Supply Chain* and the vendors. The vendor's quality must be raised to the level at which the buyer does not need to inspect incoming material, parts, or subassemblies. The vendor simply becomes an extension of the buyer's plant. Ultimately, each vendor will deliver to the plant perfect materials that need no incoming inspection. Note that this means the acceptable quality level (AQL) of incoming material reflects 100%, defect-free items. Perfection is the goal. For many years the empirical evidence from process capability studies suggested that 2 or 3 percent defective was about as good as you could get! Better quality just costs too much. For traditional mass production systems, this may have been acceptable. But in order to achieve the kinds of quality that Toyota, Honda, Sanyo, and many others have demonstrated, one has to eliminate the job shop (a functional manufacturing system) and the functional production system and integrate the quality control function directly into the linked-cell manufacturing system using one-piece flow and employee training.

14.22 Making Quality Visible

Visual display on quality should be placed throughout manufacturing facilities to make quality evident. These displays tell workers, managers, customers, and outside visitors what quality factors are being measured, what the current quality improvement projects are, and who has won awards for quality. Examples of visible quality are signs showing quality improvements, framed quality awards presented to or by the company, and displays of high-precision measuring equipment. These types of displays have several benefits. In lean manufacturing, customers often visit the plant to inspect the processes. They want to see measurable standards of quality. Highly visible indicators of quality such as control charts and displays should be posted in every department. Everyone is informed on current quality goals and the progress being made. Displays and quality awards are also an effective way to show the work force that the company is serious about quality.

14.23 Moving to World Class Quality

Quality improvement should not be dictated by a philosophy but by the quality problems that the company is facing. If the main problem is a lack of process capability, then the most effective approach is a combination of statistical methods with knowledge of the physics and chemistry of the process. There is strong anecdotal evidence that the result of process capability studies level off around a few percentage points of internally detected failures. The next order of magnitude improvement in quality performance then comes from reducing QC problem detection time by moving to manufacturing cells with one-piece flow. However, this change alone may not allow the company to achieve low parts per million (ppm) rates of defectives because human error affecting individual parts is still possible, and this is where mistake-proofing becomes important. Mistake-proofing is the strategy that takes you from a low failure rate to an extremely low failure rate like 15 ppm, or between 5 and 6 sigma in the vocabulary of the six sigma movement. Statistical methods help most when the processes are at their worst; followed by cellular manufacturing and one-piece flow to achieve world class quality levels. Therefore, the approach to quality improvement for any company should be dictated by the quality problems that they are addressing.

It is well known that quality improves as a result of converting a sequence of operations into a U-shaped cell. The effect of one-piece flow usually levels off at percentages of defects on the order of 1 % to 0.1 % or 1,000 ppm. Human error which affects individual parts is still possible and so mistake-proofing and Poka-yoke devices are needed to take you from 1,000 ppm to 15 ppm. Table 14.10 summarizes this discussion.

Table 14.10
Quality Levels and Improvement Approaches (Baudin)

Defective Parts	Quality Problem	Solution
30% - 3%	Poor Process Capability	Statistical Process Control + Process Capability Engineering
3% - 1%	Slow defect detection time	Cells, flow lines, visual controls
1% - 15 ppm	Human error	Mistake-proofing, Poka-yoke, defect prevention

The figures in the left column are interpreted as orders of magnitude and correspond to numbers typically observed in automobile companies. Once process capability is established, the plant faces a change in the nature of quality from drifts in process parameters to randomly occurring discrete events such as tool breakage, missing steps, or mislabeled components. The challenge at this point is to detect the defects quickly and react immediately. Speed takes precedent over analytical sophistication at this point. With one-piece flow designs, from the time the part leaves the input part of the cell until it reaches the output side each part moves individually from step to step without any accumulation of work between machines or workstations.

One piece flow in U-shaped cells has several positive impacts on product quality. The lead time thru a fixed sequence of operations using pull philosophies is an order of magnitude shorter than with batch and queue using push philosophies. If parts are tested immediately after processing in decouplers, negative test results can be fed back to the line before the process has a chance to damage another part. Because parts move directly to the next process step, downstream quality delivery is insured. Any defect discovered during testing in decouplers can be immediately traced back to the preceding process. Effective but simple QC checks often take the form of a *go/no-go* gage test, or what Shigeo Shingo called *successive inspection procedures.*

14.24 One Hundred Percent Inspection?
As part of their MO-CO-MOO routine, the operators check every part. These concepts are difficult for quality professionals trained in SPC to accept because they exclude both measurements and sampling. They have been trained to think that they should never use a *go/no-go* gage when you can take a measurement, because the measurement is much richer in

information. The measurements tell you more than if the measurement falls outside the upper and lower specification. It also tells you exactly where. They have been told that measuring 100% of the parts is unaffordable and the well-designed sampling plan will give you the same information at a much lower cost. In Lean Manufacturing, statistical sampling is not used inside manufacturing cells. Here is why:

1. Sampling allows defective parts to wait and accumulate between measurements which defeats the goal of detecting and responding to problems quickly.
2. Sampling often disrupts operations.

The operator's job is choreographed in a sequence of motions that is repeated with every part. When measurements are to be taken at every fifth part at one operation, and every seventh part at another; the operator's routine is not consistent. On the other hand, a check of every part can be made part of the routine and *go/no-go* gages can be engineered to make these checks fast enough so as to not slow down production. Lean manufacturing cells cannot work with processes that have even single digit percent defectives. High volume production with poor quality requires buffers between every operation simply to protect the downstream operations from the fluctuations in the output of the upstream link. When the main quality problem is one of operator error, then clearly the next frontier of improvement is their training and education. However, if the main problem is the inability of a process or machine to hold tolerance consistently, expending resources on mistake-proofing would be like lining up the deck chairs on the Titanic to aid lifeboat entry. Conversely, statistical methods are not much help in the prevention of errors that occur once or twice a month or are a result of operator error.

14.25 Conclusions

This Chapter has focused on the methodologies, tools and procedures needed to improve quality. There are, of course, may other issues that need to be addressed that are beyond the scope of this Chapter. The middle ground where one-piece flow plays a central role in quality assurance is the key to zero defects and defect prevention. However, flow lines and manufacturing cells are rarely mentioned in the literature on quality control even though their ability to markedly improve quality is well known.

The impact of superior quality is market penetration, increased sales and increased profits. Competition necessitates the need for superior quality. Due to competition in the free markets both here in the United States and abroad, there is every reason to expect that proper use of Quality Control methods will not only force major improvements in products and processes, but will also result in higher bottom line profits. The methods presented in this Chapter are relatively easy to use. They do not require extensive training and education in probability, statistics, or experimental design, so they can be quickly grasped and employed. However, proper use and interpretation of quality control methodological results need extensive training in basic statistics. While it is true that most engineering graduates in the United States have little if any formal exposure to probability, statistics, and experimental design concepts and methods, the small set of QC methods presented in this Chapter should still prove to be extremely effective. Product designers must seek the least cost methods to insure the quality of the desired functional characteristics. We again repeat the belief that this is a major area of impact for the new breed of Industrial Engineers that we call the Lean Engineer.

Review Questions

1. What is the main idea of IQC?
2. What does TQC stand for and when and where did it get its start?
3. Who was Shewhart and what was his contribution to quality control?
4. What is the difference between variance produced by an assignable cause and that produced by unknown causes?
5. Explain the two types of errors that can occur in making decisions based on samples.
6. From Figure 14.1, verify the calculation of the sample averages, the range values, the average of the 25 samples and the average range.
7. Is the sum of all the individual measurements divided by 125 equal to the sum of all the sample averages divided by 25? Discuss.
8. Explain what autonomation is all about.
9. In control charts, why is an alpha error made so small?
10. Take the data in Figure 14.1 and make a histogram out of it. Discuss in terms of comparing the process to the specifications.
11. What information does a run chart show that a histogram does not?
12. Explain the function of a Pareto Diagram.
13. What is Pareto's law?
14. What is a cause and effect diagram used for?
15. How are $\bar{X}$ and R charts used to determine the capability of a process? Using the data from Figure 14.1, calculate Cp for this process. Discuss.
16. Outline the steps in a PC study.
17. What are the five basic steps in a six sigma project?
18. How does a Taguchi method reduce the variance in a process?
19. Why aren't Taguchi methods more widely used in industry to reduce variation in processes?
20. How do SQC, OPF in lean manufacturing and Six Sigma compare in terms of percent defectives?
21. What is the difference between a run chart and a control chart? Why don't we show tolerance values on a control chart?
22. What can be done before the startup of a manufacturing cell that will minimize variance in the process capability?
23. Who has the best ideas about how to improve quality in the cells?
24. The originators of control charts believed that the variability in a process was inherent. What did they mean by that?

Problems

Problem1

Develop a brief description, with a quality based example of each of the 7 tools for quality and be ready to present one of them in class.

Problem 2:

Develop process flow diagram for the construction of a subway sandwich.

Problem 3:

Control charts are to be constructed for a machining process turning the outside diameter of a cylinder. The diameter has specifications of 4.5 ± 0.050 centimeters.

a) If $\sum \bar{X} = 157.85175$ centimeters and $\sum R = 2.18750$ centimeters for 35 samples of size n=5, calculate the centerlines and control limits for the control charts.

b) Assume that the charts show that the process is in good statistical control. What percent of the parts produced will be within tolerances if this process continues under a constant system of common causes and can be approximated by the normal distribution?

c) What percent will be within specifications if the process were to be centered at the prescribed nominal of 4.500 centimeters?

d) Graphically Show the situations in parts b and c in terms of distributions of the individual measurements.

Problem 4

Samples of 4 units were taken from a manufacturing process at regular intervals. The diameter of a bored hole is measured using an inside vernier caliper. The average ($\bar{X}$) and R is computed for each sample. After 25 samples of size 4 were taken, the following results were obtained: k=25, n=4.

	X Bar Chart	Range Chart
Upper control limit	UCL =625	UCL=37.5
Average or Mean	$\bar{\bar{X}} = 614$	$\bar{R} = 16.5$
Lower Control Limit	LCL=602	LCL=0

All the points on the $\bar{X}$ and R chart fell within the control limits and no unusual trends were observed.

The specifications were given as 610 ± 15; USL = 625, LSL =595

Questions to be answered:

a) Is the process centered with respect to the specification limits?

b) With the process centered at $\bar{\bar{X}}$, what percentage of product would you expect to fall outside the upper specification limit? What percentage would you expect to fall outside the lower specification limit? How did you calculate σ'?

c) What percentage of product would you expect to fall outside the specification limits if you were able to center the process exactly on the nominal?

d) What might be involved in centering this process, assuming that the process is turning on a lathe?

e) What assumption are you making about the distribution for the calculations you performed in parts b and c?

f) What is the process capability as measured by C_p? Is this good or bad? Discuss in the light of the control chart findings.

g) What is the relationship between the upper control limit on the X bar chart and the upper specification limit?

h) What does C_{pk} calculation take into account that the C_P calculation does not?

i) What is the standard deviation of the $\bar{X}$ values about $\bar{\bar{X}}$? Call this standard deviation σ_x. Does it have another name?

j) In using the control charts, one makes decisions about the process. If you are the person running the chart, you are going to decide if the process is in control or out of control, based on the sample information. What kinds of errors can you make in this decision? How often do you make these errors?

Problem 5

The variation in a critical dimension produced by a certain discrete part manufacturing process can be described by a normal distribution with mean equal to the nominal (the set point of the process) and standard deviation of 002 inches.

    a)   Determine the natural tolerance limits of the process.

    b)   Suppose the designed specified the actual tolerance for the output to be ±0.005 inches and the process is set at the nominal as specified by the design engineer, determine the proportion of product that will not meet the specified tolerance.

    c)   If the actual tolerance as specified by the design is ±0.005 in. but the process is set at a value that is 0.001 inches above the nominal as specified by the designer, determine the proportion of the product that will not meet specification.

    d)   Specify the required accuracy and precision of a measuring sensor that might be used in an on-line inspection system for this process. What is the standard deviation that characterizes the measurement distribution?

    e)   Calculate the Capability Index C_p for the conditions described in part b and the Capability Index for the conditions described in C where the process mean has shifted, C_{pk}

Problem 6

For figure 14.1, sample data is found that was used to plot the $\bar{X}$ and R charts. (a) Determine the control limits for the charts and comment on the process. Is the process in control? (You will have to look up some constants to do this task). (b) Redo the initial calculation for the first 11 samples of size 4 dropping out the data on line 4 of the sample measurements. Then assume you have just taken the data point at 12:00 on 6/10. What do you decide about the process? What kind of error can you make? What is the chance you will make this error?

Problem 7

To determine the present status of a piston-turning operation where the surface roughness is an important characteristic, it has been decided that a process capability study should be performed. The surface roughness has specification of 105 ± 10 microinches. Use the data shown in the table, collected in samples/subgroups of size n=5 to perform this study.

Sample	$\bar{X}$	Range	Sample	$\bar{X}$	Range
1	108.2	13	16	103.6	8
2	105.8	27	17	115.4	24
3	104.0	24	18	103.2	34
4	104.6	16	19	108.0	71
5	106.6	22	20	108.2	57
6	111.4	32	21	100.6	26
7	104.6	12	22	98.2	73
8	102.2	23	23	96.4	47
9	111.0	22	24	101.4	26
10	113.4	18	25	105.6	49
11	102.2	23	26	91.4	68
12	108.0	26	27	116.6	52
13	111.2	22	28	113.6	53
14	103.2	31	29	112.4	42
15	100.4	6	30	99.6	37

Chapter 15
Integrated Production and Inventory Control

Kanban is a tight suit that does not fit on an obese body.   *Taiichi Ohno*

15.1   Introduction

Production Control strives to answer the following questions: Where is material needed, where will it come from, when is it needed and how much is required? Inventory Control is the effort to know and determine how much material is at a given location in the manufacturing system at a given time. The integration of these two system level control functions into the production and management system is the topic of this Chapter.

Traditional methods for production and inventory control evolved out of the work of F. W. Taylor, Henry Gantt and Frank Gilbreth, the original Industrial Engineers, who developed planning and scheduling strategies for the job shop. These methodologies push material through the plant according to a master production schedule (MPS) and material requirements planning (MRP or Manufacturing Resource Planning (MRP II). We will refer to this class of software as simply MRP for convenience (Figure 15.1). These systems use forecasted demands and an economic order quantity (EOQ) calculation to release orders to the shop floor. Dates and quantities for raw material purchasing and delivery are determined by computer algorithms and MRP break-back structures. Inventory in the plant is very hard to track and control because the MRP tools are of a planning nature, with little or no ability to actually manipulate or control the inventory. There is often a lack of real-time capacity planning or capacity availability.

The lean methodology for production and inventory control uses a pull system, known as Kanban (card or signal in Japanese), which responds to actual and forecasted demand by delivering parts and products only as they are needed, just-in-time (JIT). Manufacturing and supply chain functions are all triggered by Kanban. The order release is to final Mixed Model assembly, and not the first workstation. In this way, the materials produced and used in production are controlled by the downstream customer (either internal or external) who pulls parts from suppliers (upstream) only as needed. This is accomplished by sending empty containers to be refilled, with production authorized by either single card or double card Kanban control.

Inventory is strictly controlled due to the nature of the Kanban links that connect the work cells. These links contain a specified number of containers, each with a Kanban card, that hold a precise quantity of parts. Thus, the amount of inventory in the link is known at all times and is determined by the number of Kanban cards and the number in each standard container. WIP can easily be increased or decreased by just by changing the number of containers (Kanban cards).

15.2 What is Kanban?

The Kanban system was originally conceived at the Toyota Motor Company as part of the Toyota Production System (TPS). Some have equated Kanban with TPS itself, but this is not accurate. The TPS is a manufacturing system that is managed and controlled by Kanban. The JIT philosophy at Toyota led to the development of Kanban as a way to control the movement of material in the system and minimize inventory, while ensuring that the system could meet customer demands and use JIT production. For this reason, Western manufacturers initially called the TPS system a JIT system.

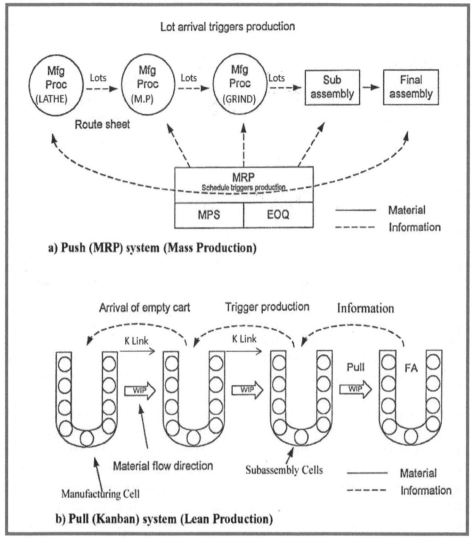

Figure 15.1
Comparison of a Push Production Control System to Pull (Kanban)

Since it was designed with JIT objectives in mind, a Kanban system requires a lean manufacturing system to be truly effective. A company cannot expect a newly implemented Kanban system to deliver dramatic results in a traditional job-shop environment. However, a company that has completed the first five steps of the redesign of a manufacturing system according to the philosophy outlined in Chapter 8 will be adequately prepared to implement a Kanban system.

When Harley Davidson implemented lean manufacturing in the early 1980's, they reduced the number of people working in the production and inventory control area from 9 to 3. The Kanban pull system was fully integrated, which means it functioned within the manufacturing system almost automatically, calling for parts as needed. Any routing sheets were eliminated, but there was no confusion about where the parts needed to go or when they need to go there, or how many to send. All of this information was built into the system.

The Kanban system is unique in that control information moves in the opposite direction of material movement. That is, downstream usage directly controls upstream production rates. Now here is an important distinction to be made about materials in the linked-cell manufacturing system. Material within a cell is called stock-on-hand (SOH). Material between cells in the links is work-in-process (WIP) inventory. When you hear the term zero inventory, this means no inventory in the manufacturing cells.

A Kanban is usually a rectangular card enclosed in a vinyl sack attached to a container of parts. Sometimes the cards are made of plastic so that they can be reused. The containers are designed for each component (part) type and hold a precise quantity (generally small) of that part number. The prerequisites of implementing a pull system with Kanban control are:

- **5- S Basics**
- **leveled and balanced production**
- **Mixed Model final assembly**
- **U shaped, manufacturing/subassembly/assembly cells for one piece flow**
- **Standard work and standard operations**
- **Reduction of setup times**
- **Quality Control with TPM (Zero defects, No breakdowns)**

Toyota is by no means the only company to implement a pull system using Kanban cards to control production and inventory. Many companies, foreign and domestic alike, have jumped on the Kanban bandwagon. Some have even developed hybrids of push and pull systems, including CONWIP constant WIP systems. Many companies have experienced tremendous success in reducing inventory while shortening delivery times. Of course, with this ever-increasing popularity a number of technical papers have surfaced criticizing Kanban performance for a variety of reasons. Where were these critics when MRP systems were destroying throughput times and dramatically inflating inventory? However, in almost every case, the Kanban criticism has arisen from an implementation which was done on a piecemeal or limited bass; thereby destroying the basic tenets of integrated Kanban. Most companies have yet to embrace the total concepts of lean manufacturing, and have failed to take the steps necessary to redesign the entire manufacturing system. To those who think that lean manufacturing is nothing more than a Kanban system, the following anecdote (liberally interpreted by the author Shigeo Shingo) is offered.

Now you might think that the Toyota Motor Company is just a company wearing a smart suit (referring to Kanban), and you want to buy such a suit for your company. However, if you only buy the Kanban subsystem, you soon discover that this suit will not fit your obese, fat body (your manufacturing system) and chaos soon results. (Shingo, 1981)

15.3 How Does A Kanban System Work?
The Kanban system (as well as any pull system) has the unique aspect of an information network that flows backwards from all external and internal customers, through the manufacturing

system, and ultimately to the company's suppliers. The system is relatively easy to design and implement. It usually consists of rectangular cards that are attached to specially designed, standardized containers. Each container is specifically designed for one part type, and it holds a precise quantity of parts. All containers have the same capacity, and a specific number of them are placed into what is known as a link. The link allows containers to move in a circuit, linking either two work cells or adjacent workstations. The Kanbans create a manual, visual system that can be easily understood by every worker on the factory floor. This allows the workers to develop a trust in the system, and this trust is necessary for any system to be accepted and successful.

15.4 Types of Kanban Systems

There are many production and inventory control systems that claim to fall under the Kanban heading. Often, any type of system that enforces pulling of parts and material will be entitled as such. In this Chapter, however, three systems will be examined. The first is known as the Dual-Card System, which as the title implies, requires two types of Kanban cards; a withdrawal Kanban (WLK) and a production authorization Kanban (POK). The second system, a Single-Card Kanban, is similar to the Dual-Card system but doesn't require the production ordering Kanban. Finally, the third system is a hybrid of push and pulls systems. This system, known as CONWIP (or Constant Work-in-Process), is not a classical Kanban system, but has similar functions and even uses some form of Kanban card or signal.

15.4.1 Single-Card System

The Single-Card Kanban System is simpler though similar in function to the Dual-Card System. It is actually a combination of push and pull production control strategies. The manufacturing portion of the system is based on a push philosophy. A daily schedule is given to each cell, and production follows this schedule rather than waiting for the signal from a production-ordering Kanban.

A withdrawal Kanban (Figure 15.2) is used to pull parts from an upstream cell only as needed by the downstream cell. This is the only Kanban needed, and it circulates between the upstream stock point and the downstream cell. No input stock points are needed, since withdrawn containers are delivered directly to the downstream cell, as downstream demand requires. Containers move according to the following pattern: Assume that Workstation1 is supplying parts to Workstation2. Once Workstation2 empties a full container, an authorization (Kanban card) is given to move the empty container to Workstation1. The arrival of an empty container at Workstation1 is a signal for Workstation1 to produce enough parts to refill the empty container. Meanwhile, the Kanban card is attached to a full container at Workstation1.This authorizes the movement of the full container to Workstation2. The full container is placed in the Workstation WIP storage (queue), and the pattern repeats itself. More specifically, the Single-Card System (Figure 15.3) functions according to the following rules: The single Kanban card is called a Move Authorization Card (MAC).

1. The MAC from a container just emptied at a downstream workstation or a cell placed in a cell *Kanban Collection Box* (KCB).
2. A material handler arrives at the KCB and moves an empty container with the Kanban card to the designated upstream production cell.

3. The Kanban card is attached to a full container of parts and the empty container placed in the output WIP area to be filled.
4. Finally, the WLK is attached to a full container and moved to the requesting downstream workstation.

............The pattern now repeats itself.

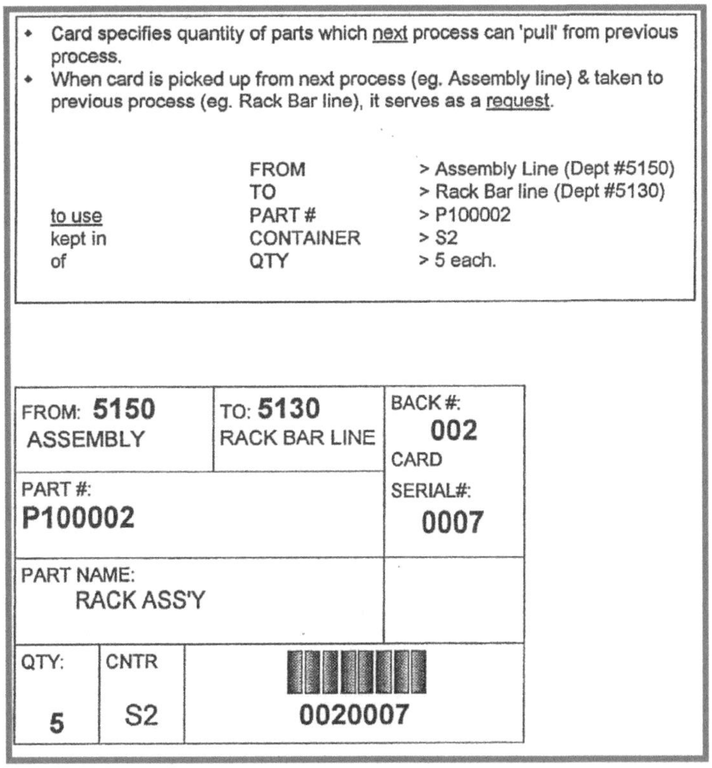

• Card specifies quantity of parts which <u>next</u> process can 'pull' from previous process.		
• When card is picked up from next process (eg. Assembly line) & taken to previous process (eg. Rack Bar line), it serves as a <u>request</u>.		

	FROM	> Assembly Line (Dept #5150)
	TO	> Rack Bar line (Dept #5130)
<u>to use</u>	PART #	> P100002
kept in	CONTAINER	> S2
of	QTY	> 5 each.

FROM: **5150** ASSEMBLY	TO: **5130** RACK BAR LINE	BACK #: **002** CARD
PART #: **P100002**		SERIAL#: **0007**
PART NAME: RACK ASS'Y		

QTY: **5**	CNTR **S2**	▌▌▌▌▌▌ **0020007**

Figure 15.2
Kanban Card in a Single-card Kanban system

The Single-Card Kanban System is solidly rooted in a JIT philosophy. The containers, for example, are all part specific; have a standard capacity; and are found in specific numbers within the links. In addition, the cells/workstations are all designed with quick setups in mind allowing small lot sizes in mixed model production to be delivered in each container. This simple system places a cap on the number of active containers in use at any one time. Figure 15.3 provides a schematic of a single card system actually being used in a rack and pinion assembly system.

One advantage of the Single-Card System is it is simple to implement. This system also relieves clutter and confusion around the downstream input area, as the need for an input stock point has been eliminated.

Once this system has been mastered, a Dual-Card System will often be developed adding a production authorization card.

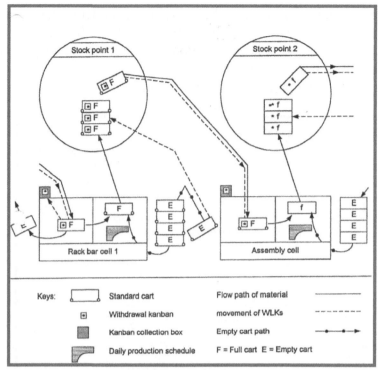

Figure 15.3
Single-card Kanban Links Rack Bar Cells to Rack and Pinion Sub-assembly Cells

---

Kanban RULES
Single card Kanban

1. Parts are transported in a specified number of containers and in specified quantities (a).
2. All parts in a container must be same.
3. All parts in a container must be of good quality.
4. A full container should have the exact number of parts as specified on the Kanban card. (a = number of parts in container)
5. A Kanban card must accompany every full container.
6. Containers must be stored only in designated areas
7. Kanban cards and containers must be picked up and delivered between departments by a material handler.
8. Parts must be made or delivered only when requested by a Kanban card.
9. When the first part is removed from a full container, the Kanban card must be immediately removed and placed in the return post.
10. The team leader should be immediately informed of any discrepancies in the Kanban system.

Figure 15.4
Rules for a Single-card Kanban System

15.4.2 Dual-Card Kanban System

The two types of Kanban cards that are used in a dual-card Kanban are: (1) the production-ordering card (POK) that signals an upstream cell or process to produce a certain part and (2) a withdrawal Kanban (WLK) that serves to link two cells or processes. Figure 15.5 shows examples of these two Kanban cards. Figure 15.6 shows an example of a WLK as used by the Toyota Motor Manufacturing Company.

Store Shelf no. *N* Item back no.	Preceeding process
Item no. *Serial # of part*	Rack bar cell
Item name *Rack bar*	
Car Type *Camry*	Subsequent process
Box capacity / Box type / Issued no. *a* / / *3 of 11*	Rack pinion assembly

Figure 15.5-a
Example of a Withdrawal Kanban (WLK) in the Manufacture of Rack Bars.

Store Shelf no. **M** Item back no.	Rack cell
Item no. *Serial # of part*	
Item name *Rack bar*	
Car Type *Camry*	

Figure 15.5-b
Example of a Production Ordering Kanban (POK) in the Manufacture of Rack Bars

There is precisely one POK and one WLK for each container. They identify the part number, container capacity, the previous cell, the next cell or process, and other information. A WLK specifies the type and the quantity of a part number that a downstream process can withdraw from the upstream process. A POK specifies the type and quantity of the part that the next cell or process must produce. The beauty of this system is that it is simple and visual and the users understand how it works. Therefore, the users trust the system. How many understood how their MRP (or ERP) system works? How can a company use a system (that controls the life blood of the company) that no one understands on the shop floor? So another aspect of the Kanban system is that everyone who uses the system understands how the system works.

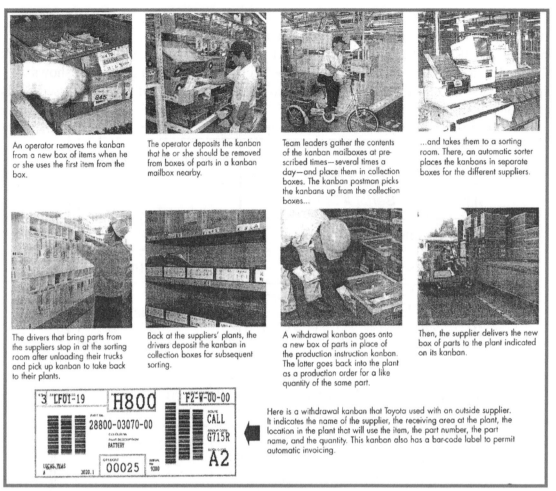

Figure 15.6
Withdrawal Kanban and Movement of Cards and Inventory for an Outside Supplier
(Courtesy Toyota Motor Company)

How a Dual card Kanban Works

Figure 15.7 demonstrates the container and Kanban flow patterns for two cells. In this example, Cell1 supplies parts to Cell2 as well as to other cells in the plant. Cell1 is serviced by stock point M, and Cell2 is serviced by stock point N. Between the cells, material is moved in standard-size carts or containers. Each container holds the same number of parts. The carts or containers move in a link (or on a circuit) from Cell1 to its stock point, then to the stock point of the next process. The material then moves into cell II for processing. Empty carts return to Cell1 for refilling. In summary, the WLKs circulate between the output side of Cell1 and the input side of Cell2, just as they did in the single card Kanban system.

POKs move between the stock point for Cell1 and the supplying work Cell2. For each container, there is one POK and one WLK. Arrival of an empty cart is the signal to the manufacturing cell to make only enough products to refill that cart. Partially full containers are not allowed. The example shown in Figure 15.7 begins at start.

1. At stock point N, a full container of parts is being moved into Cell1. The WLK is detached from the container and placed in the Kanban collection box for stock point N.
2. The most recently emptied container in cell II is transported back to stock point N where a WLK is attached.
3. The WLK and the empty container are transported back to stock point M. (Stock point N and M are not usually side by side; they may be in different parts of the plant or in entirely different buildings). The WLK is detached from the empty container and attached to a full container of parts. The full container with the WLK is returned to stock point N. (This is the withdrawing of material from the upstream cell by the downstream cell or assembly area.)
4. The full container (the one just removed from stock point M) had a POK attached to it. This POK was detached from the full container and placed in the collection box for Cell1. Then, and only then, can the container be removed from the stock point M and transported to stock point N.
5. Periodically. POKs are removed from the collection box and placed in the dispatch box for Cell1. These POKs become the dispatch list for Cell1, controlling the order of parts manufactured in the cell. These jobs are performed in the original order that POKs were received at stock point M.
6. The parts that are produced in Cell1 are placed in the empty containers taken from stock point M. The POK is attached to each container after it is filled. The container is then transported back to stock point M to be withdrawn by a downstream process when needed.
7. This sequence is repeated many times during the day. Parts are produced as needed, that is, as withdrawn from the upstream cell.
8. Cell1 may produce parts for cells or assembly lines other than cell II. All the other users operate in exactly the same manner. The carts are usually color-coded to prevent confusion.

Dual Card Kanban Rules

Rule 1: The downstream cell or process should withdraw the needed products from the upstream cell or process according to the information provided on the WLK (the needed quantity at the necessary time). While this may sound like motherhood, it is actually the realization of JIT: Parts shouldbe withdrawn "just-in-time" as they are needed, not before they are needed and not in larger quantities than needed. The key points for enforcing this rule are as follows.

1. Any withdrawal without a WLK is prohibited. This prevents a large accumulation of excess inventory at the stock point that supplies the downstream process (stock point N in the example). The number of Kanban's in the system is tightly controlled. (This will be discussed later as part of inventory control.) The withdrawal of any parts without a WLK undermines the control element of a Kanban information system.
2. Any withdrawal greater than the number of Kanban's should be prohibited. This is necessary for the same reason given above.

A Kanban (either WLK or POK) should always be attached to the physical
product or its container or should reside in a collection box. This key point
eliminates the possibility of there being any unaccounted inventory in the
system. If the number of Kanban's in the system and the number of parts per
container are strictly controlled, the amount of in-plant inventory can be quickly
determined and monitored at any time. This is the backbone of the
Kanban information system.

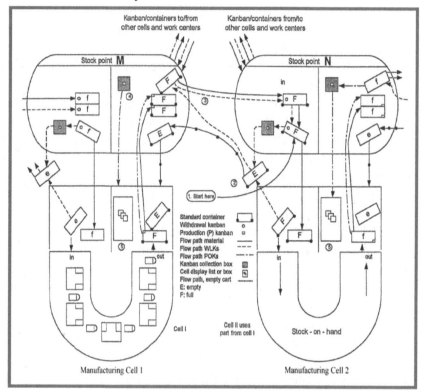

Figure 15.7
The Dual-card Kanbans Full containers From Cell 1 to Cell 2 and Returns Them Empty.
(Schonberger, 1983)

Rule 2: The upstream process should produce products in quantities withdrawn by the
downstream process or cell, according to the information provided by the POK. This is the
complement of Rule I. If parts are withdrawn *just-in-time*, then by complying with Rule 2,
the parts will be produced "This is the cornerstone of lean manufacturing that utilize the JIT
philosophy. The main points of this rule are:

1. Production greater than the number of Kanban's must be prohibited. This
   prevents a large accumulation of excess inventory at the stock point for the upstream
   process (stock point M in the example) and eliminates the possibility of
   unaccounted inventory in the system.
2. When various kinds of parts are to be produced in the upstream process,
   manufacturing should follow the ordinal sequence in which each kind of Kanban has
   been delivered. This helps to ensure that each type of part will be ready and
   available at the upstream process stock points whenever it is_ needed a t the
   downstream processes or cells.

Rule 3: Defective products should never be conveyed to the downstream process. If there is a defect, the line or cell should stop and immediately try to determine what corrective action should be taken.

It is necessary to strictly enforce zero defects and no rework to-achieve a truly effective Kanban system. Based upon Rules 1 and 2, the parts are produced and withdrawn in the necessary quantities at the necessary time. If a defective part is sent to the downstream process that operation may have to stop if there is no extra inventory in the WLK loop to replace the defective part. If a defect does occur, then the upstream processes are stopped until the problem is corrected.

Rule 4: The number of Kanban's can be gradually reduced in order to improve the processes and reduce waste. This rule conveys the fact that inventory can be used as an *independent control variable*. The level of WIP inventory is controlled by the number of Kanban's in the system at any given time. This number is initially the result of a management decision. Many companies opt for setting the initial inventory level in the link at about half of the existing level. The initial number of Kanban's can be computed by the Equation 15.1

$$K = \frac{(DD*L + SS)}{a} \qquad\qquad (15.1)$$

Where:

> K = Number of Kanbans or number of carts *(K also equals the number of POKs or the number of WLKs)*
>
> DD= Expected demand of parts, per unit time (per day)
>
> L= Lead time (i.e., processing time + delay time + [lot delay and process delay] + conveyance time)
>
> a=Container capacity
>
> SS =Safety stock

Note from Equation 15.1 the maximum inventory level *(M)* is expressed as:

$$M= Ka= DD \times L + SS \qquad\qquad (15.2)$$

The demand for parts is usually the daily demand, leveled from the monthly demand. The lead time takes into account the time needed to process a container of parts, which is the time to change over the cell and to process other items in the family, plus the time to convey a container to the usage point, plus any delay times. Delay times include lot delay and process delay. Lot delay takes into account the fact that the first part produced cannot be conveyed (to the next cell) until the last item in the lot is produced. Smaller lots reduce the lot delay time but increase both the number of containers and the material handling requirements. Process delay accounts for stoppages due to machine tool failures, broken tools, defective parts, and other manufacturing problems.

Suppose the cell is making a family of four parts, A, B, C, and D. Obviously there must be enough carts (or containers) in the loop of part A so that downstream processes do not run out of part A while the cell is making parts B, C, or D. If the manufacturing lot sizes are even moderately large, it may be necessary to create a multipart storage area, called a *Supermarket*, to buffer any downstream production while the upstream supplier produces a full cycle of parts A-D. This is characteristic of Mixed Model Final assembly.

Here is an example from Honda, in Marysville, Ohio, where they build the Accord. They have one stand of large presses that stamp out the sheet metal body parts for all the four-door Accords. This stand of presses manufactures 24 different sheet metal body parts in runs of 300 parts. (This is one day's supply of right rear doors). The presses are changed over in about 10 minutes for the different parts. Obviously, enough right rear doors must be stamped out to last through the production cycle. Honda is constantly working to reduce the time required to change over these large presses from one part to another because this will permit them to further reduce the size the run.

If the *lead time* (L) is relatively small and the demand per unit time (DD), is relatively constant, then the policy variable for safety stock can be small, resulting in a smaller inventory level. Therefore, the number of Kanbans can be smaller. In practice, this policy variable is expected to approach zero. Lead times are reduced by reducing or eliminating setup times. Another formula for the number of cards is:

$$K = \frac{DD(L_W + L_P) + SS}{a} \tag{15.3}$$

Where:

K= Number of cards; all move cards or all production ordering cards

DD= The daily demand use rate per day, derived from the level final assembly schedule, the daily master production schedule or by a similar method.

L_W= Waiting time, or withdrawal card cycle time: the amount of time for one move card to make one complete circuit between the source point of the part and the use point of the part.

LP= The processing time necessary for a production ordering card to make a complete cycle through the work cell producing the part. That includes all the time for setup, running, inspections, material preparation, and the extra time necessary for the cell to maintain a balanced running operation and keep the outbound stockpoint supplied for pickup.

a= The number of units of the part placed in each standard container.

SS= An allowance for extra time (safety stock to cover the problems and delays). Naturally, management would like to minimize the value of SS

Once the number of parts per container is calculated, this determines the number of containers required. Working with the number of containers per day maybe more convenient. Decomposing this formula, we obtain;

$$K = \frac{DD(L_W + L_P) + SS}{a} \tag{15.4}$$

Where:
    K= Number of containers required
    L_w = Withdrawal card cycle time
    Lp = Production ordering card cycle time
    SS= Safety Stock allowance

However, it may be more convenient to analyze how many cards are required by the transport process (withdrawal cards) and by the production process (production cards).The demand for parts is usually the daily demand, leveled, or leveraged over a daily or shift amount, from the monthly demand. The lead time takes into account the time needed to process a container of parts, including the time to change over the cell, process other items in the part family and convey a container to the usage point and return.

Rule 5: If there is no Kanban card, there will be no manufacturing and no transfer of parts. The WLK Kanban card, as a production control device, should always be attached to the carts (or containers) unless they are in transit within the cell to order production. This rule reveals the visual control nature of the Kanban card. The key manufacturing information is readily at hand. The removal of the Kanban card prevents a cart from being transported and used.

Rule 6: Kanbans should be flexible enough to accommodate about 10% change in demand without disturbing production. Flexibility means the system can respond to changes in demand. There are three cases where Kanban can be used to fine-tune production with respect to changes in demand.

    1.   The first case is the result of changes in the product mix of the final assembly delivery dates and of small changes in quantities. There is no change in the daily total production load. The production schedule need only be revised *for final* assembly-the schedule for all the upstream processes will be revised automatically by transferring the Kanban's.

    2.   The second case is the result of small, short-term fluctuations in the daily production load, although the monthly total load remains the same. Only *the frequency* of Kanban movement will increase or decrease. The number of Kanban's tends to be fixed despite the variation in demand.

    3.   This third case is the result of seasonal changes in demand or of increases and decreases in actual monthly demand. The actual number of Kanbans in the system must be recomputed (increased or decreased) and the production lines must be rearranged (the cycle time must be recomputed and the number of workers in the cells must be changed accordingly).

These rules must be strictly followed for the Kanban system to be an effective management information system (MIS) that also can control the inventory process (level of WIP). Kanban, as described here, is a relatively simple manual information system. Limitations for its use are as follows.

    1.   Goods must be produced in whole discrete units. Obviously, Kanban is not applicable to continuous-process industries such as oil refineries and breweries.

2. Kanban should be a subsystem of an L-CMS system, using JIT philosophy. The use of a Kanban system without the JIT philosophy makes little sense because Rules 1 and 2, regulating the use of Kanban, require the manufacture and withdrawal of the necessary arts in the necessary quantities at the necessary time.

3. The prerequisites for a L-CMS (the design of the manufacturing system, standardization of operations within the cells, and the smoothing of production) must be implemented before an effective pull system can be implemented.

4. The parts included in the Kanban system should be used every day (high-use parts). Kanban provides that at least one full container of a given part number is always available. There is not much inventory if the contents of the full container are used up the same day they are produced.

A modified version of the WLK is used to reorder raw materials (see Figure 15.8). Parts are withdrawn from a delivery lot of 500 in containers of 100. When the stack of containers reaches the material requisition Kanban, this Kanban is used to requisition a coil of steel for the process (press #10 preceded by a shear). When the signal Kanban is revealed, it is taken to the Kanban post at press #10 and placed in the queue next to the material requisition Kanban. These two Kanbans combine to instruct the worker(s) at press #10 to make 500 left doors in containers of 100. The Kanbans are reinserted in the stack of containers as shown in Figure 15.8.

15.4.3 Material Ordering Kanbans (MOK's)

There are many different types of Kanban cards which have been developed to control a wide range of operational components from the Supply Chain to final delivery. One of the more important type of Kanban control is to authorize the movement of materials and work-in-process by the Material Handling Organization. An example of a material handling Kanban is shown in Figure 15.8.

A special kind of withdrawal card is often used to get material is shown in Figure 15.9. The information on the card is very similar to what is needed on a withdrawal card except that the card has detachable pieces that go to the user's accounts receivable department as shown in Figure 15.10. A two-truck system is shown, but other similar systems have to be developed for vendor programs.

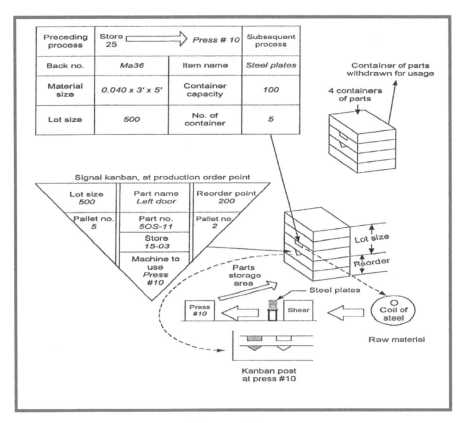

Figure 15.8
Material Ordering Kanbans

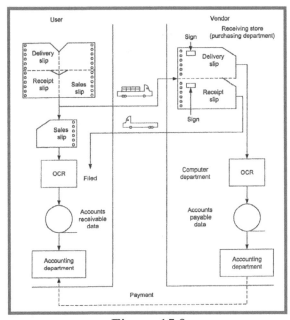

Figure 15.9
Movement of Kanban Cards between Plant and Vendor (Courtesy of Monden, 1983)

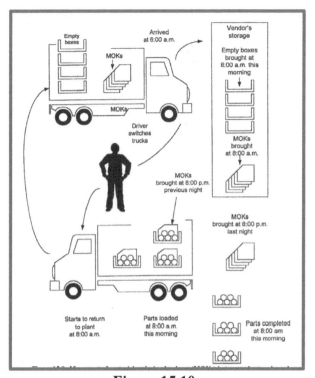

Figure 15.10
MOK Voucher Movement for a Two-truck System (Courtesy of Monden, 1983)

Here is how this simple system works.

1. At 8:00AM., a truck delivers MOKs and empty containers to the vendor. Upon arrival at the vendor, the truck driver hands the MOKs to the vendor's store worker. The driver switches to a truck loaded with parts produced to the requirement of
2. MOKs brought at 8:00PM the night before. He drives back to the user's company and delivers the parts to the correct location within the plant.
3. At 8:00 PM, the empty truck is returned to the vendor. More MOKs are given to the vendor and the full truck is taken back to the plant. One day's supply is carried in two trucks.
4. If the truck can be rapidly loaded and unloaded, only one truck is necessary. Many material transporting companies have already developed such systems.

15.5 Constant Work-in-process (CONWIP)

The CONWIP system focuses on maintaining a constant level of WIP within a specific manufacturing cell. Inside the cell WIP is allowed to move freely between machines. CONWIP is actually a push/pull hybrid. The system is not a true Kanban system, but it will normally use a card system to control WIP. However, CONWIP has been a popular topic in recent literature that discusses pull systems in general. For this reason, CONWIP will be examined and qualitatively compared to the classical one and two card Kanban systems previously discussed. CONWIP has as its basic goal to control WIP like Kanban, but it fails to make use of a pull philosophy both inside a cell and between cells. When a container of products leaves final assembly for the external customer, this signals the introduction of a new container of raw material at the beginning of the manufacturing system. Once the material has been introduced, production continues without waiting for withdrawal from the downstream cells. When finished product leaves final assembly, a new batch of raw material is introduced to take its place in the system. Either a Lean Kanban pull system can be used to link final assembly to the order release entry point, but order release is often done by using push order release systems.

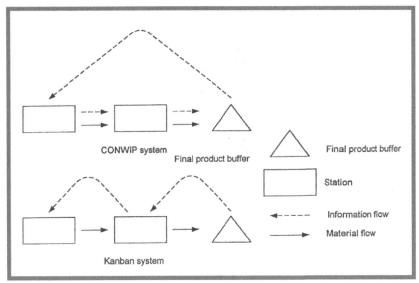

Figure 15.11
Information flow in CONWIP and Kanban Systems (Gstettnew and Kuhn)

Although CONWIP is not a true Kanban system, it can use cards which are attached to containers of products leaving final assembly, and these cards can trigger upstream cell production. The cards then act like travelers which follow the products through the manufacturing and assembly process until final assembly is completed (Figure 15.11). At this point, the product is shipped to the customer, and the card returns to the beginning to be attached to a new container and start again. The containers are all of standard size, although they are not part specific like those in a traditional Kanban system. The total number of containers in the: ,system is maintained at a fixed level. The Magic of CONWIP is that it places a WIP Cap on the entire system and within each cell.

15.5.1 Advantages of the CONWIP System
There are some advantages associated with using a CONWIP system rather than the typical Kanban system for controlling production and inventory.

1. There is no blocking. The buffers between the stations can conceivably hold the entire inventory of the system, so it is not necessary to stop and wait on the downstream cell to *ask* for a container. The station only stops producing if the raw materials upstream are disrupted.
2. It is very simple to Control. Although the Kanban system can be easily understood by all, CONWIP is an even simpler way to control inventory. The only variable is the total WIP in each cell. With Kanban, not only is the total WIP a variable to be controlled, but the WIP between each cell is controlled independently as well.

3. CONWIP works well where there is a large variety of parts. With a Kanban system, at least one container of every part produced in the plant must be held in WIP. If the number of parts is very large, then the amount of WIP can easily become excessive even if only one or two containers of each part type is available. Since the CONWIP containers are not part specific, there is no need to have one for every part type in the system. So, the total number of containers, and thus the WIP, is not affected by the number of part types the system is capable of producing.

4. Unbalanced lines can be handled well by CONWIP. A true Kanban system will have difficulty functioning in a facility that has not balanced the processes with the final assembly Takt times. There will always be some cell that prevents all of the others from producing due to its slower cycle. However, the individual cells in a CONWIP system are not subject to withdrawal from downstream. Thus, the slow cell in an unbalanced line does not prevent other cells from functioning.

5. Bottlenecks near the end of the line are not devastating. If a cell near the final assembly in a CONWIP system experiences delays, this does not necessarily affect all of the upstream cells. They are still receiving raw materials from upstream, so they are able to continue production due to the pushing of material through the system.

15.5.2 Limitations of CONWIP

Just as there are several advantages in using a CONWIP system, there are also some disadvantages. Some of these are listed below.

1. CONWIP is inferior to Kanban if the lines are balanced. Much of this has been devoted to the contention that if the first five steps of LCMS have been implemented (which involves leveling, balancing and synchronizing); then a Kanban system will provide optimal control of production and inventory. CONWIP cannot match Kanban performance under these conditions, as several authors have proven through simulation.

2. Performance is poor if bottlenecks occur near the beginning. Due to the fact that material is pushed from its introduction at the beginning through the system to final assembly, a bottleneck at the start of this process can really delay all of the subsequent cell's production.

3. CONWIP can experience routing problems. Since the parts are often *pushed* through the factory, a routing sequence must be set. This is not a problem if all of the parts will travel to the same cells. However, if some parts do not require certain processes, then a routing control methodology may be required. This is reminiscent of the traditional MRP push systems in the job shop.

15.6 Integrated Inventory Control

The most powerful analogy presented in the Japanese literature is by Shingo (1981), which are the now famous and widely published rocks in the river graphic (Figure 15.12). In this simple analogy, rocks are equated to problems and the river is inventory (material) moving through the plant. The level of the river is equivalent to the work-in-process (WIP) inventory that flows through the factory, just as the river flows between its banks. When the river level is high the rocks, which represent hazards to safe navigation, are _ covered. (See Table 15.1 for a listing of

494

problems and traditional solutions.) Now it might be asked; *isn't that good? Hasn't inventory traditionally been used to circumvent the problems of poor quality, machine tool breakdowns, long setup times, shortages of parts, and many other deficiencies in the manufacturing or production systems?* True....but there are errors in this way of thinking. Covering up the problems is the wrong approach .All inventory is wasteful and expensive to carry. The greater the inventory in the system, the longer will be the Throughput time by Little's law. Low-cost, high quality manufacturing will never be achieved when WIP levels are high. Besides, the inventory, if not controlled, may suddenly drop because of factors outside of one's control, creating flow problems in the system at the most inopportune time and throwing the entire plant into disarray.

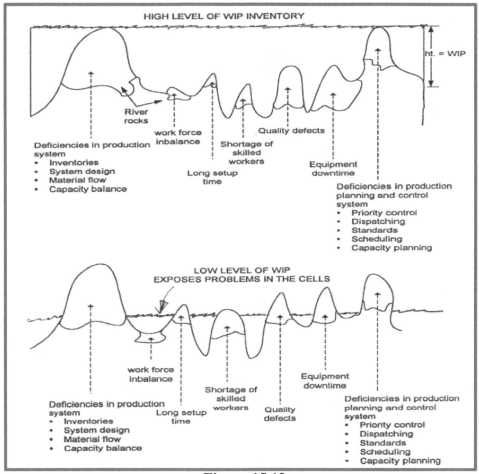

Figure 15.12
Lowering WIP Uncovers Production Problems

Table 15.1
Problems and solutions in the workplace for mass production

- **Problems in the workplace (rocks in the river)**

  Machine failures (wait for repairs)
  Bad raw materials (poor incoming quality)
  Tool failure (fractured, work, or missing tools)
  Workers absent or late
  Changeover from one part to another very long
  Waiting for parts
  Waiting for material handling
  Waiting for inspector/setup/or maintenance person

- **Typical Solutions for the mass production system**

  Lots of inventory and buffer stock
  Backup machines or material- handling equipment
  Super machines (large and expensive automation) with short cycle times
  Extra tools and materials
  Extra repair parts
  Extra workers (expediters and dispatchers)
  Elaborate information systems (Computerized)
  Robotize and automate (expensive)

15.7 Inventory: An Independent Control Variable

The Lean, Linked Cell Manufacturing System (L-CMS) alters the nature of inventory completely, changing it from a dependent variable in the classical push system to an *independent control variable* in the pull system Taking the rocks in the river analogy a step further, the river has volume of flow that represents capacity. River flows are described in cubic feet per .minute. The materials that flow in continuous processing manufacturing such as chemical plants and refineries are also described in ft^3/min. Continuous processing manufacturing systems found in refinery operations represent the prototypical continuous flow system, except that they are very inflexible. In order to make the discrete parts flow like water or oil, the Lean engineer must accomplish four key tasks: (1) Reduce the setup time and lot sizes, (2) eliminate all defective products and rework,(3) eliminate all unscheduled machine breakdowns in the system and (4) redesign machines so that they can be easily moved around and execute multiple simple tasks. Inventory movement and levels is an independent variable that can be described as follows:

- River flow rate (ft^3/min) = depth x width x velocity
- Depth of river = Amount of WIP
- Width of the river (ft.) = Number of manufacturing systems (No. of final assembly lines.)
- Length of river (ft.) = Distance materials are traveling through the system.
- TPT = Throughput time

- Velocity (ft./min) = speed = Length/TPT

Traditional Mass manufacturing systems release parts into production with little regard for the level of inventory. *In L-CMS, the depth of the river is also controlled.* That is, the inventory levels are deliberately raised or lowered, even though lowering the inventory level will expose problems. When this happens, the emerging problems are then solved. In some cases, it may be necessary to restore inventory to near its previous level to ease the pain to us or our customers, then the problem(s) which have been uncovered are attacked and eliminated. The *rocks* are removed from the river and WIP levels are reduced until another *rock* (problem) is uncovered.

How is the inventory level controlled? By the use of Kanban. Kanban allows for specific amounts of inventory to be added or extracted from the flow, thus changing the depth. The flow rate can be changed by altering the production rate or the number of workers in the manufacturing cells. The production rate in units per hour is the inverse of Takt in hours per unit. Cycle time (Takt goal) is not fundamentally a control variable, unless overproduction is initiated to buffer demand uncertainty.

Returning to the river analogy, the river level can never be lowered completely to the riverbed, because flow stops completely. *There is no such condition as zero inventory.* A certain minimum amount of inventory must always be in the system. The power of the analogy is that the more rocks are removed, the lower the river can be run without destroying the system.

In the same way, the level of WIP between cells reflects the progress that has been made in removing setup time, eliminating defective products, eliminating machine breakdowns, and standardizing the cycle times (eliminating the variability). The closer the system gets to perfection, the lower the WIP level can be and still flow smoothly.

15.8 Using Kanban as a Control Variable

Kanban cards are directly related to container size, and containers carry a specified amount of inventory. The dual-card Kanban system previously described provides a very tight control on work in process. Dual-card Kanban systems have a unique productivity improvement feature that neither push systems nor single-card Kanban systems possess. Foremen or supervisors have the authority to remove Kanban from the system to reduce inventory and thus expose problems. To do this, they do not have to remove a container from the system; all they need to do is remove a pair of Kanbans from a full container. The container cannot be moved without a Kanban attached to it. Even though workers and foremen are upset when the removal of inventory from the system causes delays in the schedule, this gives them a chance to uncover problems in the upstream cell. The inventory can be restored if necessary by reinstating the cards while problems are solved. Once a solution is in place, the inventory level can be lowered again and another round of problem solving begins. Productivity and quality are continuously improved. This feature makes the dual- card Kanban system very effective.

The dual-card system is particularly effective for small-lot, mass production of complex assembly items in which there is a potential for delays caused by the compound effects of:

1. A large wide variety of parts being produced
2. Multiple points of usage for some parts

3.   Multiple stages of manufacture/ final assembly

In order to avoid running out of parts when delays occur, huge buffer stocks are normally carried in the traditional mass production system. However in lean manufacturing, the dual-card pull system signals the manufacture of each part to match the up-and-down output rate of downstream production stages. The inventory between the cells is continually reduced by the removal of card or equivalently containers. Referring back to Figure 15.7, suppose you begin with 11 carts between the two cells. Further suppose that each cart holds 20 parts. (K=11 and a = 20). The maximum inventory between the two cells is therefore 220 parts. The removal of a single Kanban (cart) lowers the maximum inventory level to 200 parts. If no problems occur, another cart is removed. The process continues until finally no more carts can be removed without serious delays or downstream starvation. Let us assume that this occurs when there are only six carts remaining. (This number will often depend on how close one cell is to another and on the length of setup times between different parts).

Now the number of parts in each cart can be, say, cut in half (10) and the number of carts restored to 11. Thus, the level of inventory is about the same, but the frequency of production is increased. The flow is smoother (smaller lots produce smaller demand spikes in the system). The problem of setup time reduction will become immediately apparent since cutting the lot size increases the setup frequency. SMED must be undertaken immediately. Once setup time has been reduced, inventory levels can again be reduced by withdrawing Kanban cards.

This method of inventory control is very effective because it is operated entirely by the people who run the manufacturing system. Eventually, the users get the inventory level as low as possible without causing major disruptions to the manufacturing system. By Little's Law, Throughput time is also significantly reduced. The system will now usually require some workstation redesign and more automation to go to the next .level. This kind of system reconfiguration is usually easy to justify.

15.9 Kanban Pull System Compared to MRP

In this section, various features of the pull system will be compared to MRP, inventory levels, and production philosophy. A Kanban controlled pull system will now be be compared to a traditional push system.   At initialization, Kanban should be a *manual* information system utilizing cards. This helps familiarize everyone with the operational philosophy. Capital investment in a Kanban system alone will not be large compared to the costs of changing the manufacturing system itself. On the other hand, MRP systems are computer-based systems. Because the existing manufacturing system and MRP are so complex, MRP is *not* manageable without computer assistance. The capital investment in an MRP system is expensive: Usually in hundreds of thousands of dollars.

MRP systems take a long time to implement. At a conference in Canada, a material manager who had just spent two years implementing the company's MRP system was asked how many people were in his company: *About 500 workers and 200 others.* When asked how many of those people actually *understood* the MRP system, he paused and estimated three. How can any system function well when only three people understood how it is supposed to work? People will not respond well to a system they do not understand. Material is the lifeblood of the manufacturing system. To not understand how it is controlled is bad management practice. It will lead to distrust and ignoring computerized instructions.

In addition, MRP systems almost always rely upon built-in EOQ calculations. Input data (such as holding costs) are difficult to estimate, and even worse the fundamental assumptions in EOQ calculations are nonrealistic; such as constant demand. EOQ calculations usually produce large lot sizes. It is better to have fixed small quantities and vary the frequency of the parts than to have large batches of either production or move parts. The MRP system was developed for planning in the job shop, not for control of the material moving through the plant.

A dual-card Kanban system is truly a *pull* system of parts ordering and control. The ordinal production schedule is issued *only* to the final workstation on the final assembly line, and not to any other cell or workstation. The production schedule for each of the preceding processes is determined by the transfer of the parts and withdrawal Kanban cards linking the processes in the system. Therefore, parts are actually pulled through the system from the end of the line to the start.

The single-card Kanban system is a combination of a push-and-pull parts ordering system. The manufacturing aspect in a single-card Kanban system is a push system because parts are produced ·according to a daily production schedule rather than for immediate needs as in the dual-card Kanban system. Coupled with this push system for manufacturing, is a pull system for deliveries. Parts are delivered using withdrawal Kanban's only as they are needed by the downstream processes.

MRP is a pure push system of order release and control. A push system is simply a schedule-based system in which a multi-period schedule of future demands based mostly on forecasting techniques is prepared. The computer then breaks down the schedule for manufacturing and develops a production schedule for each work center based upon the master schedule. Parts are then released to the system based upon a projected due without regard to the actual system status. Any connection between the planned schedule and reality may not exist.

Companies using pull Kanban systems have less delay or lead time between parts manufacture and use, so they usually carry only hours or minutes' worth of material in inventory. An MRP system carries days, weeks, or months' worth of material in inventory because the parts are produced to cover the demand for a longer planning period

A Kanban information system is a logical element in a lean manufacturing system. The elimination of setup time makes small lot sizes economical. Making all lots equal in size and as small as possible by redesigning the manufacturing system, standardizing the operations, balancing workloads, and smoothing the manufacturing system are all integral parts of a true pull system.

An MRP system produces parts in *large* lot sizes in order to cover the demand of a single period. This system has *not* adopted the concept of small economical lot sizes through eliminating setup time and streamlining the operation. In MRP, lot sizes vary considerably, so production cannot be smoothed.

The crucial factor of determining which information system to use is the ease of associating requirements for parts with the schedule of end products. Figure 15.13 shows the relationship between the ease of associating the requirements and the type of information system used. This will affect the size of the inventory level, as discussed earlier.

The major distinguishing factor between a pull system and an MRP system is the ability of the former to accurately associate component part requirements with the end products schedule. The dual-card Kanban system is able to accurately associate these two things because it manufacturers and withdraws parts according to the actual needs of the system to meet demand requirements. Production control is truly integrated into the manufacturing system. The MRP system does not have as high a degree of association or integration because of error introduced into the part requirements as a result of changes in the actual end-product schedule. Although MRP correctly calculates the part requirements by precisely associating them with the master schedule of end products, long lead times and large lot sizes erode the close association between the part requirements and the end-product schedules. The way to make MRP a truly effective information system would be to reduce setup time to make small lot sizes more economical. This would reduce the error inherent in the MRP part requirements calculation. However, if the setup time is reduced in order to make MRP more effective, MRP is not needed for material control in lean systems.

Lean manufacturing system control techniques serve primarily as tools that workers can use to keep the production process moving toward its goal of stability and continuous improvement. *Production leveling* and the *Kanban card systems* keep production and inventory levels predictable, controllable and capped.

Kanban control methodologies help operators monitor operations and spot problems in the line or defects in the products. Once discovered, problems on the line or quality issues are spotted immediately and are resolved immediately.

15.10 Supply Chain Management (SCM)

There is currently a lot of activity in supply chain management. However, unless you have achieved lean operations/production, you don't have a chain of lean-linked suppliers; you simply have a network of suppliers, so this strategy should be called supplier network management. As part of the last two steps in building a L-CMS, you must put your production system (product design, manufacturing sales, marketing, distribution, purchasing) in order. This requires that you have integrated your internal manufacturing system, and then selected the best vendors to be sole source for a component or subassembly.

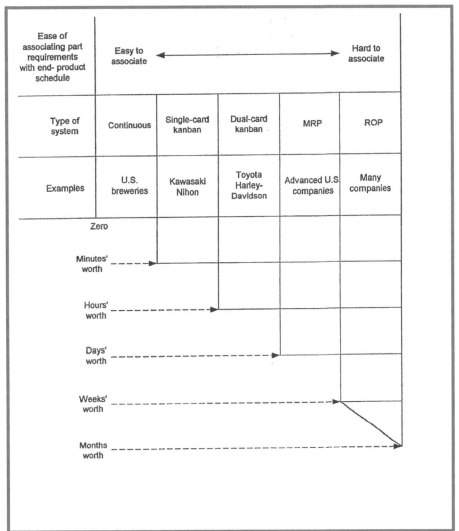

Figure 15:13
Manufacturing Inventory Systems
(Courtesy of Schonberger, 1982)

Lean monitoring and trouble-spotting tools include:
- Management by sight
- Line stop and other techniques such as hourly checks, first, middle and last inspection display and
- Control sheets, check sheets and process documentation.

The Toyota system was created by trial-and-error evolution within an environment where the language would not permit them to construct a written communication system (information system) that could control a large, complex manufacturing system. So Toyota developed a manufacturing system that was simple to operate and control with a very simple information system, one now known as Kanban. This is a pull system of production and inventory control and it is unique. Kanban uses visual cards for information transfer and inventory control. The manufacturing system is redesigned such that only the final assembly line needs to be scheduled. What could be simpler? In fact, the long-range goal of the system is to eliminate the need for

Kanban by directly linking the processes. This means that the output from each cell goes directly to the next cell. Note that this defines *what* the cell or process is to make and *when* to make products. Hence, one of the system's operational goals is the *gradual elimination of its primary information and control mechanism;* the Kanban cards.

Thus, the proper path to get us to the paperless factory of the future is evident. We must begin today to eliminate the manufacturing system called *the production job shop* and traditional Mass production lines. These must be replace with U-shaped, cellular manufacturing systems linked with Kanban. We also must redesign our flow line manufacturing systems so that they have the same characteristics as the cells; particularly regarding flexibility (ability to react quickly to changes in demand or changes in design) and one piece flow operation.

The trick is to begin to implement integrated manufacturing production systems concepts immediately in every plant. Notice we said integrated, not CIM (computer-integrated manufacturing), or CAD/CAM, or FMS. **Do not** computerize the existing system or migrate to expensive, inflexible, highly automated super machines, advanced manufacturing technologies (computers, robotics or expensive material handling systems (automated guided vehicles, automatic storage racks ASRS, etc.) Constructing large batch-building, push production_ job shops will lead to automated economic dinosaurs. These are expensive solutions, and even if they eliminate a direct laborer or two, the gains will be negligible. The correct approach is *to radically change the manufacturing system,* get rid of the push mentality, implement linked-cells and then computerize, robotize, and automate to solve production problems.

In conclusion, the Kanban system accurately associates the part requirements with the end-product schedule. Part usage and manufacturing in all the upstream processes are determined by the actual need of the end-product assemblies through the transfer of withdrawal and production Kanbans. Then parts are manufactured by design to be *just-in-time* (the necessary products, in the necessary quantities, at the necessary time).

The lean manufacturing system, being manual at the outset, is easily understood by all its users, something that cannot be said for many of the computerized systems used in most modern manufacturing facilities. All employees understand how their actions influence the entire system and that they can make the system better.

Large lot sizes, long lead times, and changes in the schedule make it difficult for an MRP system to associate accurately the part requirements with the end-product schedule.

Toyota developed a manufacturing system that was simple to operate and control with a very simple information system, one now known as *Kanban.* A pull system of production and inventory control, Kanban uses visual cards for information transfer and inventory control. The manufacturing system is redesigned such that only the final assembly line to needs to be scheduled. What could be simpler? The long-range goal of the system is to even eliminate the need for Kanban by directly linking the processes. This means that the output from each cell goes to only one customer. Note that this defines *what* the cell or process is to make. Moreover, one of the system's operational characteristics is the gradual *elimination of its primary information document,* Kanban cards and the resulting WIP inventory reduction. Thus, the proper path to obtain the paperless factory of the future is evident. The mass production manufacturing system using the job shop can be replaced with cellular manufacturing systems linked via the Kanban subsystem.

If problems are discovered, they are immediately evident on the affected production line(s) and corrective action can begin. First, middle, and last piece inspection helps determine when problems occur. MO-CO-MOO rules using integrated decouplers can be implemented to identify the source of any problem.

> •*Control charts or SPC (statistical process control) sheets* are put on display for reference, especially by the operators. Run charts ensure specifications will be met by detailing the particular operation. At the same time, they give the operator a sense of the patterns of process development and point out the effects of machine adjustments.
> •*Check sheets* are another tool used in lean production. The most common is the *machine check* list which aids operators in the correct startup procedures for the machines. Check sheets are also on display for all to view and also provide help with maintenance when the machines are down. They can be used in any function and enhance the proper application of established procedures or rules.
> •*Process Flow sheets or* standard operation sheets are the most common control tool. It provides all necessary information for the actual machining or assembly of a part. For example, they will outline speeds and feeds, material, sequence of operations, operator safety, tooling and gauging information, part number, and part name. A drawing of the part is also included. Also, standard operation sheets are positioned above each operation and provide stability and standardize the process so specifications can be met consistently. They are also used to train new operators. Leave nothing to chance or memory. Make it easy to train and retrain the operators and prevent them from making errors.

15.11 Lean Supply Chain Management (L-SCM)

Lean manufacturing is a system built on integrating suppliers into a program of continuous, long-term improvement. The company and its vendors work together to reduce lead times lot sizes, unit cost and inventory levels while improving quality. Both the vendor and the company become more competitive in the world marketplace. Once the company has implemented a lean manufacturing system, it must be extended to the vendors. This requires that the company develop a program that educates and encourages suppliers to develop a manufacturing system that produces superior quality; at the lowest possible cost; delivers on time; and is flexible. (Schoenberger's Four Horsemen). The vendors essentially become remote cells from which materials and subassemblies are withdrawn just as they are pulled from cells within the plant.

In the job shop, purchasing has the following characteristics:

> •Multi-sourcing (many vendors for the same item)
> •Weekly/monthly/semiannual deliveries
> •Long lead times (weeks or even months)
> •Large safety stocks`

- •Quantity variances
- •Late/early delivery times
- •Inspection of all incoming materials
- •Inconsistent packaging
- •Expediting

Lead time reflects the amount of time you must allow between the time you order a component and when you actually receive it, so that it arrives on time at the right place. See Figure 15.14. Expediting, an activity to be avoided, means the component did not arrive on time (or it got lost or was defective or *anything could have happened)* and someone has to go find it and get it moving within the system. Expediting is truly one of the great wastes in the mass production system.

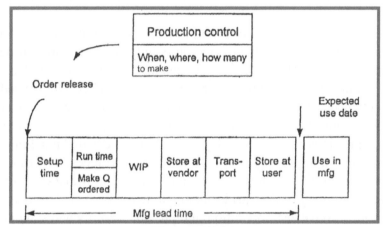

Figure 15.14
Manufacturing Lead Time Reflects How Far in Advance You Must Release an Order to the Supplier so That it Arrives Just-in-Time

The companies that use multiple sources do so for the following reasons:
- •Hedge against vendor problems (strikes, defects)
- •One vendor cannot handle all the work
- •Vendors compete with each other over price.

Now let us contrast these characteristics and functions with lean manufacturing supply chain management.

15.11.1 Lean Purchasing in the Lean Manufacturing System
The lean supply chain must be designed, transformed and operated to manufacture raw materials exactly in sync with the main manufacturing lean facility. The following is a list of desired objectives and goals.
- •Develop real partnerships with your suppliers
- •Minimize safety stock
- •Specify/ Demand low lot size deliveries
- •Structure frequent and on-time deliveries
- •Bypass incoming quality inspection by enforcing *quality at the source*
- •Use standard packaging and containers (recycle packing matrials)
- •Eliminate expediting

15.12 Characteristics of L-CMS Purchasing

Single Sourcing. One vendor, the best vendor, will be selected to be the sole source for each part, component, or subassembly used by the company. This reduces the variability between parts (improves the quality), since all the parts are coming from the same manufacturing process or system. This is a key part of the MO-CO-MOO philosophy. At every step in the process there is only one source, one machine, one set of tooling. This approach replaces the strategy of multiple vendors competing against each other. The adversarial relationship between the vendor and the customer is eliminated. Most importantly, it is easy to identify the source of every problem.

The advantages of Single Source are:

- Resources of the buyer can be focused on selecting/developing/monitoring one source rather than many
- Volume buys are higher, leading to lower cost
- Vendor is more inclined to do special favors for the customer since the customer is a large account
- Tooling dollars are concentrated in one source rather than many (save money)
- Easier to control and monitor for superior quality

Here are *Summit Polymers' six basic principles of quality.* This company is a second tier supplier to Toyota, supplying components to the supplier who builds the dashboards.

1. Do it right the first time
2. Never pass a defect on to the next team member (downstream internal customer)
3. Have a vigorous pursuit of continuous quality
4. Standardize work which ensures consistently correct results
5. In problem solving, always look for the root cause (asking why 5 times) not a temporary solution. --
6. Fix it now

15.13 Eight Best Practice Goals

1. Long term contracts. The company and the supplier develop long-term contracts (18- to 24 months) that enable the vendor to take the long range view and plan ahead. Contracts are renegotiated every 6 to 12 months with very short lead time.

1. The lean company supplies updated forecasts every month (good for 12 months).
2. The lead company commits to long-term quantity and eventual excess material buyout.
3. Delivery is specified by mid-month for the next month.

Advantages of Long-Term Contracts

- Builds schedule stability. No "jerking" up and down of vendors' built schedule; smooth increase or decrease.
- Better more frequent communication between buyer and vendor
- Better visibility. The vendor sees one years' worth of forecasted needs as soon as the company sees it, instead of a limited lead-time view.

- Less paperwork. Fewer (none, if possible) change orders to run through the system, which is good for the company and the vendor.
- Inventory reduction, initially at the lean company, and later as the lead times are reduced.

2. *Use Frequent Delivery of Raw Materials and Components*

The vendor will be expected to deliver materials to the company daily or weekly, depending on the type of part or subassembly. Most parts can be categorized according to an *ABC analysis*. An *A part is critical*, high-cost parts, and is usually one per product. For cars, this would be the engine/transmission, seat sets, steering gear, dash board and so on. *C parts* are low-cost, but numerous; *B parts are in between*, but are critical for other reasons. For example, many companies classify bulky parts like packaging materials and sheet metal as B parts. They may not be very expensive but they take up a lot of space.

3. *Mandate and Sustain 100% Good Quality*

The vendor is taught how to implement cellular manufacturing so that the vendor can deliver the correct quantity, on time, with no incoming inspection. In short, become a JIT vendor performing the MO-CO-MOO methodology.

4. *Provide Engineering Aid to All Sub-contractors*

Often the vendor will be a much smaller company, unable to afford engineering expertise in the manufacturing and quality areas. The vendor and the customer work together to improve the vendor's manufacturing processes, productivity, and quality. The customer should visit the vendor's plant at least once a year (more often if there are problems), and the vendor should visit the customer so that the vendor sees how the components are used in the customer's products.

5. *Use Local Sub-contractorss When Possible*

While it is not absolutely necessary (or even possible) that all the vendors be located close to the customer, the closer the vendor is to the customer or the company, the easier it is to provide the customer with daily deliveries. Every day material spends being transported adds to the amount of stocked material inventory.

6. *Instigate Freight Consolidation Programs*

Materials from vendors can be consolidated onto one truck for transportation to the customer. If a company has three vendors in the same town who deliver daily, one truck and driver can pick up daily from each vendor and deliver to the customer.

7. *Use Standardized Containers*

This means the containers for the parts being delivered to the lean company are standardized in terms of the size of the container and the number of parts. Half full containers are not shipped. Since the number of containers and the number of items in the container are easily seen, everyone knows how much inventory is available, so it is unlikely that the lean company runs out of critical components even though they are sole-sourced. Naturally the vendor takes precautions to ensure that they never shut the lean company down due to a lack of parts or subassemblies. For example, at Johnson Controls, Inc., who supplies seats for Toyota Camry in Lexington, there is an area in the plant called the lockup. This area contains a 24 hour supply of seats for the Camry. The seats are stored behind a large chain link fence. Only the plant manager has the key to the gate lock.

15.14 Kanban Summary

Control techniques serve primarily as some of the tools that workers can use to keep the production process moving toward its goal of stability and continuous improvement. Production leveling and the Kanban card system keep production and inventory predictable, controllable and WIP capped. Other techniques and aids help workers monitor the operations and spot problems in the line or defects in the products. Once highlighted, these troubles on the line or imperfections in the results should be tackled right away and at the site. These monitoring and trouble-spotting tools include: Management by sight, line stop and other techniques such as hourly checks; first, middle and last inspection; display and control sheets; check sheets; and process documentation.

In conclusion, the Kanban system accurately associates the part requirements with the end-product schedule. Part usage and manufacturing in all the upstream processes are used to determine the actual need of the end-product assemblies through the transfer of withdrawal and production Kanban cards or carts. These parts are manufactured Just-In-Time for the necessary products, in the necessary quantities, at the necessary time.

The lean manufacturing system, being manual at the outset, is easily understood by all its users, something that cannot be said for many of the computerized systems used in the job shop. All employees understand how their actions influence the entire system and that they can make the system better.

Large lot sizes, long lead times, and changes in the schedule make it difficult for an MRP system to associate accurately the part requirements with the end-product schedule. The inexpensive Kanban system can achieve better estimates of part requirements than computer-based MRP systems because it is used in a simpler manufacturing system, where lean manufacturing means improved quality, lower costs, reduce inventory, and true production and inventory control.

15.14.1 Kanban Rules

The rules for operating a Kanban pull system are few and straight forward.
1. Finished product is kept within visual contact of the producer whenever possible.
2. The producer owns the finished product until it is claimed by the internal customer, i.e., the next operation.
3. The customer gets what is wanted, when it is wanted, carrying light loads and making frequent trips. (The JIT mantra).
4. The producer is responsible for packaging product in the agreed way. (Standard recycling containers).
5. The customer is responsible for transportation, and for recording the transfers when necessary.
6. The producer must never let the customer run out of product, nor allow the inventory to grow beyond established limits (Ka = maximum inventory in the link)
7. The aim for the entire process is to run at a steady rate, all operations together.

507

Review Questions

1. What are the goals of production control?
2. What is the goal of inventory control?
3. What does Kanban mean, and how did it originate?
4. What is the unique aspect of the Kanban system?
5. What are two types of true Kanban systems? Briefly discuss the differences between them.
6. How does POK work with the WLK to perform the key production control functions?
7. What information is given on a WLK vs. a POK?
8. What is the difference between the dual card Kanban and the single card Kanban?
9. Write the design rule which determines the number of carts that you would need in a Kanban link, defining each item in the equation.
10. Write the alternative design rule, explaining the differences between it and the simpler version.
11. What does CONWIP stand for? Briefly describe how it differs from the 2 Kanban systems?
12. What is the primary difference between a push system and a pull system with regard to inventory control?
13. Briefly explain how inventory is controlled in a Kanban system.
14. Briefly explain Shingo's "Rocks in the River" analogy.
15. List some benefits of the Kanban system for production and inventory control over the traditional MRP system.
16. What is different about the supplier to a lean production system than the supplier to a mass production system?
17. In the dual-card Kanban system, do both the POK and WLK cards have the same information? Why or why not.
18. You are a supplier and have a Kanban system. Your customer needs 200 parts every day. Your worst case lead time is 2 hours and your carts hold 40 parts, how many Kanbans would you use as an initial starting point? Use a 10% safety stock.
19. What 3 questions does Production Control seek to answer?
20. What is a MPS, an EOQ, a MRP and how are they related?
21. What are the pre-requisites to implementing a Kanban or pull system?
22. What is the major difference between dual card and single card Kanban?
23. What is the basic information on a single card Kanban?
24. The Kanban shown in Figure 15.5 connects the _____ to the Rack and Pinion sub assembly area use _____ Kanban's with each container holding _____ parts.
25. What information is on the WLK that is not on the POK and why?
26. In the design rule for Kanban, what are the time elements that are in the lead time (L)?
27. What is the difference between lot delay and process delay?
28. Explain the effect on TPT by implementing CONWIP, using Little's Law.
29. What is the connection between Ohno's *Rocks in the River* analogy and the Kanban methods used to control inventory?
30. What is a MOK?

31. Do sub-assemblies that are synchronized to final assembly use Kanban cards? Why or why not?
32 The Lean Engineer manages many Kanban links and is continuously trying to remove inventory from them. What major measurable parameter is this effecting?

Thoughts
and Things.......

Chapter 16
Integrated Preventive Maintenance (IPM), Reliability and Continuous Improvement

If you are coasting, you are either losing speed,
Or you have peaked and you are going downhill   J.T. Black

16.1 Introduction

Time must be allotted in the manufacturing schedule for checking and maintaining the equipment and the people. Unless all the elements in the manufacturing system are properly designed and maintained, breakdowns will occur and disrupt the flow of products. Some manufacturing processes need to run continuously, but such processes already have many of the elements of lean manufacturing. Usually the problem with continuous processes is long setup times (long changeover times when the system changes materials), and all the maintenance except for emergency service is performed during these shutdown periods. The people and equipment in the lean manufacturing system must be ready to produce what is needed when needed. This may not be possible when a tightly balanced system can never catch up if slack time is not provided for catching up or for maintenance. The key is to develop a less-than-full-capacity schedule for the cells that includes time for maintenance. Repairs made under pressure may not be done well, leading to further downtime for re-repair, tinkering, and adjusting.

With *integrated preventive maintenance (IPM),* operators are required to become more aware of the behavior of their equipment and its routine problems. Chief among these problems is process drift (loss of stability or loss of aim or accuracy), so IPM is clearly linked to IQC. Getting the process centered and then maintaining the aim or accuracy of the process is different from reducing the variance and then maintaining the process spread (variability or precision). Finally, keep this thought in mind – if you are not maintaining or improving a manufacturing process or system, then it is degrading (layman's restatement of the second law of thermodynamics). The manufacturing strategy here is to design methods to check people and train people to check machines.

16.2 An 4-8-4-8 Schedule = LTFC

Integrated preventive maintenance covers the maintenance of machine tools, work holding devices, cutting tools, and personnel. This function is integrated into the daily regimen of the plant floor. Preventive maintenance responsibility is shifted from the maintenance department to the operators. The operators daily prepare and use machine tool checklists, much like the checklists pilots use to check out the aircraft before takeoff. We do not want any machines to crash in the plant during the eight-hour shift. Workers are also responsible for most of the routine machine tool maintenance. The maintenance department still does major machine overhauls and takes the lead role in the event of major breakdowns.

For a lean factory to operate effectively, the entire plant is run on a 4-8-4-8 schedule. The four-hour time blocks between the two eight-hour shifts allow for maintenance or unavoidable long setups. In addition, the eight-hour shift can begin early or run over when needed without disrupting the next shift. Therefore lean manufacturing systems are designed to be run at less-than-full-capacity (LTFC) so that there is breathing room to keep everything and everybody up and running 100% of the time. See Table 16.1.

Table 16.1
Less than Capacity Scheduling

Less than Full Capacity Scheduling or Under-capacity scheduling		
1.	Schedule	Meet the schedule every day. Breakdowns and maintenance don't cause stoppages elsewhere.
2.	Quality	Don't let schedule cause errors and defects thru haste.
3.	Marketing	Can increase output on short notice to full a hot order; buys time to add staff if necessary.
4.	Operators Jobs	The operator's No.1 Job is making parts – direct labor (DL) but if there is no DL to do (Schedule met or Stoppage) then operator performs indirect labor (IL). IL is often mental: improvements in quality, productivity and equipment. Mental work may be operators' most important contribution! You may even want to focus the reward system on mental contributions (instead of an output)
5.	Automation	With Automation, the operator may be 100% indirect (e.g., quality checker, trouble-shooter); JIT eases transition from direct to indirect labor.

Machine life and tool life are further enhanced by operating the equipment at reduced speeds or reduced production rates. This concept is totally foreign to most American factories. The idea of less-than-full capacity operation suggests less-than-100-percent utilization of equipment. What we should do is worry about effective people utilization and let the machine utilization be whatever is needed to meet the demand. However, we want the machines to always be ready to be utilized.

16.3 Scheduling Preventative Maintenance

Scheduling preventive maintenance (PM) is a task typically assigned to the lean engineer. Production supervisors believe their manufacturing processes should not be shut down simply for PM, yet higher-level PM must be performed on schedule by the maintenance department. For this reason PM should be flexible within certain limits. Scheduling PM should be between the eight-hour shifts, in the four-hour time blocks, or on weekends if necessary. This presents some inconvenience for the maintenance engineers and specialists, however. Equipment can also be used on an alternate basis when overhaul maintenance is required. One machine can be removed from the cell and replaced with another, so that the necessary overhaul can be performed.

The pace of the manufacturing system in the entire plant is synchronized to the Takt time, and production rates (cycle times, etc.) for a specific piece of equipment or cell are determined by the system needs. The machines are not run flat out. Furthermore, the entire eight-hour shift is not scheduled unless that is required to meet the daily production needs. Some time (15 minutes) at the start and end of each shift is allotted for routine repair and maintenance. If the entire eight-hour shift is scheduled, this is a serious problem and steps should be immediately taken to correct this situation. In this way the equipment lasts longer and provides higher reliability. The idea is not to overtax the machine tools, the people, or the cutting tools. People are less likely to make mistakes and machines less likely to break down if they are not pressured.

Suppose you prepared a race car to run the Indy 500 but instead of running the car at 200 mph, you operate it at 100 mph -just sufficient to get you where you needed to go on time with no waiting when you get there. The car will run much longer before it breaks down. The long distance race is won by the steady and consistent performer.

Placing a cushion of time at the end of each shift (the shift is seven and one-half hour long) allows the line to be shut down for quality circle meetings. This helps make quality part of the operators' job. The first 10 or 15 minutes of the shift are dedicated to maintenance checks, machine warm-up, oiling the equipment, checking tools, and the like.

This methodology also gives additional flexibility to respond to changes in product demand because of the wide latitude it affords the cells. In order to increase the production rate, one can add workers to the cells and increase the operation rates of the machines when necessary.

16.4 What PM Means

Preventive maintenance (PM) is designed to preserve and enhance equipment reliability. A correctly integrated PM program will provide a significant increase in production capability throughout the entire production system. The ideal PM program will prevent failure of all equipment before it occurs. The operator is trained (Yes, more operator training) and empowered to maintain his machine(s), schedule the routine maintenance and keep the area clean and neat.

16.5 Value of Preventive Maintenance

People not exposed to preventive maintenance question its value. They believe that it costs more for regular downtime and maintenance than it costs to operate equipment until repair is absolutely necessary. However, one should compare not only the costs but the long-term benefits and savings associated with IPM. Without IPM, the following costs are likely to be incurred:

Costs of Poor Preventive Maintenance

- Lost production time resulting from unscheduled equipment breakdown
- Variation in the quality of products due to deteriorating equipment performance
- Decrease in the service life of the equipment
- Safety-related accidents due to equipment malfunction
- Major equipment repair and lost production time

Long-term benefits

- If maintenance is a primary responsibility, operators are more familiar with the equipment, the way it operates, and it potential problems.

- Processes are in better control through IPM's machine and tool records, producing better quality.
- Quality, flexibility, safety, reliability, and production capability are improved.
- Reliable equipment permits reduction in inventory

Long-term effects and cost comparisons undoubtedly favor preventive maintenance. A carefully designed and properly integrated program requires a positive managerial attitude that will set the pace for a successful program.

16.6 Internal Preventive Maintenance Involves the Internal Customer

In lean manufacturing, the operator is responsible for the machine/equipment in the manufacturing cell. This philosophy encourages the operator to take responsibility for the maintenance, operation, and performance of the equipment. The operator should be trained in recognizing the machinery's optimal performance. They should be trained to trouble shoot the equipment when changes in its performance are detected. Regular lubrication and cleaning should be a normal part of an operator's daily routine. If an operator is responsible for the repair of the equipment in the cell, that individual becomes more sensitive to the care and maintenance the machines may require. When the machines break down, it reflects on that operator's performance.

Machine operators should be trained to observe their equipment and to respond to their observations. If a piece of equipment need special attention and the operator cannot perform· the level of maintenance necessary, he or she should see that the proper maintenance specialist comes to the machine. Though preventive maintenance, production operators become more conscious of their performance and take pride in their work.

Operators need to realize the importance of an orderly and clean workplace and equipment. Routine cleaning familiarizes the operator with the equipment. If a person is given responsibility for a piece of equipment, he/she will take pride in its operation. Process drift (loss of stability or loss of aim or accuracy) is a chief problem that operators need to watch for. This task is closely linked to IQC (integrated quality control), and is something that an operator should be able to identify. Housekeeping becomes a ritual in their everyday job performance. Routine cleaning familiarizes the operator with the machine, making it easier to understand the details involved in its operation. When operators have a sense of responsibility rather than just a duty to perform a specific task, they become enthusiastic about their jobs.

IPM emphasizes the significance of executing the correct procedures needed to operate all equipment in a manufacturing cell. When an operator runs the manufacturing cell in an incorrect manner, the irregularities in the processes will stand out. Problems are readily observed and may be prevented at earlier stages of the process. In summary, operators trained and equipped to perform CLAIR (tasks like: Clean, lube, adjust, inspect and repair).

16.7 Role of Maintenance

In large companies there will always be a need for designated mechanics but their role must change. If operators are doing a good job of trying to avoid emergency shut downs, the maintenance crews should be able to concentrate on more constructive tasks. There are some tasks like welding or repair of complicated equipment that operators could not be expected to learn. If the maintenance employees have been reduced significantly due to a company's dedication to lean manufacturing, then the maintenance personnel who are left should be able to devote themselves to the role of troubleshooter with diagnostic and technical skills. Maintenance personnel should be multi-skilled. They must be able to perform electric and mechanical tasks very well as well as be trained in some operating skills. They must be able to be productive at all times

16.8 Lean Engineering and Maintenance

Lean Manufacturing engineers are responsible for the design, build, test, and implementation of manufacturing equipment. Maintainability should be considered in the design or selection of equipment for purchase. Simple, reliable equipment that can be easily maintained should be specified. In general, dedicated equipment can be built in-house better than can be purchased. Many companies understand that it is not good strategy simply to imitate or copy the manufacturing technology from another company and then expect to make an exceptional product using the same technology as the competitor. The company must perform research and development on manufacturing technologies as well as manufacturing systems in order to produce effective and cost efficient products. However, an effective, cost efficient manufacturing system makes research and development (R&D) in manufacturing technology pay off. Once you have implemented the L-CMS you can begin to design, build and install your own machines into the manufacturing cells. Those who have taken this step into lean manufacturing find the following advantages.

- The cell (and the machines within it) is designed for the system so the cell operates on a MO-CO-MOO basis and the machines can meet the cycle time, where CT =Takt time for the system.
- It is flexible, allowing rapid changeover for existing products and rapid modification for new products or model changes
- It has unique processing capabilities, unique machine tools
- It has maintainability/reliability/durability· ·
- Equipment and methods are designed to prevent accidents (safety).
- It is easy to operate, has a fail-safe design (ergonomics).

The equipment is designed and developed with priority on the internal customer factors even though the factors affecting the external customer are the highest priority of manufacturing engineering. Although many plants lack the expertise to build machines from scratch, most have the expertise to modify equipment to give it unique capabilities. Modification of the equipment to prevent reoccurrence of a breakdown requires that management assign the highest priority to this work. The most skilled maintenance personnel must be given this task so that breakdowns in the L-CMS are eventually eliminated.

16.9 Total Productive Maintenance (TPM)

The Japanese imported preventive maintenance from the United States over 50 years ago. Since that time they have improved and extended it into what is now labeled total productive maintenance (TPM). Seiichi Nakajima defines the developmental stages of TPM as follows:

1950s: Preventive Maintenance -establishing maintenance functions

1960s: Productive Maintenance -recognizing importance of reliability, maintenance, and economic efficiency in the plant design.

1970s: Total Productive Maintenance-achieving productive maintenance efficiency through a comprehensive system based on respect for individuals and total employee participation (Nakajima, 1988).

1980s: Total Productive Maintenance (TPM) a concept developed by the Japanese in the 1960s from maintenance practices they learned from the United States during the 1950s, evolves as an equipment management strategy that involves all hands in a plant or facility in equipment or asset utilization. TPM is to productivity as flexibility is to competitiveness. Without it, corporate survival is questionable. It is a critical step in the lean strategy.

1990s: IPM is a key step of lean manufacturing system design (Blach, 1990). TPM involves the following five elements, often called the five pillars by Nakajima.

16.9.1 Nakajima's 5 Pillars of Preventive Maintenance

Eliminate the "six big losses" and thereby improve the effectiveness of the equipment.

Uptime Losses

1. Equipment failure - breakdowns of machine tools, material handling devices, fixtures, etc.
2. Setup and adjustment - from exchange of dies in molding machines and presses, and tooling exchanges in machine tools.

Speed losses:

3. Idling and minor stoppages- due to the abnormal operation of sensors, blockage of work on chutes, and the like
4. Reduced speed- due to discrepancies between specified and actual speed of equipment

Losses due to defects:

5. Process defects -due to scrap and rework
6. Reduced yield-from machine start-up to stable production.

16.9.2 Develop an autonomous maintenance program.

This means that the operators are involved in the daily maintenance of the equipment. The seven steps of autonomous maintenance are:

1. *Initial cleaning*: Clean to eliminate dust and dirt mainly on the body of the equipment: lubricate and tighten: discover problems, and eliminate their causes.
2. *Countermeasures at the source of problems*: Prevent the causes of dust, dirt, and spattering of liquid, improve those parts of equipment

that are hard to clean and lubricate: reduced the time required for cleaning and lubricating.

3. *Cleaning and lubrication standards*: Establish standards that reduce the time spent cleaning, lubricating, and tightening (specify daily and periodic tasks)."

4. *General inspection Procedures*: Follow the instructions in the inspection manual. Use quality circle members to discover and correct equipment malfunctions.

5. *Autonomous inspection*: Develop and use autonomous inspection check sheets."

6. *Orderliness and tidiness*: Everything has its own place and is easily found
   - Develop inspection standards for cleaning and lubricating
   - Develop standard record keeping practices
   - Develop Standards for parts and tool maintenance

7. *Document maintenance problems and their solutions*: Develop a company policy and goals for maintenance; increase the regularity of improvement activities. Record the mean time between failures (MTBF), analyze the results, and design countermeasures."

These steps are based on the five basic principles of operations management. In the Japanese literature, they are known as **the five S's**: *Seri, Seiton, Seiso, Seiketsu*, and *Shitsuke*. A rough translation of the five S's means organization, tidiness, purity, cleanliness, and discipline.

- Develop a scheduled maintenance program for the maintenance department. This is usually done in cooperation with industrial engineering. The leveled schedule greatly helps the development of a regular maintenance program.
- Increase the skill of operators and the maintenance personnel. The operators should work with the maintenance people at the time PM work is done on their equipment, discussing problems and solutions. Part of the operator's job is to keep records on the performance of the equipment, so the operators must learn to be observant.
- Develop an equipment management program. A record of the use of machines and tools denoting how much they were used and who used them.

16.10 Zero Downtime

The goal of preventive maintenance is simply to eliminate all machine breakdowns and troubles. In short, zero down time and 100% on-demand utilization. Preventive maintenance and its benefits – zero machine downtime - will not be realized without the efforts of the entire organization -operators, management, maintenance crews, and other support people. Operator involvement is critical. Zero down time, or 100% on-demand utilization, can only be assured by preventing breakdowns and problems before they happen, not by the firefighting mode of maintenance. The problem in many cases is that maintenance crews are busy fixing sudden breakdowns, while operators think that machine breakdowns are not their responsibility. *Factors leading to machine breakdown*

Machine factors:

- Dirty and oily machine
- Filled oil pan
- Leaks around the machine
- Overheating motor
- Vibration of motor
- Vibration of machine
- Inspections not routinely performed

Operator factors:

- Lack of concern for dirty machines
- Lack of training to conduct simple maintenance
- Lack of knowledge of the machine
- Not asking for help when problem is anticipated
- Considering production more important than maintenance of machine
- Lack of knowledge of when or how inspection occur

Repair crew factors:

- Not questioning "why" when problems occur to find root cause
- Failure to coach operators on simple maintenance
- Concerned only for urgent, major machine troubles
- Accepts that machines will eventually break down
- Looks only for improvement to be made through new machines

16.10.1 Major Causes of Machine downtime

- Basic machine requirements -housekeeping, tightening bolts, oiling, etc. - not performed
- Incorrect operating conditions - speeds, pressure, torque, temperature, etc.
- Employee skills maintenance crews, operators, etc. - not performed
- Deterioration --tooling, fixtures, gears, bearing, etc.
- Design – material, tooling, speeds ,feeds, etc.

16.11 Integrated Preventive Maintenance (IPM) in Lean Production

In the lean system, preventive maintenance includes another concept, Total Preventive Maintenance (IPM). *The goal of IPM is to achieve preventative maintenance through employee participation in maintenance activities.* IPM is similar to Integrated Quality Control (IQC), where employee involvement is a key to success in developing quality operators to meet customer requirements.

Lean production builds high quality into the production process itself, rather than achieving it later through inspections and repairs. Preventive maintenance catches trouble at the source, rather them letting things go wrong and then fixing each problem as it occurs. At the very foundation of TPM is partnership. TPM focuses on partnership between the manufacturing people, the maintenance department, manufacturing and design engineering and technical services to improve the overall equipment effectiveness.

16.11.1 Benchmarking Your Company

Prior to the development of an IPM plan, benchmarking or recording the current status of your organization is a good idea. This evaluation will include more than just the maintenance organization. It should embrace all parts of the organization involved in the manufacturing maintenance, design, and purchasing of the assets of the company. The most important benchmark for IPM is the current state of the equipment as measured by the percentage uptime.

Benchmarking provides the starting point from which to calculate improvement. As such, it also provides the means to demonstrate to your management that you are making progress, saving money, and improving productivity. Benchmarking also shows the production/maintenance teams that they are making progress which keeps interest and enthusiasm levels high.

Benchmarking involves documenting as much current data on equipment as possible. From the current status you can show management where the equipment was in terms of productivity, where it is, and where the trend patterns are pointing.

One of the first steps in benchmarking should be an analysis of amount of additional capacity the company could get from their machinery if they improved its uptime from where it was at baseline say, an 85-percent goal. Such data could include quality information of the equipment, time data, maintenance downtime histories, setup times, and procedures. Other documentation might include photos showing the physical condition of the equipment at program start to compare with similar photos after TPM implementation.

16.11.2 Piloting

Selecting a pilot area is an important step in the IPM sequence. Rather than attempting to implement IPM plant wide right from the start, most lean engineers start with the pilot or interim cell composed of machines from the job shop, using target pieces of equipment identified as critical to the linked-cell manufacturing system.

16.11.3 Preventive Maintenance

When considering the maintenance component of the organization, the first focus should be on a preventive maintenance program. Without a good preventive maintenance program, the equipment is never maintained at a level sufficient to ensure that the organization has assets capable of producing a world-class product or service.

16.11.4 Predictive Maintenance

Implementing a predictive maintenance program as the next step established a formal method for monitoring equipment and ensuring that wear trends are documented. In this way, equipment can be overhauled, with worn components changed, before a failure occurs. Table 16 .1 shows a typical inspection form for hydraulic and pneumatic system. Such forms must be developed for belts, chains, and general equipment, respectively.

16.11.5 Computerized Maintenance Management System

The subsequent improvement in the organization's ability to service the equipment is followed by an increase in the amount of data used to perform failure and engineering

Table 16.1
A Typical Inspection Check Form for a Hydraulic and Pneumatic System

Hydraulic Inspections	Item	OK	Needs Repair
Hydraulic Pump	Proper oil flow?		
	Proper pressure?		
	Excessive noise?		
	Vibration?		
	Proper mounting?		
	Excessive heat?		
Intake filter	Clean?		
	Free oil flow?		
Directional Control Valves	Easy movement?		
	Proper oil flow?		
Relief Valve	Proper pressure?		
	Excessive heat?		
Lines	Properly mounted?		
	Oil leaks?		
	Loose fittings?		
	Damaged piping?		
Pneumatic Inspections	**Item**	**OK**	**Needs Repair**
Compressor	Proper oil flow?		
	Proper pressure?		
	Excessive noise?		
	Vibration?		
	Proper lubrication?		
Inlet filter	Clean?		
	Free air flow?		
Directional Control Valves	Easy Movement?		
	Proper air flow?		
Muffler	Proper air flow?		
	Proper noise reduction?		
Lines	Properly mounted?		
	Air leaks?		
	Loose fittings?		
	Damaged piping?		

analysis. This highlights the need for a computerized database for tracking and trending equipment histories, planned work, the preventive maintenance program, the maintenance of spare parts, the training and skill levels of the maintenance workers, etc. Systems commonly used for this are called computerized maintenance management systems (CMMS), a comprehensive relational database accessible to the entire organization.

16.12 Continuous Improvement (CI)

As the IPM program matures, the company will see how much support maintenance provides other advanced programs. Lean manufacturing implementation is enhanced, since now the equipment runs when it is scheduled and produced at the rate it was designed to operate. IQC is enhanced, since the equipment is stable and produces a quality product every time it operates. Employee involvement (EI) programs are matured, since most of the problems the employees are currently having relate to the equipment. This makes their responsibilities easier to meet and leads to more motivated workers concerned with maximizing their competitive strengths.

16.12.1 Standardization and Continuous Improvement

The objective of standardization is to develop the daily routine so that continuous improvement can occur. Clear standards are critical to continuous improvement. Without standards, the internal customers have no goals by which to judge their work. Without standards, operations and methods will fall back to the old ways. The result will be continual firefighting, not continual improvement. Standards must be set and followed. For every deviation from the standard, the problem must be identified and must be eliminated. If possible, improve the standard so the same problem does not recur. In addition, as more jobs become standardized, there will be less confusion. Training of new operators becomes easier- which is extremely important where people rotate jobs often and work on multiple machines. In short, standards make jobs easier.

Standard Work

Standard work or standard operation is a way to achieve maximum performance with minimum waste in the cells. Standard work is composed of three elements.

1. Cycle time - The time between completion of the last component or product and completion of the next product. (CT = 1/PR)
2. Work sequence- The sequence of work performed by the internal customer.
3. Stock-on-hand (SOH) - The standard amount of work that is currently in process necessary to conduct smooth operations in the cell.
4. Standard work is not simply something management demands and operators comply with.
5. It is implemented to involve management and operators in the development process.
6. Finally, anyone should be able to follow the instructions of standard work. The goal is that a new operator should be able to master the standard work after three days of training on a process.

Table is similar to the Standard Operations Routine Sheet, but this sheet provides visual con

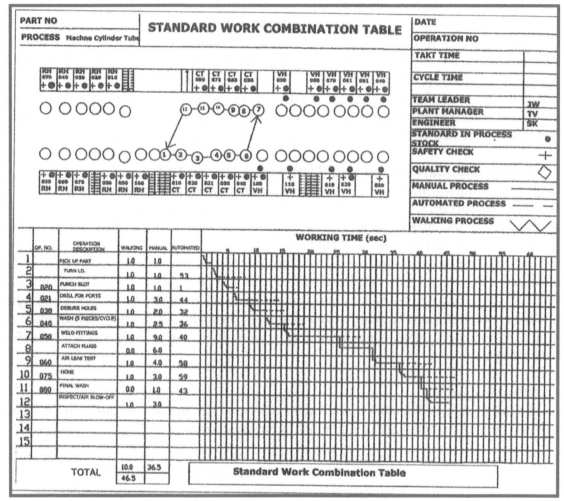

Figure 16.2
Standard Work Combination Table (Toyota Suppliers)

Standard Work Charts

A standard work chart, Figure 16.3, has two functions. It allocates an operator's areas of responsibility within a work cell or line. Second, it serves to develop standard work based on cycle time.

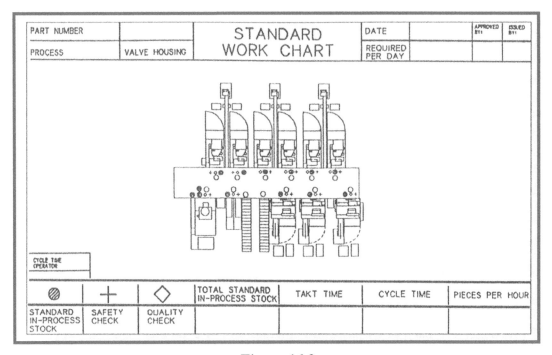

Figure 16.3
Standard Work Chart

Steps to Standardization

Standard work can be pursued through the following steps:

1. Standardize current work practices with assistance of engineers, team leaders, and operators.
2. Locate problem areas.
3. Solve the problem and develop new improved methods.
4. Implement the new methods.
5. If the new methods prove satisfactory, develop new standard work.
6. Require operators to write out the standard work combination tables with the assistance of supervisors.
7. Start the process over again.

In most cases, to the degree every person practices standards and improvement activities to eliminate waste; to that degree will the factory's competitive edge be improved.

The development and improvement of standard work involves everyone, especially the operators. It is said Toyota practices improvement activities to reduce operator work time by as little as half a second. Standardization improves the foundation.

16.13 Continuous Improvement and Employee Involvement

Machine failures are associated with the way people think and act. Operators, maintenance crews, and other support people should be trained to understand how their roles interact and what they must do to support one another.

Operators should learn how to perform routine maintenance on machines, be instructed in proper operating procedures, and develop an awareness of the signs associated with early machine deterioration. Maintenance people should learn how to assist production people with routine activities, readjustments, taking corrective actions, and increasing maintenance skills.

By involving every employee, zero machine trouble can be achieved. Employee involvement assures early detections, and early correction, of abnormal machine conditions. The clear implication of IPM is that continual skill development and training is essential to increase the abilities of all involved.

16.13.1 Continuous Improvement and the Five S Principles

Toyota developed the five S principles to describe in more detail what is meant by proper housekeeping. Here are the five S principles, or maxims:

1. Seiri ("Say-ree")- (Sifting)- Organization
Keep only the minimum of what is needed for a task and discard anything else.
Analyze what is available for the task, determine what is required to complete the task, and discard what is unnecessary. Anything extra is wasteful. For example, having extra tools, materials, pencils, and paper, is waste and should be eliminated.

2. Seiton ("Say-ton") -(Sorting) - Arrangement
Once the minimum requirement is determined, there must be "a place for everything and everything in its place."
Assign a location for all essential items, often using shadow boards. Make the work place self-explanatory so everyone knows what goes where. Thus, eliminate the confusion and the lost time associated with hunting for items out of their proper place.

3. Seiso ("Say-so")- (Sweeping)- Cleaning
Every tool should be clean and in proper working condition.
Once the work site is organized and arrangements are completed, tools must be kept clean so they can be easily obtained and used with no fumbling or lost time. If something goes wrong, a backup tool should always be available, in proper working condition, and stored exactly where it can be readily found.

4. Seiketsu ("Say-ka-sue") - (Spic-and-Span) - Hygiene
The working environment should be as clean as possible.
Hygiene usually complements the other aspects of detailed housekeeping. Effective organization and work arrangement is reinforced by keeping the entire area as clean as possible.

5. Shitsuke ("Shay-sue-ka"} - (Strict) - Discipline
All operational activities should be pursued with strict discipline.
These 5 general principles must be followed and become part of the daily routine. From observation, it seems the fifth S has been the hardest one to follow. Things start out organized, arranged and clean, but over time the workplace becomes messy. This is the second law of thermodynamics in action on the factory floor. All systems degrade with time unless maintained. Daily discipline will greatly enhance the Five S tools. As we have said in different ways, the

mark of good organization and effective work site arrangement is that nothing is hidden-everyone knows where to find or place items. The Five S's are basic principles for identifying problem areas and waste. But lean production depends on everyone's active involvement. Thus, every member of the factory must follow the Five S principles before results will be noticed and sustained on a daily basis.

16.14 Kaizen Activities Suggestion System
Kaizen ("KY-ZAN") means continuous improvement. It is the constant search for ways to improve the current situation, involving all functions of the factory.
Here are the specifics of continuous improvement, or Kaizen

1. The Team Concept
In lean production, the team is a key vehicle through which continuous improvement is achieved. Usually a production team is formed to work on problems in its specified area of the factory. Along with the production workers, it may include engineers, maintenance support and anyone else the team members might need to solve a specific problem.

Anyone can make a suggestion for improvement-and many suggestions the team implements without approval from anyone above it in the organization. The central point is to give the team the responsibility and the authority to make decisions.

The team leader for each area is the moderator. He or she has veto power over any suggestion. If problems are of a greater scale, the team calls on other teams or departments, the team may employ the use of the Seven Tools of Quality-discussed in Chapter 14 - to gather and present facts about the problem. Team meeting usually occur before or after the shift with overtime paid.

2. Monthly Engineering Meetings
The monthly engineering meets are another type of Kaizen activity. Engineers, who are usually from manufacturing, present the problem that they and the production team are tackling. The seven tools of quality are often used in a specific problem presentation format.

The engineer makes the presentation to a group of engineers, management, and production team leaders. (The presentation uses as many visual aids as possible.) After the presentation, the group is asked to suggest possible solutions. The engineer and the group will decide what options to try the next month. Through this exercise, the engineers and others in the group learn how to become problem solvers. This process is not intended to improve a person's presentation skills or writing skills. The idea is to teach everyone how to use analytical tools to solve problems systematically.

The Lean Engineer must demonstrate the ability to use these tools every day, not just for a presentation. They make it a habit to solve problems in a systematic way. As a result, a systematic, written history of the problem exists. Written histories are called *story boards* or *A3 reports* and they include the various implementations for improvement, and the current status of the situation (Chapter 17). As with the other elements of lean production, it takes practice and discipline for this process to become a daily problem solving habit. The goal, however, is not simply to go through the motions of lean production. The goal is a new, more effective and challenging way of thinking and acting.

3. The Suggestion System

The suggestion system also involves everyone in the factory. It is another mechanism to involve everyone in the production process - and to send the message that employees contribute through their minds, not just their hands.

If suggestions are not given through a team, they can be given through a formal cost reduction program. Ideas are submitted to a cost reduction coordinator who in turn presents the idea to a cost reduction team. If the suggestion can be handled by a production team, it will be given to them for review and implementation.

If a detailed study is required, a cost reduction team member will be assigned the project for additional study.

The suggestion system and, indeed, all the Kaizen activities, encourage continuous improvement through employee involvement It is a system designed to let improvements happen quickly without bureaucratic red tape. It is a system which underscores the principle that employees should be given the responsibility to suggest improvement and the authority to implement those suggestions.

16.15 Reliability

> *Reliability - **The probability of a system or device continuing to function over a given period of time and under specified conditions***

An essential element in any lean manufacturing system is system reliability. Lean systems have serial, linked processes and by definition do not have redundant components or duplicated machinery. Therefore, every component must operate properly every time. In order to accomplish this goal of continuous availability, one must consider three different, but related areas. These are the hardware systems, the software systems, and the workforce. Each of these systems can be studied independently, but the greatest improvements are realized when all three are related.

16.15.1 Reliability as a Concept

First defined by Webster's in 1816, reliability was originally: *The extent to which an experiment, test, or measuring procedure yields the same results on repeated trials.* This definition has evolved since then and now implies complete confidence in and reliance upon a system to the point that no alternate provisions are made. It now encompasses ideals such as consistency, repeatability, maintainability, and robustness. Analyzing a system requires that reliability be measured using some standardized method.

In every factory, different people have different views of reliability. For example, the plant manager would view reliability as the relative frequency, severity of unanticipated events, and consequence of the events. The area team leader might view it as the uptime versus the downtime of a process. The lean engineer would see reliability from the technical point of view with the classical *bathtub plot* of failure times. (Figure 16.4)

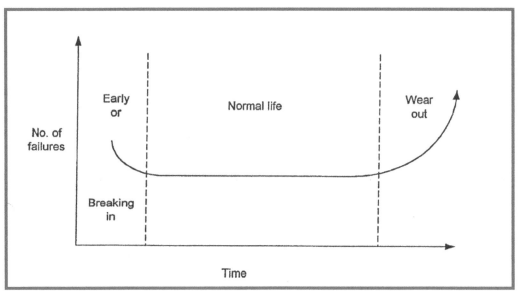

Figure 16.4
The Reliability Bathtub Curve

16.15.2 Reliability Studies

Typically when studying a system, both the "number of faults" and the "time between faults" are analyzed. The "number of faults" is a determination of the number of faults observed in a system during a fixed time interval during which the machine is operational. The "time between faults" is a measure of the time from one fault to the next.

Defining a fault is a critical step in the process. Often faults are defined as those events that bring an entire process to a halt, but this definition is not the best. A better method of defining faults would be any event that was not an anticipated event. (The application of this is further explained in the software system discussion.) The second step in the process is to record data and then statistically analyze it. The final step is to consider the faults in the system and try to relate the symptoms exhibited in their root causes. Asking *why* 5 times is a method to determine the underlying cause of a failure. Trying to fix it immediately is different than making changes based on the symptoms.

16.15.3 Hardware Reliability

Hardware systems are defined as those machines that are necessary to produce the product. These can be anything from a simple screwdriver to an automated welding machine on a factory floor.

Identifying faults in hardware systems is often easy since when a machine fails it often has a direct and immediate impact on the product or process. Hardware systems can fail many ways. The machine may fail completely and stop running, or a period of operational unavailability due to corrective maintenance may take place. A component failure in the machine or a power failure may cause the machine to be down. The challenge in hardware reliability lies in the identification of the root causes. This is because of the often complex relationships that are present in the plants.

Before any real improvements can be made to the hardware in a manufacturing cell, the shift from crisis management and reactive maintenance to planned maintenance and service must occur. This is essential and may seem intuitive, but a recent survey of over 70 manufacturing plants in a variety of industries revealed that over half of the maintenance work performed was reactive. That is, over half of the work done to keep machines running was done after the machine stopped running. It is then obvious to see that reliability of hardware is one of the areas that are in desperate need of improvement across the nation.

The best way to improve machine or hardware reliability is to decrease the number of machines that are used to do the same jobs. In manufacturing cells, this means the elimination of duplicate machinery. This may require switching to a two-shift usage of the machine. The advantages of this transition are threefold. First, when an error is found, then only one machine needs to be corrected. The same is true of improvements of the process in that only one machine is required to be replaced or upgraded. The second advantage is that it is easier to identify the source of the error and the machine associated with the fault if there is only one source for each process. The third advantage is not a reliability improvement, but rather just a general improvement to the process as it simplifies the paperwork required identifying the source of a fault once an error is found.

Another way to improve machine tool reliability is by slowing down the process. Suppose you have a manufacturing cell with 8 machines. The cycle time is 60 seconds. The third machine (lathe) has a processing time of 30seconds. For a lathe turning operation, processing or machining time is calculated by:

$$MT = \text{Length of Cut} + \text{Allowance /feed (ipr)} \times N_s$$
$$N_s = 12V/\pi D = \text{rev. per min}$$
$$V = \text{cutting speed (ft/min)}$$
$$D = \text{Diam. of work piece.}$$

Reducing the cutting speed will increase the MT but this will not change the CT for the cell so long as $CT > MT$. Therefore, a second way to improve hardware reliability is to increase the processing time (MT). One way to do this is to decrease the operating speed of a cutting tool where tool life is governed by the following equation:

$$V\,T^n = C$$

V is the operating speed of the tool. T is the tool life, and both n and C are constants related to the particular tool material. From the equation, it is clear that as cutting speed of the tool (V) is decreased, the tool life (T) will increase since the other variables are constant for each tool material/ work material combination. As tool life increases, the time between failures will increase, thus improving reliability. This is an example of a solution that requires almost no additional monetary investment but often represents significant monetary savings

A third way to improve the equipment reliability is through the implementation of successive checks following each step of a process. Inspecting at each station will allow identification and correction of the causes of errors before they become masked by other processes.

In order to improve hardware systems, the operators must have detailed knowledge of how the equipment works and be focused on the reliability of the process. Lean manufacturers are consistently better than their competitors when it comes to hardware reliability. A lean manufacturer allows the operator, someone with the required level of knowledge, to be directly involved in design development and improvement of the hardware. This is in contrast to the manufacturer that has a *specialized* or *focused* study team making recommendations, since this group may not be able to acquire such knowledge.

Reliability also requires that the operator be involved in the maintenance of the equipment. Operators do not necessarily need to be able to fix every problem, but more frequently need to be able to call upon an advanced technology when a more complex problem develops. Maintenance should work with the operators to return the machine to its original condition. Remember that the goal should always be to restore the equipment to a like-new condition, rather than just restoration of operational capability.

But remember this, computers are unreliable but humans are even more unreliable. Any system that depends on humans is unreliable (a Murphy's Law).

16.15.4 Software Reliability

Software including data and logic programs that only control the machine tools and material handling devices may also include the collection, distribution, and analysis of data essential to the production process. Software may be anything from the code that controls an automated welder to the software routines used to read barcodes on inventory cards. All software should enhance and simplify processes and operations because computers should do the routine work for man. However, software is not very flexible and always has *bugs* in it. The bugs or faults of a software system arise from both the errors in logic or the code, or program, and the faults of the computer hardware that interfaces them to their environment Those errors directly relate to human errors while creating the code or designing the system components. Frequently, this is due to some overlooked condition in the original software plan or an improperly analyzed operating environment.

When studying software reliability it is important to track not only the failures of the system, but also its faults. This is important because not every error will require the production process to come to a complete halt, but it may cause repeated errors, which will slow the system. To return to the bar code example, suppose you are scanning in tool parts. If the system scans in a two-inch drill bit as a three-inch drill bit, that does not bring the system to a halt, but could result in surplus inventory, or it could cause a machine failure if the 3-inch tool crashes into the work piece during rapid transverse or it could cause a part defect (hole 1-inch too deep!). Therefore, it is important to always track faults and NOT failures. However, just because the system does not *crash* does not mean it is error free. Software errors can hide in code and appear long after the system has gone into service.

Frequently overlooked is wrong information so that the input to a system is not correct. This can be because an input device failed or the input was inappropriate. An input device failure could be anything from a smeared lens on a bar code reader to a worn cable on a keyboard. An inappropriate input could result from a mislabeled container. An inappropriate output or response occurs when the inputs to a system are correct, but the outputs are incorrect. This reveals an error in the code.

Code reliability is highly dependent upon the time allowed between code development and its use in the field. Time must be allowed for the review and testing of the code. Obviously, the more the code can be reviewed; the errors can be identified and corrected.

Reliability of software will depend on whether you are creating the code, or simply modifying an existing code for another application. Suppose you have a code used to scan bar codes. If the program has been used to scan bar codes in a painting facility, and now you want to scan bar codes in the tool area, the modified code will have better reliability than a new one developed solely for the tool area. Because the software is at least a second-generation program, it has already undergone one complete round of the program debugging, so it is already free of some of the errors that caused faults in the first- generation of the code.

16.15.5 Workforce Reliability

Workforce reliability refers to the internal customers working in the manufacturing process keeping the process flowing continuously. A reliable workforce is a goal of almost every industry, frequently studied and theorized about. Every manager has their own theory on how to keep people on the job and productive. How did the Japanese raise from the ashes of WWII to become a great country? Was it simply by inventing better manufacturing process technology or the lean manufacturing system? Of course not! They excelled out of necessity and effort. The lean manufacturing system design provided the right environment for success.

Here are *Ed Adam's 5 P's for the people in manufacturing systems.*

1. **Purpose** - *Do my efforts always complement the task?*
    - ➢ Make certain that everyone in the organization knows intuitively what the purpose is.
    - ➢ Be the best in the world at what we do.
    - ➢ Have a design, cost and quality that permits us to pick our market (i.e., INTEL)
2. **Passion** - *How much am I really, really holding back?*
    - ➢ Have a sincere, deep and unrelenting desire to pursue our purpose.
    - ➢ Elevate the intensity, hours and planning - amount and method
    - ➢ Hire only those willing to do the same
    - ➢ Be accountable for motivating and requiring our people to do the same.
3. **Pragmatism** - *Could I explain it to my grandmother?*
    - ➢ Learn to use brains and not money
    - ➢ Eliminate waste- don't plan it forever, start today
    - ➢ Refuse to delegate work that does not add value
    - ➢ Stop useless reporting
4. **Participation** - *Should I ask my subordinates for advice?*

> Ultimately, people are our only asset, with them we can do much. Without them we can do nothing. One individual's efforts are always limited by the contribution of others.
> People must be educated, trained, respected, and fairly paid.
> We must arrange responsibilities so that it gives an opportunity to make a meaningful contribution to the organization's success. Without this our self-respect suffers. Without self-respect we cannot serve others and they cannot serve us.
> Use every employee in a humane and challenging manner. Everyone is paid to think and their thoughts are valuable. They understand this even if we do not.

5.   **Principles -** *Is our daily routine predictable?* (See Table 16.3)
> Standardize all activities as much as possible. (See Table 16.4)
> Train all involved personnel in those standard practices
> All rules are made to be broken - rule of last and first
> Only break rules after good participative investigation
> Institutionalize continuous improvement

Guiding Principle- *Our life is more important than our work.* We improve our work in order to add value to and enhance our life, not to escape from it. Try applying the 5 P's to our private life and see how we are challenged to live better, more fully, more creatively and more usefully to ourselves and others.

Author's note: Ed Adams is a design/manufacturing engineer who designs machine tools, work holders, decouplers, Poka-yokes and more for manufacturing cells for lean manufacturing vendors.

16.16 Shift Work

The L-T-F-C schedule calls for a second 8 hour shift. Perhaps one area that can often be improved in Lean manufacturing systems is the use of shiftwork in the process. According to Bureau of Labor Statistics approximately 15.5 million Americans work non-traditional (either rotating or fixed) shifts.

The reasons to be concerned about shiftwork may not be obvious at first. *The workers often report that they feel isolated* from management and the rest of the organization on evening/night shifts. A major problem is that they experience a loss of contact with friends and family. Another reason that shiftwork is so demanding is the fact that certain routine tasks that must be accomplished by the worker can only be done during normal business hours. Some examples of

Table 16.3
Training, Rewards, Involvement, Motivation for Internal Customer

Training, Rewards, Involvement, Motivation for Internal Customer	
Tell all workers that part of their job is to make signs and manuals covering:	
Job	what it is; how it is done
Machine tool	how to keep it running right and avoid activities
Team activities	its successes, its current performance, its concerns
Results	
Worker (and foreman) has primary responsibility for job descriptions and training, rewards, involvement and motivation. (Personnel department has secondary responsibility.)	

Table 16.4
Housekeeping, Station Design (Arrangement) and Cleanliness

Housekeeping, Station Design (Arrangement) and Cleanliness	
	Operator-centered
	Tie to JIT implementation - so reasons are clear (not just harassment)

	Precise arrangement to . . .
	  * Eliminate search time (human travel)
	  * Cut material travel
	  * Cut tool travel

	Absolute cleanliness for . . .
	  *Quality
	  *Safety
	  * Long machine & tool life
	  * Making problems visible

such tasks are routine car maintenance (such as oil changes), grocery shopping, and even doctor's visits. This causes most workers to return to daytime schedules on their days off.

In addition to this, the number of routine tasks that must be completed does not vary depending on the amount of free time the worker has. This means that the worker must still do X hours of household chores, and Y hours of routine tasks each week no matter how much time is spent at work. Therefore, extra work hours add to fatigue and allow less time to rest.

The result of this, if left unchecked is an increase in errors on the job as well as safety violations. The workers ability to concentrate on tasks is significantly reduced. Sick days increase as the worker drains his energy, his general health declines, he may become ill, and fail to make the required office visit to receive proper medical care. In addition, activities that were missed, such as a child's school play, can cause additional stress at home. These are the typical variables in the complex workforce equation for reliability.

Several characteristics help classify shiftwork. The first is characteristic of a shift is whether it is a permanent shift or a rotating shift. This refers to whether the same workers are always on the second (or third) shift, or if they rotate from days to nights periodically. Each type has its own advantages. Permanent shifts allow the person the opportunity to adjust to the non-traditional work schedule. Rotating shifts do not disrupt the traditional schedules except during brief periods.

The second characteristic used to classify shiftwork is the *speed and direction* of the rotation, if applicable. The rotation between shifts is said to be in a forward direction if a worker moves from the day shift to the evening shift. It is backwards if they move from nights to evenings, evenings to days, or days to nights (in a 3 shift cycle). The speed of rotation refers to the amount of time the worker has to adjust to the schedule shifts. This typically varies from a half day to half a week.

The third characteristic of shiftwork is the *work-to-rest ratios*. This is important when determining how alert a worker will be. As the ratio increase (more work, less rest) the alertness of the worker decreases as expected.

The final distinguishing characteristic is *how the worker views shift stability*. This means, does the worker see it as regular and/or predictable, or does he see it as sporadic. Workers typically adjust better to shift rotations when they can prepare for them.

Some ways to improve shift work are to avoid permanent night shifts and keep consecutive night shifts to a minimum. Avoid several days of shiftwork followed by 4 to 7 day mini-vacations. Try to plan free weekends and keep the schedule regular and predictable. In addition, ensure good inter-shift communication by holding organizational and planning meetings during all shifts, not just the day shift. The 4-8-4-8 schedule discussed earlier utilizes the 4 hour shift break during the day for meetings.

16.17 Implementation of Integrated Productive Maintenance

Here are the 12 steps involved in developing and implementing productive maintenance program (Nakajima).

Step 1:
> ➢ Announce top management's decision to introduce IPM.
> ➢ State IPM objectives in company newsletter/emails.
> ➢ Place articles on IPM in company newspaper/emails.

Step 2:
> ➢ Launch an educational campaign.
> ➢ For managers, offer seminars/retreats according to level.
> ➢ For internal customers, provide slide presentations.

Step 3:
- ➢ Create organizations to promote IPM.
- ➢ Form special committees at every level to promote IPM.
- ➢ Establish central headquarters and assign staff.

Step 4:
- ➢ Establish basic IPM policies and goals.
- ➢ Analyze existing conditions.
- ➢ Set goals
- ➢ Predict results

Step 5:
- ➢ Formulate master plan for development.
- ➢ Prepare detailed implementation plans for the five foundational activities

Step 6:
- ➢ Hold a TPM kickoff event.
- ➢ Invite external customers, affiliated and subcontracting companies

Step 7:
- ➢ Improve effectiveness of each piece of equipment.
- ➢ Select model equipment.
- ➢ Form project teams

Step 8:
- ➢ Develop an autonomous maintenance program.
- ➢ Promote the Seven Steps
- ➢ Build diagnostic skills and establish worker procedures for certification

Step 9:
- ➢ Develop a scheduled maintenance program for the maintenance department.
- ➢ Include periodic and predictive maintenance.
- ➢ Include management of spare parts, tools, blueprints, and schedules

Step 10:
- ➢ Conduct training to improve operation and maintenance skills.
- ➢ Train leaders together·
- ➢ Have leaders share information with group members.

Step 11:
- ➢ Develop initial equipment management program.
- ➢ Use MP design (maintenance prevention).
- ➢ Use start-up equipment maintenance.
- ➢ Use life cycle cost analysis.

Step 12:
- ➢ Perfect IPM implementation and raise IPM levels.
- ➢ Evaluate for PM prize.
- ➢ Set higher goals.

16.18 Summary

Increase Reliability for 3 components of Lean manufacturing
- ➢ Hardware -the machine tools that are necessary to produce the product

> Software- the data and logic systems that control the machine tools and collect and distribute data essential to the control of the manufacturing system
> Workforce Reliability- the people required to actually operate the machinery within the manufacturing system

Promote Reliability from 3 different viewpoints
> Project Managers.- up time vs. down time
> Plant Managers- relative frequency and severity of unanticipated events
> Equipment Manufacturer- financial cost, engineering effort, and reputation costs associated with a new technology

Recognize the basic Do's and Don'ts of Reliability

Do .....

1. Understand failures caused by interactions
2. Be aware of program timing
3. Understand what customers expect
4. Use common systems and methods
5. Communicate with other engineers
6. Track vendor performance
7. Design for easy assembly
8. Consider external influences
9. Regard reliability as a moving target

Don't .....

1. Invent during product development process
2. Focus too narrowly and forget your component interactions
3. Think your job is complete is everything is running smoothly

Follow Proven Software Guidelines

From the International Journal of Quality and Reliability Management, Volume 14

Think about *faults* rather than *failures* when considering the reliability of a system. Faults can be wrong information, disconnected line, system crash, or just inappropriate responses.

Workforce Guidelines

From DHHS (NIOSH) Pub. No. 97-145

- Avoid permanent (fixed or non-rotating) night shifts
- Keep consecutive night shifts to a minimum
- Plan free weekends
- Keep schedule regular and predictable

From lEE Solutions June 1998

- Ensure good inter-shift communication
- Hold meeting during all the shift

Standardize Support Functions

- In the lean production factory system, all support functions reinforce the core principles of stability and continuous improvement through employee involvement.
- Standardization of machines and work routines lends stability to the workplace and production process. Preventive maintenance - a habit of mind which prevents

break downs in machinery also assures a stable, predictable production environment.

- The *Five S Principles*, for keeping work places clean, orderly and always ready at the command of the worker smooths the road to effective work.
- On the stable platform created by all support functions, *Kaizen* activities and a *Suggestion System* draw the minds and energies of the workforce into the production process. Using active teams, employee suggestions and a systematic approach to problem solving, stability allows continuous improvement through people.

Over 50 years ago, the Japanese learned about preventive maintenance from the United States. They have since developed and improve the preventive maintenance program into what they presently call total productive maintenance. The objective of integrated preventive maintenance is to prevent failure of all equipment before it actually occurs. The objective of total productive maintenance is to increase productivity, improve efficiency, increase the percentage of time equipment is able to operate, and minimize the number of required steps by keeping it simple. The program requires heavy involvement with production workers. This is IPM.

Review Questions

1. Why is the Lean Manufacturing system designed to run at Less-Than-Full-Capacity (LTFC)?
2. What are the benefits of Preventive Maintenance
3. Discuss five pillars of IPM.
4. What are the factors that lead to machine breakdowns? List a few causes.
5. What is the importance of piloting in IPM?
6. What are the three elements of standard work, why is SOH so important in standard work?
7. What are the functions of standard work charts?
8. List five principles of Continuous Improvement and explain.
9. Explain the "Bath Tub" plot of failures.
10. "Any system that depends on humans is unreliable!" Explain.
11. List Ed Adam's five P's for the people in the manufacturing systems.
12. What are Nakajima's 12 steps for developing and implementing an integrated maintenance program

Chapter 17
Lean Tools

Lean tools can help the company achieve level production, resulting in dramatic reductions in throughput time while improving cost, quality and flexibility -the four horsemen of the L-CMS.
J.T. BLACK

17.1 Introduction

This Chapter covers some of the more important lean tools that might be used to design, build and control a Lean manufacturing system. Some have been previously mentioned and some have not. Lean engineers use these tools to design, analyze and control the lean production system.

KAIZEN FOR CONTINUOUS IMPROVEMENT			
PULL / IPC / Kanban	MANUFACTURING, MACHINING AND ASSEMBLY CELLS		TPM
JIDOKA/ QUALITY / 6 SIGMA/ TAGUCHI	POUS	SMED	VISUAL CONTROL
STANDARD WORK	POKA – YOKES DECOUPLERS	A3 REPORTS	TEAMS
5S SYSTEM	PROCESS FLOW MAPPING VSM	JUST IN TIME/ MMFA / Takt TIME	

Figure 17.1
Lean Tools Used to Operate the Toyota Production System (TPS)

Value Steam Mapping

Value stream mapping began with Toyota, who for many years used a methodology called information and material flow diagrams, which were developed from traditional process flow diagramming techniques. Mike Rother and John Shook used their knowledge of this practice to create a simple way for managers to see the flow of materials to which value was being added. This was called *value stream mapping* and was introduced in 1998. (Rother and Shook, 1998). See Chapter 8 for details.

17.2 Process Flow Charts

Many different process mapping techniques can be used to document and analyze processes including group technology, operation process charts, flow process charts, man-and-machine process chart, two-handed charts and value stream mapping, discussed earlier in Chapter 8.

If a picture is worth a thousand words, then the major benefit of process flow mapping has been defined. Process maps illustrate the essential details of a process in a way that written description cannot. To begin a process mapping project define the process as a series of tasks or sequences of operations and that when linked together, can turn inputs (raw materials) into output (finished goods and products). It is a way for factories to organize their resources (physical elements) to accomplish their goals (meet customer needs). Furthermore, a process must have a beginning and an end and all work is part of a process that ends with a customer.

17.2.1 Conducting Value Stream Mapping

- The first step in a process analysis is to decide which process or family of process to investigate. The fact-finding phase of an investigation often part of a kaizen event. Outlines of the questions to be asked and the actions to be observed should be known before anyone is contacted. Department managers or supervisors should be informed of the objectives of the process along with the operators. More and better data result from extending similar courtesies during the investigation.
- Be certain the operators are involved and the objectives of the investigation.
- Their contribution will lead to the success of the study.
- Solicit and encourage suggestions by the operators.
- Be courteous and complimentary.
- Do not criticize or correct the way the workers are doing anything. The mission at this stage is to find facts, not to correct faults.

17.2.2 Process Flow Symbols

Early renditions of process flow charts had many symbols for plotting activities. Since that time the number of basic symbols has decreased but the format variations for process charting are almost innumerable. And this situation is probably healthy, because customized applications can make charts much more useful for particular situations such as computer procedures or forms flow analysis. Standardized symbols are found in *Operation and Flow Process Charts, ASME standard 101*, ASME, New York, 1947. The following standardized symbols will be used here.

◯	Operation -intentional changes in one or more characteristics
⇨	Transportation -a movement of an object or operation that is not an integral part of an operation or inspection.
☐	Inspection -an examination to determine quality or quantity
D ▽	Delay -an interruption between the immediate and next planned action. Storage -holding an object under controlled conditions.
⊡	"Combined" -Combining two symbols indicates simultaneous activities. The symbol shown means an inspection is conducted at the same time an operation is performed.

It is important to distinguish between an operation process chart which follows the routing and acts performed on an object, and an *operator* process chart which shows the sequence of activities performed by a person. The two types should not be combined. Three typical formats for process charts designed to serve different purposes are shown in Figure 17.2. The actual charts represented by the illustrations would provide entries for additional details.

Type and Purpose	Characteristics	Illustration	
Single column - to study the detailed steps in a relatively simple process	Charts are often drawn on printed forms; processes are shown by connecting appropriate symbols; space is provided for additional data	Symbols	Description
			Invoice in mail room
			Determine addressee
			To addressee's secretary
			Placed in action tray
Multicolumn - to analyze detailed steps in the flow of work that is quite complex	Horizontal lines show operational areas; symbols are entered to show activities of the process	Operator	Activities
		Mail Clerk	
		Secre-tary	
		Man-ager	
Layout diagram -- to improve the layout by avoiding unnecessary steps	Lines indicate the travel and symbols show the activities on a layout often drawn to scale	Mail room	Manager
			File
			Secretary

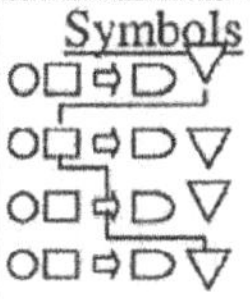

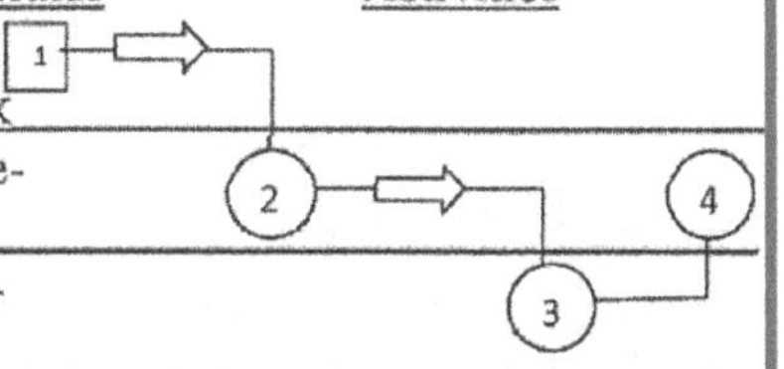

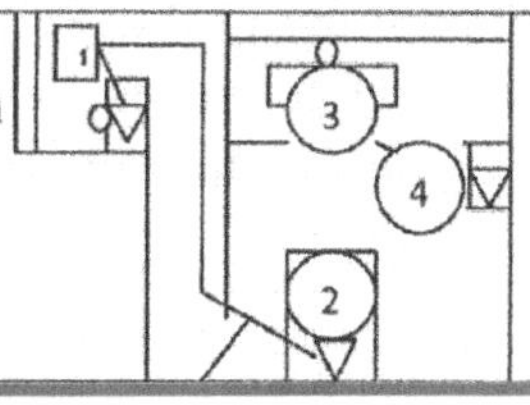

Figure 17.2
Typical Formats for Process Flow Charts

Each process flow activity is represented by a charting symbol and placed in the swim-late of the function that does it. Here is what to do to become skilled at process mapping.

- Elicit process-related information using various techniques, such as interviewing, facilitation, observation, and written procedure.
- Ask the appropriate questions to capture activity steps decision points, and pertinent information.
- Software charting tools are available to check to see if symbols are properly aligned and evenly distributed.
- Redesign the process maps to minimize excessive crossover of lines.
- Have the employees write down the steps in the process (since they know it best). If there are multiple people doing the same thing, have each of them write down the process steps independently and then compare all the process maps. This will develop a shared but consistent view.
- Capture the right level of detail and describe steps using the minimal number of effective words.

539

Process mapping is not the end product but is usually the first step. The information derived from process mapping should be used with other lean tools. Process mapping is not a method to determine process costs and savings. Although process maps can capture costs as secondary information, costs are not the main focus. Process mapping can be used for benchmarking a process but will not reveal the industry benchmark. Engineers would still need to see what other organizations are doing to compare and discover best practices. Process mapping is not a tool to analyze organizational structure. They do not describe organization charts. Figure 17.3 provides an example of a process flow chart.

Here are a bakers- dozen tips on how best to create a process map.

1. Clearly identify and define the process or sub process to be reviewed. Failures to do this could result in the wrong group of tasks being identified as a business process and targeted for change. The risk is that you may implement improvements or make system changes that do more harm than good.
2. Understand the objective for mapping a process. For instance, will the process map be used to identify bottlenecks? Will it used to see which steps are redundant?
3. Frame the process under review. Establish what's in and out of scope.
4. Be aware that the processes have inputs, outputs, and enablers. An example of an input could be an invoice; an output could be a payment check sent to a supplier. An enabler is a factor that helps a process achieves intended results. For example, information technology is a common enabler, and an information system enables a process by automating or supporting a step or managing workflow.

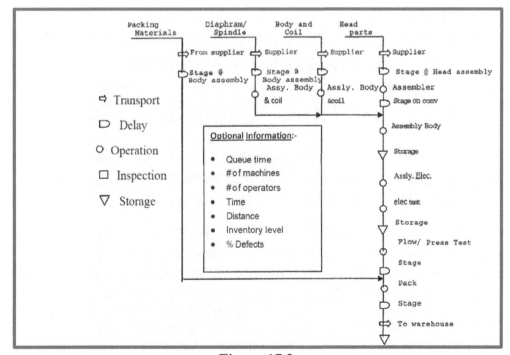

Figure 17.3
Process Flow Chart for a Head and Body-coil Sub-assembly

5. Show the key decision points. These important steps are the most difficult to capture because they tend to be mental steps as opposed to tangible steps.
6. A process map without decision points usually lacks the essential details required to use an effective analytical tool.
7. Created process maps in facilitation session that involve people from multiple functions and disciplines as much as possible. This unearths many good process improvement suggestions. Fine-tune facilitation skills to cope with human anxiety and potential conflicts. Team building exercises and a bit of humor help.
8. Pay attention to the level of detail you capture. The primary steps you need to show are the ones that add value, transport the work item, or introduce a delay. If it gets done, show it. While constructing the process map, constantly ask, "Did I capture enough information to make me understand the critical elements of the process?" If the answer is yes, then you most likely have a completed map and should consider stopping.
9. Observe and listen to reactions from the subject matter experts, project team members, or client stackers. If you hear the term *Aha* a lot and see happy faces then you most likely created a useful process map. If not, keep working at it.
10. Avoid mapping the *to-be* and concentrate on mapping the *as-is*.
11. Don't hesitate to ask challenging and probing questions.
12. Take the initiative to suggest an effective action work for an activity step or a phrase for a decision point if the subject matter expert, project team member or client stakeholders are struggling.
13. Practice, practice and more practice. Try to experiment with various process mapping techniques and master the one you're most comfortable using.

The benefits of process mapping are:
- Process maps provide a visual picture. Employees are often surprised by the complexity of the process and are excited to see the whole process or system for the first time.
- Employees are more willing to participate in the change initiative and introduce new ideas if they participate in process mapping
- Process maps aid in critical business communication, problem solving, and decision-making processes. The process steps demonstrate connections and sequences that allow immediate location of any element of the process, which can be big time saver. Process maps illustrate the important characteristics of a process, such as: How many people and departments are involved? How much redundant steps-hand-offs, verification, approvals -are there? What are the bottlenecks and inefficiencies along the critical paths? Does technologies serve the process or does the process or does the process serve the technologies? Is this process executed in the best way?

Limitations of Process Flow Mapping:
- Process maps are very detailed. Hence it may be difficult to show all the data on a

particular process map.

- It is more useful for identifying wastes in a particular process, as it is a micro level activity. It does not show the overall picture of the factory.
- An experienced person is required for process flow mapping, because of the minute details involved
- Process flow mapping does not show information flow, as in value stream mapping.
- It is only company specific.

Value stream mapping (Chapter 8) and process flow mapping are both very effective tools in implementing the lean manufacturing methodology. Value stream mapping can be used for identifying sources of waste in an organization and help a company to visualize the current state of progress. This helps to make a plan of the improvements that are required to be made and thus strategize on the future state development. It is a qualitative tool to measure progress. Value stream mapping helps the management to formulate and answer a specific set of questions which are used for future development and improvement.

Process flow modeling is also extremely effective in identifying all the minute sources of waste within the processes or a factory. It draws an exact picture of what the process is and provides a detailed understanding of the different factors that decrease the value added time in a process.

17.3 Lean Six Sigma

Lean/Six Sigma engineers gain experience as keen listeners and problem-solvers, not strictly Lean tool masters. They use creativity and leadership to guide challenging projects to successful outcomes. Training levels for six sigma were discussed in Chapter 14.

Lean/Six Sigma methodology includes a variety of tools such as: Brainstorming, Affinity Analysis, Survey, Process Mapping, Pareto Chart, Fishbone Diagram, Kaizen, Kanban, Value Stream Mapping, 5S, Setup Time Reduction, Mistake Proofing, Point of Use Storage, X bar and R Chart Control Charts for process capability, Taguchi Methods, Confidence Test, ANOVA, Regression, Design of Experiments, Monte Carlo, Correlation and Multi-Variable Chart.

The drive towards superior quality has led to the introduction of a set of variance reduction tools called *Taguchi methods* for improvement in products, product design, and processes. Basically, SPC looks at processes and control, the latter loosely implying improvement. Taguchi methods, however, span a much wider scope of functions and include the design aspects of products and processes, areas that were seldom if ever formally treated from the quality standpoint. Another threshold has been reached in quality control, witnessed by an expanding role of quality in the production of goods and services. The consumer is the central focus of attention on quality, and the methods of quality design and control have been incorporated into all phases of production.

17.3.1 Taguchi Methods Incorporate the Following General Features

- Quality is defined in relation to total loss to the consumer or society from less than-perfect product quality. Methods include placing a monetary value on quality loss. Anything less than perfect is waste.
- In a competitive society, continuous quality improvement and cost reduction are necessary to stay in business.

- Continuous quality improvement requires continuous reduction in the variability of the product-performance characteristics. Taguchi methods can be an alternative approach to making a process capability study. The Taguchi approach uses a truncated experimental design. This is called an *orthogonal array,* and determines which process inputs have the greatest effect on process variability (for example, precision), and which have the least. Inputs with the greatest influence are set at levels that minimize their effect on process variability. In other words, Taguchi methods seek to minimize or dampen the effect of the causes of variability and, thus, reduce total process variability. See Chapter 14 for an example.

Taguchi methods are more than just mechanical procedures. They infuse an overriding new philosophy into manufacturing management, basically making quality the primary issue in manufacturing. The manufacturing world is rapidly becoming aware that the consumer is the ultimate judge of quality. Continuous quality improvement toward perfect quality is the ultimate goal. Finally, it is recognized that ultimate quality and lowest cost of a manufactured produce are determined to a large extent by the manufacturing processes and the manufacturing system (integration of the product and the process)

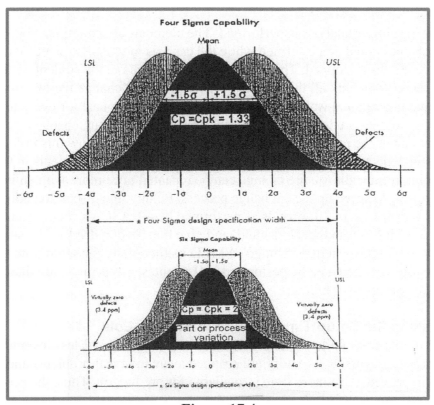

Figure 17.4
To Move Six sigma Capability from Three Sigma Capability Requires the Process Variability be Greatly Improved (σ reduced).

With the advent of six-sigma philosophies and Taguchi Methods, a new understanding of quality has emerged. Process variability can be reduced and controlled. The noise level of a process can be reduced by exploring the nonlinear effects of the products (or process) parameters on the performance characteristics. Using the six sigma tools, process variability can be reduced, as shown in Figure 17.4 where a process with 3 sigma capability has been improved to a 6 sigma level. Sigma is a unique kind of measurement in that it changes in size (decreases) as the variability in the process capability decreases while the upper and lower specification limits stay the same size. The limits are product design specifications or tolerances added/subtracted to a nominal size dimension.

17.4 SMED

The following exercise introduces students to the basic concepts of the rapid exchange of tooling and dies, including setup-reduction philosophy and techniques. This methodology was developed by Shigeo Shingo, an IE working at Toyota for Taiichi Ohno. He called it SMED, an acronym for Single Minute Exchange of Dies.

The class is led to a parking lot and a student is asked to drive a car up to the group and stop. The student is asked to change the rear tire while the rest of the class observes. Students are instructed to document tire-changing work, breaking the job down into tasks or elements, and timing each element. The procedure is videotaped. This changeover task usually takes 15-20 minutes, provided the student can find the necessary tools.

Then the following simple analysis is performed. The elements are examined and students determine which ones could be performed while the car was moving and which can only be performed while the car is stopped. This analysis usually results in the student getting the spare tire out of the trunk along with all the necessary tools needed to change the tire, and laying them out on the ground in a rational way [good housekeeping means a place for every tool and every tool in its place). Now the student drives up to the area again, gets out, changes the tire and drives off in about half the previous time. The discussion centers on how this task can be further improved. The students will have lots of suggestions for simple improvements in tools or the jack. At this point, if time allows, tire change can be repeated once more with an improved jack (electric power jack) to speed up the exchange process. Finally, the discussion concludes with a video of a changeover time of 20 seconds as performed by a NASCAR racing team (to change four tires, fuel a vehicle, clean the windshield, and give the driver a drink). The students have now performed the first two steps of Shingo's SMED methodology for setup time reduction. SMED really means the setup can be performed in 9 minutes, 59 seconds, or what Shingo meant by single minute -less than 10 minutes.

17.5 Decreasing the Setup Time Results In Smaller Lots

Building in small lots costs far less per unit. Quality improves because less inventory needs to be managed and stored. Layers of inventory are stripped away to reveal problems and throughput time is greatly improved. Also more time is available for production. Thus, the economics of setup elimination are persuasive.

So, in lean manufacturing, as setup time is reduced, products can be built in the smallest lot sizes possible at a rate dictated by the daily demand and the Takt time. As the unit cost approaches the variable cost, material costs tend to dominate the variable cost per unit. The delightful news is that eliminating setup is not some complex, sophisticated undertaking. Setup reduction requires knowing some simple rules and applying good operations and methods analysis along with a bit of good common sense. Clearly, it could have been done long ago, but most companies failed to see the need.

17.6 Organizing to Eliminate Setup Time

Most setup problems are related to materials, manufacturing processes and systems, and management practices. Contrary to popular opinion, labor is usually a minor factor. Many companies have used a team approach rather than individuals for setup reduction efforts. The recommended approach combines a team, including the internal customers, trained in Single Minute Exchange of Dies (SMED) fundamentals. (Shingo, 1985). The team tackles difficult setup problems, develops standard procedures, and trains operators in SMED. The follow sequence of steps is recommended:

1. Select a full-time project leader who believes in the lean manufacturing philosophy and setup reduction.
2. Select a project team to do the initial setup reductions. The setup improvement team typically includes some or all of the following people: the operators, the lean engineer, a design engineer, setup workers, a toolmaker, a consultant (who has experience in setup reduction), a foreman or supervisor, a manager from the project area, and a union leader. Keep in mind that a team of four or five dedicated individuals will be able to be more nimble than a larger team.
3. Hold a series of informational meeting with management, supervisors and foremen, and all workers including the union committee. These meetings must emphasize that the setup reduction program will result in faster but more frequent setups and that the workers will be responsible for doing setups in the cells. These meeting must explain what is to be done, why it is to be done, who is to do it, and how it is to be done.
4. Select specific areas of the plant for the pilot projects. This may be a collection of machines, processes, and operations already organized into manufacturing cells. As soon as machines have been arranged into cells, setup problems will have to be addressed. Harley-Davidson formed cells specifically to reduce setup time so Harley-Davidson was able to obtain immediate increases in capacity while eliminating many long setups. The initial pilot project may have long setups, scheduling problems, large inventories (work-in-process), a high inventory value, or severe quality problems.
5. Suggestions should be very welcomed. Keep the union advised and involved. Invite the union president to your team meetings. This program has nothing to hide. The sole motivation is that by reducing setup time, manufacturing runs can be shortened, inventory and costs can be reduced, and productivity and quality can be improved.

Once the team is trained in SMED and setup operations, begin specific SMED training of additional operators and setup personnel. The SMED methodologies are so simple and direct that everyone can do it. Besides, the company does not have time to wait until the team gets around to all the machines in the plant. Therefore, the attack on setups must be company-wide.

17.7 The Project Team

Regardless of the team size or makeup, every member must be well trained in setup reduction, problem solving, and have a positive attitude. The feeling that the job can be done better and less expensively is essential for the success of this initial team and the project itself. Do not neglect the factory workers. Include workers on the team on a rotational basis. They will know more about what it takes to eliminate setup time on their jobs than anyone else. The proper atmosphere is important here. This should be a grassroots program, dominated by the shop floor personnel. This is not another engineering project. A key element is that the people who developed the existing setup should not be on the team to try to improve it because they have vested interests. Problems with pride of authorship are best avoided. A fresh perspective is sometimes the best approach, particularly since we are all good critics but few among us are creators.

An alternative to the team approach is to do most of the work through existing channels. The people in engineering and tool design only review setups and try to invent solutions. This serves two purposes: (1) it generates some high quality ideas; and, (2) it introduces support areas to the idea of quick setups that can apply to other work areas. However this approach often fails to unearth many easy, low-cost, readily implemented solutions. Harley Davidson discovered that with the team approach, it was able to reduce the setup time from three hours to less than twelve minutes on the first machine line the team studied.

The temptation is to do other improvement tasks with the project team, including process improvement, changing the process sequence, quality standards, and so forth. This approach may be the easiest to take but is not the most effective. The major objectives of the setup team should be to develop and implement solutions to reduce setup time and to train all the workers and foremen in setup reduction. Therefore, by spending time establishing objectives and obtainable goals, the project team avoids trying to solve all the problems they will encounter in areas other than those directly related to setup. The initial team also avoids trying to solve all the setup problems themselves or else they run the risk of getting bogged down in one area.

To summarize, the role of the project team is to:
- Train and involve all operators, supervisors, and support personnel
- Gain experience from the worst setup projects
- Prepare plans and set priorities
- Determine installation timing
- Coordinate group efforts
- Create and maintain everyone involver's enthusiasm

17.8 Motivation for Single-Minute Exchange of Dies (SMED)

The lean production system is a serial industrial process. The Just-In-Time concept means items are made *when* they are required, in the *quantities* required, and inexpensively as possible. Minimizing inventories, synchronizing manufacturing system elements and convert all upstream subassemblies and component manufacturing to Lean cells in order to minimize work-in-process.

It is important to stimulate the need for reducing setup change time. Lean production uses *Kanban*. Depending upon the type, Kanban uses indicator cards or signs that, in addition to preventing overproduction and performing production control functions, provide information on production and transactions. Kanban also acts as a tool for forcing the gradual improvement of the manufacturing system, i.e., reduction of throughput time, fewer defects, and no late deliveries. To function this way, the Kanban must be located so that everyone can see the sequence, amount, and timing of work to be done.

Over time, the number and size of the Kanban will be decreased, and, in turn, decreasing the system's WIP. Reducing the lot size is fundamental to reducing WIP. Reducing setup times is fundamental to reducing lot sizes.

The basic approach to reducing setup times includes several key points:
- It is important to have the conviction that drastically shortened setups are possible. Dramatic reductions can be made by starting out with the attitude that tooling changes are not just a matter of removing one tool or die and attaching another. If the setup time can be reduced in one area, then managers and supervisors will gain direct experience of the improvements, making it easier to extend these improvements laterally to other areas and operations.
- The Setup process is simplified so that the workers can do it. Keeping setup changes away from machine operators merely creates a class of setup experts. Dealing with this issue is one of the principal goals of setup improvements. Workers must be directly involved in the change process.
- Locating and eliminating adjustments should be emphasized. Adjustments depend on the right touch or on luck and are not repeatable. Differences show up when different people make the same adjustments. Even the same person may take varying times to make the same adjustment on different occasions. For all these reasons, adjustments should be eliminated, replaced by settings.
- A setup changes should produce defect-free products right from the first piece after changeover. There is no logic to speeding up a setup operation without knowing if quality products can be manufactured.
- The Setup Process flows through the cell one operation at a time and is performed by the operator. Each time the operator completes a CW loop, he performs a setup on the next machine in the sequence.
- The total setup time is the time from the last good part of the previous setup to the first acceptable part in the new setup. Anything affecting that time frame is in the scope of a setup reduction program. The basics of the SMED methodology to improve tool and die exchanges are given next.

17.9 Four Stages for Reducing Setup Time
The four basic stages of a SMED, setup time reduction program are as follows.

1. Determine the existing method.
2. Separate the internal elements from the external elements where: External Setup -All operations (such as transporting dies to storage or to the machines) that can be conducted while a machine is in operation or running. Internal Setup -All operations (such as mounting or removing dies) that can be performed only when a machine is stopped.
3. Shift the internal elements to external elements.
4. Improve all elemental operations. Reduce or eliminate internal elements by continuously

improving setup. Apply methods analysis, and practice doing setups. Eliminate any on-line adjustments.

17.10 Determining Existing Practices (As-Is)

Operational analysis, using motion and time study, can be used to determine the current setup procedure. The usual objective is to improve work methods, eliminate all unnecessary motions, and arrange the necessary motions into the best sequence. The setup is broken down into short elements and activities that consume the most time. Problem-solving techniques can be applied separately to each particular activity to achieve the lowest possible time. Setup procedures are usually thought of as infinitely varied, depending on the type of operation and the type of equipment being used. Yet when these procedures are broken down into elements and analyzed from the SMED point of view, one finds that all setup operation comprise a sequence of steps. Traditional setup changes the distribution of time similar to that as shown in Table 17.1

Figure 17.5
How Setup Flows Through the Cell

Operation vs Proportion of Time Spent	
Operation	**Proportion of time**
Preparation after-process adjustment and checking of raw material, dies, jigs, gauges, cutting tools etc.	30%
Removing and mounting new tooling etc. for next part	5%
Centering, dimensioning and setting of other conditions	15%
Trial runs and adjustments	50%

Preparation, after-process adjustments, checking of materials, tools, etc., ensures that all parts and tools are where they should be and that they are functioning properly. Also included in this step is the time period after the setup is completed when the items like the removed tooling and dies are returned to storage and the process is cleaned, etc. Removing and mounting work holders, dies, tools, parts, etc. includes the removal of parts and tools after completion of processing. This also includes the insertion of work holders, tooling, parts and cutting tools for the next job. Measurements, setting and calibration refers to all of the measurements and calibrations that must be made in order to perform a production operation, such as centering, dimensioning, measuring temperature or pressure, etc.

Trial runs and adjustments means that adjustments are made after a test piece is machined and checked against the specifications. The greater the accuracy of the measurements and calibrations in the preceding step, the easier it will be to make these adjustments.

The frequency and length of test runs and adjustment procedures depend on the skill of the setup people. The greatest difficulties in a setup operation lie in adjusting the equipment correctly. The largest proportion of time associated with trial runs is derived from these adjustment problems. So, manufacturers should seek to eliminate trial runs and adjustments from existing setups. Because the existing setups can be quite long, videotaping of two or three setups is very helpful for later review and analysis. Having the worker, as a team member, review the videotape will reveal much waste in the existing setup, even without carrying out an operations analysis.

Typically, about 20 to 50% of the time is spent in adjusting (making a part, checking it, adjusting setting, repeating) and 10 percent of the time in locating and securing the part to be manufactured and its associated tooling.

A motion analysis of the process using a stopwatch is the next best approach. Such an analysis, however, takes a great deal of time and skill. Another possibility is to use a work-sampling study. The problem with this option is that work samples are precise only where there is a great deal of repetition. Such a study may not be suitable when few actions are repeated. A third useful approach is to study conditions on the shop floor by interviewing workers and having them explain what they do in a step-by-step manner. Even though some consultants are advocates of an in-depth continuous production analysis to improve setup, informal observation of the setup and a discussion with operators often suffices.

17.11 Separating Internal from External Elements

Internal, or mainline, elements refer to setup actions that require the machine to be stopped. External or off-line elements refer to actions that can be taken while the machine is running. The most important step in implementing SMED is distinguishing between internal and external setup actions. While everyone will agree that preparation of the dies and work holders, performing maintenance, and gathering tools should not be done while the processes are stopped it is absolutely amazing to observe how often this is the case.

If an honest effort is made to treat as much of the setup operation as possible as external setup, then the time needed for internal setup, steps performed while the machine is off, can usually be cut 30 - 50 percent. Mastering the distinction between internal and external setup is most direct route to achieving SMED. Internal and external elements must be rigorously separated. Once the process is stopped, the worker should never leave it to handle any part of the external setup. As part of the external setup, the die, tools, and materials should be ready for insertion into the process while it is still working on the job. Any modifications or repairs to tooling or dies should have been made in advance. In the internal setup, only removal and insertion operations should occur operations that absolutely have to be done with the process stopped.

The exchange of the old die for the new die can be facilitated by having dies lined up ready for insertion when the press is stopped. To enhance the change, roller conveyors can be used for staging and changing the dies, as shown in Figure 17.5. A typical sheet metal die-exchange procedure is as follows (an actual method to be standardized and practiced):

549

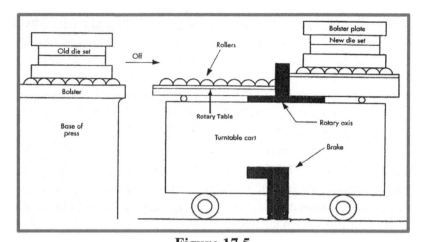

Figure 17.5
A Die-exchange Cart that Holds Both an Old and New Die Set on a Rotary Table

- Detach the old die from the bolster plates on the machine.
- Push the table cart over to the press and secure the table next to the press with the brake or stopper.
- Push the old die onto the table.
- Rotate the table and unload the new die onto the bolster.
- Pull the table away from the press and attach the new die to the machine, using the same set of bolts

Standardization of the height of all the dies going into a press and the length of bolts used to secure the die to the press bed greatly reduces the internal setup time. The die is precisely located on the press bed every time by the method shown in Figure 17.6 and 17.7. The die set is modified, that is standardized, by adding female locator plates to the bottom bolster plate. The male locator plate is permanently mounted on the bed of the machine (press). This locator established the X-Y position of every die set every time a die is placed in the press.

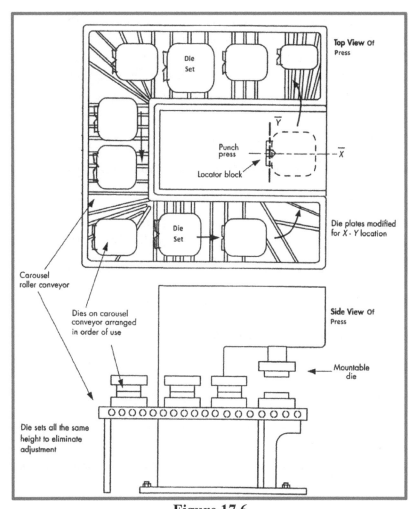

Figure 17.6
Punch Press Equipped with a Carousel Conveyor for Quick Exchange of Dies. All Dies are Modified to a Standard Height for Standardized Location on a Press Bed.

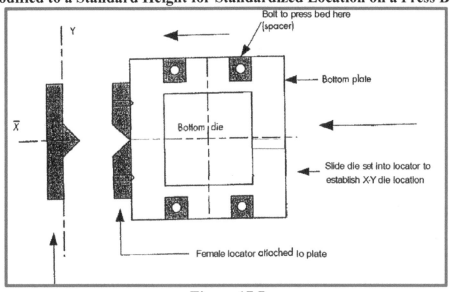

Figure 17.7
A Locator Fixed on the Press Bed

The same ideas can be applied to metal-cutting machines, as shown in Figure 17.8. Suppose a vertical drilling or milling machine is scheduled to process four jobs on four different jigs and fixtures. Each setup consists of removing the old jig or fixture, then installing and aligning the new fixture to the spindle of the machine. In this example, the four jigs are mounted on a turntable, and each is aligned to the spindle when rotated and locked into position with the spring stops. This is an example of one touch exchange of dies (OTED).

A higher-level solution might involve a turret mill with an oversized table. Remember that machines in cells process families of parts. Reducing the variety of parts coming to the machines permits you to modify the machine so that setup times can be eliminated altogether.

17.12 Converting Internal Setup to External Setup

One of the most important concepts in reducing setup time is converting internal setup operations to external operations. Chief among the elements that can be readily shifted from internal to external are:

- Searching time (trying to find the correct die, looking for the right tools, carts, fixtures, nuts, bolts, etc.
- Waiting time (waiting for cranes, carts, skids, or instructions)
- Setting time (setting dies, tooling, fixtures, etc.)

If an activity can be safely carried out when the machine is running, then it can be shifted to the external setup. An example is preheating metal molds for a die-casting machine, using the waste heat of the furnace from this machine, before inserting them into the machine. This means that trial shots, often needed to heat the dies to the right operational temperature, can be eliminated and the production run starts sooner. The addition of temperature sensors on the dies could tell the operator when the dies are at the right temperature for good molding.

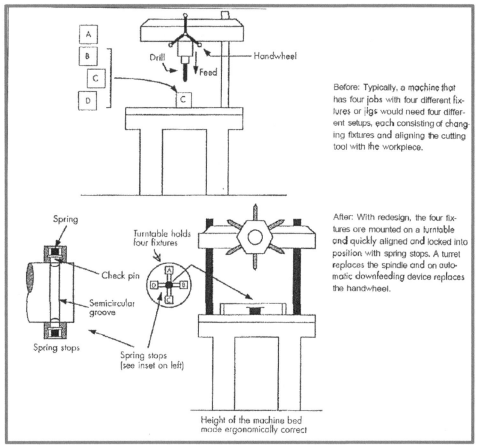

Figure 17.8
Modification of the Machine Tool in a Lean Cell to Process a Family of Parts with Rapid Changeover

The external operations for preparing the dies, tools, and materials should be made into routines and standardized. The internal die exchange should also be standardized. Such standardized operations should be documented and posted on the bulletin board for the workers to see. Instruct the workers to practice setups during slack times to master and improve the routine method. Make sure the best setup times are posted for all to see. While a defect-free exchange is the primary goal, it is important not to lose sight of safety during the process.

17.13 Reduce or Eliminate Internal Elements

Eliminating or reducing the internal elements in the setup time cycle will directly affect the setup time. In the exchange of dies, for example, the process of changing (adjusting) the shut height of the punch press often takes 50 percent to 70 percent of the total internal setup time. This activity is considered essential to the proper setup of the machine and often requires highly skilled personnel. However, the entire activity can be eliminated by standardizing the shut height of the press. Liners and permanent spacers are added to the die set so that altering the stroke of the machine is never necessary. If the sizes (and shapes) of all the dies or fixtures are completely standardized during the tool design phase, setup times will be shortened tremendously. Standardization however can be an expensive long-range solution if started after the fact in a mature factory.

17.14 Intermediate Work Holders

Another way to reduce setup time is use the intermediate work holder concept. To wit, work holding devices are designed so that they all appear the same to the machine tool. This usually requires the construction of intermediate jigs or fixture plates to which the job or fixture is attached. The dies, jigs, or fixtures are all different sizes. But the plates are identical.

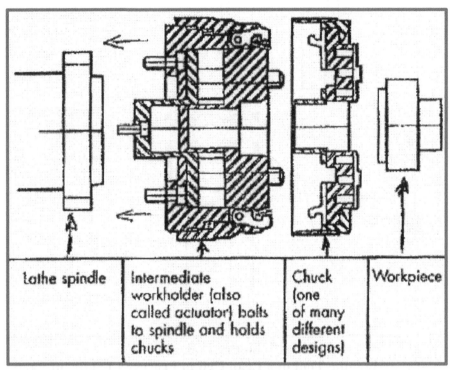

| Lathe spindle | Intermediate workholder (also called actuator) bolts to spindle and holds chucks | Chuck (one of many different designs) | Workpiece |

Figure 17.9
Example of an Intermediate Work Holder for a Lathe, Permitting Rapid Exchange of Chucks. (Courtesy Sheffer Collet Co.)

Work holding devices are designed so that they appear the same to the machine tool or process. This usually requires construction of an intermediate work holder such as jig or fixture plates to which the work holder for the part is attached. Jigs and fixtures are different, but plates are identical. Work holding devices, such as an intermediate jig, can help to quickly achieve one-touch setups. Figure 17.10 shows a solution where chunks of many different configurations are quickly mounted on an intermediate jig that remains attached to the lathe spindle. A key to reducing setup time is eliminating adjustments. A significant difference exists between setting and adjusting. The channel selector sets the television to a channel. A thermostat for the house sets the temperature. If settings can be manipulated, that tinkering or fine-tuning is called adjusting. Eliminate adjustments whenever possible to facilitate simpler tool and die. If the height of the base plates is standardized, the same fastening bolts, nuts, and tools can be used for all the dies. Bolts are the most popular fastening devices in tool and die mounting. A bolt fastens at the final turning of the nut and loosens at the first turn. Therefore, only a single turning of the nut is really needed. Many quick-acting fasteners have been designed to take advantage of this fact.

17.15 Applying methods Analysis

One of the least expensive ways to improve setup times is by applying methods analysis to examine in detail the methods of the internal setup. Methods analysis techniques are the subject of many basic texts and handbooks. The secret is to teach these basic methods to all the operators so that everyone is looking for ways to reduce setup time and improve the process. This is part of making the operator multifunctional able to do many things other than "just run the machine and make parts." Methods analysis helps operators to eliminate unnecessary tooling movement, reduce manual effort, eliminate extraneous walking, etc.

A large punch press or large molding machine will have many attachment positions on its left and right sides as well as on its front and back. The setup actions for such a machine can take one worker a long time. However, methods analysis can lead to the development of parallel operations for two persons, eliminating wasteful movement and reducing internal setup time. Reducing setup time from one hour to 10 minutes means that the second worker would only be needed for 10 minutes during the internal exchange. Setup specialists perform many of the external setup operations and assist the machine operators in setup actions.

17.16 Standardize Methods and Practice Setups

Dies, tooling, fixtures, part design, part specification, and methods are standardized. Once a standardized setup method has been achieved, it must be documented by the workers. This means that workers are asked to write down, step by step, the setup procedure for each machine (within a cell). The write-up is compared to the standard to see if the worker is doing what should be done. Extra and missing steps will become apparent. Some manufacturers have the setup teams practice setups during slack periods of the working day to further reduce the internal setup time.

Eliminating adjustments from the setup operation is a critical step in reducing internal time. Using spacers on die sets in a die setup eliminates the need for adjusting the shut height on a press. Shut height is never changed for this family of dies. However, situations always occur that require the machine to be reset. Even then, the number of actual setting positions needed on most machines or operations is usually quite limited, especially in cells. Setting is an activity that should be considered independently of adjustment. This can be accomplished by instrumenting the machine as necessary to permit the reestablishment of initial (or previous) setup conditions without any trial and error. The use of digital readouts or limit switches, for example, expedites resetting the machine without adjustment or fine tuning. The setup conditions should be determined, recorded, and marked so they can be readily reproduced time after time. A record of speed, feed, and depth of cut should be posted along with data on temperature, pressure, and the like. Step-function setting, like the push buttons on a car radio, can eliminate adjustment.

Molding machines typically require a different stroke for the knockout punch, depending on the size of the die being used. The stroke of the machine is halted with a limit switch. To find exactly the right stroke position, an adjustment (movement of the limit switch) is always necessary. A molding machine put into a cell environment requires only five positions for the limit switch. Instead of the one limit switch, five limit switches can be installed, one in each of the five required positions. A simple electric circuit is rigged to send electric current only to the limit switch that needs to be activated. As a result, the need to adjust limit switch position is eliminated (Figure 17.10). The mechanism is left alone, and only a function switch is changed to accomplish the change in setting. No adjustment of the limit switches is ever needed because they are not moved. Machine-tool manufacturers usually do not know the application of their products for a particular company, so they provide machines with continuously variable positional setting. Machines placed in cells, however, have limited applications, so the adjustment process is converted to settings or stops, often with templates or digital readouts to accomplish settings without adjustment.

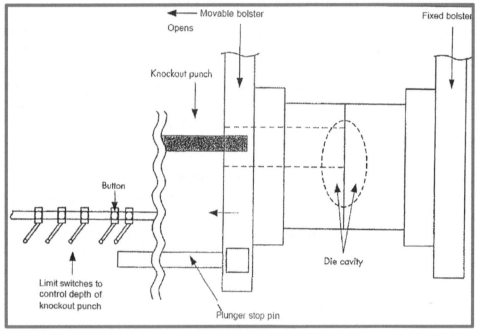

Figure 17.10
Installing Limit Switches at all Required Positions Eliminates Knockout-stroke Adjustments. (Shingo, 1981)

17.17 Abolish Setup

The final step in setup reduction is to abolish setup entirely or have it done automatically, but this is usually an expensive solution. Some ways to abolish setup have already been suggested, but here are two additional approaches that can eliminate setup:

- Redesign the product so it is uniform and uses the same part in various products.
- Produce the various parts at the same time. This can be achieved by two methods:

In the first method, the parts are processed in parallel using less expensive, slower machines. For example, an arbor press instead of a large punch press is placed in a welding cell to provide a simple bending function. Each worker handles a small arbor press as well as other welding jobs in the cell. This press has a small motor and can perform the same function as a heavy punch press. If several presses of this kind are available, they can be used in parallel and dedicated to producing one type of part at low cost. Multiple versions can be made available to produce a limited variety of parts.

The second method uses the *set-of-parts* system. For example, in the single die of the punch press, two different shapes of parts A and B are produced as a set, punched at the same time, and

556

then separated. No changeover is ever done. This requires that parts A and B be needed in the same quantities. Honda does this with doors, producing the front left and right doors simultaneously. For additional discussion and examples on eliminating setup times, refer to the books on group technology that describe elimination of setup as a natural outgrowth of the formation of part families.

17.18 Summary for SMED

Reduction or elimination of setup time is a critical step in converting any manufacturing system to a lean and flexible system. This effort is usually one of the first that the lean engineer at the company will be able to undertake. The results are immediate and obvious, but this does not mean that the setup reduction program is a short-term project.

Table 17.2 shows the results of a setup reduction program at one company where it took three years to reduce setup times to less than 100 seconds for 62 percent of the setups. Getting to this level of setup time reduction will typically come in phases. The first phase will require little capital expenditure, and solutions can be achieved in a relatively short time. Reductions of 20 percent to 30 percent are typical. No analysis other than videotaping is required providing everyone gets involved. This is a methodology that involves workers. The objective is to improve setup incrementally, until it is eliminated or economically prohibited.

Table 17.2
Setup Time Reduction Results at One Particular Company

Setup Time Reduction Results Over a 3 Year Period of Time			
Setup Time	Year 1	Year 2	Year 3
60 minutes	30%	0	0
30-60 minutes	19%	0	0
20-30 minutes	26%	10%	3%
10-20 minutes	20%	12%	7%
5-10 minutes	5%	20%	12%
100 seconds -5 minutes	0	17%	16%
100 seconds	0	41%	62%

The second phase involves operation analysis; minor modification to dies, tools, fixtures, machines, and procedures; and some modest expenditure. Again, benefits of 30 percent to 50 in setup time reduction can be achieved in a relatively short time.

The third phase may involve methods analysis, design changes, standardization of dies, tools, parts, machines, operations, and procedures. Large capital expenditures may be required, and complete conversion to very rapid setups may take years to achieve. Benefits of 10 percent to 40 percent may be expected.

For manufacturing cells, setups of 10 minutes are probably achievable in the first phase, 1 minute in the 2nd phase, and 10-20 seconds in the 3rd phase where the machine tools are custom built for the cells. The point to remember is that every time lot size is decreased, the need to reduce setup time will be felt. Setup is an interruption of the process. Work-in-process inventory between the cells is protects downstream processes from upstream problems. The rapid exchange of tooling in the manufacturing cells is the key to flexibility.

17.19 The Paperless Factory of the Future

Dinosaurs ruled during the age of reptiles. The biggest dinosaur, the swamp monster *Brachiosaurus,* was too large to support its own weight on land. It needed the buoyancy of water to stand. It might be said that manufacturing systems have their *dinosaurs* and their *endangered species*. The principal one is the production job shop, a manufacturing system kept afloat (by the ingenuity of people and by oceans of inventory), long after it has reached a size and complexity that should have caused its utter collapse. Clearly the time has come for the invention of a new manufacturing system. What motivated Taiichi Ohno, then vice-president of Toyota, to develop this system, and why does the system have the characteristics that it does?

After World War II, Japan was known as a nation that made junk. The Japanese wanted to develop full employment in their country through industrialization. To do this, they needed to learn how to build quality products-products other nations would buy. They thought that we in the United States knew how to do this, so they learned about quality control from us (the now-famous trips of Deming and Juran). Initially our quality gurus taught quality control techniques to the engineers and managers. Then the Japanese did something very different, something not done in the U.S. factories. Most surprisingly, the workers were taught quality control techniques. Next, the workers were given the responsibility for quality and the authority to stop the processes if something went wrong. Toyota and its suppliers began to develop early versions of linked-cell manufacturing systems for the manufacture of (families of) parts. In these early cells, they learned to eliminate setup and to work together to eliminate defective parts. The workers became multi-process, so that they could operate different processes. The MO-CO-MOO methodology became the operational standard. But these events do not explain why this unique system was so successful.

The answer probably lies in the very nature of the Japanese language. In his seminars, Richard Schonberger tells of not being able to find a typewriter on his travels to various Japanese companies. As a language, Japanese is very difficult to use for written communication. The classical, large mass production system requires a very sophisticated information system to deal with the complexity of the manufacturing system and its interface with the production system. Since Taiichi Ohno could not change the language, he had to find a way to simplify the manufacturing and production systems to eliminate the need for written communication replacing it with visual or automatic signals. He began to eliminate all kinds of unnecessary functions which he called waste. In contrast, manufacturers in the United States tried to computerize and optimize .the functions.

17.20 Visual Control or Management by Sight

The objective of management by Sight (or visual control) is to provide an easy method to exercise control of the plant and provide quick feedback by simply using your eyes to view the status of operations. Whenever an abnormal condition exists, the system will provide a signal requiring that timely corrective action be taken. Management by sight calls for all signals to be active changing and provide up-to-date information. Visual control causes all employees to get out into the plant on a regular basis to exercise control. Anyone at any time can go to "the shop floor and view the conditions. When successful fewer reports will need to be sent through the organization (thereby reducing paper flow.) There are two kinds of management by sight, *Information on Displays* and *Work Place Organization through the Shop.*

17.20.1 Information on Display

For visually displayed information to be effective in controlling plant operations it must be updated continuously and be flexible, able to change with daily operations.

Here are some of the kinds of items displayed in the plant in implementing a lean system.

- Cleanliness Control Board
- Control Charts
- Job Training Chart
- Housekeeping
- Evaluations Machine Checklists
- One point lessons
- Poke-Yoke Maps/Sheets
- Process Flow Charts or Value Stream Maps
- Scrap Tally Sheets
- Set-Up charts
- Standard Work Combination Tables and Production Capacity Sheets
- Team Kaizen Project Reports

17.20.2 Workplace Organization through the Shop

Here are some ways to properly identify items on the shop floor so that abnormal conditions can quickly be seen.

Visual Control at Standard Operations

- Items such as gauges, meters and valves can be marked to indicate normal operating conditions.
- Color coding can be used for gauging and meters, while valves may be tagged to indicate the normal position.
- The responsible person's name and telephone number may be posted so that anyone finding a problem can immediately report it.
- When equipment is moved from one plant to another, visual preparation ensures installation can be completed quickly and easily at the point of destination

Visual control should be made a part of every employee's daily operations so everyone can be involved in spotting abnormal conditions. Visual identification can and should extend to all areas related to maintenance throughout the plant, not only to equipment.

17.20.3 Visual Control for Housekeeping and Work Place Organization

When people store tools wherever they wish, the obvious result is confusion and lost time. Housekeeping procedures should be developed and followed. For example, if a tool is missing at the end of the day, everyone should be made aware of it and be involved in finding it. Management by sight highlights the abnormal conditions: Taped areas, tool display boards with proper identifications, and color coding increase the visual control. With this control operating, housekeeping and work place organization are more efficiently monitored by simply walking onto the shop floor.

17.20.4 Visual Control for Storage Area

The same principles apply with identification of any area where inventory, machines, equipment, containers and scrap bins are kept. The following are examples.

- If these areas are taped, color coded or partitioned off, they can become useful tools in determining problem areas. The idea is *A place for everything, and everything in its place.*
- If the scrap tub is missing from a machine, it is quickly noticeable because an empty area is taped on the floor.
- A forklift parked in an area not identified is not located in the proper position.
- Work-in-process containers found in an unidentified area belong in another location in the factory.
- If inventory is missing, or too much is in the designated location, here is a visual *flag* signaling that there may be a problem.

Management by sight forces thought about the location and function of all items located in the shop area and provides instant visibility for tools and resources being used.

17.20.5 Visual Control to Convey Information

This area focuses on items that convey information either by their mechanical or electrical function.

- Andon Board -Used by operators to signal the occurrence of a problem. The andon board lights to signal that assistance is needed.
- Scrap Bins .Red metal bins divided into days of the week. An attached scrap tally sheet keeps a one month history of the reasons and quantities of scrap.
- Clean Stands -Yellow stands designed to hold one piece of product and indicate when the cleaning solution in the parts washer should be changed. After washing, the part is put on display along with the results of the cleanliness check. Upon reaching a specified unacceptable level of cleanliness, the solution is changed before unacceptable parts reach assembly.
- First, Middle, Last Part Samples -High quality samples of each machined part is taken near the start, middle and finish of each shift. Sample parts are displayed to show the quality status at the point of inspection.

Visual control enables factory operations to be more tightly linked, with improved communication and better problem solving routines. It is one more set of techniques which underscores the driving force of continual improvement in the lean production system.

Management by sight also reinforces another central theme of lean production in the factory: for effective management employee and team involvement is critical.

17.20.6 Line Stop Concepts

Line stop is a fundamental lean production factory control technique. It refers to stopping the production line when a problem occurs, identifying the problem and regaining the flow as soon as possible and then resolving the problem so that it will not occur again. *Jidoka* is pronounced Ji-doy-ka. The Japanese refer to the concept of line stop as Jidoka. It means literally to make machines intelligent, capable of determining if a line should be stopped. The objective of line stop is to give operators the authority to stop the process any time a problem occurs.

However, it is difficult to implement. It takes discipline to respond to problems quickly. And it takes commitment from top management to shut down machines and production line if necessary.

17.20.7 Line Stop with Machines and Line Operations

The line stop concept applies to machines and line operations. With machines indicator lights help the operator detect abnormal occurrences in the machines. On the Line, the operator pushes a trouble button to get the attention of the team leader. If the problem can be solved in time, the line may keep moving; if not, the line will stop. An andon light is normally used to signal these occurrences. A yellow light may be used when requesting assistance with a problem and a red light for stopping the line if the problem cannot be resolved quickly. Buzzers or music are often used in conjunction with the andon lights to enhance visibility. In brief, line stop is one more technique encouraging continuous improvement. Line stops should not be feared, but encouraged in order to expose, and solve, problems.

17.21 Other Control Techniques

A number of other techniques and tools help in the planning and control of a lean production factory system. Here are a few of them:

17.21.1 Hourly Check

In addition to Poka-yoke devices to spot and prevent defects, hourly checks of the product further eliminate the possibility of passing defective work to the next downstream process. A buzzer sounds once per hour to trigger 100 percent inspection on critical processing dimensions. All parts are 100 percent visually inspected. The hourly check reinforces the previous efforts of the quality system and pin points problems at their source.
Sample Size, n = 2 or 3.

When the operators are not checking parts in the MO-CO-MOO cells, the quality control folks do a sampling inspection where the first, middle, and last part of each shift is taken from the line and is inspected in the inspection lab to ensure that specifications are being met. These pieces are then put on display to show that quality parts are being produced on that shift. If problems are discovered, the details are relayed to the production line for corrective action. First, middle and last piece inspection also helps determine when a problem occurred. If a problem was found on the last piece inspection at the end of the shift, but not found earlier in the shift, the first two pieces can be analyzed again to ensure the problem only occurred near the end of the shift. These imperfect parts can then be isolated for corrective action.

17.21.2 Control Charts, Check Sheets, and Other Process Documentation

Various documents are also used to help control the process
1. *Control charts or SPC (statistical process control) sheets* are put on display for reference,

especially by the operators. Control charts and run charts were discussed in Chapter 14 They give the operator a sense of the patterns of process development and point out the effects of machine adjustments.

2. *Check sheets* are another tool used in lean production. The most common is the *machine check* list which aids operators in the correct startup procedures for the machines.

3. Check sheets are also on display for all to view and also provide help with maintenance when the machines are down. They can be used in any function and enhance the proper application of established procedures or rules.

4. A *Process sheet or standard operation sheet* is the most common control tool. *It provides all necessary information for the actual machining or assembly* of a *part.* For example, they will outline speeds and feeds, material, sequence of operations, operator safety, tooling and gauging information, part number, and part name. A drawing of the part is also included.

5. Standard operation sheets are positioned above each operation and provide stability and standardize the process so specifications can be met consistently. They are also used to train new operators. Leave nothing *to* chance or memory. Make it easy to train and retrain the operators and prevent them from making errors.

17.22 Group Technology

Group technology (GT) offers a systems level methodology for the reorganization of the functional job shop manufacturing system, restructuring the job shop into manufacturing cells. In a manufacturing facility, component parts of similar design, geometry, or manufacturing sequences are grouped into families. Machines can then be rearranged into groups or manufacturing cells to process the family of components. This manufacturing sequence defines the arrangement of machines and processes in the cell. Finding families of parts is one of the first steps in converting the functional job shop into a cellular manufacturing system. There are a number of ways to accomplish this. Judgment methods using axiomatic design principles are, of course, the easiest and least expensive, but also the least comprehensive. Eyeball techniques work for small manufacturers and restaurants, but not in large job shops where the number of components may approach 10,000 and the number of processes reaching 500 or more.

17.22.1 Production flow analysis (PFA)

PFA uses the information available on route sheets. The idea is to sort through all the components and group them by a matrix analysis, using product routing information (see **Figure 17.11**). This method is simple, inexpensive, and fast, but still more analytical than tacit judgment. PFA is a valuable tool for systems reorganization problems. For example, it can be used as an *up-front* analysis, a sort *of before-the-fact* analysis that will yield some cost/benefit information. This would give decision makers some information on what percentage of their product would be made by cellular methods, what would be a good first cell to undertake, what other analysis method would work best for them, how much funds they may have to invest in new equipment, and so forth.

In short, PFA can greatly reduce the uncertainty in making the decision on reorganization of the factory floor. As part of this technique, an analysis of the material flow of the entire factory is performed, laying the groundwork for the factory's new linked cell layout. The use of PFA to identify the elements of the first cell permits a company to implement that first cell without waiting until all the parts in the plant have been coded by comprehensive coding systems.

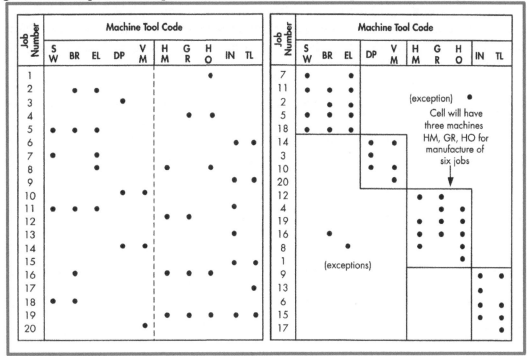

Figure 17.11
Information On Routing Sheets Suggest Part Families

The PFA method helps the lean engineer identify potential cells and answers the following questions regarding their existing equipment.

- How many cells could be developed for its existing equipment list?
- For each cell, which machine and products could be assigned to which cell?
- Which machines must be purchased because they cannot be duplicated in all the cells?
- Which routings do not match the machine composition of even one cell?
- Which products' routings may require two or more cells?
- Are there products that can be produced in anyone of several cells?
- Will the cells be designed to be flexible and accommodate future changes in part mix, demand volumes, and new parts?
- Should the company design a lean cell layout with machines designed or built for cellular layouts?

17.22.2 Coding/ Classification Methods

Many companies converting to a cellular system have used a *coding/Classification (CIC)* method, which is more comprehensive and time consuming than PFA. Under the C/C systems there are design codes, manufacturing codes, and codes that cover both design and manufacture. *Classification* sorts items into classes of families based on their similarities. It uses a *code* to accomplish this goal. *Coding* is the assignment of symbols (letters, numbers, or both) to specific component elements based on differences in shape, function, material, size, and manufacturing processes.

563

No attempt to review C/C methods will be made here. Coding/classification systems exist in bountiful number in published literature and from consulting firms. Most C/C systems are computer-compatible, so computer sorting of the codes generates the classes of parts families. The system does not find groups of machines. If a code is based on design data, errors in forming good manufacturing families will occur.

Whichever C/C system is selected, it should be tailored to the particular company and should be as simple as possible so that everyone understands it. It is not necessary that old part numbers be discarded, but every component will have to be coded prior to the next step in the program, finding part families. This coding procedure is costly and time-consuming, but most companies opting for this conversion understand the need to perform this analysis.

17.22.3 Other GT Cell Design Methods
Other methods used to form cells, involve the following tasks.

1. Finding a key machine, often a CNC machining center, declaring all parts going to this machine a family, and moving machines needed to complete all parts in family around the key machine. Often it will be prudent to off-load operations from the key machine to other machines in the cell.
2. Building a cell around a common set of components like gears, splines, spindles, rotors, hubs, shafts, and so on. There are "natural" families of parts that will have the same or similar sequences of processes.
3. Building a cell around common set of processes. For example, drilling, boring, reaming, key setting, and chamfering holes makes up a common sequence.
4. Building a cell around set of parts to eliminate the most time-consuming element in setups between parts being made in the cell. This was the approach used by Harley-Davidson.
5. Picking a product or products, then designing a linked-cell manufacturing system beginning downstream with final assembly line (convert final assembly to mixed model) and moving upstream toward subassembly; and finally to component parts and vendors.

Part families' will not all have the same material flow and, therefore, they will require different designs (layouts). In some families, every part goes to every machine in exactly the same sequence. No machine is skipped, and no backflow will be allowed. This is the purest form of a cellular system. Other families may require some components to skip some machines and some machines to be duplicated. However, cellular backflow is still not allowed except under extremely unusual circumstances such as very low volume, high process times, and experienced cell workers.

GT cells have the following shortcomings.
1. They may have to accommodate both high-volume commodities and rarely-made spare parts or prototypes in the same line.
2. Operator job design (ergonomics) is not considered, and the machines are not modified for operator productivity, safety and ergonomics.
3. The design work is often done by engineering without any input from production (i.e. the operators)

17.23 Summary

This Chapter has covered some of the lean tools not discussed elsewhere in the book. The Chapter (the list) is never complete but keeps in mind; lean is all about changing the manufacturing system design and not just implementing lean tools into your old system. The TPS is a system designed to eliminate waste, where production is leveled, defects eliminated and products match consumer preferences.

Review Questions

1. Briefly discuss the importance of Takt time?
2. What are bottlenecks in a sequence of processes and what is the basic design rules associated with them?
3. What are the major differences between VSM and process flow mapping?
4. When might you do process flow mapping instead of VSM?
5. Why is throughput time such a great metric?
6. In lean manufacturing, what happens to the EOO equation as setup time (cost) is reduced?
7. Outline the 4 stages of SMED methodology developed by Shingo.
8. In replacing flat tires, what are elements that are standardized for almost all cars?
9. In the example of changing a tire of a car, what are the examples of internal Vs. External elements?
10. In a die changing operation for small presses, what are some of the die features that can be standardized or modified to help reduce setup time?
11. For tire changing, what are some ways you could reduce the internal elements?
12. What is an intermediate jig or workholders? Name an everyday example.
13. Explain how setup flows through the cell when the cell is changing over from Part A to Part B.
14. SMED has many advantages for the lean factory. What is having the operators doing the setups the key to the SMED methodology? See previous question.
15. Suppose you have a cell where the operator is moving counter clockwise (CCW) while the parts are moving clockwise (CW), Will "flow through" setup still work in this design or do you have to stop, clean out all the A parts, perform setups for B parts and then restart the cell?

Thoughts
and Things... ...

Chapter 18
Sustaining the Lean Enterprise

David S. Cochran, Ph.D.
Associate Professor and Director: Center of Excellence in Systems Engineering
Indiana University-Purdue University, Fort Wayne Indiana

"For now we see through a glass darkly; but then face to face: now I know in part, but then shall I know even as I am known." I Corinthians 15:12 (KJV)

18.1 Introduction

As we advance into the 21st Century, the next Revolution is the wide-scale understanding and practice of systems thinking and enterprise system leadership, design, and engineering. Perception becomes reality in complex systems even when perception has obvious counter examples. The process of thinking differently can drastically change perception. This Chapter is concerned with how to achieve collective agreement about the purpose of a system, and describes a process for designing and implementing systems that are sustainable.

Sustainable systems require the people within an enterprise to all be *on the same page*, communicating intention and measuring results based on this shared understanding. This Chapter is intended for people who are already practicing lean, and is a guide to a way of thinking about lean that will make lean sustainable beyond the initial implementation.

The approach of Enterprise Design and Leadership communicates Lean as both a system design and a rigorous approach to learning and improvement. The Enterprise Design and Leadership approach applies equally well to manufacturing and non-manufacturing environments. This Chapter describes an approach to sustaining the lean enterprise, with the intention of inspiring the reader to define and achieve effective Enterprise Systems.

18.2 What is the Meaning of Lean?

What's in a name? If you ask thirty people in a room what lean means, there will typically be twenty or even thirty *different* answers. The result of implementing the Toyota Production System is an enterprise system that is lean. The Toyota Production System (TPS) is the name of an integrated Manufacturing and Enterprise System (Black, 1991) that is both self-sustaining and self-organizing. What Toyota called the Production System, we now call the Enterprise System.

Is it possible to sustain something when there are so many different points of view about what *it* is? Many people now say that the goal of lean is to *eliminate waste*, without Mr. Ohno's original meaning of waste elimination in the context of an enterprise system design (Ohno). When the definition of lean is described as the *elimination of waste*, it must be described in the context of improvement of the system rather than the single operation (Rother and Shook). One need to sustain lean in today's environment is to treat lean as a system design of the entire enterprise. The system design must foster effective communication and clarity in purpose for the organization. Leaders must use the system design to create an environment for learning and continuous improvement.

One definition of leadership is that a leader has primary responsibilities to drive out fear and to eliminate ambiguity (Montgomery). To eliminate ambiguity requires effective and also simple-to-use methods of communication. The adage; *if you don't know where you are going, you'll get there* often applies to the management and control of large systems and complex enterprise management. Effective communication requires defining, understanding, and creating shared agreement about system purpose, cooperatively, with the people who work within an enterprise.

A sustainable enterprise system design includes not just the manufacturing system design, but also product development (Lenz and Cochran, 2000), supply chain, human resource management, information technology, performance measurement, and management accounting. While many of the lean practices present in industry today target manufacturing systems, ignoring the other aspects of the enterprise is akin to improving single operations; an objective that does not meet the original design of lean. Design of the enterprise system requires communication of enterprise purpose, called *functional requirements* (FRs), and creating *physical solutions* (PSs) as means to achieve that purpose. It also requires creating an environment for learning and improvement, and a methodology for determining performance measures and visual feedback that reinforces the improvement of the system design.

The professions of Industrial Engineering and Systems Engineering have a lack of definitional clarity, rigor, and relevance when it comes to describing and designing systems that are *Lean*. This is probably why the definition of the Toyota Production System, which is an Enterprise System Design, is called *Lean;* and the recognition of Lean is only associated with the elimination of waste. Nothing could be more wrong. The principles of Lean Engineering promise to change the very future of Industrial and Systems Engineering: Particularly the way in which manufacturing and enterprise systems are designed, integrated and operated. Sustaining this type of system is one of the most difficult of human challenges, and requires a framework to not only design such systems but to do research about them as well. For one reason, this requires self-reflection about what a Lean system design actually involves.

18.3 Enterprise System Design

Enterprise system design may be thought of as a hybrid of the attributes of a complex system and a classically engineered system. Complex systems engineering deals with making decisions in which the outcome is unknown. In contrast, the objective of classical engineering is to make outcomes predictable, reliable and stable (Braha, 2006).

A system may be defined as an assemblage of subsystems, hardware and software components, information, and people, designed to perform a set of tasks to satisfy specified functional requirements and constraints (Suh, Cochran and Lima, 1998; Suh,2001). A system in this sense is certainly complex in that it includes the thoughts and actions of human beings, which invariably results in unpredictable and often unknown outcomes.

To lead an enterprise system design that is sustainable requires everyone within an organization to first have a common mental about the purpose of an enterprise (Senge,1990). Since an enterprise is a system, and since systems, by definition must be designed to perform tasks to satisfy its purpose in terms of its functional requirements, then it is first necessary that the people within an organization are able to communicate that purpose with each other in an effective manner.

Axiomatic Design is one such way to communicate the purpose of an enterprise([Suh, Cochran and Lima, 1998). A functional requirement (FR) may be used to define what an enterprise must accomplish. For example, an FR could be:

FR: Deliver No Defects to the Customer

Or, consider an alternative statement, FR_{alt}:

FR_{alt}: Do Not Advance a Defect to the Next Operation

Based on the definition of the FR, the physical solutions (PSs) to these two FRs are very different. The first FR statement is much easier to accomplish than FR_{alt}. A PS of final inspection before shipping will achieve the FR. However, achieving FR_{alt} must be done differently since PS_{alt} must consider manufacturing process designs for each operation that are self-checking. Dr. Shingo called PS_{alt} source inspection (Shingo, 1986). Sustaining an enterprise requires that the people who work in the enterprise system first agree on the functional requirements of the enterprise that will sustain it and then do and improve the work that is necessary to achieve the functional requirements.

As noted before, the Toyota Production System (TPS) is an Enterprise System Design. An enterprise can be a service company, a manufacturing company, a hospital or network of hospitals. An enterprise can be a government agency. The key is that the team defines the system boundary and then agrees upon the set of FR's that the system must accomplish.

18.4 System Design and Engineering of a Lean Enterprise

In reviewing Dr. Blanchard's work about Systems Engineering, the word *system* is used to describe the arrangement of physical elements to achieve an intended *purpose* (Blanchard, 2008). This definition does not include the people who design the system in the first place. What if we were to re-define the system boundary to include the people who design and create the system? Enterprise System Design is the term that describes the inclusion of these people, the system designers, inside the system boundary.

What if the thinking process and mental models of the people who define *system purpose* are part of the system itself? Let us call the system design activities that embrace this point of view the *Enterprise System Design* or simply *Enterprise Design*. This lexicon not only involves the operational system, but the people who both design and operate the system. In for-profit businesses, this is at the heart of *Company Design*. With government agencies, the service sector and with not-for-profits, the approach is the *Enterprise Design*. This term is created based on the realization that the people who define the purpose of a system are part of the system itself.

Enterprise Design attempts to achieve stable, predictable and reliable results in a classical engineering sense. However, the people who organize, develop and design the system and expect the design to work must work within the domain of complex engineered systems – because the result of design decisions is not known in advance.

A system must have a clearly defined system boundary and purpose (Figure 18.1). An enterprise system is designed to achieve customer needs and should be designed to be stable and predictable. At the same time, people internal and external to a system are either consciously or unconsciously determining how a system will operate. For this reason, the Enterprise Design should be developed to achieve enterprise purpose in a stable manner, while at the same time human decisions about the design of the enterprise are best understood as a complex system because the results and consequences of many design activities are not known *a priori* – in advance of the actual decision making.

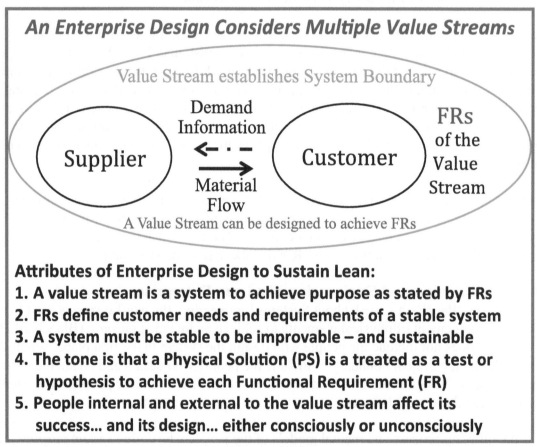

Figure 18.1
Enterprise Design is a Complex System Trying to Create a Stable System

Since the Enterprise Design is supposed to meet customer needs and achieve stable, predictable results at the same time, the decision making process to create the Enterprise Design can be classified as a complex system. Complex systems exist when the decisions that a person makes or group of people make has unknown outcome.

The fact that a design decision has an unknown impact underscores the importance of having a language to express and guide how people communicate when developing an Enterprise Design. A language for design assists people to express and communicate their shared purpose and to agree on the physical means to achieve shared purpose. For example, companies that have tried and failed at lean may view *it* as a tool set that will fix their problems. To sustain an enterprise that seeks to be lean requires the people in an organization to work together to create a shared vision of purpose, and to create and do the work that is necessary to sustain the enterprise as a system design.

For these reasons, to sustain the Lean Enterprise it is important to add the use of Axiomatic Design as a language to express and to communicate the design of enterprise systems (Cochran and Pincham, 2009) that augments the multiple enterprise system purposes that the value stream mapping process (Rother and Shook, 1999) attempts to convey and the Plan-Do-Check-Act learning loop process attempts to refine and improve (Deming, 2000).

Enterprise Design uses a language for the system design of each value stream that expresses system purpose in the form of Functional Requirements (FRs) and hypothesized Physical Solutions (PSs) to achieve purpose (Figure 18.2).

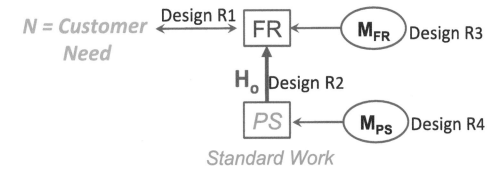

Figure 18.2
A Language for Enterprise Design

The first design relationship **R1** relates customer need with the *Enterprise Functional Requirement* (FR). For example, if the customer need is to receive a leather sofa in one week after it is ordered and the customer wants a custom blue leather sofa with bun feet and not Queen Anne styled feet, then the FRs of an enterprise are:

FR1: Produce sofas in one week or less order to delivery time
FR2: Produce sofas in the color that the customer wants
FR3: Produce sofas in the style that a customer wants

Design relationship **R1** does not define *how* the physical system achieves the desired relationships. Design relationship **R2** methodologically defines the physical implementation of the enterprise system, as a *value stream* to achieve the stated FRs. The value stream system in this case includes the ordering processing, manufacturing, shipping, and delivery subsystems. **R3** and **R4** illustrate that the definition of performance measures occurs *after* the FRs and PSs are defined by the enterprise system designer or the *Lean engineer*.

Lean is the term that is now used to describe the results of an Enterprise Design, known as the Toyota Production System (TPS) (Womack et al, 1990). To address the sustainability of the Lean Enterprise and the Enterprise Design used to create the Lean Enterprise, a system design team might decide to eliminate the term *lean (and the use of all other Japanese buzz-words for that matter)* to express and to characterize the system-design relationships. Collective agreement about the enterprise purpose as functional requirements (FRs), and the means to achieve purpose as physical solutions (PSs), and how a group of people improves the work to achieve system purpose is even more difficult than the use and application of lean tools. It is the tone of the people within an enterprise toward each other and their work relative to the achievement of the value stream FRs that becomes key to sustainability of an Enterprise Design (Cochran, 2006). A team that works together on an enterprise design must have common understanding and definitions of terms that *they* use – this understanding leads to a common mental model about the system design (Senge, 1990).

It is more important that a team have collective understanding about what they want to accomplish (FRs) and the means of achievement (PSs), than whether they are *right* about their design. For example, the way people cooperate with each other and their tone with each other ultimately ensures long-term sustainability of an enterprise and its design. After all, the ability to adapt to change is the basis for survival of any enterprise. An advantage of a language to design and communicate an Enterprise FRs and PSs is that it helps to avoid the tendency to speak to each other in the form of *solutions* -- tools, processes, or procedures. In a typical organization, these tools, processes or procedures are often times mandated for implementation by the executive leadership. These decisions are often based upon superficial logic or perceived needs, rather that sound, analytical reasoning.

In many cases, the use of tools as the *solution to a problem*, used robotically without thinking about their use, results in the improvement of isolated operations instead of the system. The use of Kaizen events (and Six-Sigma tools) often addresses isolated, visible problems that must be solved. However, the trap is that the Kaizen events (and Six-Sigma tools) often result in spectacular unit operations improvements…but do not improve the achievement of the Enterprise Design as a system. The promise of Enterprise Design is that the achievement of FRs truly does result in lower total cost and increased sales for an enterprise. Point Kaizen and Six-

Sigma improvement projects that are typically isolated to operation improvement may appear to flow-down to the bottom line – to flawed management and accounting models (Johnson and Bröms, 2000).

In order to sustain an enterprise that is designed to be *lean*, the implementer must understand and be clear about the system FR that the lean *tools* are attempting to achieve. This situation is practically always true for people within organizations implementing lean, but also applies to the implementation of Six-Sigma, and other programs that require systemic change (Hopp and Spearman, 2008).

Another key issue in sustaining an enterprise that seeks to be lean is that corporate training and educational programs are often perceived to be the end of a journey, rather than the beginning of a journey to achieve the enterprise FRs. Lean training and programs are necessary, but not sufficient to sustain lean within organizations. The reason is that lean is the *result* of a system design, not the system design itself which means that the objective of a lean enterprise is to achieve its FRs (purpose), with minimum cost and with waste reduced in the context of system (i.e., value stream) improvement.

A system design must state both the FRs (purpose) and the PSs (means) to achieve system purpose. The term *Lean* results from the observation of the Toyota Production System by John Krafcik (SM '88), an MIT student, who noted that the result of implementing the Toyota Production System was to *become Lean (*Womack et al, 1990). Krafcik had observed that the result of Toyota's tools: standardized work, *andon* boards, single piece flow cells, multi-functional workers, mistake proofing, pull production, resulted in lower inventory; the consequence of which he called *Lean Thinking*

Taiichi Ohno's model of waste reduction is in the context of achieving enterprise design purpose. He states that: *True efficiency improvement comes when we produce zero waste and bring the percentage of value added work to 100 percent* (Ohno, 1988). Ohno and his brilliant engineering support personnel relentlessly pursued waste (*Muda)* reduction.

Waste reduction as Ohno writes takes place in the context of a system which includes the thoughts and actions of people, instead of treating people as units of production – equivalent to machines. He said; *In the Toyota Production System, we must make only the amount needed and manpower must be reduced to trim excess capacity to match the needed quantity* (Ohno, 1988). Waste reduction is done in the context of the *system,* and the work of people and machines is designed to operate at the pace of customer demand.

The *7 Wastes* as stated by Mr. Ohno are shown in Figure 18.3 (Ohno, 1988).

If Lean is described as the elimination of waste, but out of context from the Enterprise Design; operations are improved but not the system itself. Shigeo Shingo's seminal book, *A Study of the Toyota Production System from an Industrial Engineering Viewpoint*, points out that an operation can be improved but might have no real impact on the overall process… or *system* (Shingo, 1989).

Mr. Ohno elevated the needs of people to the highest status in the company. For example, Mr. Ohno said; *Hiring people when business is good and laying them off when business is down, is a bad practice* (Ohno, 1988).

> **"The Preliminary Step toward application of the Toyota Production System is to identify wastes completely:**
>
> 1. **Waste of Overproduction (too much, too early)**
> 2. **Waste of time on hand (workers waiting on machines)**
> 3. **Waste in Transportation**
> 4. **Waste of Processing itself (unnecessary processing)**
> 5. **Waste of stock on hand (unnecessary inventory)**
> 6. **Waste of movement (motion)**
> 7. **Waste of making defective products"**

Figure 18.3
Taiichi Ohno's Description of the Seven Wastes

In traditional assembly line mass production systems, each worker is usually assigned to only one machine or workstation doing repetitive, short cycle time work (Cochran and Dobbs). This kind of work does not respect the skills and ability of the worker to identify and resolve problems and treats people as a unit of work, rather than the system's most valuable asset. This type of design is not sustainable, because people become bored, and when improvements need to be made they usually have no authority to improve the system. People are isolated to one operation, making it virtually impossible to identify quality problems that span multiple workstations

Systems must be designed to *sustain* technological change and *support* design modifications. Furthermore, any reduction in the waste of motion at a machine would result in the worker waiting longer at the machine. This type of improvement is improvement of the operation, but it does not reduce total operator processing time or material flow time in system. In contrast, in the cellular arrangement of machines when motion waste is reduced, the *system design itself facilitates improvement*. The fact that machines are designed such that a worker loads a part and can walk away from the machine means that if enough improvements are made, a cell can be run with fewer workers at the same Takt time. The point is that motion improvement is made in the context of system improvement – while respecting the knowledge and capabilities of the people who work in the system.

A second example illustrates why improving an isolated operation may not always reduce waste for a system. A large automotive supplier automated the transport operation with the use of an Automated Guided Vehicle (AGV) to move parts from one department to another in their facility. This did not improve system performance measures since neither operation was a bottleneck or critical operation. This is a clear case of where the operation of transport (transportation…one of Toyota's 7 wastes) was improved, but not the system itself. A systemic improvement would eliminate the need for transport altogether.

How do we avoid the problem of making improvements to an operation instead of to the system value stream? Ohno was very specific about enterprise purpose when he stated what Just in Time (JIT) means to: *Make only what is needed, when it is needed, in the quantity that is needed* (Ohno, 1988). Ohno established the enterprise purpose, one pillar, called JIT of the Toyota Production System house model (Figure 18.4).

The enterprise design language expresses JIT in the form of FRs for each value stream design to accomplish. To achieve the FRs stated by the JIT pillar, the physical implementation of the work, the PS, is to balance the work of all people and machines in the value stream to be less than are equal to the actual pace of customer demand – to the downstream final customer cell or operation – called the *Takt time*.

The Takt time is the available production time per shift divided by the average demand per shift. The Takt time is the *heartbeat of a lean production system* (Rother and Shook, 1999). It is important to note that the *Takt* time is a PS, a physical solution, to achieve the FR, to produce the customer consumed quantity every time interval – related to JIT to: *Make a product only when needed, in the quantity that is needed* (Ohno, 1988).

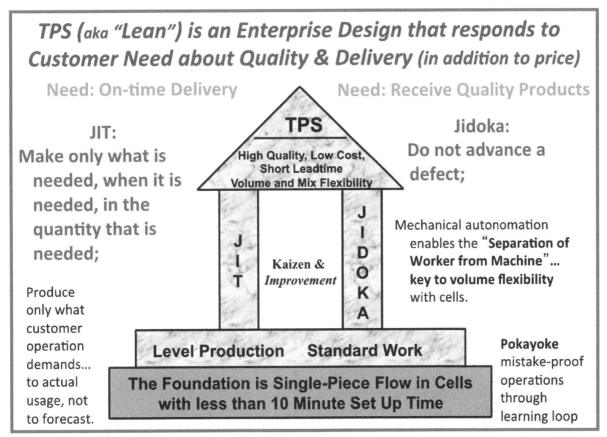

Figure 18.4
The TPS House Model: Pillars Define the Purpose (FR's) of an Enterprise System

To summarize, system improvement and sustainability requires a language for enterprise design because there must be shared agreement and shared understanding about what a system is supposed to accomplish and how it is accomplished. The Enterprise Design Language in Figure 18.2 expresses the design relationships.

18.5 Sustaining the Enterprise Design

Since the objective of Enterprise Design is to implement systems that are sustainable, the language for Enterprise Design that is necessary to communicate and to create a sustainable system ensures that people are on the same page and that there is collective agreement about the FRs and PSs of a system. Success by design is the result of mapping PSs to the FRs. This means that the system designers must identify a PS that is intended to achieve an FR. An FR defines the purpose of an enterprise as a system and is written to describe what the system must achieve to be successful. When an FR is not achieved a system cannot be successful. FRs do not define solutions; the PSs can and do. PSs define how to achieve the FRs. Furthermore, the PSs clarify and define how people do the work and/or implement tools to achieve the FRs. This definition leads to defining the standard work – which is the standard and *normal* way of doing all work. Collective agreement is codified and standardized through the creation of an Enterprise Design to communicate a design and to create a mental model of the system.

Dr. Cochran established the Production System Design Laboratory at the Massachusetts Institute of Technology (MIT) to create a way to express the design of systems. One result of this research is the development of the Manufacturing System Design Decomposition (MSDD) shown in Figure 18.5 describing the Functional Requirements (FRs) and Design Parameters (DPs), later renamed to Physical Solutions (PSs), of an effective manufacturing system (Cochran, Link, Reinhart and Mauderer, 2000) .

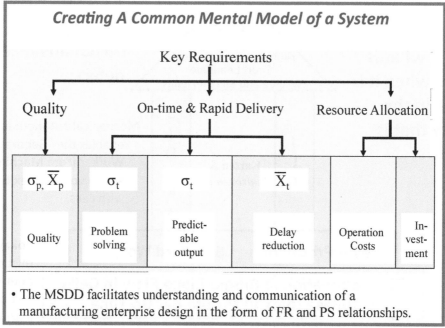

Figure 18.5
Manufacturing System Design Decomposition (MSDD)

Investment managers know how to allocate capital and human resources to sustain a manufacturing system because of the MSDD (see Figure 18.5 and Cochran, Duda, Linck, & Arinez, 2001). By having all managers in agreement about the FRs of the system, there is no need to do traditional "cost justification" to spend money to do "the right thing" – because the right things, the PSs, are already defined. Furthermore, the right thing to do doesn't have to be debated by management because the process of creating a system design map requires mangers to collectively agree on the FRs and PSs for their specific system design. Each system design map is custom made by each business' leadership team and then tested and refined by all people in the enterprise.

Collective agreement is imperative for an organization to function. Think of collective agreement as defining the mission (FRs) of the enterprise and also the methods, strategy and tools (lumped together as PSs) to achieve the mission. FRs define what the business, system or enterprise must accomplish to exist. The PSs define the *tool(s)* to achieve desired purpose.

Figure 18.6 illustrates the Enterprise System Design Map (ESDM) that serves as a template for enterprise implementations. The ESDM extends the MSDD (Figure 18.5) to include product design and system engineering objectives. The Enterprise System Design Map is the master plan for implementing a system and communicates the logic and thinking behind that plan.

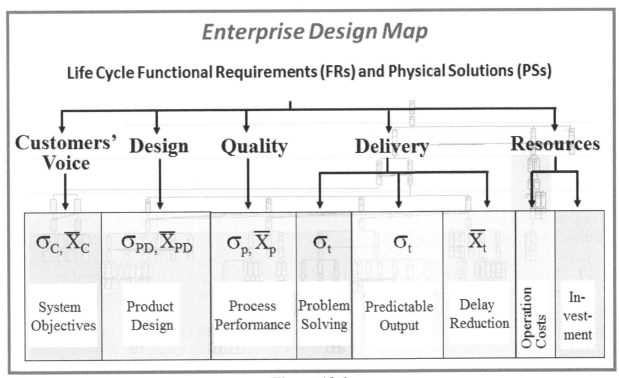

Figure 18.6
Enterprise Design of a Systems Life Cycle

The following eight steps serve as a framework for managing an enterprise that is designed to be lean.

1. Leadership by Design. The General Manager of an organization must first be committed to the design/re-design of the enterprise and to lead the design.

2. Define the Standard Work as the *Normal Condition of all work*, even if it is not the best or optimum. Eventually, as the standard work is improved, it becomes a *Vehicle for Continuous Improvement*, and forms the basis for sustaining an Enterprise Design.

3. Define the Enterprise Value Streams defined within the system boundaries, and align all work to achieve these Value Streams.

4. Define System Purpose in terms of Functional Requirements FRs of the Value Stream – Based on innovative ideas designed to delight the external and internal customer; define FRs that meet basic internal and external customer needs.

5. Understand the Functional Requirements FRs of the system very well before determining the physical solutions (PSs) and implementation. Treat a Physical Solution (PS) as a hypothesis to achieve an FR. Develop the experimentation mindset. Choose PSs so that the design is acceptable. Each must be implemented with standard work.

6. Build physical models of the work (PSs) to achieve the FRs in a safe environment.

7. Derive Performance Measures of a System after understanding the FRs and PSs of the system and after validating that a PS does in fact achieve an FR.

8. Align all costing information to each of the enterprise value streams; convert / allocate overhead costs to each value stream. Invest in the resources necessary to achieve the FRs of the Enterprise design.

These are 8 key principles which must be followed to achieve success. They are so important that we will discuss each one in detail.

1. Leadership by Design. The General Manager of an organization must first be committed to the design/re-design of the enterprise and to lead the design.

Sustaining an enterprise builds on its initial design. When the General Manager of an organization is committed to the re-design of the enterprise, it is remarkably easy to facilitate change.

Once the system is setup, it is much easier to cooperatively agree on new system FRs and PSs, since the Enterprise Design becomes visible, and everyone in the organization understands the commitment of senior leadership to systemic change.

2. **Define the Standard Work, the *Normal Condition of all work*, even if it is not the best or optimum. Eventually, as the standard work is improved, it becomes a *Vehicle of Continuous Improvement*," and forms the basis for sustaining an Enterprise Design.**

To sustain an enterprise, once the executive management team is committed to the change, first requires an organization to get a baseline understanding of the way that work it is done, even if the work is not done well. When an enterprise has not defined the content, sequence and timing of all work, it is, by definition, *not possible* to define a normal operating condition (Spear and Bowen, 1999).

Defining the normal operating condition enables the workers to see and understand the abnormal condition, which is the basis for sustainability of any system that involves the work and thoughts of people within an enterprise. Senior leaders form teams to continuously improve the work. The team leader develops the initial standard work definition and implements a process for improving the work with the team members. Typically, the span of control of the team leader is about 7 team members.

The Standard Work Procedure may use visual management techniques to ensure that all team members understand the way work is done and to identify when an abnormal condition occurs. A team leader – team member structure works for all work in a value stream – for both manufacturing and non-manufacturing work (including the work of professionals within a value stream). Work is work and the way that it is done must be written down and improved once the organization that is based on the Enterprise Design is put in place to do so. The very way that work is to be done is documented and written down for all engineering and design activities. Standard work and standard work practices are not just for shop floor activities!

The frequency of change and variation from standard work practices determines the frequency of the *check* to improve the Standard Work. Typically, this is done on a daily (or weekly) basis during a 10-minute *stand-up meeting*. A stand-up meeting facilitates the improvement. The team identifies the work to improve, who is responsible for completing the task, and the completion date. Once the team improves the work, the team leader issues the new Standard Work Procedure.

Ken Kreafle, a former TMMK Quality Manager (Kreafle, 2001), called Standard Work a *Record* of Problems Solved, although in reality it is a reflection of how to execute a job cheaper, more effectively or in less time. As the Standard Work is improved, it is able to address the problems that the Enterprise Design has encountered thus far. While striving to never have any problems is an unrealistic task, ensuring that problems are not repeated is attainable and is key to sustaining the Enterprise Design.

3. **Define Enterprise Value Streams and align all work to the defined Enterprise Value Streams.**

The term *Value Stream* is now almost universally accepted as characterizing work that people do to create the flow of material and information through any enterprise (Womack et al, 1990). A value stream extends the customer-supplier connections that are present within the manufacturing system – to customers upstream and downstream of the manufacturing system. Additionally, the concept of a Value Stream applies to flow in non-manufacturing environments.

The Value Stream concept may be used to break barriers between systems to define the system boundary. For example, the idea of an integrated enterprise and manufacturing system depends only on the definition of the manufacturing system vis-à-vis the enterprise system. The Value Stream concept can burst the idea of a separate and severable manufacturing system from the enterprise system. The Value Stream concept may also apply to University Education by breaking down barriers between departments. An Enterprise Design may consist of several value streams as illustrated in Figure 18.7 (Brennemen, 2013).

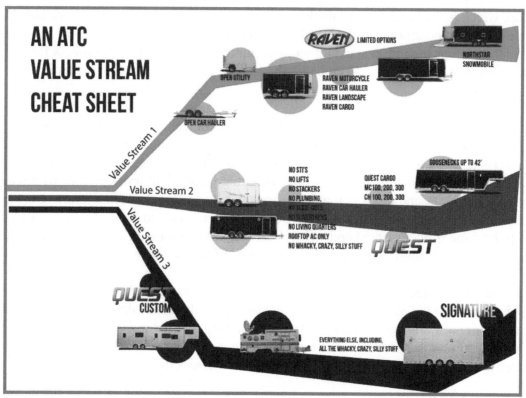

Figure 18.7
An Enterprise Design Consists of Multiple Value Streams

4. **Define System Purpose in terms of Functional Requirements FRs of the Value Stream**

The Value Stream approach creates a *de facto* representation of the system boundary. Fundamental to systems engineering is the concept that a system must have a defined boundary. Either one is *in* or *out of* a Value Stream. Therefore, either one is within or outside of the system boundary established by a Value stream.

Principle 4 acknowledges that the Value Stream is a necessary, but not sufficient condition for a sustainable enterprise design. Why? Because the definition of a system is that it must achieve Functional Requirement(s). The way that a Value Stream is defined is to do what is necessary to create enterprise value. Done properly, a Value Stream defines the Functional Requirements to create enterprise value and defines the functional requirements of a system to be stable.

An enterprise system has many sub-systems. With the use of the system design language, it is possible to develop Enterprise Design Maps that may serve as templates that define the general FRs and PSs for any enterprise or business, whether it is private, service or government. An Enterprise may adapt an existing system design Map's FRs and PSs to meet their specific needs. For example, manufacturing systems (the MSDD) or product delivery systems (product design + manufacturing) and/or financial reporting, accounting and performance measurement systems may have a generalized set of FRs and PSs that are applicable to many different products or services (Cochran, 2006).

The Enterprise Design Map provides a guideline for new system design and implementation. For this reason, a company that wants to become lean may use pre-existing Enterprise Design Maps to understand successful system designs that others have implemented, or the system design and leadership team may customize the PSs to their situation as they invent new technologies and procedures. Likewise, the people within an enterprise may add new FRs (or modify existing FRs) to adapt an existing Enterprise Design to their situation as market conditions change.

Success by design is the result of mapping PSs to the FRs. The FRs define the purpose of the system and are written to describe what the system must achieve to be successful. When an FR is not achieved, a system cannot be successful. The PSs clarify and define how people to achieve the FRs do the work. This definition leads to defining the standard work for the Enterprise Design – which is the standard way of doing things to achieve enterprise FRs. Collective agreement is codified through the creation of the Enterprise Design. The Enterprise Design is the master plan for implementing a system and communicates the logic and thinking behind that plan.

The Enterprise Design map assists in driving out fear in an organization since it requires collective agreement and clear articulation of the enterprise FR - PS relationships and corresponding performance Measures (M_{FR} and M_{PS}).

When an enterprise designer defines the FRs of the future state value stream (Rother and Shook, 1999), the FRs drive the physical solutions (PSs) of the value stream implementation. For example, consider the *Seven Manufacturing Enterprise Design FRs* illustrated in Figure 18.8 (Cochran and Barnes, 2013). The first six FRs must be achieved to create a stable system.

FR7 reduces the time in system through both design and continuous improvement. Compare Figure 18.8 with Figure 18.4, which illustrates Toyota's *House Model*. The house model does not define Enterprise PSs. Instead, it defines what the system must achieve in terms of enterprise purpose – the pillars of the model are the FRs of Toyota's Enterprise Design. FR1 and FR2 are derived from the Just in Time (JIT) pillar. FR3 is derived from the Jidoka pillar (Figure 18.3).

FR1 meets one portion of JIT, to produce the quantity consumed by the downstream customer. Takt time is defined as the average pace of customer demand. To achieve FR1 requires designing each operation to produce at or less than the Takt time for each operation (cell, plant or work loop). This design solution is the PS1 to achieve the corresponding FR1.

Manufacturing Enterprise Design FRs

FR1 – Produce the customer-consumed quantity every shift (time interval) – **from JIT**

FR2 – Produce the customer-consumed mix every shift (time interval) – **from JIT**

FR3 – Do not advance a defect to the next customer of your work – **from Jidoka**

FR4 – Achieve FR1, FR2, FR3 in spite of variation – **Robust Design**

FR5 – When a problem occurs accomplishing FR1, FR2, FR3, identify the problem condition immediately and resolve for the long-term – Controllable Design

FR6 – Provide a safe, healthy, ergonomically sound environment – **Fundamental, must have**

FR7 – Produce product with the the least time in system, **once FR1-6 are achieved – Cost / Continuous Improvement**

Figure 18.8
Creating a Stable and Sustainable Enterprise Design

Likewise, FR2 is to produce the variety or mix of product that a customer operation (or cell, or plant) consumes, once they consume it. PS2 achieves FR2 by pacing and leveling production demand information to the upstream producer of work to you (which requires decreasing the run size quantity through setup time reduction). With the ideal case of run size one unit. To accomplish a run size of one unit requires one touch setup, meaning zero changeover time from one unit to another (i.e., red product to blue).

The concept of Jidoka is to not advance a defect to the next customer in the Value Stream. The PS3 to achieve Jidoka was codified by Shigeo Shingo's methodology of source inspection and mistake proofing called Poka-yoke (Shingo, 1986).

The system designers must accomplish the 7 FRs with the least resources. The achievement of an enterprise FR must not be compromised in the near term to only gain what might appear to be a cost advantage. For example, pulling out the "cost" to implement quick changeover will compromise the ability to achieve FR2. Taking out quick changeover may be a short-term cost savings, but will most certainly result in a long-term total cost increase. FR2 must be accomplished and not be compromised; management must invest in the methods and tools that minimize and preferably, eliminate, changeover time.

The role of leadership and management becomes to make the necessary investments in the equipment, tools, people -- training and work methods -- and material and information flow practices at the lowest total cost while achieving all enterprise design FRs. Leadership nurtures the spirit of improvement of the system to achieve the FRs and not of single-point operations. FR7 is listed last, since the system must be put in place to first achieve FR1-6.

FR7 fosters long-term sustainability through continuous improvement and additional cost reduction. Think of the analogy of a river with a high water level. The rocks are hidden. The water level is analogous to inventory level. PS7 is the systematic reduction of Standard Work in Process (SWIP) inventory to expose the rocks (the variation) that still exists even though the system has been put in place to be stable by achieving FR1-6. If a stable system has not yet been put in place to deal with the problems caused by variation, there is no benefit to exposing the variation. For example, a facility put in a single-piece flow cell to manufacture catalytic converters. One day, the manufacturing engineering that had the only knowledge in the plant about how to properly execute the changeover of a welding machine that welded the two cat shells together was sick. No one knew how to reset the machine – what do you think happened?

Instead of stopping work because the single-piece flow cell had indicated a problem existed, the workers were still being measured on how many parts were produced – so the workers did what was logical to them – to keep producing and to pile up inventory *inside of the single-piece flow cell*. The lean team did *not* implement the standard work necessary to facilitate rapid (and autonomous) changeover; FR2 was not achieved. Furthermore, FR5 was not achieved – *a process had not been set up to identify and resolve problems in a pre-defined way*. No alarms sounded, no line or area managers were contacted. Instead, people did what they always had done, pump out the product the best way they could and hope that the problem was discovered when their shift had ended.

The consequence of implementing lean, half way, as a set of tools to impress people will never work: It causes more grief than benefit and can actually be counterproductive. To sustain an enterprise that seeks to be lean requires a completely new way of working together to engage people in collective agreement about the work that people want to accomplish together and the way that they accomplish it (Cochran, 2006; Rother, 2011). Enterprise design makes the distinction between the purpose and objectives of a system from the technologies, tools and work methods. There are at least two stages of agreement: what we want to accomplish and how it is accomplished.

In summary, *all FRs of an enterprise design must be achieved*. The enterprise design FRs represents collective agreement about what a system *must* accomplish. The FRs do not represent a wish list; they are not *stretch* objectives, and they do not represent spam of any kind. The challenge for the people in an enterprise who choose a set of FRs is how to achieve the FRs for the least cost. It is not questioned whether the FRs will be accomplished once there is collective agreement to achieve them. This is a prime job responsibility of what we have called a *Lean Engineer*.

18.6 Stable System Design

For any system to be stable, in general terms, a system design must meet the Fundamental Customer FRs derived from JIT (FR1 and FR2) and Jidoka (FR3), and achieve the following Four System Design Principles:

Principle 1: Design System to minimize process variation σ_p to achieve the target process mean $\overline{x_p}$ minimize the variation of a part's total time in system σ_s^2, and to minimize the average total time in system (Throughput time). Figure 18.9 illustrates the total time or flow time through a system as a single part's time out of the system (t_{out}) minus a single part's time into the system (t_{in}). Imagine putting a red dot on a single part and recording its time in and time out; t_{out} – t_{in}, gives the total time in system for the red-dot part, this is sometimes called the lead-time or throughput time. The average time in system, $\left(\overline{X_t}\right)$, estimates the average time for a part to move through the system. FR7 states to reduce the average time in system, once FR1 through FR6 are achieved.

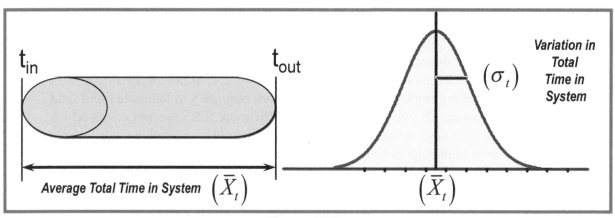

Figure 18.9
Illustration of Total Time and Variation in Total Time in System

Principle 2: Design System to be Controllable. To control a system requires feedback at pre-determined time intervals regarding the state of the system. FR5 expresses the design requirement to first know that a problem condition exists and then to do something about it is a pre-defined way. The response can be lack triage, as the problem persists an escalation occurs from team leader, group leader, value stream manager and finally to plant manager.

Principle 3: Design System to be Robust. A system that is designed to be robust achieves its primary mission, in spite of variation (problems). A design team must first ask what its primary mission FRs are, i.e., FR1, FR2, and FR3… but a system may not be limited to this set of three FRs. The Standard Work in Process (SWIP) inventory which is pulled based on the Takt time from a pace setter enables a system to be robust to variation.

Principle 4: Design System to Facilitate Active Learning and Leaders who are System Designers. A system that rewards improvement encourages people to speak up about problems that exist. Leaders have the responsibility to encourage people to suggest improvements and to reward improvement. Leaders must drive out the fear among people to voice concerns. A stable system design is defined by the FRs and PSs associated with the above Four System Design

Principles, and it is also a design which achieves the Fundamental Customer FRs. For a manufacturing system to be stable, it must achieve the 6 FRs and PSs of system stability.

FR1, FR2, and FR3 are derived from the Toyota House model and represent fundamental customer needs of superior quality products, on-time delivery of a variety of products.

FR4 and FR5 also address whether the delivery of a product is on time. FR4: Achieve FR1, FR2, FR3, in spite of variation. FR4 ensures that a product is delivered on time in the presence of variation. One solution is that PS4 can be stated as:

> *PS4: Customer pulls from marketplace, – Standard Work in Process (SWIP)*

FR5, on the other hand, applies the controllability principle to the enterprise design. FR5 states:
> FR5: *When a problem occurs accomplishing FR1, FR2, FR3, identify the problem condition immediately and resolve for the long term.*

The system is able to produce product on time when it is able to detect and resolve problem conditions as rapidly as possible. The solution is to sound a visual and/ or audible alarm and resolve the problem in a pre-defined way.

> *PS5: Alarm followed by a Pre-Defined (Standardized) Problem Resolution Process*

To control a system first requires feedback. This means that a system must be designed to sound an alarm at the moment a problem occurs. However, providing the alarm signal to the engineers is not enough. The engineers must have a pre-defined troubleshooting and problem resolution process (PS5) that leads to the long-term mitigation of the problem condition.

FR6 is stated as:
> FR6: *Provide a safe, health, ergonomically sound work environment*

The physical solution is the work environment and standardized work procedures for safety.
> *PS6: Health and Safety Standard Work Procedures*

18.7 The Unit Cost Equation and Stable System Design

If Lean is as easy as some claim, why is it so hard to achieve? This section addresses that penetrating question. Fundamentally, it is because the existing organization has its own set of PR's that have been largely defined by accounting procedures. This is nowhere more evident than the misguided practice of driving the system with a unit cost equation. The practice of applying the Unit Cost Equation to management decisions is best characterized as *Mass production*. In contrast, the consequence of applying TPS was to become *Lean* in the book, *The Machine that Changed the World* (Womack et al, 1990). Lean was described as the *result* or consequence of implementing the TPS in the United States with U.S. workers. The book's research conducted through the Massachusetts Institute of Technology (MIT) International Motor Vehicle Program (IMVP) showed that American Workers within the umbrella of TPS could produce cars of equivalent cost and quality as Japanese workers in Japan within the TPS umbrella. This profound finding disclaimed the myths that American workers were overpaid and slothful as claimed by Big 3 pundits at the time. Womack and his coworkers demonstrated that the difference in quality and cost were due to the *System* and not the workers.

The unit cost equation defines minimum total cost as the sum of the minimum unit costs ($/unit) at each operation i; for i =1, 2,......M operations..

Minimum Total Cost $=\sum\{min[Unit\ Cost(Op_i)]\}$ for i = 1 to M   18.1

Unit Cost$(Op_i)=\{$ Labor Hours(Op_i)*Wage Rate + Material\$$(Op_i)$ + Overhead\$$(Op_i)\}/N_i$

Where: N_i= Average number of parts produced per hour at operation i
And: Overhead\$$(Op_i)$ = Labor Hours(Op_i)/ Labor Hours(total) * Total Overhead\$

For managers tasked to run operations according to the terms of this equation, the goal is to minimize the unit cost ($/unit) of each operation as if each operation is an isolated stand-alone entity whose inputs and outputs *do not affect and are not affected by* other operations as illustrated by Figure 18.11 (Cochran, 2006).

> Incorrectly, the goal of the unit cost equation is to minimize the cost of each operation (op), independently and without regard for other operations.

$$Unit\_Cost(Op_i) = \frac{^{\$}DL(Op_i)\ +\ ^{\$}MTL(Op_i)\ +\ ^{\$}OVHD(Op_i)}{N}$$

...where $^{\$}OVHD(Op_i) = \frac{DL\ Hours\ (Op_i)}{Total\ Plant\ DL\ Hours}\ (^{\$}OVHD_{total})$

N =number of parts produced per hour; op =operation; DL =Direct Labor; MTL =Material; OVHD =overhead

> The result is to <u>compromise system objectives</u>, while assuming optimal decisions.

Unit Cost Equation <u>Falsely Assumes</u>:

$$Total\_Cost(system) =$$
$$\sum_{i=1}^{n} Unit\_Cost(Op_i)$$

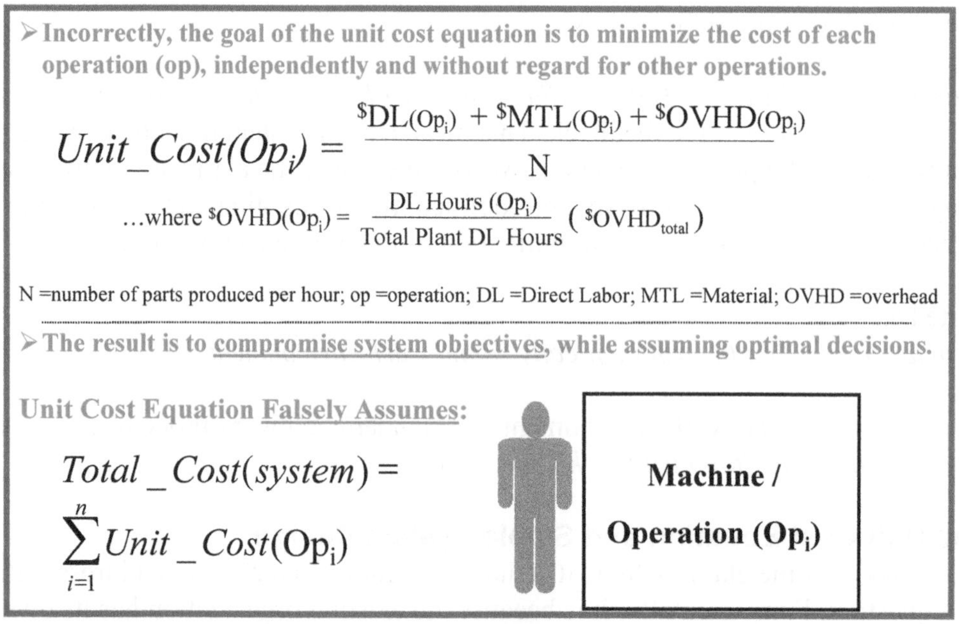

Machine / Operation (Op$_i$)

Figure 18.11
The Unit Cost Equation Focuses on Individual Operations and Not the System

When Ford Motor Company began using this equation to define the design intent / purpose of plant operations, *they implicitly changed the organization's primary FR* from meeting customer needs (i.e., producing cars based on actual customer demand) to minimizing each process (single operation) unit cost as defined by Unit Cost Equation. Therefore, by seeing the company's operations through the lens of the Unit Cost Equation, Ford Motor Company viewed sales volume fixed within certain bounds, and not as an *input parameter* to dynamically drive manufacturing. In reality, however, volume of sales is a reflection of how well customer needs are met in terms of the design, quality, reliability, variety and availability of options, delivery speed, cost, service, and support of a product offered in the marketplace.

The correct system design requires use of a system design language (FR's and PS's) which reflect collective agreement about system objectives and how to achieve those objectives. Dr. David Cochran has called this methodology *Collective System Design* or CSD (Cochran, 2006). The CSD methodology may be used to diagnose the design of an existing system, by using the CSD language to identify the currently practiced (the system's purpose being what it does) as reflected by the actual FRs and PSs that were implemented. The *incorrect* and mis-leading system FRs and PSs *implied* by the Unit Cost Equation are:

FR1_misled: Increase Speed (N) of Operation *i*   *PS1_misled: High Speed Machines*
FR2_misled: Decrease Direct Labor Content   *PS2_misled: Automate the Operation*
FR3_misled: Decrease Direct Labor Wage Rate   *PS3_misled: Low Wage Environment*

Managers and engineers, who apply the Unit Cost equation to make investment decisions purchase high-speed, automated machines and arrange those machines in departmental (operation-focused) plant layouts. The equation points managers and engineers to also source products to low-wage countries, to achieve FR3.

A typical plant layout that results from the application of the Unit Cost equation is shown in Figure 18.12 (Cochran and Dobbs, 2001b).

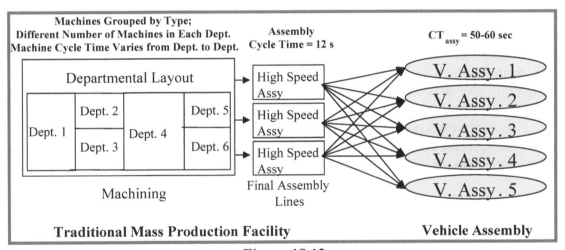

Figure 18.12
The Unit Cost Equation Results in Cost-Independent Workcenters

In contrast, Enterprise Design may be used to define the FRs necessary to create stable flow and output to the customer to meet actual demand. Enterprise design principles also address the design of a system to be robust to achieving system FRs in the presence of variation, and self-diagnosis and correction of problems.

In *Relevance Lost,* H. Thomas Johnson and Robert S. Kaplan identified the allocation of overhead as a major concern with this type of unit cost equation; they proposed Activity-Based Costing (ABC) as the solution to this problem (Johnson and Kaplan, 1987). Johnson's 1992 book, *Relevance Regained*, refuted the idea of using ABC to control operations costs, as Johnson came to realize that the work itself creates the cost and that an improved overhead allocation schema does not, in itself, improve the work within a system (Johnson, 1992). In 2000, Johnson took his insights one step farther in *Profit Beyond Measure* by arguing that numbers and accounting information cannot be used to control operations. The work or PS's by the people within a system affect the outcomes (FR's) of the system, and a system is much greater than the sum of its parts (Johnson and Bröms, 2001).

The unit cost equation seduces and misleads engineers and managers into thinking that they are making the right decisions; while in fact, blindly driving off a cliff. Non-systems thinkers still claim that the problem with the Big 3 is direct labor cost. The proponents of this theory seem to forget that if a product is deemed inferior by the free market place, people will not buy it.

On the other hand, any product delivery system designer knows that a company must produce products that people want to buy, this is FR0:

    FR0: Produce Products that People Want to Buy.

An Enterprise Design integrates product development and manufacturing. Toyota's Physical Solution (PS) to achieve FR0 is the product development process.

    PS0: Product Development Process

Some of the attributes of this process include systematic freshening of a vehicle's design (3rd year after new product introduction) and new product introduction every 5 years. The Big 3 do not follow this systematic approach. Instead, for example, an investment trade – off decision is typically made in which the profit to develop and introduce a new vehicle is compared to the declining profit (due to declining sales) of an older vehicle. The decrease in market share of the Big 3 is due to the product offering itself. It should be noted that the U.S. Automaker's new vehicle prices to the market are less than comparably equipped international nameplates. The claim that market share has been lost due to higher labor costs, therefore, is not tenable – but results from the thinking of the Unit Cost Equation.

The Unit Cost Equation tacitly assumes that reducing labor cost for each operation will reduce total system cost. Yet, to understand any system, requires voicing the FRs --- moving the unspoken to spoken or making implicit knowledge explicit. The unspoken practice of Detroit may in fact be to produce cars to achieve high utilization, *regardless* of whether customers are actually buying the vehicles. This practice is the result of the *engrained Detroit DNA*, called the Unit Cost Equation. The Unit Cost Equation assumes that a piece-part of a system can be optimized as an entity that is independent of the other parts of a system around it. For example, this way of thinking is analogous to a person's left leg independently deciding to grow without bound and not in proportion to the right leg and deciding to consume all resources from the body without concern for the consequence of its actions on the rest of the body. Yet, the Detroit DNA as mis-led by the Unit Cost Equation implements systems with this poisonous way of thinking.

18.8 A Factory Re-Design Example

A team put together a new plant layout to control material flow, only to find that the management accounting practice driven by the unit cost equation in the plant rewarded

Department Managers to earn standard labor hours rather than achieving the true manufacturing system and sub-systems correct FRs:

FR1: Produce the customer consumed quantity per time interval

FR2: Produce the customer consumed mix (variety) per time interval

The tools (PSs) of the lean manufacturing system were implemented; but the performance measurement and management accounting subsystem were not changed as part of the *lean* implementation. Since the performance measurement and management accounting subsystems were not changed, managers were driven by FRs implied by the management accounting system. The FRs of the management accounting system drove the wrong behavior.

The implied FRs were not aligned with the true Enterprise Design FRs: FR1 and FR2 above, which define the functional purpose of the manufacturing system in meeting the product delivery needs of the customer.

5. Understand the Functional Requirements FRs of the system very well before determining the physical solutions (PSs) and implementation. Treat a Physical Solution (PS) as a hypothesis to achieve an FR. Develop the experimentation mindset. Choose PSs so that the design is acceptable. Each must be implemented with standard work.

Step 5 emphasizes the importance of knowing the FR, the *Why* of Enterprise Design before jumping to the PS; which is the *How*.

Again we ask: If Lean is so easy, why is it so hard? It may be because we jump to solutions – before understanding the real purpose of work.. the *Gemba* (Brennemen, 2013; Cochran, 2007).

The approach of sustainable Enterprise Design is that the enterprise FR must be more than simply saying:

FR: Make Money

With an enterprise design approach, we can start with this high-level FR, but immediately the design approach requires us to define the PS, physical solution to achieve the FR. Each PS is implemented through standard work. Standard work defines what people must do to be successful. Standard work represents the best way of doing work within the organization at the present time. It represents organizational wisdom and is the record of solved problems. Standard work not only applies to routine work tasks; it also may be used to define how people respond in a crisis situation and correct a problem condition. Standard work ensures clear and unambiguous definition of roles, responsibilities and measures of people within an organization.

Senior leaders must work with their management team to define *how* the enterprise will make money. Furthermore, the notion of M_{FR} defines measures on the FR, which define how much money, and on what sales volume. It is ineffective for a senior leader to simply say: *Make money, I don't care how you do it.* The general PS to achieve the FR is the Enterprise Design.

The PS is the Hypothesis to achieve the FR. The Enterprise Designer then defines the FRs of the Enterprise Design, for each different Value Stream.

The enterprise design enables a leadership team to quickly focus on the, *what* (the objectives) and the *how* (means) of the system; and not on financial performance measures alone to run the enterprise (Johnson and Bröms, 2001).

6. Build physical models of the work to achieve the FRs in a safe environment.

When a team builds a physical model of the work to achieve the FRs of an Enterprise Design, it facilitates shared agreement and understanding. A safe environment means that the work may be done without fear of failure.

Figure 18.10
Teams Building a Physical Model of the Work to Achieve the 7 FRs

7. Derive Performance Measures of a System after thoroughly understanding the FRs and PSs of the system.

A common trap is to define performance measures before establishing the FRs. Performance measures should be used to reinforce the achievement of system FRs. Figure 18.2 illustrates that performance measures can be associated to the FR, called M_{FR}, to measure whether system purpose is achieved. Performance measures can also be associated with a PS, called M_{PS}. A measure on a PS determines whether the work for a PS is being done or not. Even though a M_{PS} performance may be high, it should be remembered that a PS is only a hypothesis to achieve the desired enterprise purpose that is expressed as the system FR. Doing a PS well and having a good M_{PS} score does not ensure that an FR is achieved. For example, if the FR is:

FR: Predict First-Year College Grades

If the PS is: *PS: SAT I Test*

A high score on the SAT, the M_{PS}, does not always ensure academic preparedness. In fact, predictive validity of the SAT, the M_{FR}, with respect to first-year, explains only 4% of the variation in first year college grades (Fairtest, 2007).

8. Align all costing information to each of the enterprise value streams; convert / allocate overhead costs to each value stream. Invest in the resources necessary to achieve the FRs of the Enterprise design.

The benefit of defining an enterprise that consists of multiple value streams is that all indirect and overhead costs may be allocated and/or converted to direct costs associated with each value stream. As the time in system is reduced (FR7), total cost is reduced.

18.9 Moving Forward

I recall touring the Georgetown KY plant with J. Black. He took me to the side after we had heard a tape that Toyota was playing for the plant visitors. He asked me: *David is it the system design or the culture that makes TPS successful*? A few years ago, he said: *Toyota had said in the demo tape that the success at Toyota was its culture…. The people. But on that day, Toyota's tape said that its system was the reason for its success.* So which is it? ….. or is it both?

In my mind I began to examine the real secret in how Taiichi Ohno had managed to create the most successful automotive manufacturing system in the world. Was it the people, the system design or the system itself? How *did* the Toyota Production System (TPS) change the world? Could its success be applied elsewhere?

The Machine that Changed the World seems to have been the first to use the term, *Lean* to describe TPS; the true impact of the book is that it disproved the notion at the time that TPS was successful because the Japanese were better workers than Americans (Womack et al., 1990]. Prior to this landmark publication, engineers and social scientists both believed that the *culture* in Japan made it possible to produce vehicles that were superior to those made in America. This book disproved the notion that the superior results were due to the worker. This book provided evidence that Dr. Deming's assertion that success (or failure) is due to the *system* and not the *people*. Idealistic notions of cultural superiority do not make a better mousetrap. Avoid the, *we need to change the culture* trap at all costs. It is the system design and the properly defined value stream FR's that will constantly achieve the purpose of the system or enterprise.

18.10 Working with Enterprise Systems

People create systems. The key to unlocking Enterprise Design is to realize that the people doing the creating of *something … anything…* are in a system themselves while creating new systems.

Miguel Ruiz is his book the *Four Agreements* says that each person lives in his or her own *dream* (Ruiz, 19970. Practically speaking, this means that each person sees reality in his or her own way. The profound implication is that people can and do view a system in their own way.

The *engineers'* mindset is that anything and everything can be designed. This is a different point of view than others who are trained to *observe* what a system does and *execute* system goals. Observation skills are required to improve engineering designs. But observers may not be designers and designers may not be observers. Each person has his or her own prism and worldview. For example, one school of thought describes; *the purpose of a system should not be different from what it does* (Lockton, 2012). The system is a *living organism* that constantly evolves. This description appears opposite to the precepts of engineering which focuses on designing something.

If we take an iPhone as an example; *is the purpose of the iPhone what is does? Is the iPhone's existence simply its manifest reality?* Are there design intentions that were not realized? Was the purpose of the iPhone's existence to serve society or to make money or both? We don't know. Only by asking the designers would we know.

To *fulfill* a person's dream, we must ask characterize and *understand* their dream. We cannot simply assume that their dream is our perception of reality… and that there perceptions and understanding of a system is ours… this is called checking our assumptions at the door – and is the reason why a language of Enterprise Design is necessary. Properly executing design and implementation is the goal of a *Lean Engineer* and the purpose of a *Lean system*.

18.11 Summary

The principles expressed in this Chapter advocate the use of a structured language to communicate a system design and to make good design decisions, and consistently execute those decisions. Peter Senge has written about the learning organization: John Shook, Mike Rother and the Lean Enterprise Institute emphasize the importance of lean being practiced as a PDCA learning loop: and Steve Spear asserts that all work must be done with the mindset of a hypothesis to achieve an intended result. The learning organization described by Peter Senge just might be fully achieved through a properly coordinated and executed design procedure led by a Lean Engineer.

As much as possible, the Language for expressing system FRs, PSs and defining relevant system performance measures must be combined with the understanding of the importance of shared and collective agreement about system FRs and the implementation of PSs through an experiential PDCA learning loop that fosters the mindset that work is improvable and part of shared learning.

The discussion has gone full circle; success is all about the way that work is done in the context of the Enterprise Design --- which was initially the foundation of Industrial Engineering. The traditional Industrial Engineer *has* to come to terms with either changing, or being replaced by someone or something else. This book is about **Lean Engineering**, which is a testament to this fact.

Review Questions

1. Why must a system be designed to be stable before it can be sustained?
2. What are the 7 FRs of a stable manufacturing enterprise design?
3. What are the 7 PSs of a stable manufacturing enterprise design? (Note: there can more than one PS implementation.)

4. Describe:
    a.) The significance of each of the 8 steps of the enterprise design framework to sustain an enterprise that is designed to be lean.
    b.) Discuss the benefit of including each of the 8 steps in your enterprise design, and
    c.) Provide an example of the cost of not including each of the 8 steps in your system design?
5. Define and Describe the *Four Stable System Design Principles*
    1. Design to achieve Fundamental FRs to minimize $(\bar{X}_{tt})$ (σ_{tt}) $(\bar{X}_{p})$ and (σ_{p})

    2. Design system to be Robust
    3. Design system to be Controllable
    4. Design System to Facilitate Active Learning and Leaders who are System Designers
6. Write the mathematical expression of the unit cost equation and define the variables. Does the unit cost equation apply to systems or operations? What are the implied FRs and PSs of the Unit Cost Equation and state your reasoning.
7. Toyota discovered that a vehicle could by manufactured with linked-cellular manufacturing systems so that the Unit Cost (cost to produce per vehicle) does not decrease significantly beyond 100k units per year. Explain why Toyota's knowledge about this point is a factor in the bankruptcy of GM and Chrysler in 2009? Hint: Think about the consequence of using the unit cost equation to make platform, investment and production decisions.
8. Why is the implementation of Stable Manufacturing System Design a pre-requisite to long-term, sustainable improvement?
9. Why does a Stable Manufacturing System Design reduce cost over the long term?
10. Why do operation-focused motion *kaizen* events typically *not* reduce cost over the long term? What does Toyota put in place before doing motion Kaizen?
11. Why does separating the worker from a machine in a cell require Poka-yoke devices?
12. What is the reason for separating workers from the machines in a cell
13. Why is excess work / motion the shadow of an unstable system design?
14. Define what Standard Work means and why it is important.
15. What is the "unit Cost Equation"? Why does the use of this equation destroy the basic concepts of Integrated Lean systems?

Thoughts
  and Things........

Chapter 19
Advances in Mixed Model Final Assembly
Kavit Antani

Some minds are like concrete, thoroughly mixed up and set Unknown Author

19.1 Introduction

Since the days of the famous Model-T, product requirements and thereby the requirements of production systems have changed significantly. Customers want a unique product that meets their needs and fulfills their desires. This along with tremendous competition in the marketplace has driven the need for mass customization. Companies can no longer afford to offer a few standardized products and expect to be profitable and gain market share through mass production. In the automotive industry, product variety is driven by globalization of the marketplace, increasing demand for technology-integration, and changes in energy costs. For example, on the BMW 7-series sedan, there are so many options to choose from, that 10^{17} unique combinations can be manufactured on the assembly line. To enable such a diversified product range, without jeopardizing the advantages of an efficient manufacturing process, specialized flow lines called mixed-model final assembly lines are utilized. In a mixed-model final assembly line, the use of cross-trained operators and relatively simple tooling leads to a significant reduction in setup time. This enables the assembly of different products in mixed product sequences on the same line with a lot size of one. There is usually a common base product which can be customized by selection of optional features from a standard list of options.

19.2 History

In the late 1790s, Eli Whitney developed the first mechanized assembly line to manufacture muskets. Henry Ford later adopted this concept to produce the Model T in 1913. In the early days of automotive manufacturing, perhaps a large group of skilled workers built a vehicle together, one at a time. Ford's assembly line made division of labor possible but it was based on a large lot size. A conventional assembly line used to build a single product type is shown in Figure 19.1. At each station, some operations are done to add value to the product.

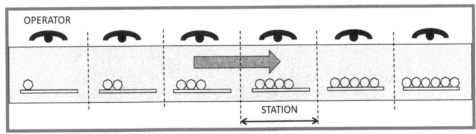

Figure 19.1
Conventional Assembly Line building a Single Product

The goal of mass production was to lower the unit cost by distributing the depreciation cost of specialized equipment and tooling over a large number of identical parts and by reducing the number of changeovers. In the early days of mass production, a lower selling price made products more accessible to common man, thereby creating a favorable environment for mass production of relatively identical products with very little variety. It was a seller's market in the early days of automotive manufacturing. The manufacturers assumed that whatever they built would get purchased and they used a traditional push system. Model variety was purposely avoided by Henry Ford. All cars had essentially the same color, configuration and standard features. This enabled worker specialization and standard work patterns based upon traditional time and Motion study. The two biggest downsides of this system were excess inventory and lack of flexibility.

The solution to the problem of producing cars in a variety of styles, options and configuration was pioneered by General Motors Corporation, and reached its zenith some 50 years later when a new production system was pioneered by the Toyota Motor Corporation. This system was called the *Toyota Production System* (TPS). The Toyota Production System is based on four basic principles: (1) Elimination of waste, (2) Just-In-Time delivery, (3) Separation of worker from machine and(4) Pull production. These principles resulted in the ability to build customized products with a lot size of one on what is known today as the *Mixed-Model Final Assembly* (MMFA) line.

19.3 Mixed Model Final Assembly

In the case of automotive assembly, the process starts in the Body Shop by Spot-Welding various sheet metal panels to the base chassis. This body undergoes various cleaning and corrosion protection processes and is called a Body-In-White (BIW) at this stage in the manufacturing process (Figure 19.2).

This is usually where the sequencing of various models takes place based on both demand and product mix. Sequencing will be addressed separately in the following sections of this Chapter. The Body-In-White then goes through the multi-layer painting process and then arrives on a conveyor to the Final Assembly department of the automotive plant (Figure 19.3). Final Assembly department is usually a maze of several sections of smaller assembly conveyors connected by transfer stations (Figure 19.4). The painted body structure which is no more than a basic skeleton structure of the vehicle enters the assembly line on one end and out comes a finished vehicle on wheels at the other end, ready to be road tested and loaded on a delivery truck.

Figure 19.2
Body-In-White (Image Source: BMW)

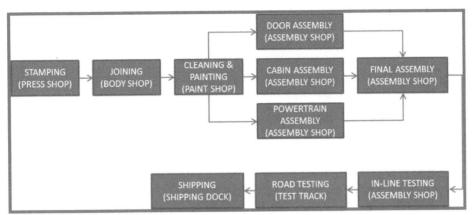

Figure 19.3
Block-diagram showing automotive assembly steps

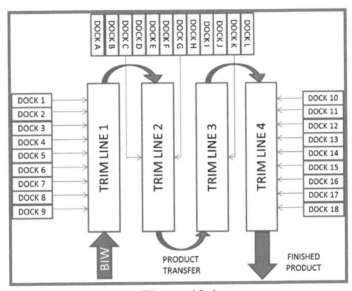

Figure 19.4
Block Diagram Showing Product Flow and Raw Material Supply from Dock Doors

Each segment of an assembly line (Called a *Trim Line*) may have 20 to 25 stations, each of which may contain multiple work-zones or workstations where operators assemble components to the vehicle body. All conveyors are synchronized and move at a constant pace determined by the *Takt-time*. Takt Time is defined as the ratio of total number of available time units for production to the total number of products (vehicles) to be made; dictated by the demand of the end user.

$$\text{Takt Time} = \frac{\text{Total available time per shift (sec.)}}{\text{Total product demand per shift}} \qquad (19.1)$$

For example, Takt time for a major automotive assembly plant in the United States is 101.25 seconds. The Takt time represents drop-off rate, which means every 101.25 seconds, one vehicle rolls off the end of the assembly line. This Takt time is calculated as follows:

$$\text{Takt Time} = \frac{9 \text{ hours x } 3{,}600 \text{ sec.}}{320 \text{ vehicles per shift}} = 101.25 \text{ sec.} \qquad (19.2)$$

It is important to note that the denominator in this equation should be based on the rate at which the vehicles are sold to the final customer to maintain a Lean Supply Chain.

In order to increase profits, the least number of operators should be utilized to produce the required number of vehicles. For example, if 50 operators on an assembly line can produce 320 vehicles, but through some improvement projects, if the same 50 operators now produce 350 vehicles, this results in overproduction of 30 vehicles, compared to the actual demand. Overproduction is a fundamental waste. The goal should be to produce the required 320 vehicles per shift with the least number of operators.

At each station, multiple assembly operators work in their respective zones of the vehicle as it travels on a constantly moving conveyor. The vehicle may be raised, lowered or tilted using programmable logic controllers to allow the operators to assemble components to the vehicle with the least ergonomic stress. Each associate has a set of tasks to be completed within the available Takt time. This task assignment results in a non-polynomial (NP) hard problem called the Mixed-Model Assembly Line Balancing Problem which will be addressed separately in the Enabling Systems section of this Chapter. At the end of the assembly segment, there may be a quality check station that focuses on critical assembly characteristics associated with the tasks completed in that segment. A detailed explanation of the systems used for Quality Control and Defect Prevention in MMFA will be provided in a separate section in this Chapter as well. At the end of the assembly segment, the vehicles get transferred across the logistics aisle to the next segment to continue the assembly process. This process continues along a serpentine sequence of conveyors until the entire assembly is complete. After the assembly is complete, the vehicle is driven on its own power to the testing area. All vehicles get driven into a booth that has a Dynamometer. The vehicle gets accelerated to its maximum rated speed and a series of tests are done on the Dynamometer. Finally the vehicle goes through a Road Test which includes driving it on a specially developed surface prior to certifying the vehicle for final delivery.

19.4 A Real-World Example of MMFA

The following data shows a typical range of values for key metrics that will give readers a perspective of a real-world Mixed-Model Automotive Assembly line:

a. Takt time: 60 to 125 seconds
b. Assembly Takts (multiple per station): 355 to 450
c. Labor per vehicle (Joining, Paining, Assembly) = 25 to 30 hours
d. In-process dedicated quality-checks = 8 to 10
e. Available labor hours per shift = 8 to 10 hours
f. Vehicle built per shift (9 hours) = 260 @ 60 s. and 540 @ 125 s.
g. Takt time utilization = 92% - 96%
h. Number of mixed base models assembled on a line = 2 to 3
i. Number of variants of each base model assembled = 20 to 25
j. Number of optional sub-assemblies per variant = 300

Note: Takt time utilization for a given workstation or zone of work is the ratio of the sum of task times assigned to the assembly operation to the Takt time determined by customer demand. This ratio is a measure of Line efficiency measured by the ability to produce according to line Takt time at each Workstation or zone of work. A perfect ratio would be 1.0. This metric will be explained in greater detail in the Assembly Line Balancing section of this Chapter.

Multiple base models may be assembled on the same assembly line. For example, a base model can be a small 5-seater SUV (Sports Utility Vehicle) and it can be assembled alongside another base model which can be a 7-seater large SUV. Each of these base models can have multiple variants such as Left-Hand Drive / Right-Hand Drive, choice of 2.5 liter gasoline engine / 3.0 liter larger gasoline engine or a turbo-charged diesel engine, or it can be a market specific variant that meets regulations of a certain country where the vehicle will be shipped. There can be 20 to 25 different variants per base model in a modern automotive assembly plant.

Option content refers to the possible option choices that customers have when they configure the vehicle. For example, one may be able to choose for up to 7 different roof-rails for a vehicle, depending on the selected variant. The multiple base models, their variants, and the associated option content make Mixed-Model Final Assembly a challenging multi-disciplinary problem.

19.5 Key Enabling Systems

Mixed-Model Final Assembly is feasible because the following key enabling systems function seamlessly in the background.

19.5.1 Just-In-Time (JIT) Component Deliveries

The goal of Just-In-Time deliveries is to have the required components at the required time in the required quantities in order to prevent accumulation. One of the fundamental pillars of the Toyota Production System is Waste Elimination. Although inspection, transportation, and storage of inventory are required elements of a manufacturing process, only the actual processing is value added. JIT aims at eliminating one of the primary waste sources which is storage of raw material and finished goods. Mixed-Model Final Assembly reduces or practically eliminates finished goods accumulation because the vehicles are produced just-in-time, directly based on the customer orders and the same strategy is applied to the raw material receiving side.

Most automotive assembly plants that have implemented JIT deliveries have to place a tremendous focus on schedule and capacity. Some companies such as BMW allow customers to place a completely customized vehicle order through a web-based configurator. These orders are then sequenced in the form of a production plan and broadcast to the respective vendors. On the other hand, some assembly plants operate on a sales forecast but divide their schedule into several layers. The top level master schedule is based on extensive market survey to estimate a relatively approximate demand for each model type. This high level planning is used to plan capacities in the plant and raw material suppliers. This estimate is given to the plants and vendors between 60 to 90 days in advance and firmed up usually 30 days before the planned production date. The firm numbers are used for the second level (weekly) and third level (daily) planning. A final leveled schedule is sent to the final assembly line which drives the demand using a Kanban system.

Instead of the long final assembly lines (Figure 19.4), the newer plants have a layout like the fingers of a hand (Figure 19.5). Just like airports would like to maximize the number of available gates, this floor layout allows the assembly plant to have a significant number of dock doors all along the various fingers for JIT deliveries of sub-assemblies and components directly at the point-of-use on the assembly line. This minimizes the need to move racks over long distances from the dock doors to the point-of-use, using forklifts.

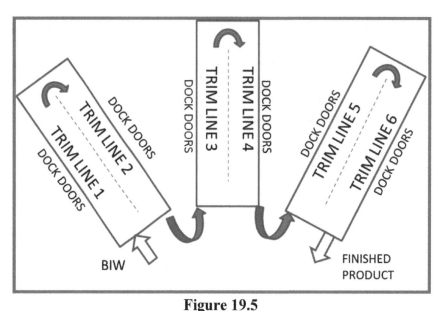

Figure 19.5
Finger-Layout **provides significantly greater number of dock doors for JIT part supply**

19.6 Assembly Line Balancing (ALB)

The decision problem of optimally partitioning (balancing) the assembly work among the stations with respect to some objective is known as the assembly line balancing problem (ALBP). Work on methods to develop a decision support system for Assembly Line Balancing system at a major automotive assembly facility in the United States, is ongoing. A summary of the key concepts of Assembly Line Balancing, challenges with the current state-of-the art, and potential opportunities to improve will now be discussed.

In today's mixed-model automotive assembly plants, several hundred models of a common base product can be manufactured using a transfer lot size of one. An assembly line consists of (work) stations $k = 1,\ldots,m$ arranged along a conveyor belt or a similar mechanical material handling equipment. The work-pieces (jobs) are consecutively launched down the line and are moved from station to station. At each station, certain operations are repeatedly performed with reference to the cycle time (maximum or average time available for each work-cycle). Manufacturing a product on an assembly line requires partitioning the total amount of work into a set of elementary operations which we will call *tasks* $V = \{1, \ldots ,n\}$. Performing a task j takes a task time t_j and requires certain set of machines and/or skills of workers. Due to technological and organizational conditions precedence constraints between the tasks have to be observed. These elements can be summarized and visualized by a precedence graph. It contains a node for each task, node weights for the task times and arcs for the precedence constraints. Figure 19.6shows a precedence graph with $n = 9$ tasks having task times between 1 and 9 (time units).

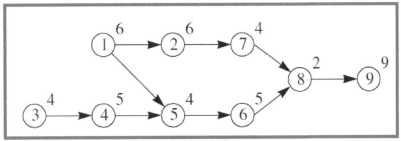

Figure 19.6:
Precedence Graph

For example, the precedence constraints for Task 5 mandate that its processing requires that Tasks 1 and 4 (direct predecessors) and 3 (indirect predecessor) be completed. The other way Task 5 must be completed before its (direct and indirect) successors 6, 8 and 9 can be started.

Any type of ALBP consists in finding a feasible line balance, i.e., an assignment of each task to a station such that the precedence constraints and further restrictions are fulfilled. The set S_k of tasks assigned to a station $k = 1, \ldots, m$ constitutes its *station load*, which is the acccumulated task times $t(S_k) = \sum_{j \in S_k} t_j$ is called *station time*. When a fixed common cycle time c is given, a line balance is feasible only if each station time is less than or equal to c. If $t(S_k) < c$, the station k has an idle time of $c - t(S_k)$ time units in each cycle. If both, number of stations and the cycle time can be altered, the line efficiency E is used to determine the quality of a balance. The line efficiency corresponds to the productive fraction of the line's total operating time t_{sum} and is typically defined as $E = t_{sum} / (m.c)$. Since the total idle time is equal to $t_{sum} - (m.c)$, a maximization of E also minimizes idle times.

Among the family of ALB problems, the best known and best-studied is certainly the SALB (Simple Assembly Line Balancing) problem. Although it might be far too constrained to reflect the complexity of real-world line balancing, it nevertheless captures its main aspects and is rightfully regarded as the core problem of ALB. In fact, vast varieties of more general problems are direct SALB extensions or at least require the solution of SALB instances in some form. In any case, it is well suited to explain the basic principles of ALB and introduce its relevant terms. A comprehensive review of SALB and its solution procedures is provided by Scholl and Becker(2006). SALB problems are based on a set of limiting assumptions.

- (S-1) Mass-production of one homogeneous product.
- (S-2) All tasks are processed in a predetermined mode (no processing alternatives exist).
- (S-3) Paced line with a fixed common cycle time according to a desired output quantity.
- (S-4) The line is considered to be serial with no feeder lines or parallel elements.
- (S-5) The processing sequence of tasks is subject to precedence restrictions.
- (S-6) Deterministic (and integral) task times t_j.
- (S-7) No assignment restrictions of tasks besides precedence constraints.
- (S-8) A task cannot be split among two or more stations.
- (S-9) All stations are equally equipped with respect to machines and workers.

With regard to real-world line balancing, one of the strongest simplifications of SALB is certainly caused by assumptions (S-2) and (S-9). Empirical surveys stemming from the 1970s and 1980s revealed that only a very small percentage of companies were using a mathematical algorithm for configuration planning at that time. The apparent lack of more recent scientific studies on the application of ALB algorithms indicates that this gap still exists or even has widened. Boysen et al.(2007) took the first decisive step to resolve this problem by developing a consistent, authoritative classification of ALB problems including all relevant constraints and objectives. A uniform classification enables practitioners to compare their individual problem settings with those covered by research and to single out suitable solution techniques. Furthermore, future research challenges can be identified by structuring the existing body of literature according to the classification scheme.

The classification scheme is constructed as follows. Any ALB problem will at least consist of three basic elements: A precedence graph which comprises all tasks and resources to be assigned, the stations which make up the line and to which those tasks are assigned and some kind of objective to be optimized. Accordingly, the presented classification will be based on those three elements which are noted as tuple $[\alpha \mid \beta \mid \gamma]$, where:

α precedence graph characteristics;
β station and line characteristics;
γ objectives.

19.6.1 Manual Assembly Line Balancing

This example is based upon a major automotive assembly plant and a pilot line where the electrical harnesses, floor insulation, and curtain head airbags get assembled into the vehicle. As part of the current (conventional) Assembly Line Balancing project, based on the change in model mix, the assembly team reviews the work distribution and changes it as needed, on a monthly basis. The current manual process of reviewing the various operations and rearranging the tasks to improve the average utilization of the operators is labor intensive. To baseline the current line balancing process, two line balancing workshops were held. These workshops included the following steps:

- Generate a visual display of all Takt's in the assembly line,
- Analyze tasks that will exceed the cycle / Takt time based on projected volume of vehicles,
- Re-balance each Takt while ensuring that tooling / station / work zone constraints are not violated,
- Calculate the line utilization metrics,
- Conduct trial runs to verify feasibility, and
- Finalize the proposed line balance.

In each workshop, a cross-functional team was comprised of 5 experienced individuals from assembly, Industrial Engineering, training, and quality departments. During the course of such a line balancing workshop that is typically done two times per year, for each assembly line, each participant focuses 100% on the work content evaluation and line balancing process. Tasks are manually arranged until the team reaches consensus on the organization and then line trials are conducted. Similarly, on a monthly basis, the line gets re-balanced to account for the volume changes that have been forecast for the following month. This exercise is usually done on a smaller scale than the workshop described above, and includes 2 experienced associates who conduct the planning and analysis in one day followed by line trials for two shifts. Although sounds simple, this process relies heavily on the knowledge of the participants and during the workshops it was evident that several constraints that should have been taken into account were not easy to remember while making decisions manually, thereby re-iterating a strong need for a decision support system.

19.6.2 Precedence Constraints

In addition to balancing a new assembly line, an actual running line has to be re-balanced periodically or after changes in the production process or the production program have taken place. Balancing means assigning the tasks to the stations (workplaces) based on, among other things, the precedence graph. In the automotive industry, typical information and planning system contains the description of tasks including their deterministic task times (derived for example by a motion-time measurement MTM approach), the current assignment of tasks to Takt's and the execution sequences of tasks within each Takt. However, no precedence relations are documented, not to mention an entire precedence graph. The huge manual input and the multitude of tasks (up to several hundreds or even thousands) prevent manufacturers from collecting and maintaining precedence relations.

This absence of documented information on precedence relations is the main obstacle in applying well explored theoretical assembly line balancing methods in practice. In practice, planning, balancing and controlling assembly lines are based on subdividing the production processes and, hence, the assembly lines into segments. Each segment is managed by a dedicated human planner, who becomes an expert for this part of the system. Though some software systems provide a component for automatic line balancing, the planners mostly balance their segments of the line by manually shifting tasks from one station to another, because precedence data is not available or existent data is not reliable. This is a very time-consuming and fault-prone job, which is solely driven by the experience and knowledge of planners. By appending the plans of succeeding line segments, the entire production plan is developed.

In the literature, some research work related to deriving precedence information can be found. However, most researchers do not aim at constructing precedence graphs but to find feasible sequences. Such concepts can be categorized in two main classes: on the one hand, there are manual and automated approaches that are intended to detect all feasible sequences. On the other hand, genetic algorithms and case-based reasoning procedures are applied to search for a few good sequences. The earliest approach to find feasible assembly sequences was introduced by Bourjault (1984). During his question-based procedure, an expert has to decide on the feasibility of assembly actions. However, as the number of liaisons between parts (connection points that correspond to mounting operations) rises, the number of questions grows exponentially. De Fazio and Whitney (1987) modified Bourjault's approach and reduced the number of questions to two times the number of liaisons. While both methods are based on the assembly of the product, Homen de Mello and Sanderson (1990) pioneered the disassembly approach. The basic idea is to enumerate all possibilities to disassemble a product by examining cut-sets of the liaison graph. The liaison graph is an undirected graph that contains parts as nodes, where two nodes are adjacent if they are connected in the assembly (by some mounting operations). Due to high error-proneness and the huge amount of input needed, manual approaches fail to work for highly complex products like automobiles. Therefore, researcher's interest turned to automated methods.

Ammer (1985) used the subassembly structure of the multilevel bill of material for the outline of the precedence graph. Precedence constraints for each subassembly are automatically extracted with the help of CAD data and merged thereafter. Most automated methods to find assembly sequences are based on geometric reasoning and, thus, geometric product models (e.g., CAD models). For early automated methods based on collision analyses in direction of the orthogonal axes, refer to Romney et al (1995). and Su (2007), who expanded the collision analysis to 360 deg. around the assembly. There are certain issues that restrain the use of automated methods especially in the automotive industry: first of all, manufacturers do not necessarily possess geometric product models and often do not even have the CAD data. Furthermore, even if CAD models are available, soft components like clips and gasket rings cannot be handled. Since both manual and automated methods only work for relatively small assemblies, Bonneville et al. suggested the use of genetic algorithms. Feasibility of new generated individuals (i.e., sequences) is checked with the help of the liaison graph and expert knowledge. Accepted individuals are evaluated by means of a fitness function. This approach or other genetic methods, e.g., as proposed by Smith and Smith, basically combine a (potentially successful) heuristic meta-strategy for solving the assembly line balancing problem with one of the (manual or automated) approaches for detecting or examining precedence relations.

Thus, there is no methodological gain in finding the relations of the precedence graph. Swaminathan and Barber (1996) included a case-based reasoning (CBR) approach into assembly planning. The basic CBR approach can be described as follows: a *case-based reasoning tool* solves new problems by matching and adjusting solutions used for similar problems in the past. In order to find feasible assembly sequences, Swaminathan and Barber divided the liaison graph into cyclic substructures and the case-based reasoning tool searches for a feasible sequence for each cycle. Dong et al. used subassemblies instead of cycles of the liaison graph. Criteria and the approach to find the best suitable subsequences are described by Su (2007), and Chen et al. (2006), who developed the software *Body Build Advisor* that was used for the assembly of an automotive body.

Contrary to procedural algorithms, which manually search for assembly sequences, both genetic algorithms and case-based reasoning approaches also take older plans into account. However, a structured use of those methods requires extensive additional input. Despite these recent developments in research, requirements on the precedence graph generation procedure put by automotive producers are still not satisfied by the available methods.

	Literature	Automotive industry
1	Small number of parts and tasks	Large number of parts and tasks (several hundreds to some thousands of tasks)
2	Precedence graph from scratch (esp. manual methods)	Credited assembly plans (containing feasible sequences) for past assemblies available
3	Product is assembled in only one or a few variants (models)	Product is assembled in almost arbitrary number of models (up to several millions or even billions)
4	Geometric product models are available	Geometric product models are not or only partly available, many (simple) parts responsible for great portion of tasks are not contained in those models

Figure 19.7
Academic Research vs. Real World

These methods either need an extensive manual input, which manufacturers are not willing or able to provide, or refer to geometric product models, which are often not available. Furthermore, a precedence graph concept suited for the automotive industry must work for several hundreds or even thousands of tasks and has to incorporate all the options for assembly of specific car models; both cannot be guaranteed by the existing approaches. Automobile manufacturers regularly re-balance their assembly lines both in the short term and in the long term. Typically, actual customer orders are known in advance for about 4 to 6 weeks, which makes it possible and necessary to re-balance the line according to the current model mix (share of different models and/or options in the sales plan). This will take place daily, weekly or monthly. Of course, adjustments introduced during re-balancing will not change the entire balance, but will consist of local re-arrangements (shifting some tasks to other stations). Long-run balancing takes place in case of new models or modifications in the production process.

19.6.3 Manual Precedence Mapping

Here is an example of a pilot manual precedence mapping exercise on an assembly trim (segment) comprised of 15 assembly stations at a major automotive assembly plant where the roof rails, electrical harnesses, sub-woofer, floor insulation, and curtain head airbags etc. get assembled into the vehicle. The primary purpose of this exercise was to understand the various constraints that would need to be captured for the decision support system that would be the primary data source for the optimization algorithm / construction heuristic.

In order to understand the process instructions and the actual work content, the individuals who conducted this study underwent hands-on training on each assembly station involved on the pilot line. The key advantages of conducting this training are as follows:

1) Visualization of Process Instructions.
2) Understand precedence relationships.
3) Understand undocumented supporting tasks.
4) Gain basic understanding of additional complexity due to high option content.
5) Learn the effect of a work overload situation on operational metrics.
6) Awareness of the constraints that must be incorporated into the optimization model.

7) Experience the ergonomic impact of repetitive tasks.
8) Understand the human behavior to adapt the required task to make it less strenuous and more effective.
9) Gain input from assembly line associates based on their work experience.

After a thorough understanding of the tasks and the complexity, the precedence mapping was manually undertaken in the following manner:

1) Stage 1: Each Takt was evaluated to determine precedence relationships between tasks within each Takt.
2) A cross audit was conducted by multiple process experts to verify the precedence relationships
3) Stage 2: The scope of the mapping exercise was expanded to the entire assembly line and relationships were mapped across Takts. In several cases, entire groups of tasks were found to be predecessors of another group of tasks in a downstream Takt.
4) Data verification was done to ensure that cyclic relationships did not exist. A cycle refers to a relationship that points from task i to j and another points in the reverse direction, making j a predecessor for task i.

In addition to the basic task level precedence relationships, the following constraints were identified and recorded during Stage 1 precedence mapping:

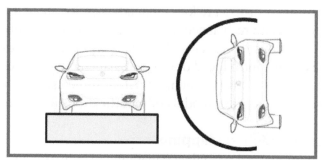

Figure 19.7:
Product Raised (left) and Turned 90 Degrees to Provide Access to Underbody (right)

1. **Product State constraint:** The *Product State* can be defined as the physical state in which the product gets presented at a certain assembly station.

2. **Assembly Zone constraint:** In the assembly of certain large products such as an automobile or a large machine, it would be critical to capture the location of the assembly operator with reference to the product while conducting the specific task. With reference to this study, the automobile would be divided into 9 assembly zones (Figure 19.8). This information needs to be captured as a constraint for each task so that the optimization algorithm takes the zone into account and prevents the assignment of multiple operators in the same assembly zone at the same time doing different tasks.

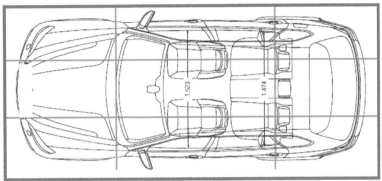

Figure 19.8:
Top View of Product Showing 9 Assembly Zones

3.   **Ergonomic constraint:** Every assembly task is assigned an ergonomic rating. When tasks are not designed appropriately in systems that depend on human operators, the system is particularly vulnerable to problems associated with worker health, production, quality, and increased training costs. It is important to capture this information as a constraint for the task distribution algorithm to be able to set an objective to maintain the average ergonomic rating for a specific assembly station below a pre-determined target.

4.   **Tooling constraint:** During the precedence and constraint mapping exercise, it is important to identify and record the specific tooling needed to execute a given task (e.g., overhead lift assist systems). Moving such capital equipment is expensive, so it should be kept at a certain station and included in the optimization algorithm as a constraint.

19.6.4 Challenges with Manual Precedence Mapping

The basic precedence mapping exercise required the identification of enabling predecessors (tasks that need to be done before commencing the successor tasks). In principle, as long as each one of those preceding tasks was completed, the dependent task could be done, as shown in the battery installation example in Figure 19.9.

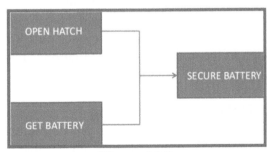

Figure 19.9:
Precedence Graph - Successor is independent of Predecessor Task Sequence

However, in reality there were tasks that needed to happen immediately after a preceding task had taken place. For example, the windshield would need to be assembled immediately after the adhesive was applied. This presents a challenge in terms of capturing the input data for the construction heuristic which is used for task distribution, as the intent is not to treat the preceding tasks as independent tasks. If treated independently, the task distribution process could potentially add several other tasks between the adhesive application and the windshield assembly operation in order to reduce idle time at each Takt. From a process requirement standpoint, this would be unacceptable. A possible solution would be to group such tasks to ensure that they get executed in a sequence which would be pre-determined based on the design or process requirements (Figure 19.10). Such grouping is considered as an Adjacency Set. Another solution is to include time relations between tasks, such that minimum and maximum separation between tasks can be enforced.

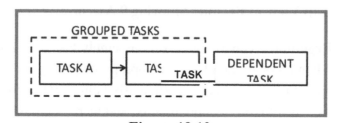

Figure 19.10
Task Grouping Required to Prevent Separate Assignment During Optimization

In summary, all these constraints were mapped for each Takt during Stage 1 of precedence and constraint mapping. Once the precedence relationships and constraints for each Takt were documented in Stage 1, the second stage included mapping the precedence relationships between tasks, across all Takts. It was important to review process sheets that had a detailed listing of every single task and the task time. The hands-on training was very beneficial to visualize the tasks and identify predecessors from stations that were not on adjacent stations.

The comprehensive precedence map which included Stage 1 and Stage 2 data was then created using Microsoft® Visio software to review the precedence relationships in a visual form. The visual representation was beneficial in highlighting circular references or *floating processes* that did not have any predecessors. In some cases, a large cluster of closely linked processes were found. This can occur when several small tasks such as individual wire connections are done at a specific station. As long as they are part of a sub-assembly (example, audio system), there was no additional advantage in trying to split each small task and set the precedence relationships to the upper level task which would in this case be, *Install the audio system*. Instead it was found beneficial to link each of these tasks as an adjacency set and link the very first one to the preceding upper level task such as the audio system installation. This ensured that the small tasks did not get fragmented during the task distribution process.

19.6.5 Precedence Relationship Learning

Many manufacturers hold yearly workshops, where experts re-balance the manufacturing lines. In order to internally charge production times and costs, all feasible plans (also called credited plans) are stored in the manufacturer's databases. This is a very valuable source of information as it implicitly contains the knowledge of planners about precedence relations (and other aspects of assembly conditions).

Recently, it has been a highly discussed topic by well-known car producers, on how to do assembly line balancing using these proven feasible plans. A first promising step in this direction was done by Minzu, et al.(1999) who noticed that the knowledge of all the feasible assembly sequences is equivalent to possessing the precedence graph. Further reflections of these ideas point out that the combination of several assembly sequences very often results in getting information on new feasible sequences. The experimental results show that it is a very effective method which meets the outlined requirements of the automotive industry much better than existing approaches.

Some of the available historical feasible sequences were found during a study performed on a pilot line at a major automotive assembly plant where the electrical harnesses, floor insulation, and curtain head airbags get assembled into the vehicle.

The novel *Precedence Graph Learning* concept is based on the fact that these past feasible sequences can be used to generate a precedence graph in a partially automated manner, as compared to the labor intensive manual process.

19.6.6 Terms and definitions

Precedence constraints can be summarized and visualized in a precedence graph. The precedence graph is an acyclic digraph $G = (V, E)$ that contains a node for each task $j \in V$ and an arc $(i, j) \in E$ for each non-redundant precedence relation which requires that task $i \in V$ is finished before another task $j \in V$ can be started. The task times t_j are allocated as node weights. The learning precedence graph concept is based on three precedence graph types that are distinguished by their *fill level*: *target graph, maximum graph and minimum graph*. As data type for storing these graphs an adjacency matrix is used. For a precedence graph with n tasks the adjacency (witness) matrix is an $(n \times n)$-matrix A, with tasks i and j. Various precedence conditions and the corresponding values of cell a_{ij} and a_{ji} ($\forall i, j \in V$) are shown in Figure 19.11. In the pilot study that we conducted at a major OEM facility, tasks within a station were not specifically sequenced, hence the precedence relationships between tasks within that station were considered as unknown, due to lack of specific evidence. Thus, a_{ij} and a_{ji} will be set to value U (which stands for unknown), as shown in Figure 19.11. Comparison of two feasible sequences generates a *Witness Matrix*. All possible combinations of codes generated while comparing another feasible sequence with existing witness matrix are shown in Figure 19.12. For a precedence graph $G = (V, E)$ with adjacency matrix A, its transitive closure is defined as the precedence graph $G^T = (V, E^T)$ that contains in E^T an arc (i, j) from $i \in V$ to $j \in V - \{i\}$ if there exits at least one path from i to j in G. The corresponding adjacency matrix is called A^T. The precedence graph that correctly describes the real production process is called *target graph G*. This is the graph that we try to approximate by generating the precedence graph through learning from past feasible sequences and by getting expert feedback, as closely as possible. The maximum graph $\overline{G}=(V,\overline{E})$ with its adjacency matrix A is a precedence graph that contains the same nodes (tasks) as the target graph and at least all the precedence constraints of the target graph, i.e., $E^T \subseteq \overline{E}^T$. We propose a simple but very effective method to construct the *maximum graph* $\overline{G}=(V,\overline{E})$ based on a set of $S \geq 1$ precedence-feasible assembly sequences (permutations) $s_1, s_2, \ldots, s_n$ of the n tasks available. All possible combinations of two tasks $i, j \in V$ are examined as follows: if there is one sequence s_n, in which task i is conducted before task j and another sequence $s_m, (n, m \in \{1, \ldots, S\}$ and $n \neq m)$, in which i is executed after j, then i and j are definitely independent as they were already executed in both orders in the past, and a_{ij} and a_{ji} will be set to value I, as shown in Figure 19.12. However, if i is executed before j in each of the sequences $s_1, s_2, ..., s_n$, it is assumed that i is a predecessor of j and a_{ij} will be set to "P" (which stands for potential predecessor which has not yet been contradicted in any past feasible sequence) and a_{ji} will be set to value P, as shown in Figure 19.11.

#	CONDITION	PRECEDENCE	WITNESS MATRIX i to j	WITNESS MATRIX for j to i
1	Tasks i and j belong to the same station in the feasible sequence	Unknown	U	U
2	Task i precedes j in the feasible sequence but not enforced by experts or contradicted	Potential but unproven	P	- P
3	Task i precedes j in one sequence AND j precedes i in another sequence	Proven Independence	I	I
4	Task i MUST ALWAYS precede j (Forced due to Process Physics / Quality)	Forced	Z	- Z
5	Task i MUST PRECEDE j per Expert AND j precedes i in a given feasible sequence	Conflict - Raise Alert!	Z	P

Figure 19.11:
Precedence Relationship Conditions & Variables

		Witness Matrix for Precedence						
		O	U	P	-P	Z	-Z	I
Old proven	O	O	-	-	-	-	-	-
feasible	U	-	U	P	-P	Z	-Z	I
sequence	P	-	P	P	I	Z	Alert	I
	-P	-	-P	I	-P	Alert	-Z	I

Figure 19.12:
Possible Combinations while Comparing Feasible Sequences with a Witness Matrix

Possibly, there are (still unknown) feasible sequences in which j is executed prior to i. Usually, the procedure will be applied in an iterative manner, starting with a single sequence (s_1) which defines the maximum graph as a single chain arranging the nodes according to the given precedence relationships observed. If, in a later iteration, a maximum graph representing s sequences is already present, the graph can be easily updated when a new sequence s_{n+1} becomes known. It is only necessary to examine if the chain representing the new sequence and the present maximum graph show contradictory precedence relations and to delete those relations from the (transitive closure of the) maximum graph. An interesting scenario can occur when a confirmed precedence relationship is marked by an expert, such that $a_{ij} = Z$"and a contradictory situation is found in a following sequence in which $a_{ji} = P$ (which suggests j is a predecessor of i). Such scenarios should be highlighted as an alert for the expert to review and confirm the precedence relationship or independence.

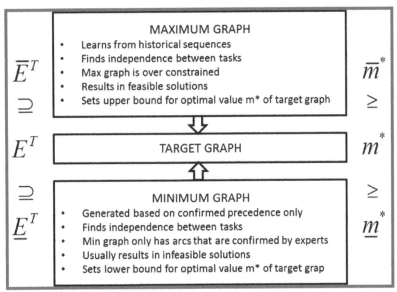

Figure 19.13:
Relationship and Contributions of Max, Min, and Target Graphs

The minimum graph $G = (V, E)$ with its adjacency matrix A is a precedence graph that contains the same nodes and task times as the target graph, but only a subset of precedence restrictions of the target graph, i.e., $\underline{E}^T \subseteq E^T$. Contrary to the maximum graph, all precedence relations described by the minimum graph are valid, whereas the independencies are temporary.
The above relations can be summarized as:

$$\underline{E}^T \subseteq E^T \subseteq \overline{E}^T \qquad (19.3)$$

The minimum graph initially contains no arcs (zero matrix $\underline{A}^T$). In order to get closer to the target graph, the minimum graph is filled with confirmed constraints, e.g., from R&D department or CAD, in a bottom up manner. All the arcs that are part of $\underline{E}^T$ are also part of E^T. However, the minimum graph generally contains too few constraints and therefore resulting sequences might be infeasible or, in other words, the sequences represented by G build a subset of the sequences represented by the minimum graph G. That is, the minimum graph ALBP is a relaxation of the original ALBP. As a result, the optimal solution with objective value $\underline{m}^*$ found via a line balancing procedure for the ALBP instance constructed from the minimum graph serves as a lower bound for the optimal solution value in the target graph. In contrast to the maximum graph, it is necessary to find an optimal solution for the minimum graph to assure that $\underline{m}^*$ is a valid lower bound. To summarize, we get the following relations:

$$\underline{m}^* \leq m^* \leq \overline{m}^* \qquad (19.4)$$

Hence, we can conclude that if the best solution found for the maximum graph and the optimal solution in the minimum graph requires the same number of stations, this value is optimal for the target graph, without knowing this latter graph. In this case, the found maximum graph solution is optimal and also constitutes an optimal solution for the target graph.

$$\underline{m}^* = \overline{m}^* \Rightarrow m^* = \underline{m}^* = \overline{m}^* \qquad (19.5)$$

Depending on the currently known minimum graph, only remaining potential arcs, i.e., task pairs (i, j) with a_{ij} or $a_{ji} = $ "P" or "$-P$" are to be considered for possibly confirming the dependency.

We are only able to examine a subset of all the potential arcs, as experts' time and knowledge are limited and databases only contain information on a part of the relations. For each considered potential pair (i, j), one of the following three situations will occur (Figure 19.14):

1. Task i turns out to be an actual predecessor of task j. Then, arc (i, j) is added to the minimum graph as a confirmed precedence relation by setting a_{ij} = "Z" (which stands for a confirmed precedence).
2. Though always being performed before task j in the yet known practical sequences, task i is identified not to be a predecessor of j which enables us to set a_{ij} and a_{ji} = "I", i.e., to add a confirmed independency to the maximum graph. This strengthens the overall approach.
3. It does not become clear, whether or not there is a precedence relationship between tasks i and j. This has no effect on either graph.

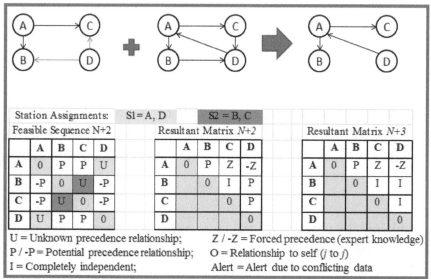

Figure 19.14:
Iterative Evaluation of Past Feasible Sequences

19.6.7 Results of the Pilot Study: Precedence Learning

Twelve past feasible sequences were used to develop a Precedence Graph. Figure 19.15 summarizes the key findings from the pilot study based on a real-world automotive assembly line with 317 tasks distributed across 15 Takts. The objective for this ALB exercise was to maximize Labor Utilization.

$$\text{Labor Utilization} = \frac{\sum_{j \in S_k} t_j}{m \times c} \qquad (19.6)$$

Where:
    j = Task that belongs to a set of tasks S_k
    t_j = Time required per task j
    m = Number of stations
    c = Takt time

Description	Value
Total # of tasks in pilot assembly line	317
Total number of staffed takts (# of operators)	15
Number of precedence relationships identified by manual mapping	293
Number of comparison codes generated per square matrix for each feasible sequence (317 x 317)	100,489
% Labor Utilization using manual precedence relationship data	96%
% Labor Utilization using resultant predence relationships using new learning technique	89%
Annual time required for manual precedence data generation & manual line balancing	1316 hrs.
Annual time required for automated precedence data generation and line balancing (including 16 hours per month for trial runs)	264 hrs.
Annual time required for expert process planner to manually update precedence graph to compensate for 7% reduced utilization	192 hrs.
Annual time saved per assembly line (approx. 300 tasks / 15 takts)	860 hrs.

Figure 19.15
Summary of Results from Pilot Study an e
Existing Automotive Assembly Line

To overcome one of the main gaps between line balancing in theory and real-world application, the application of existing line balancing approaches requires a precedence graph. The authors are aware of no automotive manufacturer possessing a precedence graph describing the entire production process of any of its models. They examined the available approaches for generating precedence graphs and found out that none of the existing methods is suitable for the conditions in the automotive and related industries. The presented learning precedence graph concept takes requirements of manufacturers into account and relies on the analysis of available feasible sequences.

This approach was tested on a real world assembly line with 317 tasks, and considered 12 past feasible sequences. It was shown that a graph constructed by learning from these 12 feasible sequences closely approximated the graph created through manual precedence mapping. In these experiments, the author found only a 7% deviation from an assembly line utilization metric which was computed by a construction heuristic using the precedence data generated by both approaches. This gap from optimal utilization can be bridged manually by process experts.

19.7 Sequencing

The primary purpose of Mixed Model Final Assembly is to manufacture products on an *on-demand* basis, but most importantly have the ability to do so with a lot size of one. As seen earlier in this Chapter, several unique combinations of vehicles can be manufactured from a vast variety of option content that is available in automobiles these days. However, despite any succession of models being technically feasible, the actual sequence influences various economic targets. Two different general objectives are impacted by the model sequence:

19.7.1 Work Overload: As explained earlier, diverging processing times of the respective models that are produced on the MMFA are averaged. In order to improve capacity utilization, the cycle time is determined such that it is observed on average over all models. As a result, some models may have a longer processing time than the cycle time and some may be shorter. Whenever multiple *option-heavy* models follow each other in direct succession, a work overload situation occurs, which means the operator is not able to complete all the tasks prior to reaching the border of the down-stream station. Such overload would typically result in a line stoppage and potentially quality defects. Work overloads can be avoided if a sequence of models is identified, where those models which are less labor-intensive alternate with the ones which are highly labor-intensive. To avoid such overloads, the mixed-model sequencing problem was introduced. Within this sequencing approach, a detailed scheduling is applied, which explicitly takes operation times, station definition, operator walk and other operational details of the MMFA line into account. This allows the work overload to be quantified exactly and, thus, be minimized.

19.7.2 Just-in-Time objectives: Different models require different options so the model sequence influences the progression of material demands over time. A Mixed-Model Final Assembly line is supported by a strong local JIT network of suppliers. For example, a truck carrying seats in the same exact sequence as the vehicles being assembled would arrive right next to the assembly station. A fork-lift would unload the rack of seats and place it directly next to the MMFA line for the operator to install it into the vehicle using a lift-assist system. An important requirement for JIT supply as stated in the literature is a consistent demand rate of raw material over time, as otherwise the advantages of JIT are lost due to the increased inventory required to avoid starving the assembly line.

These two basic objectives (minimizing work overload and leveling part usage) have been undertaken in three alternative sequencing approaches discussed in literature:

a. **Mixed Model sequencing:** This approach aims at minimizing sequence-dependent work overload based on a detailed scheduling which takes operation times, station definition, operator walk and other operational details of the MMFA line into account.

b. **Car sequencing:** This approach attempts to minimize sequence-dependent work overload in an implicit manner, to avoid the data collection effort associated with mixed-model sequencing.

c. **Level scheduling:** While the first two approaches aim at minimizing violations of capacity constraints, level scheduling seeks to find sequences which support JIT supply of raw material and component sub-assemblies to the line. In this approach, "ideal" production rates are defined and models are sequenced in such a manner that deviations between ideal and actual rates are minimized.

The rest of this Chapter describes the basic mathematical formulation of each of the above stated approaches along with their respective constraint definitions. Based on literature review, following are the basic assumptions that apply to all three approaches:

The assembly line is paced and consists of multiple stations arranged along a transportation system, typically a conveyor, which continuously moves the vehicles from station to station. The constant velocity of the conveyor belt is normalized to one.

a. Operators conduct a set of specified tasks during each cycle on the vehicle and move along with the vehicle on the conveyor. As soon as the work is done, the operators move back to the beginning of their station or until they reach the next vehicle.

b. Buffers do not exist between stations.

c. The model-mix is fixed and not subject to sudden changes.

d. Multiple models contain different options (components) and require respective set of tasks with individual processing times. Thus demands for material and the utilization of the station capacities change from model to model.

e. No interruptions in material supply or production is considered.

f. Working across station boundaries is not allowed (i.e. the station is considered closed).

g. The operators return with infinite velocity to the next work piece. This assumption is considered adequate whenever the conveyor speed is much slower than the walking speed of operators.

h. Work overload is measured by the time or space necessary to complete all work in excess of respective station's right border.

i. Work overload has no impact on the succeeding station. The model assumes that the work overload is either compensated by a utility worker or by increasing processing speed.

19.8 Mixed-Model Sequencing

The following table shows the notation used in the Mixed-model sequencing model (Figure 19.16).

M	Set of models with $m \in M$
T	Number of production cycles with $t=1,...,T$
K	Number of stations with $k=1,...,K$
O	Set of options with $o \in O$
d_m	Demand for copies of model m
a_{mp}	Demand for model m for part p
a_{mo}	1, if model m contains option o and 0, otherwise
p_{mk}	Processing time of model m at station k
c	Cycle time
l_k	Length of station
r_p	Target consumption rate of part p
r_m	Target production rate of model m
$H_o:N_o$	At most H_O out of N_O successively sequenced copies may require option O.
BI	Big integer
x_{mt}	1, if model m is produced in cycle t; 0, otherwise
w_{kt}	Work overload occurring at station k with the tth unit is processed
s_{kt}	Position of operator in station k when cycle t begins
y_{mt}	Total cumulative production quantity of model m up to cycle t
z_{ot}	1, if a rule with respect to option o is violated in cycle t; 0, otherwise

Figure 19.16
Notation for Sequencing models

All the studied mixed-model sequencing models share the following characteristics:

1. The planning period is divided into T production cycles with $t = 1, ... T$).
2. For each model $m \in M$, the demand d_m at the end of the planning horizon is given and has to be met. It follows that the sum over model demands is equal to the number of production cycles available, $\sum_{m \in M} d_m = T$.
3. The assignment decision is represented by binary variables x_{mt}, which indicate whether a copy of model m is produced in cycle t ($x_{mt}=1$) or not ($x_{mt}=0$):

$$x_{mt} \in \{0,1\} \quad \forall m \in M; \quad t = 1,...,T. \qquad (19.7)$$

In each production cycle t exactly one model m is produced:

$$\sum_{m \in M} x_{mt} = 1; \quad \forall t = 1,...,T. \qquad (19.8)$$

Production must satisfy the demand for model copies d_m

$$\sum_{m \in M} d_m = T$$

Where: $\sum_{t=1}^{T} x_{mt} = dm, \qquad \forall m \in M \qquad (19.9)$

Based on these characteristics and the assumptions stated above, the objective function and constraints are listed below. The mixed model sequencing problem has been formulated in a similar manner by Yano and Rachamadugu (1991) and Bard et al.(1994).

$$\text{Objective: Minimize } Z(X,S,W) = \sum_{k=1}^{K}\sum_{t=1}^{T} w_{kt} \qquad (19.10)$$

Subject to: Constraints 20.7, 20.8, 20.9 and:

$$s_{k,t+1} \geq s_{kt} + \sum_{m \in M} p_{mk}.x_{mt} - w_{kt} - c \qquad \forall k = 1,...,K; \ 1,...,T \ (19.11)$$

$$s_{kt} \sum_{m \in M} p_{mk}.x_{mt} - w_{kt} \leq l_k \qquad \forall k = 1,...,K; \ 1,...,T \qquad (19.12)$$

$$s_{kt} \geq 0, \quad w_{kt} \geq 0 \qquad \forall k = 1,...,K; \quad 1,...,T \qquad (19.13)$$

$$s_{kt} = 0, \quad s_{k,T+1} = 0 \qquad \forall k = 1,...,K \qquad (19.14)$$

The Objective function minimizes the total work overload. Constraint Equations 19.11 ensure that processing of a model copy in cycle $t + 1$ by station k cannot start before the station has completed the preceding unit in cycle t. Work is restricted to the stations' border by Equations 19.12 and the non-negativity constraints for s_{kt} in Equation 19.13. Constraint set 19.14 ensure that the line is in an initial state prior to and at the end of the planning horizon.

Bard et al.(1994) have introduced a preliminary framework for characterizing mixed-model sequencing problems, which is, however limited to a small subset of operational characteristics discussed in the literature. Any mixed-model sequencing problem will at least consist of three basic elements: Operational characteristics of the stations, characteristics of the assembly line as a whole and an objective to be followed. Accordingly, the classification is based on these three elements, noted as a tuple $[\alpha, \beta, \gamma]$, where α refers to the characteristics of stations, β refers to the characteristics of the assembly line, and γ refers to the objectives.

19.8.1 Sequencing

Instead of a detailed scheduling of work content, car scheduling considers the succession of work intensive product options like sunroof installation, navigation system installation, etc. in order to avoid work overload. A set of product options can be subject to sequencing rules, which restrict the maximum number of occurrences within a subsequence of a certain length. The car sequencing problem then seeks to find a sequence of models which meet the required demands for each model without violating the given sequencing rules. Car sequencing originally stems from the practical applications in the automobile industry and was first formulated by Parrello et al (1986) Their approach can also be applied to mixed-model assembly problems in other industries.

The sequencing rules are typically of type $H_o : N_o$, which means that out of N_o successive models only H_o ma contain the option o in order to avoid work overload. Such rules are derived from considering the capacity situation at the stations expressed by Drexl and Kimms (2001) as follows:

Assume that 60% of the cars manufactured on the line need the option "sunroof". Moreover, assume that five cars (copies) pass the station where the sunroofs are installed during the time for the installation of a single copy. Then, three operators (installation teams) are necessary for the installation of sunroofs. Hence, the capacity constraint of the final assembly line for the option "sunroof" is three out of five in a sequence, or 3:5 for short.

Thus, the car sequencing problem can be formulated as a constraint satisfaction problem, based on Constraints, 19.7, 19.8, 19.9, and 19.15.

$$\sum_{t'=t}^{t+N_o-1} \sum_{m \in M} a_{mo}.x_{mt'} \leq H_o \quad \forall o \in O; \quad t = 1,...,T - N_o + 1 \tag{19.15}$$

Additionally, recent publications investigate the applicability of combinatorial optimization techniques to solve instances of car sequencing. In these cases, the constraint satisfaction problem has to be transformed into an optimization problem. Such an optimization problem has the advantage of finding a model sequence which minimizes violated rules whenever a solution without violations is not existent. Boysen and Fliedner (2007) proposed an optimization model based on binary variables z_{ot} and Objective Function 19.16:

Objective function:

$$\text{Minimize } Z(X,Y) = \sum_{o \in O} \sum_{t=1}^{T} z_{ot} \tag{19.16}$$

Subject to Constraints 19.7, 19.8, 19.9 and 19.17:

$$\sum_{m \in M} \sum_{t'=t}^{t+N_o-1} a_{mo}.x_{mt'} - \left(1 - \sum a_{mo}.x_{mt}\right).BI \leq H_o + BI.z_{ot} \quad \forall o \in O; \quad t = 1,...,T \tag{19.17}$$

19.8.2 Level Scheduling

The third and final problem type in Mixed-Model Sequencing research is Level Scheduling. Level Scheduling has recently received significant focus in research and practical applications. This approach aims at evenly smoothing the material requirements induced by the production sequence over time, so that a just-in-time supply of material is facilitated and safety stocks are minimized. For that purpose, each material receives a (theoretical) target consumption rate, which is determined by distributing its overall demand evenly over the planning horizon. Hence, a sequence is sought where actual consumption rates of materials are as close as possible to target rates.

Consider a set M of models each of which consist of different parts ; p with $p \in P$). The production coefficient a_{mp} specifies the number of units of material p needed in the assembly of one unit of model m. The target consumption rate r_m per production cycle is then calculated as follows:

$$r_p = \frac{\sum_{m \in M} a_{mp}.d_m}{T} \quad \forall p \in P \tag{19.18}$$

Together with the integer variables y_{mt}, which represent the total cumulative production quantity of model m up to a cycle t, the part-oriented level scheduling LS^P can be modeled as follows:

$$LS^P : \text{Minimize} \quad Z(X,Y) = \sum_{t=1}^{T} \sum_{p \in P} \left(\sum_{m \in M} a_{mp} \cdot y_{mt} - t \cdot r_p \right)^2 \qquad (19.19)$$

Subject to:

$$y_{mt} = \sum_{t'=1}^{t} x_{mt'} \quad \forall t = 1, \dots, T \qquad (19.20)$$

This objective function aims at minimizing the sum over all deviations of actual from ideal cumulative demands per production cycle t and part p.

In real-world applications, where products may consist of thousands of different parts, the resulting problem instances of the above objective function are barely solvable to optimality. Accordingly, researchers propose a class of simplified approximate models, which, under specific prerequisites, are claimed to be sufficient to level part usages without explicitly considering the materials contained in those products. The objective of these model-oriented level scheduling problems is to achieve a constant production rate rm for each model m: $r_m = d_m / T \forall m \in M$. Thus, the objective function given in Equation 19.19, is replaced by the following objective function:

$$LS^M : \text{Minimize} \quad Z(X,Y) = \sum_{t=1}^{t} \sum_{m \in M} \left(y_{mt} - t \cdot r_m \right)^2 \qquad (19.21)$$

Subject to 20.7, 20.8, 20.9, and 20.20

19.8.3 Conclusion

This section of the Chapter gives a brief overview of the three major approaches for sequencing mixed-model assembly lines. Based on the literature review and practical experience with Mixed-Model Final Assembly of automobiles, the author finds an apparent lack of empirical research evaluating the goodness of fit of the mathematical models developed by academic researchers in real-world applications.

19.9 Summary

This Chapter gives a brief overview of recent advances in dealing with Mixed-Model Final Assembly systems, which are key enabling systems that make the Lean manufacturing system seamless and versatile. A review of the analytical problems associated with Assembly Line Balancing and Sequencing of products in MMFA was also presented. Manufacturers in a wide range of industries face the challenge of providing a high level of product customization at a very low cost while maintaining a lean supply chain of components and finished goods. This requires the implementation of efficient Mixed-Model Final Assembly systems. MMFA offers a rich area for academic researchers to work closely with industrial partners to bridge the gaps between academic research and real-world requirements for planning and implementation of this unique manufacturing system.

Thoughts
  and Things.......

Bibliography

Albin, S. Approximating a Point Process to a Renewal Process: Superposition Arrival Processes to Queues. *Operations Research, Vol. 30* , 1133-1162.

Allen, A. (1990). *Probability and Queuing Theory* (Second Edition ed.). Academic Press.

Ammer, E. D., 1985, "Rechnerunterstützte Planung Von Montageablaufstrukturen Für Erzeugnisse Der Serienfertigung." IPA-IAO Forschung Und Praxis, **81**.

Askin, R., & Strandridge, C. (1993). *Modeling and Analysis of Manufacturing Systems.* John Wiley & Sons, Inc.

Bard, J. F., Shtub, A., and Joshi, S. B., 1994, "Sequencing Mixed-Model Assembly Lines to Level Parts Usage and Minimize Line Length," International Journal of Production Research, **32**(10) pp. 2431.

Bartholdi, J., & Eisenstein. (n.d.). Retrieved from http://www2.isye.gatech.edu/-jjb/bucket-brigades.html

Baudin, M. (2007). *Working with Machines.* Productivity Press.

Black, J. (1983). Cellular Manufacturing Systems Reduce Setup Time, Make Small Lot Production Economical. *Industrial Engineering* , 36-48.

Black, J. (1991). *The Design of the Factory with A Future.* McGraw-Hill.

Black, J., & Chen, J. (1994). Decoupler-improved Output of an Apparel Assembly Cell, *Journal of Appl. Manufacturing System* , 47-58.

Black, J., & Hunter, S. *Lean Manufacturing Systems and Cell Design.* Dearborn, MI: Society of Manufacturing Engineers, ISBN 2003-102-33-5.

Black, J., & Kohser, R. A. (2008). *DeGarmo's Materials & Processes in Manufacturing, 10th edition.* Wiley.

Black, J., & Schroer, B. (1988). Decouplers in Integrated Cellular Manufacturing Systems, "Jo. *Journal of Engineering for Industry, Transactions ASME, Vol. 110* , 77-85.

Black, J., & Schroer, B. (1993). Simulation of an Apparel Assembly Cell with Walking Workers and Decouplers. *Journal of Manufacturing Systems, Vol 12, No. 2* , 170-180.

Black, J., & Sipper, D. (1991). *Decouplers for Integrated Pull Manufacturing.* ASME-PED.

Black, J., Jiang, B., & Wiens, G. (1991). Design, Analysis and Control of Manufacturing Cells, PED-Vol.53. *The American Society of Mechanical Engineers* .

Blanchard, B., System Engineering Management, Fourth Edition, John Wiley and Sons, 2008.

Bourjault, A., 1984, "Contribution a Une Approche Méthodologique De L'assemblage Automatisé: Elaboration Automatique Des Séquences Opértiores." Universite De Franche-Comte, .

Boysen, N., Fliedner, M., and Scholl, A., 2007, "A Classification of Assembly Line Balancing Problems," European Journal of Operational Research, **183**(2) pp. 674-693.

Braha, D., Minai, A., & Bar-Yam, Y. (2006). *Complex Engineered Systems.* Springer.

Buffa, E., & Sarin, R. (1987). *Modern Production/ Operations Management, ISBN 0-471-81905-0.* John Wiley & Sons.

Burbridge, J. (1975). *The Introduction of Group Technology.* John Wiley & Sons.

Burgridge, J. *Production Flow Analysis for PLanning Group Technology.* Oxford, England: Clarendon Press.

Burke, P. (n.d.). The Output Process of a Stationary M/M/S Queuing System. *Annals of Math. Stat. Vol.39* , 1144-1152.

Buzacott, J., & Shanthikumar, J. (1963). *Stochastic Models of Manufacturing Systems.* Prentice-Hall.

Chen, G., Zhou, J., Cai, W., 2006, "A Framework for an Automotive Body Assembly Process Design System," Computer Aided Design, **38**pp. 531-539.

Cochran, David S. "Detaching from Management Accounting Requires System Design," Journal of Cost Management, March/April 2006, pp. 20-28.

Cochran, D. S. and Barnes, J. "Enhancement of the Systems Engineering Process in the Life Cycle with Axiomatic Design," Proceedings of ICAD2013, The Seventh International Conference on Axiomatic Design, WPI, Worcester, MA, June 2013. ICAD-2013-26.

Cochran, D. S., & Dobbs, D. C. (2001/2002). Evaluating Manufacturing System Design and Performance Using the Manufacturing System Design Decomposition Approach. *Journal of Manufacturing Systems, Vol. 20 No. 6* .

Cochran, D. S., Arinez, J. F., Duda, J. W., & Linck, J. (2001/2002). A decomposition Approach for Manufacturing System Design. *Journal of Manufacturing Systems Vol 20 No. 6* .

Cochran, D. S., Eversheim, W., Kubic, G., & Sesterhenn, M. L. (2000). The Application of Axiomatic Design and Lean Management Principles in the Scope of Production System Segmentation. *International Journal of Production Research Vol. 38, No. 6* , 1377-1396.

Cochran, D., Duda, J., Linck, J., & Arinez, J. (2001/2002). The Manufacturing System Design Decomposition. *SME Journal of Manufacturing Systems , 20* (6).

Cochran, D., Linck, J., & Neise, P. (2001). Evaluation of the Plant Designs of Two Automotive Suppliers Using the Manufacturing System Design Decomposition (MSDD). *Transactions of NAMRI/SME .*

Cochran, D. S., Linck, J., Reinhart, G., and Mauderer, M., "Decision Support For Manufacturing System Design (MSD): Combining A Decomposition Methodology With Procedural MSD," Proceedings of The Third World Congress on Intelligent Manufacturing Processes and Systems, MIT, Cambridge, MA, June 2000, pp. 9-16.

Cochran, D. S. and edited by Pincham, W. H., "Enterprise Engineering, Creating Sustainable Systems with Collective System Design – Part I," The Journal of RMS (Reliability, Maintainability, Supportability) in Systems Engineering, Winter Journal, 2009.

Crosby, P. *Quality if Free.* New York, NY: McGraw-Hill Inc.

Curry, G., & Feldman, R. (2009). *Manufacturing Systems Modeling and Analysis, ISBN-978-3-540-88762-1.* Springer-Verlag.

Curry, G., & Phillips, D. (May 10-12, 2000). Renewal Approximations of the Departure SCV's for Batch Arrivals and Service Processes'. *International Conference on Modeling & Analysis of Semiconductor Manufacturing.* Tempe, Arizona.

Dar-El, E. M. (Dec 1973). MALB-A Heuristic Technique for Balancing Large Single-Model Assembly Lines. *IIE Transactions .*

De Fazio, T. L., and Whitney, D. E., 1987, "Simplified Generation of all Mechanical Assembly Sequences," IEEE Journal of Robotics and Automation, 4, pp. 640-658.

Dennis, P. (2002). *Lean Production simplified.* Productivity Press.

Drexl, A., and Kimms, A., 2001, "Sequencing JIT Mixed-Model Assembly Lines Under Station-Load and Part-Usage Constraints," Management Science, 47(3) pp. 480-491.

Feldman, R., & Valdez-Flores, C. (1996). *Applied Probability and Stochastic Processes.* Boston: PWS.

Feldman, R., & Valdez-Florez, C. (2010). *Applied Probability and Stochastic Processes* (Second Edition ed.). Springer-Verlag.

Gaither, N. (1980). *Production and Operations Management, ISBN-0-03-074622-1.* Orlando, FL: Dryden HBJ.

Goldratt, E. (1992). *The Goal.* Croton-On Hudson, New York, N.Y.: North River Press, Inc.

Goldratt, E. (1990). *Theory of Constraints.* Croton-On Hudson, New York, N.Y.: North River Press, Inc.

Groover, M. (2001). *Automation, Production Systems and Computer Integrated Manufacturing.* Prentice-Hall.

Gross, D., & Harris, C. (1997). *Fundamentals of Queuing Theory* (Third Edition ed.). Wiley Interscience.

Gross, D., & Harris, C. (1998). *Fundamentals of Queuing Theory.* John-Wiley and sons, Inc.

Gryna, F. M., Chua, R. C., & Defeo, J. A. (2007). *Juran's Quality Planning & Analysis for Enterprise Quality, Fifth Edition.* The McGraw-Hill.

Hall, R. (1991). *Queuing Methods for Services and Manufacturing.* Prentice-Hall, Inc.

Hall, R. W. (1982). *Kawasaki U.S.A: A case Study.* Alexandria, VA: American Production Inventory Society.

Hall, R. (1983). *Zero Inventories.* Home Wood, IL: Dow Jones-Irwin.

Harmon, R. L., & Peterson, L. D. (1990). *Reinventing the Factory, Productivity Breakthroughs b Nabyfacturing Today.* Arthur Anderson & Co.

Hendryx, S. (July/Aug 1990). Manufacturing Execution: Theory and Concepts. *AT&T Technical Journal* .

Hino, S. (2006). *Inside the Mind of Toyota.* New York: Productivity Press.

Hirano, H., & Black, J. (1988). *JIT Factory Revolution.* Productivity Press.

Homen de Mello, L. S., and Sanderson, A. C., 1990, "AND/OR Graph Representation of Assembly Plans," Robotics and Automation, IEEE Transactions On, 6(2) pp. 188-199.

Hopp, W., & Spearman, M. *Factory Physics, ISBN 0-256-24795-1.* Irvin: McGraw Hill. http://en.wikipedia.org/wiki/The_purpose_of_a_system_is_what_it_does

Hunter, S. L. (2001/2002). Ergonomic Evaluation of Manufacturing System Designs, vol. 20, No. 6. *Journal of Manufacturing Systems* .

Hunter, S., & Black, J. (2007). Lean Remanufacturing: A cellular Case Study. *Journal of Advanced Manufacturing Systems, Volume 6, Number 2, World Scientific* .

Jackson, J. (1957). Networks of Waiting Lines. *Operations Research, Vol. 5* , 518-521.

Jensen, P., & Bard, J. (2003). *Operations Research: Models and Methods ISBN 0-471-38004-0.* John Wiley and Sons, Inc.

Johnson, H. T. and Kaplan, R., <u>Relevance Lost</u>, Harvard Business School Press, 1987.

Johnson, H. T., <u>Relevance Regained</u>, The Free Press, Division of Simon & Schuster, 1992.

Johnson, H. T., and Bröms, A., <u>Profit Beyond Measure</u>, The Free Press, 2001.

Kingman, J. (1962). On Queues in Heavy Traffic. *Journal of Royal Statistics Society, Vol. 32* , 102-110.

Kreafle, K., TMMK Quality Manager, personal conversation, Georgetown, KY.

Lenz, R. K., and Cochran, D. S., "The Application of Axiomatic Design to the Design of the Product Development Organization," Proceedings of ICAD2000, First International Conference on Axiomatic Design, Cambridge, MA, June 2000, pp. 18-25.

Liker, J. K. (2004). *Becoming Lean.* Productivity Press.

Liker, J. K. (2004). *The Toyota Way - 14 Management Principles frm the world's greatest Manufacturer.*

Little, J. A Proof for the Queuing Formula L=λW,. *Operations Research , 9*, 383-387.

Lockton, D., "The Purpose of a System is what it Does (POSIWID) and determinism in design for behavior change," Brunel University, accessed working paper, available at http://danlockton.co.uk, 2012.

Marshall, K. (n.d.). Some Inequalities in Queuing. *Operations Research, Vol. 30* , 651-655.

Martinich, J. (1997). *Production and Operations Management, ISBN 0-471-54632-1.* John Wiley & Sons, Inc

Minzu, V., Bratcu, A., and Henrioud, J. M., 1999, "Construction of the precedence graphs equivalent to a given set of assembly sequences. ," Proceedings of the 1999 IEEE International Symposium on Assembly and Task Planning, Anonymous pp. 14-19.

Monden, Y. (2001/2002). *Toyota Production System.* Industrial Engineering and Management Press, IIE.

Monden, Y. (1983). *Toyota Production System, An Integrated Approach to Just-In Time.* Institute of Industrial Engineers, second edition.

Montgomery, D. C. (2001). *Introduction to Statistical Quality Control.* Wiley.

Morgan, J. M., & Liker, J. K. (2006). *The Toyota Product Development System.* Productivity Press.

Morgan, J. M., & Liker, J. K. (2006). *The Toyota Product Development System.* Productivity Press.

Nahmias, S. (2005). *Production and Operations Analysis, ISBN 0-07-286538-7* (Fifth Edition ed.). McGraw Hill, Inc.

Nakajima, S. (1990). *Introduction to TPM.* Cambridge, MA: Productivity Press, ISBN 0-915229-23-2.

Nicholas, J. (1998). *Competitive Manufacturing Management, ISBN 0-256-21727-0.* Irvin: MCGraw Hill.

Ohno, T. (1988). *The Toyota Production System.* Productivity Press, ISBN 0-915229-14-3.

Ohno, T. (1988). *Toyota Production System: Beyond Large-Scale Production.* Productivity Press.

Ott, E. R., & Schilling, E. G. (1990). *Process Quality Control, Second Edition.* Cambridge, MA: McGraw-Hill.

Parello, B. D., and Waldo C. K., Job shop Scheduling Using Automated Reasoning: A Case Study of the Car sequencing Problem, Journal of automated Reasoning, 2, 1986, pp. 1-42

Phillips, D., Ravindran, A., & Solberg, J. (1976). *Operations Research: Principles and Practice, ISBN 0-471-68707-3.* John-Wiley and Sons, Inc.

Pound, S., & Spearman, M. (Sept 5, 2007). The Great Push vs. Pull Diversion. *Industry Week.* Pyke, D., & Cohen, M. (1990). Push and Pull in Manufacturing and Distribution Systems. *Journal of Operations Management, Vol. 1*, 24-42.

Pyzdek, T. (2003). *Six Sigma Handbook.* McGraw-Hill.

Romney, B., Godard, C., Goldwasser, M., 1995, "An efficient system for geometric assembly sequence generation and evaluation. ," Proceedings of the ASME International Computers in Engineering Conference, Boston, Massachusetts, pp. 699-712.

Ross, P. J. (1988). *Taguchi Techniques for Quality Engineering.* McGraw-Hill.

Rother, M., & Shook, J., *Learning to See.* Lean Enterprise Institute.

Rother, M. and Shook, J., "Value-Stream Mapping Workshop," Lean Enterprise Institute, Version 1.3, September 2009.

Rother, M., "Toyota Kata – Time to Retire the PDCA Wedge," Presentation, Northeast Shingo Conference, Springfield, MA, 2011.

Ruffa, S. A., & Perozziella, M. J. (2000). *Breaking the Cost Barrier.* John Wiley & Sons, Inc.

Ruiz, M., The Four Agreements, Amber-Allen Publishing, 1997.

Sakasegawa, H. (n.d.). An approximation Formula for M/G/1 Queues. *Annals of the institute of Statistical Mathematics, Vol. 29* , 67-75.

Sakasegawa, H. (1977). An approximation Formula L_(q= 〚αβ〛 ^ρ (1-ρ)). *Annals of the Institute of Statistical Mathematics Vol. 29* , 67-75.

Santos, J., Wysk, R. A., & Torres, J. M. (2006). *Improving Production with Lean Thinking.* Wiley.

Scholl, A., and Becker, C., 2006, "State-of-the-Art Exact and Heuristic Solution Procedures for Simple Assembly Line Balancing," European Journal of Operational Research, **168**(3) pp. 666-693.

Schonberger, R. J. (1982). *Japanese Manufacturing Techniques: Nine Hidden lessons in Simplicity.* The Free Press.

Schonberger, R. J. (1986). *World Class Manufacturing: The Lessons of Simplicity Applied.* The Free Press.

Schonberger, R. J. (1996). *World Class Manufacturing: The Next Decade.* The Free Press.

Schonberger, R. (1986). *World Class Manufacturing.* New York: The Free Press.

Schroer, B., & Black, J. (1993). Simulation of an Apparel Assembly Cell with Walking Workers and Decouplers. *J. Manufacturing System* , 170-180.

Sekine, K. (1990). *One-Piece Flow: Cell Design for Transforming the Production Process.* Productivity Press.

Sekine, K., & Arai, K. (1992). *Kaizen for Quick Changeover.* Productivity Press.

Shingo, S. (1985). *A Revolution in Manufacturing: The SMED System.* Cambridge, MA: Productivity Press.

Shingo, S. (1989). *A study of The Toyota Production System.* Productivity Press.

Shingo, S. (1988). *Non-Stock Production - The Shingo System for Continuous Improvement.* Productivity Press, Inc.

Shingo, S. (1986). *Zero Quality Control: Source Inspection and the Poka-Yoke System.* Productivity Press.

Shingo, S., A Study of the Toyota Production System From an Industrial Engineering Viewpoint, Productivity Press, 1989.

Shingo, S., & Translated By Dillon, A. P. (1987). *The sayings of Shigeo Shingo, Key Strategies for Plant Improvement.* Productivity Press, Inc.

Sipper, D., & Buffin, R. (1997). *Production Planning Control and Integration.* McGraw-Hill, Inc.

Skinner, W. (May/June 1974). The Focused Factory. *Harvard Business Review*, 113-121.

Sly, D. (n.d.). Retrieved from www.proplanner.com

Spear, S. and Bowen, K., Decoding The DNA of the Toyota Production System, Harvard Business Review, September – October 1999.

Spearman, M. (n.d.). Retrieved from www.factoryphysics.com

Sugimori, Y., kusunoki, K., Cho, F., & Uchikawa, S. (1977). Toyota Production System and Kanban System: Materialization of Just-in-Time and Respect-for-Human System. *International Journal of Production Research*, 553-564.

Suh, N. P. (1992). Design Axioms and Quality Control. *Robotics and CIM, Vol. 9 Number 4/A, Aug-Oct*, 367.

Suh, N. P., Cochran, D. S., & Lima, P. C. (1990). "Manufacturing System Design" CIRP Annals Manufacturing Technology. *Paris, France: Institute for Production Research. Vol. 47, No. 2*, 627-639.

Suh, N. P. (1990). *The principles of Design.* New York, NY: Oxford University Press.

Suh, N. P. Axiomatic Design: Advances and Applications, Oxford University Press, 2001.

Su, Q., 2007, "Applying Case-Based Reasoning in Assembly Sequence Planning," International Journal of Production Research, **45** pp. 29-47.

Suzaki, K. (1987). *The New Manufacturing Challenge.* The Free Press.

Swaminathan, A., and Barber, S., 1996, "An experience-based assembly sequence planner for mechanical assemblies," IEEE Transactions on Robotics and Automation, Anonymous **12,** pp. 252-267.

Taha, H. A. (2003). *Operations Research, ISBN 0-13-032374-8* (Seventh Edition ed.). Prentice Hall, Inc.

Thomopoulos, N. (Oct 1967). Line Balancing - Sequencing for Mixed Model Assembly. *Management Science*.

Whitt, W. (1983). A Queuing Network Analyzer. *Bell Systems Technical Journal Vol. 62 No. 9* , 2779-2815.

Whitt, W. (1984). Departures from a Queue with Many Servers. *Math of Operations Research Vol 9* .

Whitt, W. (1984). open and Closed Models for Networks of Queues. *AT&T Bell Laboratories Technical Journal, Vol. 63, No.9* , 1911-1979.

Whitt, W. (1994). Towards Better Multi-Class Parametric Decomposition Algorithms for Open Queuing Networks. *Annals of Operations Research, Vol. 48* , 221-248.

Winston, W. *Operations Research: Applications and Algorithms.* Thompson, Brooks and Cole, ISBN 0-534-38058-1.

Womack, J. P. (2005). *Lean Solutions.* New York, NY: Simon & Schuster.

Womack, J. P. (2006). *Value Stream Mapping.* Manufacturing Engineering, SME.

Womack, J. P., Jones, D. T., & Roos, D. (1991). *The Machine That Changed the world.* Harper Perennial.

Womack, J., & Jomes, D. (1996). *Lean Thinking.* New York: Simon & Schuster.

Yano, C. A. and Rachamadugu, R., 1991, Sequencing to Minimize Work Overload in Assembly Lines with Product Options, Management Science, **37**(5) pp. 572-586.

*Thoughts
and Things.......*

Index

A

B

C

D

E

V

value stream mapping (VSM), 257
variance, 5, 24, 26, 29, 35, 41, 57, 79, 82, 95, 97, 98, 100, 105, 115, 122, 126, 127, 131, 132, 133, 165, 286, 293, 294, 297, 298, 301, 302, 303, 304, 309, 310, 311, 312, 314, 317, 322, 325, 331, 332, 343, 345, 381, 382, 383, 384, 386, 387, 388, 389, 390, 391, 398, 402, 407, 413, 418, 420, 427, 429, 431, 434, 435, 436, 437, 441, 448, 454, 456, 460, 461, 462, 464, 465, 474, 511

W

waste elimination, 6, 26, 30, 82, 83, 566
withdrawal Kanban, 480, 483, 499
work holding devices, 66, 72, 149
work in process, 22, 34, 55, 313, 321, 497
workholding devices, 511
workstation, 2, 14, 24, 27, 49, 50, 53, 78, 79, 101, 141, 147, 148, 149, 154, 155, 156, 158, 159, 160, 162, 165, 169, 171, 182, 191, 204, 282, 289, 290, 291, 297, 298, 304, 313, 315, 325, 326, 331, 333, 335, 345, 347, 383, 384, 402, 421, 422, 423, 424, 428, 429, 431,458, 470, 477, 480, 481, 498, 499

X

x-bar chart, 436, 437, 448, 450, 456

Z

zero defects, 20, 26, 42, 44, 52, 70, 78, 82, 85, 99, 132, 172, 203, 251, 278, 283, 372, 373, 433, 435, 438, 457, 468, 473, 487
Zero downtime, 517

Thoughts
and Things.......

Glossary of Important Lean Concepts and Terms

Andon: A visual control device that indicates the status of a machine, line, or process. Frequently, audible alarms or warning messages accompany Andon status lights as a secondary method of communicating a problem has arisen.

Autonomation: Autonomation refers to the autonomous control of both defects and machine shutdowns See also Jidoka.

Axiomatic Design Principles: Used in the design of manufacturing cells and systems. It involves determining the customer needs and transforming them into functional requirements, design parameters, and process variables.

Balancing: Assigning work to each workstation such that every workstation and worker works the same amount of time. Balance is dictated by task assignments and Takt time.

Benchmarking: The process of establishing progress or comparison, particularly competitive products and competing methodologies.

Cause and Effect Diagram (a.k.a. fishbone diagram): A method of defining an occurrence of an undesirable event or problem. The effect is the fish head. Contributing factors, or causes, are the fish bones attached to a backbone and the fish head.

Cell Balancing: Assigning workers and work to Lean cells such that every cell is working to the same drumbeat or Takt time requirement.

Changeover: Process of converting a line or machine from running one product to another.

Continuous Process Improvement (CPI): Continuously improving and updating a process.

Control Chart: One of the primary techniques of statistical process control. It plots sample statistics (measurements) of a quality characteristic for the process versus time (or the sample number).

Decouplers: Decouplers physically separate (or decouple) one station or machine from another. They are placed between two workstations or sub-cells to improve capability and flexibility in a cell. Decouplers also provide the JIT pull cell with *Make One, Check one, and Move One On* capability. Decouplers can have multiple functions including part transportation and inspection.

Defect: The undesirable result of an error in a process, and one of many types of waste in a system. Defects are often expressed as either yield of good parts or as Defects per Million Opportunities (DPMO).

Deming Cycle: The Deming Cycle, also known as the Plan-Do-Check-Act (PDCA) cycle is a standardized system for making gains in quality and conducting continuous improvement.

DMAIC: Acronym for a five step problem solving process process; Define, Measure, Analyze, Improve and Control.

Downtime: A period of the time during which a device or operation is malfunctioning or inoperative causing the system to be inactive.

Error Proofing: Designing a potential failure or cause of failure out of a product or process (a.k.a. Poka-yoke)

External Customer: A person or organization outside your organization who receives the output of a process. Of all external customers, the end-user should be the most important.

Five S's: The five Japanese terms designated for maintaining an efficient and organized workspace. The terms are Seiri (sorting), Seiton (straightening or setting in order), Seiketsu (standardizing), Seiso (sweeping or shining), and Shitsuke (Sustaining).

Flexibility: A production system's ability to respond and adapt to internal and external changes, such as changes in product design or customer demand, in a cost-effective and timely manner.

Flowchart: A graphic representation using symbols to show the step-by-step sequence of operations, procedures, or activities. See Value Stream Mapping

Gemba: The place where value is being added to a product

Gemba Walk: Searching out Muda or places where improvements can be made

Heijunka Box: A visual scheduling tool that strives to level production to the mixed model final assembly.

Histogram: A a graphical representation that shows how many observations fall in certain intervals. It is used to estimate the probablity distribution function of a variable for a large set of data.

Integrated Quality Control (IQC): Everyone in an organization has a complete understanding of quality control, its methods, benefits and expectations.

Internal Customer: The people, machines, or processes being supplied with the products or parts made in preceding work areas.

Jidoka: A cornerstone of the Toyota Production System, Jidoka or autonomation prevents the production of defective products, eliminates overproduction, and preventing problem recurrence.

Jig: A work holding device used to locate a work piece with respect to the tooling during manufacture or assembly.

Job Shop: A manufacturing system where machines are functionally grouped and incoming jobs can be processed through the system using route sheets.

Just-In-Time Manufacturing: JIT is the lean production manufacturing system design which minimizes inventory and operators by having the right materials at the right place at the right time.

Kaizen: The Japanese word for improvement.

Kaizen Events: Executed by a team of workers to facilitate improvement in processes, methodologies and technology insertion

Kaizen Blitz: An intense, short term Kaizen event to execute radical or immediate change

Kanban: A pull system for production and inventory control system used in lean manufacturing, provides the links that connect cells and control inventory in the system.

Lead Time: The total time a customer must wait to receive a product after placing an order. In a lean shop, lead time is independent of the processing time for individual machines.

Lean Manufacturing/Production: A manufacturing system design to eliminate waste from all of its activities and operations. Lean produces products better, faster, and cheaper than its competitors using the economies of scope.

Leveling: Leveling or smoothing of production involves making the final assembly line into a Mixed Model subassembly or Mixed Model final assembly line to produce an even distribution of products every hour, every shift, every day.

Line Balancing: Making the amount of work, and thus the time it takes to perform the work, as equal as possible (in terms of time) at each Workstation.

Linked-Cell Manufacturing System(L-CMS): Manufacturing system design using multifunctional workers in cells that are linked by Kanban to other cells.

Little's Law: A law that shows the relationship between the work-in-progress inventory (WIP), the throughput time (TPT), and the production rate (PR) for the manufacturing system; WIP = TPT x PR

Make One-Check One-Move One On (MO –CO- MOO): Perform a step in the process; Check the product to ensure that step was done correctly; Move on to the next step.

Make-to-Stock: Intermediate product to sustain smooth, level flow or to buffer uncertainty (See WIP)

Make-to-Final Inventory: Finished product temporarily stored to satisfy on-time customer delivery.

Mass Production: Large-scale manufacturing with high-volume production and output; implies pre computer-era methods, with departmentalized operation and reliance on "economies of scale" to achieve low per-unit costs.

Materials Requirements Planning: Using software, material planning is accomplished through evaluating the Bill of Materials (BOM), Inventory Data, and the Master Schedule in order to stimulate replenishment of materials to be consumed and present purchase orders (PO's) for future materials needed.

Muda: Japanese term for any human activity that absorbs resources or time, but creates no real value.

Mura: Japanese term for *unevenness*. It refers to an uneven flow of parts in the Toyota Production System.

Muri: Japanese term for the act of overloading an area or otherwise executing unreasonable work.

One-Piece Flow Production: A production system where, through the development of a product-oriented layout and operators capable of handling multiple processes, parts are produced and transferred one piece at a time.

Pareto diagrams: Pareto diagrams styled after the Pareto Principle are a type of bar chart or histogram which displays the frequency with which a particular phenomenon is occurring relative to the occurrence of others.

Pareto's law: The law, often stated as 80/20 rule, developed by an Italian economist in the 1800s based on the principle that the vast majority of an end result (wealth, cost, quality problems, etc.) is determined by a small percentage of a group (the number of people, items, etc.).

Pokayoke (also Poka-yoke): Japanese for *mistake-proofing* or *defect prevention*. Mistake-proofing devices made by designing parts, processes, or procedures so that mistakes physically or procedurally cannot happen.

Preventive Maintenance: Activities done to insure the operating availability of a production resource and its ability to meet process specifications. They are done on a pre-scheduled basis, or based on the identification by a monitoring system of conditions that may cause future breakdowns.

Process Failure Mode and Effects Analysis (PFMEA): An analytical technique used by a manufacturing responsible engineer/team as a means to assure that, to the extent possible, potential failure modes and their associated causes/mechanisms have been considered and addressed.

Process Sheet: A set of manufacturing instructions for a specific batch, lot or run that describe the operating parameters and settings for the equipment and facilities used, and any associated tooling or supplies.

Product Life Cycle: The processes, costs and revenues associated with a product from its initial creation to its abandonment, and often categorized by the stages of introduction, growth, maturity and decline.

Product Mix: The number of individual products produced or sold by an organization. Sometimes involves the concept of *sequencing*.

Quality: The characteristic of a product or service to satisfy stated or implied needs.

Reliability: The ability of a machine, device or component to operate as it was designed to operate over a specified period of time without failure or entering a failed state.

Repeatability: Capability of a process, tool or procedure to consistently and repeatedly yield the same measureable result.

Root Cause Analysis: A problem-solving approach to reduce or eliminate recurrence of the same problem whereby the underlying cause of a problem is first identified and only then is the corrective action or solution designed.

Seven quality tools: Tools used for process improvement that are usually: Cause and Effect Diagrams, Check Sheets, Control Charts, Histograms, Pareto Charts, Scatter Plots and Run Charts.

Seven Wastes: (According to Taiichi Ohno) These wastes are: Overproduction, Idle Time, and Unnecessary Material Handling, WIP or Excess Inventory, Excess Processing, Wasted Motion Poor Quality.

Six Sigma: An improvement program using statistical analysis tools and directed Kaizen events to reduce Standard Deviation from a spread of 6-sigma to 12 –Sigma between the natural Lower Tolerance Limit and the natural Upper Tolerance Limit.

Standard Operations Routine sheets: Diagrams that show all of the tasks of a job including the sequence of operations and times, walking time, task time, the tooling needed, the stock on hand and the motion of the operator.

Taguchi Methods: A set of methods and statistical Design of Experiment tools to both enforce Design Centering (Meeting Targets) and to reduce variance (6-Sigma)

Takt Time: In a Lean Production System, Takt time is the time between production of consecutive items in a production system. Takt time is the drumbeat or heartbeat of

Lean production, and is defined as the available working time divide by demand over a specified period of time.

Throughput Time: The time needed for a product to pass through a machine, Lean cell, final assembly or an entire manufacturing system.

Type I error: The error of deciding that a process is out of control, when it is actually not making any defects. The probability of this occurring is given by Alpha (α).

Type II error: The error of deciding that a process operating in an acceptable manner, when it is actually out of control. The probability of this occurring is given by Beta (β).

Value Added Work: Any activity that actually transforms raw materials into a finished product.

Value Stream Mapping: A technique used in Lean manufacturing that maps the flow of material and data, and associated time requirements, from initial supplier to end customer for a given business process. VSM seeks to identify non-value added activities.

Variability: The extent to which the data points are spread from the mean.

Variety: Not the same as variability. Variety is a necessary component to sell finished product in different design configurations to satisfy the customer.

Withdrawal Kanban: A card used that specifies the kind and quantity of product which a manufacturing process should withdraw from a preceding process.

Work in Progress (WIP): All materials in the production process once they are withdrawn from the store until they are stored as finished goods. Wip is sometimes classified as (1) Parts waiting for value added transformation, (2) Parts waiting plus parts undergoing value added transformation, (3) Only those parts held between manufacturing areas or Lean Work cells, (4) Parts in intermediate or final storage, and (5 All of the above). Toyota called WIP *The root of all evil*.

Zero defects: The goal of a Lean system is to have no defects. Zero Defects is the utopian goal of Lean manufacturing and is necessary to operate a Lean system..

Zero Inventory: Made popular by Robert Hall's *Zero Inventories*. Hall meant target of zero WIP inventory between manufacturing areas or Lean work cells..